The Application of
Charge Density Research
to Chemistry and Drug Design

NATO ASI Series

Advanced Science Institutes Series

A series presenting the results of activities sponsored by the NATO Science Committee, which aims at the dissemination of advanced scientific and technological knowledge, with a view to strengthening links between scientific communities.

The series is published by an international board of publishers in conjunction with the NATO Scientific Affairs Division

A	**Life Sciences**	Plenum Publishing Corporation
B	**Physics**	New York and London
C	**Mathematical and Physical Sciences**	Kluwer Academic Publishers
D	**Behavioral and Social Sciences**	Dordrecht, Boston, and London
E	**Applied Sciences**	
F	**Computer and Systems Sciences**	Springer-Verlag
G	**Ecological Sciences**	Berlin, Heidelberg, New York, London,
H	**Cell Biology**	Paris, Tokyo, Hong Kong, and Barcelona
I	**Global Environmental Change**	

Recent Volumes in this Series

Volume 248—Lower-Dimensional Systems and Molecular Electronics
edited by R. M. Metzger, P. Day, and G. C. Payavassiliou

Volume 249—Advances in Nonradiative Processes in Solids
edited by B. Di Bartolo

Volume 250—The Application of Charge Density Research
to Chemistry and Drug Design
edited by George A. Jeffrey and Juan F. Piniella

Volume 251—Granular Nanoelectronics
edited by David K. Ferry, John R. Barker, and Carlo Jacoboni

Volume 252—Laser Systems for Photobiology and Photomedicine
edited by A. N. Chester, S. Martellucci, and A. M. Scheggi

Volume 253—Condensed Systems of Low Dimensionality
edited by J. L. Beeby

Volume 254—Quantum Coherence in Mesoscopic Systems
edited by B. Kramer

Series B: Physics

The Application of
Charge Density Research
to Chemistry and Drug Design

Edited by

George A. Jeffrey

University of Pittsburgh
Pittsburgh, Pennsylvania

and

Juan F. Piniella

Universitat Autonoma de Barcelona
Bellaterra, Barcelona, Spain

Springer Science+Business Media, LLC

Proceedings of a NATO Advanced Study Institute on
the Application of Charge Density Research
to Chemistry and Drug Design,
held April 17–27, 1990,
in Sant Feliu de Guixols, Costa Brava, Spain

Library of Congress Cataloging-in-Publication Data

NATO Advanced Study Institute on the Application of Charge Density
  Research to Chemistry and Drug Design (1990 : Sant Feliu de Guixols,
  Spain)
    The application of charge density research to chemistry and drug
  design / edited by George A. Jeffrey and Juan F. Piniella.
        p.   cm. -- (NATO ASI series. Series B, Physics ; v. 250)
    "Proceedings of a NATO Advanced Study Institute on the Application
  of Charge Density Research to Chemistry and Drug Design, held April
  17-27, 1990, in Sant Feliu de Guixols, Costa Brava, Spain"--T.p.
  verso.
    "Published in cooperation with NATO Scientific Affairs Division."
    Includes bibliographical references and index.
    ISBN 978-1-4613-6645-4   ISBN 978-1-4615-3700-7 (eBook)
    DOI 10.1007/978-1-4615-3700-7
    1. Drugs--Design--Congresses. 2. Drugs--Structure--Congresses.
  3. Charge density waves--Congresses. I. Jeffrey, George A., 1915-
  . II. Piniella, Juan F. III. North Atlantic Treaty Organization.
  Scientific Affairs Division. IV. Title. V. Series.
  RS420.N37 1990
  615'.19--dc20   91-11799
                                                            CIP

© 1991 Springer Science+Business Media New York
Originally published by Plenum Press New York in 1991
Softcover reprint of the hardcover 1st edition 1991

ORGANIZING COMMITTEE

H. Fuess

Institüt of Kristallografia
Universität
Frankfurt/Main, Germany

J. A. K. Howard

Department of Inorganic Chemistry
University of Bristol
Bristol, United Kingdom

G. A. Jeffrey[†]

Department of Crystallography
University of Pittsburgh
Pittsburgh, Pennsylvania, USA

F. K. Larsen

Department of Chemistry
Aarhus University
Aarhus, Denmark

M. Nardelli

Instituto di Chimica Generale
Universita di Parma
Parma, Italy

J. F. Piniella[†]

Departament de Cristallografia
Universitat Autonoma de Barcelona
Bellaterra, Barcelona, Spain

[†]Co-Directors

PREFACE

In the past twenty years, the X-ray crystallography of organic molecules has expanded rapidly in two opposite directions. One is towards larger and larger biological macromolecules and the other is towards the fine details of the electronic structure of small molecules.

Both advances required the development of more sophisticated methodologies. Both were made possible by the rapid development of computer technology. X-ray diffraction equipment has responded to these demands, in the one case by the ability to measure quickly many thousands of diffraction spectra, in the other by providing instruments capable of very high precision. Molecules interact through their electrostatic potentials and therefore their experimental and theoretical measurement and calculation is an essential component to understanding the electronic structure of chemical and biochemical reactions.

In this ASI, we have brought together experts and their students from both the experimental and theoretical sides of this field, in order that they better understand the philosophy and complexity of these two complementary approaches.

George A. Jeffrey
Department of Crystallography
University of Pittsburgh
Pittsburgh, Pennsylvania 15260 USA

CONTENTS

LECTURES

General Considerations on Methods for Studying Molecular Structures
    and Electron Density Distributions................................. 1
  M. Nardelli

The Past and Future of Experimental Charge Density Analysis.......................... 7
  P. Coppens and D. Feil

Determination of Atomic and Structural Properties from Experimental
    Charge Distributions.. 23
  R. F. W. Bader and K. E. Laidig

Electrostatic Properties of Molecules from Diffraction Data............................. 63
  R. F. Stewart

X-ray Diffraction and Charge Distribution. Application to the Electron
    Density Distribution in the Hydrogen Bond..................................... 103
  D. Feil

Electron Density Models: Description and Comparison................................. 121
  C. Lecomte

Experimental Requirements for Charge Density Analysis 155
  R. H. Blessing and C. Lecomte

Necessity and Pitfalls of Low-Temperature Measurements............................ 187
  F. K. Larsen

The Use of Synchrotron Radiation and Its Promise in Charge Density Research....... 209
  P. Coppens

The Use of Neutron Diffraction in Charge Density Analysis............................ 225
  H. Fuess

Structural Chemistry and Drug Design.. 241
  G. Gilli and P. A. Borea

Correlation of Crystal Data and Charge Density with the Reactivity and
    Activity of Molecules: Towards a Description of Elementary Steps
    in Enzyme Reactions 287
  G. Klebe

Charge Density Studies of Drug Molecules.. 319
  E. D. Stevens and C. L. Klein

POSTERS

Charge Density Distribution of Dimethyl-1,3-amino-4-uracil 337
    F. Baert, A. Guelzim and V. Warin

Structure and Electron Density of *Trans*-dichloro-bis(creatinine)platinum(II)
    Dihydrate 338
    A. M. Beja, J. A. Carvalho Paixão, J. M. Gil and M. Aragon Salgado

A Low-Temperature (23 K) Study of L-Alanine: Topological Properties
    of Experimental and Theoretical Charge Distributions 339
    R. Bianchi, R. Destro, C. Gattti and F. Merati

Electrostatic Properties and Topological Analysis of the Charge Density of
    Syn-1,6:8,13-biscarbonyl[14]annulene Derived from X-ray
    Diffraction Data at 16 K 340
    R. Bianchi, R. Destro and F. Merati

Electron Density Analysis of the Antimicrobial Drug and Radiosensitizer
    Dimetridazole at 105 K 341
    H. L. De Bondt, N. M. Blaton, O. M. Peeters, C. J. De Ranter
    and I. Kjøller Larsen

The Influence of Packing at a Molecular Level: Conformation, Geometry,
    Spectroscopy (IR, Raman) 351
    B. Bracke, B. van Dijk, C. Van Alsenoy and A. T. H. Lenstra

The Experimental Electron Density Distribution of a C–H→Co Bridged Complex 353
    L. Brammer and E. D. Stevens

Transferable Atom Equivalents. Molecular Electrostatic Potentials from the
    Electric Multipoles of *PROAIMS* Atomic Basins 357
    C. M. Breneman

An Ab Initio Study of Clavulanic Acid and the Relation to Its Chemical Reactivity ... 359
    B. Fernández and C. Van Alsenoy

UHF Calculations of the g Tensor in Metal-Carbonyles Radicals 360
    A. Grand, P. J. Krusic and R. Subra

Magnetostriction in NiF_2: Combined γ-Ray and Neutron Diffraction 361
    W. Jauch, A. Palmer and A. J. Schultz

The Conformation of Some Neuromuscular Blocking Agents 362
    H. Kooijman, J. A. Kanters and J. Kroon

A Structural Comparison of Potential HIV Inhibitors:
    2',3'-Didesoxinucleosides and 2',3'-Didesoxicarbonucleosides 363
    R. A. Mosquera, B. Fernández, S. A. Váquez and
    M. A. Ríos y E. Uriarte

Experimental Observation of Intermolecular π-Electron Interactions in
    Photoreactive Crystals 365
    I. Ortmann, St. Werner and C. Krüger

The Metal-Nitroxyl *Interaction* in MNO Metallacycles (M = Cu, Pd):
    An Ab Initio SCF/CI Study 369
    M.-M. Rohmer, A. Grand and M. Bénard

First-Principles Theoretical Methods for the Calculation of Electronic
    Charge Densities and Electrostatic Potentials....................................... 371
    J. M. Seminario, J. S. Murray and P. Politzer

Comparison of Precise Electronic Charge Densities and Electrostatic
    Potentials Obtained from Different Ab Initio Approaches........................ 383
    J. M. Seminario and P. Politzer

Theoretical and Experimental Study of Dimerization of Substituted
    Phenylacetylenes.. 384
    F. Sevin and N. Balcioglu

Electron Deformation Density of Metal Carbyne Complexes............................. 385
    A. Spasojevic-de Biré, N. Q. Dao, P. Becker, M. Bénard,
    A. Strich, C. Thieffry, N. Hansen and C. Lecomte

The Potential Function in Short O–H···O Bonds. Experimental
    Evidence and Ab Initio Modelling 401
    F. Vanhouteghem, C. Van Alsenoy and A. T. H. Lenstra

The Effect of Conformational Changes on the Polarizability of PNA 403
    G. J. M. Velders and D. Feil

Charge Density in Adeninium Hydrochloride Hemihydrate and
    $1H^+$-Adeniniumtrichlorozinc(II): The Effects of Metal Binding
    to a Nucleobase 404
    L. M. Vilkins and M. R. Taylor

Participants 405

Index 407

GENERAL CONSIDERATIONS ON METHODS FOR STUDYING MOLECULAR

STRUCTURES AND ELECTRON DENSITY DISTRIBUTIONS

M. Nardelli

Istituto di Chimica Generale ed Inorganica della Università di Parma
Centro di Studio per la Strutturistica Diffrattometrica del CNR
Viale delle Scienze, I-43100 Parma, Italy

From the discovery of X-ray diffraction by crystals it became clear that these kinds of experiments were useful to study not only the X-ray radiation, but also the structure of the diffraction lattices, *i.e.* crystals. It was shown by W. H. Bragg,[1] that, owing to the periodical nature of the atomic distribution in crystals and the interaction of the X-ray radiation with the electrons in it, it is possible to deduce the distribution of the electron density in crystals from the intensities of the diffracted beams.

This property was first used with the only purpose of determining the distribution of the atoms in crystals and hence the geometrical structure of crystals and molecules. Indeed, if exception is made for a few simple cases in which the electron density distribution itself was used to interpret the nature of bonding in crystals, it is necessary to reach the middle of the 1960's, with the development of single crystal diffractometers and computing facilities, to have systematic studies of the electron density deformation with the purpose of obtaining an experimental description of the chemical bonding to compare with the picture given by quantum chemistry theoretical calculations.

The experimental determinations of one-electron densities[2,3] are based on intensity measurements of X-ray photons elastically scattered by crystals, and the changes observed with respect to the distribution of the valence electrons of spherical atoms (promolecule or procrystal) are interpreted as due to the formation of chemical bonds. The fundamental equations are:

$$I(\mathbf{H}) \propto |F(\mathbf{H})|^2 \tag{1}$$

$$F(\mathbf{H}) = \int_V \rho(\mathbf{r}) \exp(2\pi i \mathbf{H} \cdot \mathbf{r}) \, d\mathbf{r} \tag{2}$$

$$\rho(\mathbf{r}) = (1/V) \sum_{\mathbf{H}} F(\mathbf{H}) \exp(-2\pi i \mathbf{H} \cdot \mathbf{r}) \tag{3}$$

where $\mathbf{r}$ is a position vector in direct space, $\mathbf{H}$ a reciprocal lattice vector, I the intensity of the diffracted beam, F the structure factor, $\rho(\mathbf{r})$ the electron density and V the unit cell volume. To have reliable results it is necessary to overcome a number of experimental difficulties, the major ones being as follows:

- very accurate measurements of low temperature intensities,
- correction for absorption, extinction, TDS, multiple scattering and anomalous scattering,

The Application of Charge Density Research to Chemistry and Drug Design
Edited by G. A. Jeffrey and J. F. Piniella, Plenum Press, New York, 1991

- scaling of intensities,
- phasing of the structure amplitudes, particularly in the case of acentric structures when phases are subject to refinement,
- series termination in calculating $\rho(\mathbf{r})$,
- correct description of atomic displacements (thermal motion, disordered distributions),
- the use of spherical averaged atoms to define the promolecule (or procrystal) model is not free of criticism.

Refinements are carried out taking into account the expansion or contraction of the valence-shell with variation of atomic charge (κ refinement) and introducing aspherical atom-centered density functions (multipole and Hirshfeld refinements).

It is important to keep in mind that experimental electron densities obtained by diffraction of crystals are electron densities of the molecules in the crystals, so they experience the effects of molecular packing: van der Waals, Coulomb and hydrogen-bonding interactions.

Theoretical electron densities[3,6] are usually based on one-electron density distributions calculated by the expression

$$\rho(\mathbf{r}) = N\int |\psi(\mathbf{r}_1,s_1,\mathbf{r}_2,s_2,\cdots\mathbf{r}_n,s_n)|^2 \, ds_1 \, ds_2 \cdots ds_n \, d\mathbf{r}_1 \, d\mathbf{r}_2 \cdots d\mathbf{r}_n \qquad (4)$$

where $\psi(\mathbf{r}_1,s_1,\mathbf{r}_2,s_2,\cdots\mathbf{r}_n,s_n)$ is the n-electron wave function of the system and $\mathbf{r}_1,\mathbf{r}_2,\cdots\mathbf{r}_n$ and $s_1,s_2,\cdots s_n$ are the space and spin coordinates of the electrons, respectively. This theoretical electron density is for an isolated molecule.

The sources of errors in this kind of calculation are:

- the assumption that nuclear and electronic motions are separable (Born-Oppenheimer approximation),
- neglecting relativistic effects,
- averaging $\Delta\rho(\mathbf{r})$ with respect to the vibrational nuclear motion,
- finite expansion basis set (Roothaans' method, Slater-type orbitals, uncontracted and contracted Gaussian-type orbitals),
- correlation effects.

The *ab initio*[7] molecular orbital calculations are based on a non-parameterized treatment (even if a number of simplifying assumptions are made also in this case), while the semi-empirical[8,9] methods consider more drastic approximations. Both methods are based on the LCAO-SCF procedure for building up molecular orbitals, having the proper symmetries and giving the lowest electronic energies, from sets of atomic orbitals centered at the constituent atoms. The self-consistent field (SCF) approach takes into account the electron-electron repulsion by considering the interaction between an electron in a given orbital and the mean field produced by the other electrons in the molecule.

One problem arising in using the SCF-MO method is connected with the treatment of "open-shell" and "closed-shell" systems, *i.e.* with and without unpaired electrons, respectively. Closed-shell systems are calculated using the spin "restricted Hartree-Fock" theory, which considers each orbital doubly occupied or empty. Open-shell systems can be treated by using the "unrestricted Hartree-Fock" formalism, which considers one set of molecular orbitals for each type of spin. This formalism allows for spin polarization, *i.e.* an unpaired electron perturbs a formally paired spin, and is not limited to one electronic state as it considers also mixing in states of higher spins.

The computer programs for *ab initio* and semi-empirical calculations[10] work in the following general way. Starting from the Cartesian coordinates of the atoms, the nuclear repulsion is first calculated. Then the atomic orbitals are assigned to each nucleus. The semi-empirical programs use a set of predetermined parameters to define the forms and energies of the atomic orbitals. The *ab initio* programs calculate the various one- and two-electron integrals using a standard set of coefficients and exponents defining the orbitals (basis set). Then, both semi-empirical and *ab initio* programs produce a trial set of molecular orbitals (usually atomic orbitals in the semi-empirical case, extended Hückel in the *ab initio* case) to use at the start of the SCF calculations. The SCF equations are iteratively solved in a process ending when the electronic energy is at a minimum and the density matrix does not change. From the molecular orbitals obtained in this way, the atomic charges, overlaps, dipole moment and correlation energy are calculated. The molecular geometry is optimized by an iterative process that reduces to zero the atomic forces and minimizes the total energy.

Ab initio methods are becoming increasingly important in a variety of chemical studies including organic synthesis and reactivity, catalysis, molecular biology, polymer science, etc. Nevertheless, there are practical limitations, mainly connected with the computing time; the computational effort increases with the fourth power of the number of the basis functions just for the evaluation of the inter-electron repulsion integrals, and inclusion of correlation via configuration interaction expansion makes the calculations quickly prohibitive, even on the most powerful supercomputers. This is the reason why semi-empirical methods became so popular, but these do not frequently give results accurate enough to solve current problems of molecular geometry, conformational energy, heat of formation, etc.

Quantum mechanical methods, even at the semi-empirical level, usually require heavy calculations to give results of acceptable reliability, so it is easily understandable why other methods, like those of Molecular Mechanics (Force Field)[11-13] based on quite a different approach, have been developed. These methods were found to be very successful for practical applications, particularly in computer graphics and drug design. According to them, a molecule is considered as a collection of atoms held together by elastic forces that can be described by potential energy functions related to the structural parameters (bond lengths, bond angles, non-bonded contacts, etc.). The total steric energy, E, arising from the deviation from an ideal structure of a molecule is approximated by a sum of the different types of energy contributions:

$$E = E_s + E_b + E_{sb} + E_w + E_{nb} + \cdots \qquad (5)$$

where E_s is the energy of the bond stretching, E_b that of bending bond angles, E_{sb} that of stretch-bend cross terms, E_w that from twisting about bonds and E_{nb} that from non-bonded (van der Waals, Coulomb, hydrogen bonding) interactions.

The molecular mechanics methods, starting from a trial geometry, give an optimized geometry of the molecule through a process of minimization of the steric energy. The technical procedures used are the steepest-descent, or the pattern search, or better, the Newton-Raphson methods, making use of the analytically evaluated second derivatives of the molecular energy with respect to the geometrical parameters. The following expressions for the different types of energy contributions are those most frequently used:

$$E_s = \tfrac{1}{2} \sum_{i=1}^{N} k_i \, (d_i - d_i^\circ)^2 \qquad (6)$$

$$E_b = \tfrac{1}{2} \sum_{i<j}^{M} k'_{ij} (\theta_{ij} - \theta_{ij}^\circ)^2 \qquad (7)$$

$$E_{sb} = \tfrac{1}{2} \sum_{i<j} k''_{ij} (d_i - d_i^\circ + d_j - d_j^\circ) \, (\theta_{ij} - \theta_{ij}^\circ) \qquad (8)$$

3

where N is the total number of bonds and M the total number of bond angles, k and k' are the stretching and bending force constants, d_i and θ_{ij} are the assumed bond lengths and angles while $d°_i$ and $\theta°_{ij}$ are the same magnitudes at the equilibrium. The energy due to the changes of the ω torsion angles is expressed in a Fourier series

$$E_\omega = \tfrac{1}{2}\sum_{i=1}^{P} \left[V_1(1+\cos\omega_i) + V_2(1-\cos2\omega_i) + V_3(1+\cos3\omega_i) + \cdots \right] \quad (9)$$

usually truncated at the third term, where P is the total number of unique torsion angles and V_1, V_2, V_3 are parameters calculated from some test molecules giving calculated conformations in agreement with the experiment.

For the nonbonded potential energy, E_{nb}, usually the van der Waals, E_{vW}, and Coulomb, E_C, terms are considered (in the case of hydrogen bonding, a suitable function for this kind of interaction should be considered):

$$E_{nb} = E_{vW} + E_C, \quad (10)$$

where

$$E_{vW} = \sum_{i<j} (Ar_{ij}^{-12} - Br_{ij}^{-6}) \qquad \text{(Lennard-Jones)} \quad (11)$$

or

$$E_{vW} = \sum_{i<j} [A'\exp(B'r_{ij}^{-1} - Cr_{ij}^{-6}] \qquad \text{(Buckingam)} \quad (12)$$

and

$$E_C = \sum_{i<j} q_i\, q_j\, r_{ij}^{-1} \quad (13)$$

where q_i and q_j are the atom charges (calculated by quantum mechanical methods) and r_{ij} is the distance between atom i and atom j, the sums being extended to all the contacts under a threshold value.

Molecular mechanics calculations are becoming more and more popular, as the computational time they require is much less than that required by molecular orbital methods. In fact, with molecular mechanics this time increases only with the square of the number of the atoms, and therefore also large molecules can be fully optimized. The predicted geometries are often near to the experimental ones and this is the reason why these methods are increasingly used in studying structure-activity relationships, particularly in connection with drug design. Nevertheless, it must be clear that these methods require good steric molecular energy models and getting such models is certainly not a trivial task. Care must be taken in relating the observed distributions of the structural parameters (bond distances, angles, torsions, etc.) to the potential energy functions responsible for these distributions. Of course, molecular mechanics methods are not appropriate for studying properties where electronic effects are predominant or when excited states are invovled.

Considering the problem of the molecular structure and biological activity of drugs, the paramount importance of the experimental data concerning the structures of the active molecules and receptors becomes quite clear, and in this respect the information collected in the Cambridge Structural Database (CSD)[14,15] and in the Protein Data Bank (PDB)[15,16] is certainly of enormous value. Indeed, even if there is a tendency to substitute the calculated structures for the experimental ones in tackling problems of drug design, the experimental data remain, at least for the time being, the fundamental source of structural information. Computer programs dealing with the data from these databases and with the calculated models

4

have been developed, and research devoted to combining the structural information on active small molecules with that on the biological macromolecules is going to have a very promising future.

Acknowledgements: It is a pleasure to thank Professor Luigi Oleari (Parma) for his critical reading of the manuscript.

REFERENCES

1. W. H. Bragg, *Phil. Trans. Roy. Soc.*, **210**, 253 (1915).
2. P. Coppens, "Concepts of Charge Density Analysis: The Experimental Approach", in: *Electron Distributions and the Chemical Bond*, P. Coppens and M. B. Hall, eds., Plenum Press, New York and London (1982).
3. M. Breitenstein, H. Dannöhl, H. Meyer, A. Schweig, and W. Zittlan, "Experimental Versus Theoretical Electron Densities: Methods and Errors", in: *Electron Distributions and the Chemical Bond*, P. Coppens and M. B. Hall, eds., Plenum Press, New York and London (1982).
4. J. D. Dunitz, "X-ray Analysis and the Structure of Organic Molecules," Chapter 8, Cornell University Press, Ithaca and London (1980).
5. G. Moss, "Pseudomolecular Electrostatic Properties from X-ray Diffraction Data," in: *Electron Distributions and the Chemical Bond*, P. Coppens and M. B. Hall, eds., Plenum Press, New York and London (1982).
6. V. H. Smith, Jr., "Concepts of Charge Density Analysis: The Theoretical Approach," in: *Electron Distributions and the Chemical Bond*, P. Coppens and M. B. Hall, eds., Plenum Press, New York and London (1982).
7. W. J. Hehre, L. Radom, P. v.R. Schleyer and J. A. Pople, *Ab initio Molecular Orbital Theory*, Wiley-Interscience, New York (1985).
8. J. A. Pople and D. L. Beveridge, *Approximate Molecular Orbital Theory*, McGraw-Hill, New York (1970).
9. W. L. Jorgensen and L. Salem, *The Organic Chemist's Book of Orbitals*, Academic Press, New York (1973).
10. T. Clark, *A Handbook of Computational Chemistry. A Practical Guide to Chemical Structure and Energy Calculations*, John Wiley & Sons, New York (1985).
11. U. Burkert and N. L. Allinger, *Molecular Mechanics*, ACS Monograph 177, American Chemical Society, Washington, D.C. (1982).
12. K. Rasmussen, *Potential Energy Functions in Conformational Analysis*, Springer Verlag, Berlin (1985).
13. D. B. Boyd and K. B. Lipkowitz, *J. Chem. Ed.*, **59**, 269 (1982).
14. F. H. Allen, O. Kennard and R. Taylor, *Acc. Chem. Res.*, **16**, 146 (1983).
15. F. H. Allen, G. Bergerhoff and R. Sievers, eds., *Crystallograhic Databases. Information Content, Software System, Scientific Applications*, Data Commission of the International Union of Crystallography, Bonn-Cambridge-Chester (1987).
16. F. C. Bernstein, T. F. Koetzle, G. J. B. Williams, E. F. Mayer, Jr., M. D. Brice, J. R. Rodgers, O. Kennard, T. Shimanouchi and M. Tasumi, *J. Mol. Biol.*, **112**, 535 (1977).

THE PAST AND FUTURE OF EXPERIMENTAL

CHARGE DENSITY ANALYSIS

Philip Coppens

Chemistry Department
State University of New York at Buffalo
Buffalo, New York, 14214, USA

Dirk Feil

Chemical Physics Laboratory
University of Twente
P.O.B. 217
7500 AE Enschede
The Netherlands

TRIAL AND ERROR: THE EARLY YEARS

Though the ability to obtain charge densities experimentally from accurate X-ray data has only recently become widely accepted, the possibility to probe electronic structure was recognized almost immediately after the discovery of X-ray diffraction by von Laue in 1912, as is evident from Debye's statement (in 1915):[1]

"It seems to me that experimental study of the scattered radiation, in particular from light atoms, should get more attention, since along this way it should be possible to determine the arrangement of the electrons in the atoms".

Convinced that X-rays were scattered by electrons, Debye, with Scherrer, turned his attention to chemical bonding and thereby essentially initiated the field of X-ray charge density studies. Debye and Scherrer wondered how the valency dashes, used by the chemists to describe the bond between atoms, could be replaced by an electron model that was consistent with their results with the newly discovered powder diffraction method and with Bragg's ionization counter measurements.

The first problem addressed was the covalent bond in diamond. They first assumed the electrons to occupy the space between the neighbouring atoms as in the Lewis dot model, which had become widely used. But this model was not in agreement with the diffraction results of Bragg, since it would yield a strong 222 reflection, which is not

The Application of Charge Density Research to Chemistry and Drug Design
Edited by G. A. Jeffrey and J. F. Piniella, Plenum Press, New York, 1991

observed in the diffraction pattern. They concluded that electron clouds in the bonding regions could not be detected in diamond or other comparable solids.

A second model consisted of a spherical carbon atom with the electrons moving rapidly such as to give a constant average distribution within a spherical volume centered at the nucleus. This model gave a quantitative account of the observed decrease of the atomic scattering factor with increased diffraction angle. The dimension of the sphere derived from this decrease was consistent with the interatomic distances, and of the same order of magnitude as calculated by Bohr on the basis of his proto-quantum theory.

Debye and Scherrer then turned their attention to the origin of the bond in ionic crystals like NaCl. The then current theory assumed Coulomb forces between ions. The fact that the Bragg spectra of KCl showed the absence of 'even' reflections seemed to imply that K and Cl have the same number of electrons, or in other words that the atoms are present as monovalent ions. But the accuracy of the measurements was not deemed sufficient for such a conclusion. Moreover, in the much lighter-atom crystal of NaF, the mixed reflections were not absent. Debye and Scherrer showed, however, that a difference in thermal motion of the ions could destroy the equal scattering power of the two ions, thus evading the conclusion that NaF is not an ionic crystal. A more quantitative study on LiF, based on the ratio of the measured 'even-even-even' and 'odd-odd-odd' reflections extrapolated to zero scattering angle, led a value of 1.52 for the electron-ratio (Li+F)/(Li-F), and therefore to the conclusion that the crystal consisted of singly charged ions (Fig. 1).[2]

This result was criticized by W.L. Bragg, James and Bosanquet in 1922, who wrote:[3] "it is interesting to see whether any evidence can be obtained as to whether a valency electron has been transferred from one atom to the other or not. This may be put in another way: Can we tell from the atomic form factors whether their maxima are at 10 (for Na) and 18 (for Cl) or at 11 and 17 respectively? It appears impossible to do this; and when we come to consider the problem more closely, it seems that crystal analysis must be pushed to

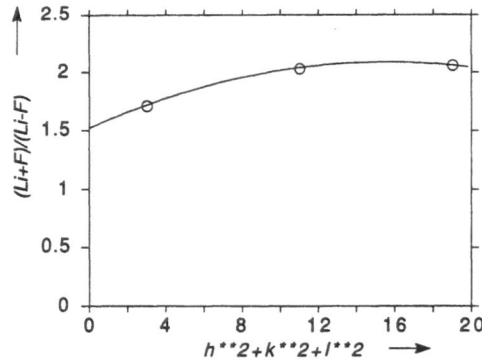

Fig. 1. The (Li + F)/(Li - F) electron ratio as a function of $h^2 + k^2 + l^2$ as plotted by Debye (ref. 2).

a far greater degree of refinement before it can settle the point. If all the electrons were grouped close to the atomic centres, and if the transference of an electron meant that one electron passed from the Na group to the Cl group, then a solution along the lines of that attempted by Debye and Scherrer for LiF might be possible. The electron distributions we find extend, on the other hand, right through the volume of the crystal. The distance between Na and Cl centres is 2.81 Å, and we find electron distributions 1 Å from the centre in sodium and 1.8 Å from the centre in chlorine. If the valency electron is transferred from the outer region of one atom to that of the other, it will still be in the region between the two atoms for the greater part of the time".

The negative implication on the feasibility of determining atomic charges persisted a long time after, as evidenced by a statement by James in his 1948 book:[4]
"any attempt to determine the state of ionisation of the atoms in a crystal is likely to fail, since the scattering factor curves will differ appreciably only at angles for which no spectra exist".

The concept of the electron *distribution* used in these early papers suggests that the experimentalists anticipated the electron probability distribution to be defined by Born in 1926 in terms of the new quantummechanical concepts!

A most interesting 1921 paper by W.H. Bragg concentrates on the bonding density implied in the observation of the 222 reflection of diamond:[5]
"Another point of interest is the existence of a small (222) reflection (in diamond). This has been looked for previously but without success. The structure of the diamond cannot be explained on the hypothesis that the field of force around the carbon atom is the same in all directions: or in other words, that the force between the two atoms can be expressed simply by a function of the distance between the centres. If this were so, the sphere, which would then represent the carbon atoms appropriately, would adopt the close-packed arrangement. As a matter of fact, each atom is surrounded by four neighbours only. It is necessary, therefore, to suppose that the attachment of one atom to the next is due to some directed property, and the carbon atom has four such special directions: as indeed the tetravalency of the atom might suggest. In that case the properties of the atom in diamond are based upon a tetrahedral not a spherical form. The tetrahedra point away from any (111) plane in case of half the atoms in diamond and towards it in case of the other half. Consecutive (111) sheets are not exactly of the same nature; and it might reasonably be expected that they would not entirely destroy each other's effects in the second order reflection from the tetrahedral plane. It is this effect which is now found to be quite distinct, though small[†] ".

Wyckoff in 1930 attempted to predict the effect of covalency on the scattering of more complicated crystals:[6]

"The 'anomalous' (222) reflection of the diamond indicates that some of its electrons may, under suitable conditions, give rise to appreciable reflections

9

which are probably not to be associated with atomic centres. In such an organic compound as urea it might be expected that similarly bound electrons of the nitrogen and oxygen, as well as of the carbon, atoms would be potential sources of 'anomalous' effects and would result in atomic scattering powers lower than those determined from ionic salts. No 'anomalous' reflections were found, but it may well be that part of the reduced scattering powers of the atoms in urea are to be ascribed to the 'homopolar' bonds with which the atoms of its molecules are held together".

It is remarkable that changes in net charges on the atoms and the effect of covalent bonding, addressed in the pioneering papers of the early days, are still the among the topics of today's research.

REALIZING THE PROMISE: THE IMPORTANCE OF TECHNICAL
BREAKTHROUGHS

The interest in charge densities persisted through the work of a a few pioneers including Brill and coworkers[7] and Goettlicher and Woelfel, who published widely quoted studies on the total charge density in diamond and silicon[8] and sodium chloride.[9] The concept of the deformation density seems to have been first introduced by Roux and Daudel in a theoretical study of Li_2;[10] and initially referred to as the "Roux" density. But difference densities obtained by crystallographic methods turned out to be amazingly flat and devoid of much meaningful structure. The reason for this was, as was realized in the middle sixties, that the least squares adjustment of structural parameters, introduced by Hughes,[11] was so successful that it accounted for all bonding features by a subtle adjustment of structural and thermal parameters. This bias was demonstrated when neutron diffraction became more generally available through the development of high flux beam reactors, in particular at Brookhaven National Laboratory. The combined X-ray/neutron (X-N) deformation map in the plane of the s-triazine molecule (Fig. 2), very clearly shows both bonding and lone pair densities.[12] Other studies on oxalic acid dihydrate[13] and decaborane[14] soon followed and confirmed the earlier results. The availability of combined X-ray and neutron data sets also led to a much better understanding of the atomic asphericity shifts. It had been generally recognized that C-H, N-H and O-H bond length from X-ray diffraction were about 0.1 Å shortened because of the non-coincidence of the electron distributions' center of gravity with the nucleus of the atom. The subject was discussed by Jensen and Sundaralingam in an article in Science in 1964.[15] For heavier atoms, however, the effect is much smaller because of the invariance or near-invariance of the inner electron shells, but still measurable when accurate X-ray and neutron data are compared.

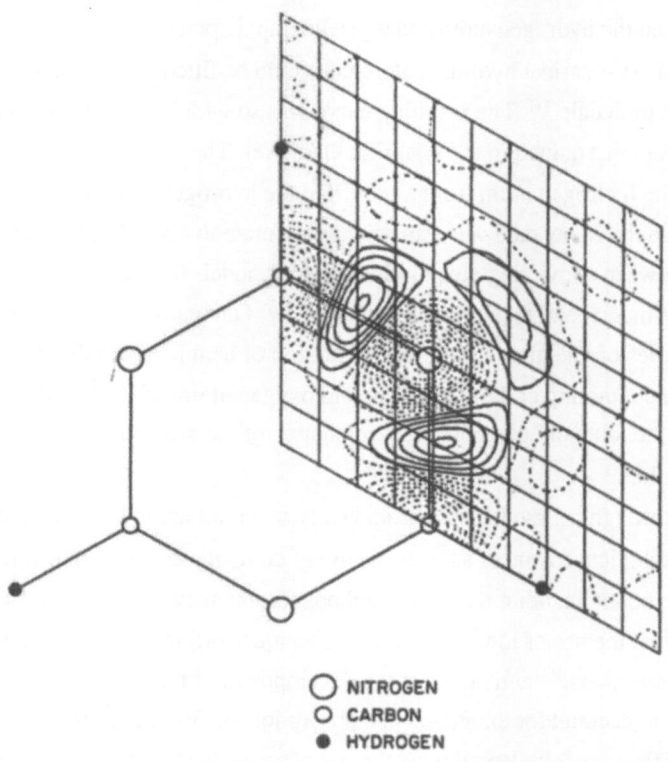

○ NITROGEN
○ CARBON
● HYDROGEN

Fig. 2. The X-N map in the plane of the s-triazine molecule (ref. 12). Contours at
0.05 eÅ^{-3}. Negative contours dotted.

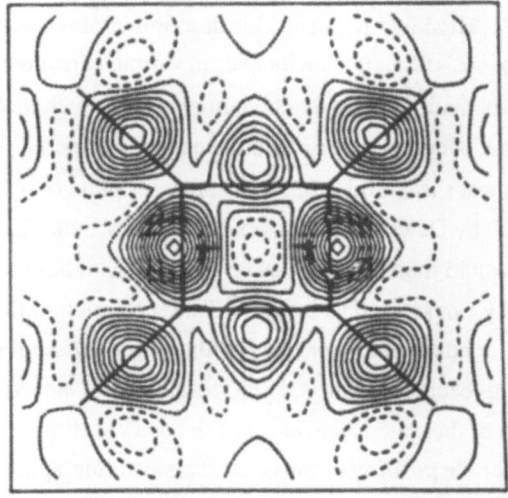

Fig. 3. The deformation density in the plane of the cyclobutadiene ring, showing bond-
bending (ref. 19). Contours at 0.05 eÅ^{-3}.

11

In 1965 Stewart, Davidson and Simpson bridged the gap between the theoretical distribution around the hydrogen atom and crystallographic practice by calculating the scattering of the best spherical hydrogen atom that could be fitted to the theoretical density for the hydrogen molecule.[16] The resulting curve was so widely used that the paper became one of the most quoted in the scientific literature! They also derived an aspherical form factor for the hydrogen atom from a best fit to the hydrogen molecule.

By this time the X-ray data were sufficiently accurate to allow Groenewegen and Feil to distinguish between competing quantummechanical models for the ammonium ion by comparing experimental and theoretical form factors.[17] Qualitative applications became possible. Examples of the latter are the demonstration of bent bonds in three and four-membered rings containing carbon, nitrogen and oxygen atoms (Fig. 3),[18,19] and the confirmation of the mutually perpendicular orientation of the π-systems in a cumulene double bond chain.[20]

The breakthroughs of the late sixties and early seventies are to be attributed to very significant technological advances such as the use of computer-controlled diffractometers with scintillation counters rather than film methods for intensity recording, the advent of neutron diffraction, the use of low-temperature techniques originally pioneered by Lipscomb and coworkers,[21] and not least the development of high-speed computers which made it possible to account for extinction and absorption and to fully analyze the recorded profiles of the reflections. It also allowed the use of sophisticated least-squares models in which the bonding effects are specifically accounted for.

THE DEVELOPMENT OF CHARGE DENSITY REFINEMENT MODELS

The increased experimental accuracy made it necessary to develop a more realistic analytical description of the electron density distribution . In 1966 Weiss used a model for the charge density in diamond consisting of the symmetry-allowed spherical harmonic functions[22] and in 1971, Hirshfeld published his description of the aspherical atom, consisting of a series of $\cos^n \alpha$ functions with axes in various directions (which are equivalent to linear combinations of spherical harmonics).[23] Such models, which do not *explicitly* account for the bonding density, as did the model proposed by McWeeny[24], ultimately proved the most successful for the general description of bonding effects. It was used in the careful work by Dawson and coworkers on silicon, and fully developed by Stewart [25] (the"pseudoatom model"), by Kurki-Suonio[26] (with harmonic oscillator-type radial functions), and by Hansen and Coppens.[27] The separation of the valence shell and inner shell(s), i.e. core, scattering is a common feature, which make it possible to parametrize the bonding features. At the simplest level this is done with only the (monopole) population of the spherical average of the outer shell (the 'L-shell projection' method[28]). Variation of the *position* of the outer shell also (the 'split atom' method), reproduced from the X-ray data alone some of the asphericity shifts observed by the comparison of X-ray and neutron results.[29] But it soon became evident that variation of charge must be accompanied by variation of the shape of the density function. This is a consequence of elementary considerations concerning electron-electron repulsions, often described as the screening of the nucleus by the other electrons, and incorporated by Slater in his rules for valence shell orbital exponents.[30] The κ parameter,[31] which multiplies the radial coordinate r, proved to be a widely applicable improvement of the scattering formalism. The simple model with two extra parameters per atom (κ and the valence population) turned out to be surprisingly successful in reproducing dipole moments for a number of molecules. The experimental relation between the net atomic charge and the kappa parameter for a number of nitrogen atoms in different environments is remarkably close to the prediction of Slater's rules (Fig. 4)! The two parameter monopole model is very computer-time efficient and therefore has been dubbed the *'poor man's charge density refinement'*.[32] But with the present abundant, and still increasing, availability of computer power *poor men* have become increasingly rare and multipole models relatively more popular, provided of course that the additional complexity is warranted by the quality of the data.

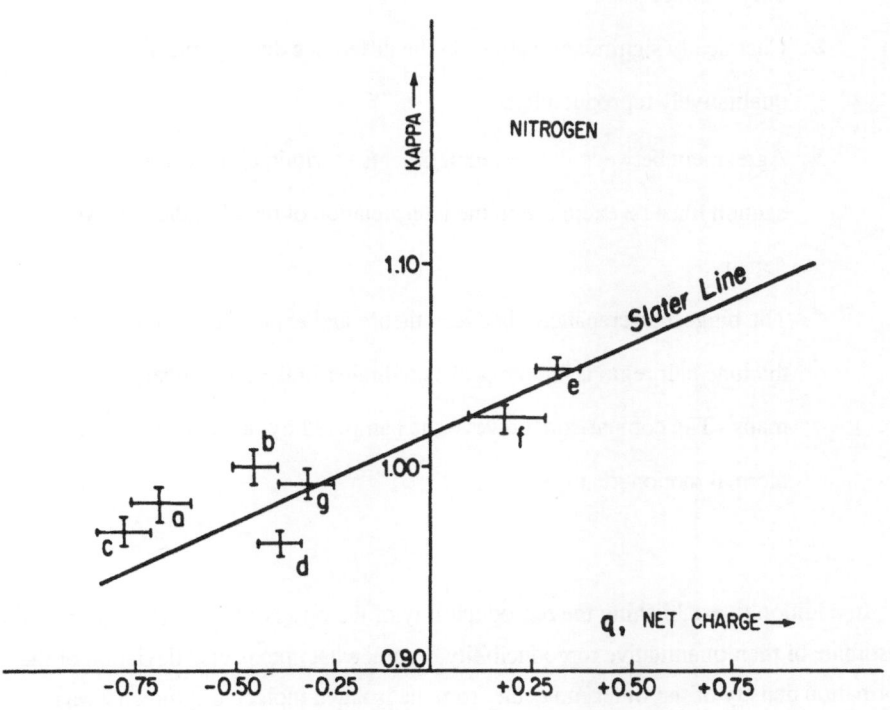

Fig. 4. Relation between kappa and net charge for nitrogen atoms in a number of structures (ref. 31).

A TEST OF REPRODUCIBILITY OF THE RESULTS: THE OXALIC ACID PROJECT

At the point where the feasibility and relevance of the technique appeared established, it became important to examine the reproducibility of the observed features, qualitatively as well as quantitatively. At the initiative of the Commission on Charge, Spin and Momentum densities of the International Union of Crystallography, a charge density determination project was organized, for which oxalic acid dihydrate was selected as a compound with good quality, easy to grow crystals with a variety of bonding features. Four X-ray and five neutron low-temperature data sets were collected in different laboratories and three different theoretical calculations on the oxalic acid molecule were performed. The conclusions obtained after extensive intercomparison of the results[33] are listed in Table 1.

Table 1. Some conclusions of the IUCr oxalic acid project (1984)

1. Positional parameters are reproducible to precisions of 0.001Å or better. Average discrepancies between some of the experiments are only 0.0005Å.

2. Chemically significant features in the difference density maps are qualitatively reproducible.

3. Agreement between the best experiments is within about $0.15e\text{Å}^{-3}$ so caution must be exercised in the interpretation of detail in the observed density.

4. The biggest discrepancies between theory and experiment occur in the lone pair regions, where peaks are higher in the theoretical maps. The comparison, however, is hampered by inadequacies in the thermal motion treatment.

In addition to establishing the reproducibility of the observed features, and providing an estimate of their quantitative reproducibility, a somewhat unexpected deviation of the deformation density of the water molecule from the isolated molecule symmetry was observed. The oxygen lone-pair density was found to be polarized into the direction of the short hydrogen bond, an indication that intermolecular effects on the density were becoming observable. It was clear that application of the methods to physical and chemical problems was entirely feasible.

SOME SELECTED APPLICATIONS

One of the most interesting applications of the charge density method is the ability to distinguish between different theoretical calculations by comparison of experimental and theoretical ground state densities. It was already known from the early work by Smith and Richardson[34] that semi-empirical and minimal basis set calculations do not do not give a proper representation of the overlap density in covalent bonds, and that at least polarized basis sets are needed to get a reasonable build up in this region in a molecule such as N_2. Comparison of experimental deformation maps on formamide with thermally smeared extended basis set HF quality maps, gave almost featureless $\Delta(\Delta\rho)$ maps, calculated by subtraction of the experimental and theoretical deformation maps.[35] Some remaining features, at the boundary of significance, could be due to the lack of electron correlation in the HF calculations, an effect investigated theoretically by V.H. Smith for the water molecule,[36] and by Becker for CO.[37]

Pendellosung measurements allow very accurate determination of structure factors. Velders and Feil[38] showed that density functional calculations on clusters of Si-atoms, together with the stockholder method to isolate a bonded atom, give ab-initio structure factors which are in excellent agreement with the observed Pendellosung values of Saka and Kato,[39] corrected by Cummings and Hart[40] for anomalous scattering.

Electrostatic interactions

With the rapid rise of molecular simulation methods, the need for reliable force fields has become urgent. The calculation of electrostatic interaction requires a knowledge of the atomic charge distribution in molecules. The size of molecules of biological importance often prohibits the use of ab-initio quantumchemical methods and, in the absence of reliable rules for transferability of local charge distributions only X-ray diffraction seems capable of yielding the required information. Berkovitch-Yellin[41] showed how crystal morphology could be derived from X-ray charges by calculating the energy of two-dimensional layers of molecules. More recently Spackman et al[42] calculated energies of molecular interactions from Bragg diffraction data which agree well with results obtained with other methods. Electrostatic interaction plays a major role in hydrogen bonding. Krijn and Feil[43] showed that quantitative agreement between observed and calculated structure factors can only be obtained by taking mutual polarization of interacting molecules and the effects of Pauli repulsion into account. The observation of increased capability for hydrogen bonding of water molecules that participate in a hydrogen bond network confirms many-body effects in water clusters described by Hermansson.[44]

Charge partitioning

An important step in relating the results of electron density studies to the main body of physics and chemistry was made when procedures were developed to partition the electron density distribution into fragments which can be associated with the concept of atoms as elementary components of the molecule or solid. Although the electron density distribution in crystals as determined by X-ray diffraction is a continuous function, the total density is dominated by the peaks at or near the nuclear positions. The importance of partitioning of the density distribution is manyfold:

- in general many concept in chemistry and physics are expressed in terms of the properties of atoms and ions, rather than as a function of the continuous distribution.

- the electrostatic interaction between molecules can be calculated as the sum of the electrostatic interactions between the constituent atoms. Since the ratio of the size of the interacting particle to the distance between the particles is much more favorable for atoms than for molecules, the use of atomic fragments requires fewer multipole moments per fragment and leads to more rapid convergence of the calculation.[45]

- as the atomic fragments may be approximately transferable between molecules, the density of larger molecular units may be constructed. This is of particular importance in the calculation of interactions between macromolecules in drug design.

A number of partitioning methods have been developed. Atomic charges and higher moments follow from the application of charge density refinement formalisms. Since the density functions generally overlap, such pseudoatoms are not strictly localized. The κ-refinement model is the simplest of such models.

Two other partitioning methods have found widespread acceptance. Bader[46] discovered a procedure based on the total electron density distribution, in which the boundaries of the atoms are planes for which the vector product of the normal to the plane and the gradient of the electron density distribution is zero. The so defined atoms satisfy a number of important quantummechanical properties, including the virial theorem. Hirshfeld[47] developed the *stockholder method* in which the deformation density is partitioned. It is based on the concept of the *promolecule*, defined as the superposition of free spherical atoms located at the molecular atomic positions. The stockholder method allocates the density at each point of the distribution in proportion to each atom's density in the promolecule, i.e. every atom gets back a share proportional to its investment. Atomic multipole moments can be obtained from the partitioned density by numerical integration. It follows from the definition that the atoms of the promolecule do not carry any charge. This is not true for the atoms in the Bader definition.[48]

Eisenstein[49] carried out an extensive comparison between the stockholder multipole moments obtained from X-ray data and those derived from a quantumchemically calculated density distribution. The agreement between the charges was good, between the dipole moments fair, while the correlation between experimental and quantumchemical quadrupole moments was marginal. However, since the basis set used was of intermediate quality, the theoretical values of the higher moments may not be reliable.

Transition metal compounds

In principle transition metal complexes are less suitable for charge density studies because of the larger number of much less perturbed inner shell electrons. But they often have unusual chemistry, of relevance in catalysis and in a large number of biological processes. They also play a key role in many materials, such as the high Tc superconducting cuprates. Furthermore, theoretical calculations for many-electron systems are often ambiguous. They require approximations, which can be tested by comparison with the experiment. The potential for applications is therefore considerable.

The first studies on larger complexes were done by Iwata and Saito on $Co(NH_3)_6C(CN)_6$ and $Co(NH_3)_6Cr(CN)_6$,[50] by Rees and Coppens on benzene chromium tricarbonyl,[51] and by Rees and Mitschler on $Cr(CO)_6$.[52] Information on metal-metal bonding was obtained from experimental and combined theoretical and experimental studies on a number of complexes containing a metal-metal bond such as Cr-Cr,[53] Fe-Fe[54] and Mn-Mn.[55] The super-short (1.879Å at 79K) bond in tetrakis (μ_2-hydroxy-6-methyl pyridine dichromium [$Cr_2(mhp)_4$] showed a large accumulation at bond center (Fig. 5). In

Fig. 5. The deformation electron density in Cr_2 (mhp)$_4$, a compound with an extremely short Cr-Cr bond (ref. 53).

most other metal-metal bonded complexes studied the bonding density becomes only apparent if molecular fragments are subtracted as demonstrated by Hall and coworkers.[56] The preferential occupancy of field stabilized d-orbitals was clearly evident in all these studies. Several years later a systematic method became available to obtain the d-orbital occupancies from the results of the multipole refinement.[57] Systematic application to a number of iron(II)porphyrins showed often remarkable agreement with theoretical values, as for example in bispyridine iron(II) tetraphenylporphyrin.[58] In other cases, such as the intermediate spin (triplet) state complexes of iron(II)phthalocyanine and iron(II)tetraphenylporphyrin the experimental results helped clear up long standing contradictions in theoretical results.[59] In systems with very closely-spaced energy levels, however, the effect of intermolecular interactions on the electronic ground state cannot be neglected. Studies of different polymorphs, and mixed molecular crystals may become useful, in particular if the experimental accuracy can be improved further.

WHAT'S NEXT ?

At present, charge density analysis is far from a routine technique. It requires careful collection of often very large data sets, and attention to experimental effects such as extinction, absorption and thermal diffuse scattering and the shape of the peak profiles. In the best data sets collected on four circle diffractometers, agreement between symmetry-related reflections of 1-2% in intensity can be achieved. Such internal consistency is a necessary condition if the results are to be interpreted with any confidence.

What can be done to reduce the effort and increase the accuracy of the results? In some cases very significant advances in accuracy have been reported. When perfect crystals are available, as in the case of silicon, special techniques (in particular the measurement of the spacing of the "Pendellosung" fringes) may be used which allow measurement of the structure factors with an accuracy of 0.1% or better.[39,40,60,61] Other techniques are being developed. For example, advances in high-speed computing have made it possible to accurately correct for extensive multiple diffraction effects in electron diffraction, and to obtain charge densities in solids by this technique, as demonstrated in the study of GaAs.[62]

Many of the factors limiting the accuracy of the experimental structure factors can be reduced or eliminated if smaller crystals and shorter wavelengths could be used. Examples are absorption, and effects that depend on the interaction between two or more beams in the crystal such as extinction and multiple reflection. It is here that the advent of synchrotron radiation may greatly expand the possibilities. It is intense, thus allowing the use of smaller crystals and/or and improvement in the accuracy of the weaker reflections, and is tunable to shorter wavelengths, in particular at beamlines receiving radiation from insertion devices. The much higher brightness of third generation sources now under development and the use of accurate area detectors can greatly reduce data collection times. The use of synchrotron radiation in Charge Density Studies is described later in this volume.[63] The technical advances now underway are such that charge density analysis can become a truly analytical technique for calibration of theoretical methods, and application to a large number of problems in Physics, Materials Science, Chemistry and Molecular Biology. It is likely that the changes in the next twenty years will be as dramatic as the onces we have witnessed in the past decades!

ACKNOWLEDGEMENTS

Support of the work in Buffalo by the National Science Foundation (CHE8711736) and the National Institute of Health (5RO1HL23884-09) is gratefully acknowledged.

REFERENCES

1. P. Debye, Dispersion of Rontgen rays, <u>Ann. Phys</u>. 46:809 (1915).
2. P. Debye and P. Scherrer, Atomic structure, <u>Physikalische Zeitschrift</u> 19:474 (1918).
3. W.L. Bragg, R.W. James, and C.H. Bosanquet, The distribution of electrons around the nucleus of the sodium and chlorine atoms, <u>Phil. Mag.</u> 44:433 (1922).
4. R.W. James, "The optical principles of the diffraction of X-rays," Oxbow Press , Woodbridge, CT (1982).
5. W.H. Bragg, The intensity of X-ray reflection by diamond, <u>Proceedings of the Physical Society of London</u> 33:301 (1921).
† It is interesting that, as recorded by James (ref. 4), the original observation of the 222 reflection by W.H. Bragg was almost certainly due to multiple scattering later discovered by Renninger.
6. R.W.G. Wyckoff, A powder spectrometric study of urea, <u>Zeitschrift fur Kristallographie</u> 75:529 (1930).
7. R. Brill, H.G. Grimm, C. Hermann, and C. Peters, Application of Rontgen-ray Fourier analysis to questions of chemical linkage, <u>Ann. Phys.</u> 34:393 (1939).
8. S. Goettlicher, R. Kuphal, G. Nagorsen, and E. Woelfel, X-ray detection of the electron density distribution in crystals (VI) Electron distribution in Si, <u>Z. Phys. Chem.</u> NF21:133-145 (1959). S. Goettlicher and E. Woelfel, X-ray detection of the electron-density distribution in crystals. (VII) Electron distribution in the diamond and Si lattices, <u>Z. Elektrochim.</u> 63:891 (1959).
9. S. Goettlicher, Der Beitrag der Thermisch Diffusen Streustrahlung zur Intensitat der Roentgeninterferenzen und die Elektronendichteverteilung im NaCl, <u>Acta Cryst.</u> B24:122-129 (1968).
10. M. Roux, S. Besnainou, and R. Daudel, The distribution of electron density in molecules I. Effect of the chemical bond, <u>J. Chim. Phys.</u> 54:218 (1956). M. Roux, The distribution of electron densities in molecules III. Influence of the nature of the molecular wave function on the representation of the effect of chemical bonds, <u>J. Chim. Phys.</u> 55:754 (1958).
11. E.W. Hughes, Crystal structure of melamine, <u>J. Am. Chem. Soc.</u> 63:1737 (1941).
12. P. Coppens, Comparative X-ray and neutron diffraction study of bonding effects in <u>s</u>-triazine, <u>Science</u> 158:1577 (1967).
13. P. Coppens, T.M. Sabine, R.G. Delaplane, and J.A. Ibers, An experimental determination of the asphericity of the atomic charge distribution in oxalic acid dihydrate, <u>Acta Cryst</u>. B25:2451 (1969).
14. R. Brill, H. Dietrich, and H. Dierks, Die Verteilung der Bindungselektronen im Dekaboran-Molekul ($B_{10}H_{14}$), <u>Acta Cryst.</u> B27:2003-2018 (1971).

15. L.H. Jensen and M. Sundaralingam, Hydrogen atom thermal parameters, <u>Science</u> 145:1185 (1964).
16. R.F. Stewart, E.R. Davidson, and W.T. Simpson, Coherent X-ray scattering for the hydrogen atom in the hydrogen molecule, <u>J. Chem. Phys.</u> 42:3175-3187 (1965).
17. P.P.M. Groenewegen and D. Feil, Molecular form factors in X-ray crystallography, <u>Acta Cryst.</u> A25:444 (1969).
18. A. Hartmann and F.L. Hirshfeld, Structure of Cis-1,2,3-Tricyanocyclopropane, <u>Acta Cryst.</u> A20:80 (1966). T. Ito and T. Sakurai, The structure and electron density of ethyleneimine quinone, <u>Acta Cryst.</u> B29:1594 (1973).
19. H. Irngartinger, Electron density distribution in the bonds of cumulenes and small ring compounds, <u>in</u>: "Electron distributions and the chemical bond," P. Coppens and M.B. Hall, eds., Plenum, New York (1982).
20. Z. Berkovitch-Yellin and L. Leiserowitz, Electron density distribution in cumulenes: Low-temperature X-ray study of tetraphenylbutatriene, <u>J. Am. Chem. Soc.</u> 97:5627 (1975). Z. Berkovitch-Yellin, and L. Leiserowitz, Electron density distribution in cumulenes: An X-ray study of tetraphenylbutatriene at 20°C and -160°C, <u>Acta Cryst.</u> B33:3657 (1977). Z. Berkovitch-Yellin, and L. Leiserowitz, and F. Nader, Electron density distribution in cumulenes: An X-ray study of the complex allenedicarboxylic acid-acetamide (1:1) at -150°C, <u>Acta Cryst.</u> B33:3670 (1977). H. Irngartinger and H.-U. Jager, Kristall- und Molekularstrukturen von Zwei Carbodiimiden: Bis(diphenylmethyl) Carbodiimid und Bis(p-methoxyphenyl)Carbodiimid, <u>Acta Cryst.</u> B34:3262 (1978).
21. W.E. Streib and W.N. Lipscomb, Growth, orientation, and X-ray diffraction of single crystals near liquid helium temperatures, <u>Proc. Nat. Acad. Sci.</u> 48:911 (1962).
22. R.J. Weiss, "X-ray determination of electron distributions," John Wiley and Sons, New York (1966).
23. F.L. Hirshfeld, Difference densities by least-squares refinement: Fumaramic acid, <u>Acta Cryst.</u> B27:769 (1971).
24. R. McWeeny, X-ray scattering by aggregates of bonded atoms. III. The bond scattering factor: Simple methods of approximation in the general case, <u>Acta Cryst.</u> 6:531-637 (1953). R. McWeeny, X-ray scattering by aggregates of bonded atoms. II. The effect of the bonds: With an application to H_2, <u>Acta Cryst.</u> 5:463-468 (1952).
25. R.F. Stewart, Electron population analysis with rigid pseudoatoms, <u>Acta Cryst.</u> A32:565-574 (1976B)
26. K. Kurki-Suonio and V. Meisalo, (1967) Ann. Acad. Sci. Fen. AVI, 1 Spherical harmonic expansions in X-ray diffraction analysis. K. Kurki-Suonio and V. Meisalo, (1967) Ann. Acad. Sci. Fenn., Ser. AVI 243, 3 Non-spherical analysis of charge density.
27. N.K. Hansen and P. Coppens, Electron population analysis of accurate diffraction data. VI. Testing aspherical atom refinements, <u>Acta Cryst.</u> A34:909 (1978).
28. R.F. Stewart, Valence structure from X-ray diffraction data: an L-shell projection method, <u>J. Chem. Phys.</u> 53:205-213 (1970).
29. P. Coppens, Determination of nuclear positions from X-ray data by a double-atom refinement method, <u>Acta Cryst.</u> B27:1931 (1971).
30. J.C. Slater, Atomic shielding constants, <u>Phys. Rev.</u> 36:57 (1930) as quoted in C.A. Coulson, Valence, Oxford University Press, 1961.
31. P. Coppens, T.N. Guru Row, P. Leung, P.J. Becker, Y.W. Yang, and E.D. Stevens, Electron population analysis of accurate diffraction data. VII. Net atomic charges and molecular dipole moments from spherical atom X-ray refinements and the relation between atomic charge and shape, <u>Acta Cryst.</u> A35:63 (1979).
32. P. Coppens, T.N. Guru Row, P. Leung, and E.D. Stevens, ACA Program and Abstracts, Series 2, Vol. 6, p. 7 (1978). "Variation of atomic charge and shape in the spherical atom approximation: The poor man's charge density refinement."
33. P. Coppens, J. Dam, S. Harkema, D. Feil, R. Feld, M.S. Lehmann, R. Goddard, C. Kruger, E. Hellner, H. Johansen, F.K. Larsen, T.F. Koetzle, R.K. McMullan, E.N. Maslen, and E.D. Stevens, Commission on charge, spin and momentum density project on comparison of structural parameters and electron density maps of oxalic acid dihydrate, <u>Acta Cryst.</u> A40:184 (1984).
34. P.R. Smith and J.W. Richardson, Electron density shifts during chemical bond formation, <u>J. Phys. Chem.</u> 69:3346 (1965). P.R. Smith and J.W. Richardson, Bonding and hybridization in the nitrogen molecule, <u>J. Phys. Chem.</u> 71:924 (1967).

35. E.D. Stevens, J. Rys, and P. Coppens, Quantitative comparison of theoretical calculations with the experimentally determined electron density distribution of formamide, J. Am. Chem. Soc. 100:2324 (1978).
36. V.H. Smith, Jr., Theoretical determination and analyses of electronic charge distributions, Physica Scripta 15:147-162 (1977)
37. P. Becker, Theoretical electronic densities: Accuracy, comparison with experiments, effect of thermal smearing, Acta Cryst. A31:S227 (1975).
38. G.J.M. Velders and D. Feil, Calculation of the electron density distribution in silicon by the density-functional method. Comparison with X-ray results, Acta Cryst. B45:359-364 (1989).
39. T. Saka and N. Kato, Accurate measurement of the Si structure factor by the Pendellosung method, Acta Cryst. A42:469-478 (1986).
40. S. Cummings and M.J. Hart, Redetermination of absolute structure factors for silicon at room and liquid-nitrogen temperatures, Aust. J. Phys. 41:423-431 (1988).
41. Z. Berkovitch-Yellin, Toward an ab initio derivation of crystal morphology, J. Am. Chem. Soc. 107:8239-8253 (1985).
42. M.A. Spackman, H.P. Weber, and B.M. Craven, Energies of molecular interactions from Bragg diffraction data, J. Am. Chem. Soc. 110:775-782 (1988).
43. D. Feil, X-ray diffraction and Charge distribution, this volume.
44. K. Hermansson, Many-body effects in tetrahedral water clusters, J. Chem. Phys. 89:2149-2159 (1988).
45. G. Moss and D. Feil, Electrostatic molecular interaction from X-ray diffraction data. I. Development of the method; test on pyrazine, Acta Cryst. A37:414 (1981).
46. R.F. Bader, this volume
47. F.L. Hirshfeld, Bonded-atom fragments for describing molecular charge densities, Theor. Chim. Acta 44:129 (1977).
48. E.N. Maslen and M.A. Spackman, Atomic charges and electron density partitioning, Aust. J. Phys. 38:273 (1985). E. Krafka, New ways to analyze electron densities in molecules, Thesis, University of Koln, Germany, 1984.
49. M. Eisenstein, Static deformation densities for cytosine and adenine, Acta Cryst. B44:412 (1988). M. Eisenstein, SCF deformation densities and electrostatic potentials of purines and pyrimidines, Int. J. Quantum Chemistry 33:127 (1988).
50. M. Iwata, X-ray determination of the electron distribution in crystals of $[Co(NH_3)_6]$ $[Cr(CN)_6]$ at 80K, Acta Cryst. B33:59-69 (1977). M. Iwata and Y. Saito, The crystal structure of hexamminecobalt(III) hexacyanocobaltate(III): An accurate determination, Acta Cryst. B29:822-832 (1973).
51. B. Rees and P. Coppens, Electronic structure of benzene chromium tricarbonyl by X-ray and neutron diffraction at 78K, Acta Cryst. B29:2515 (1973).
52. B. Rees and A. Mitschler, Electronic structure of chromium hexacarbonyl at liquid nitrogen temperature. 2. Experimental study (X-ray and neutron diffraction) of σ and π bonding, J. Am. Chem. Soc. 98:7918 (1976).
53. M. Benard, P. Coppens, M.L. DeLucia, and E.D. Stevens, Experimental and theoretical electron density analysis of metal-metal bonding in dichromium tetraacetate, Inorg. Chem. 19:1924 (1980). A. Mitschler, B. Rees, R. Wiest, and M. Benard, Electron deformation density for the "supershort" Cr Cr bond: A joint experimental and theoretical study, J. Am. Chem. Soc. 104:7501 (1982).
54. A. Mitschler, B. Rees, and M.S. Lehmann, Electron density in Bis(Dicarbonyl-Pi-Cyclopentadienyliron) at liquid nitrogen temperature by X-ray and neutron diffraction, J. Am. Chem. Soc. 100:3390 (1980). M. Benard, Electron deformation density distributions in binuclear complexes of transition metals: Computation and interpretation from ab initio molecular orbital wavefunctions, in: "Electron distributions and the chemical bond," P. Coppens and M.B. Hall, eds., Plenum, New York (1982).
55. W. Heyser, Thesis, Free University, Amsterdam. W. Heyser, E.J. Baerends, and P. Ros, Faraday Symp. Chem. Soc. 14:211 (1980).
56. M.B. Hall, Computation and interpretation of electron distributions in inorganic molecules, in: "Electron distributions and the chemical bond," P. Coppens and M.B. Hall, eds., Plenum, New York (1982). P.T. Chesky and M.B. Hall, Electronic

structure of metal clusters. 1. Photoelectron spectra and molecular orbital calculations on alkylidynetricobalt nonacarbonyl clusters, <u>Inorg. Chem.</u> 20:4419 (1981).

57. A. Holladay, P.C. Leung, and P. Coppens, Generalized relations between d-orbital occupancies of transition metal atoms and electron density multipole population parameters from X-ray diffraction data, <u>Acta Cryst.</u> A39:377-387 (1983). P. Becker and P. Coppens, Analysis of charge and spin densities, *International Tables for X-ray Crystallography* (In Press).

58. N. Li, P. Coppens, and J. Landrum, Electron density studies of porphyrins and phthalocyanines. 7. The electronic ground state of bis(pyridine)(meso-tetraphenylporphinato)iron(II), <u>Inorg. Chem.</u> 27:482-488 (1988).

59. N. Li, Z. Su, P. Coppens, and J. Landrum, X-ray diffraction study of the electronic ground state of (Meso-Tetraphenyl porphinato) Iron (II): Evidence for the effect of intermolecular interactions, <u>J. Am. Chem. Soc.</u>, In Press. P. Coppens and L. Li, Electron density studies of porphyrins and phthalocyanines. III. The electronic ground state of iron(II)phthalocyanine, <u>J. Chem. Phys.</u> 81:1983-1993 (1984).

60. P.J.E. Aldred and M. Hart, The electron distribution in Silicon. I. Experiment, <u>Proc. Roy. Soc. Lond.</u> A332:223-238 (1973). P.J.E. Aldred and M. Hart, The electron density distribution in Silicon. II. Theoretical interpretation, <u>Proc. Roy. Soc. Lond.</u> A332:239-254 (1973).

61. N. Kato and S. Tanamura, Absolute measurement of structural factors of Si by using X-ray Pendellosung and interferometry fringes, <u>Acta Cryst.</u> A28:69-80 (1972).

62. J.M. Zuo, J.C.H. Spence, and M. O'Keeffe, Bonding in GaAs, <u>Phys. Rev. Lett.</u> 61:353 (1988).

63. P. Coppens, The use of synchrotron radiation and its promise in charge density research, this volume.

DETERMINATION OF ATOMIC AND STRUCTURAL PROPERTIES FROM EXPERIMENTAL CHARGE DISTRIBUTIONS

Richard F.W. Bader and Keith E. Laidig

Department of Chemistry
McMaster University
Hamilton, Ontario L8S 4M1
Canada

SCIENCE AND CLASSIFICATION

Science is classification. A discipline begins with the empirical classification of observations. It yields to science when these observations are classified in such a way as to reflect the structure imposed on a system by the physics that governs its behaviour. The structure may be real or abstract. An example of abstract structure, as determined by the properties of Hilbert space and associated linear operators, is the classification of a given quantum state in terms of the eigenvalues of the maximal commuting set of operators predicted by quantum mechanics to be the observables for that state. In the Russell-Saunders coupling scheme for example, each atomic state is classified by the eigenvalues of the observables H, $\hat{L}^2$, $\hat{L}_z$, $\hat{S}^2$, $\hat{S}_z$ ($\hat{J}^2$ and $\hat{J}_z$). In the limiting case, prediction and classification are synonymous and the classification is based on what physics predicts can be observed.

The atomic classification of the properties of matter is an example of a classification resulting from physical structure, the structure exhibited by the distribution of electronic charge in real space. The form assumed by the distribution of charge in a molecular system is the physical manifestation of the forces acting within the system. Dominant among these is the attractive force exerted on the electronic charge density by the nuclei, a consequence of the localized nature of the nuclear charge. This interaction is responsible for the single most important topological property exhibited by a molecular charge distribution of a many-electron system -- that $\rho(r,X)$ exhibits significant local maxima only at the positions of the nuclei. This is an observation based on experimental results obtained from x-ray diffraction studies on crystals and on the results of theoretical calculations on a large number of systems.

The Application of Charge Density Research to Chemistry and Drug Design
Edited by G. A. Jeffrey and J. F. Piniella, Plenum Press, New York, 1991

The atomic classification scheme is the cornerstone of chemistry. It is based upon the observation that atoms and functional groupings of atoms make recognizable contributions to the total properties of a system. In practice, we recognize a group in a system or predict its effect upon the static and dynamic properties of the system in terms of a set of properties assigned to that group.

Because the charge density ρ exhibits local maxima at the nuclei, recognizable atomic forms are created within a molecular charge distribution. These forms are so dominant in determining the structure of the charge density that their individual properties make characteristic contributions to the properties of the total system. Thus the atomic basis for the classification of chemical properties could and did evolve as the working model of chemistry before its underlying physical basis was known.

This classification is predicted by physics. The atomic forms defined by the topology of the charge distribution of a molecule or a crystal are open systems with boundaries defined in real space. Electronic charge and momentum can be exchanged across the boundaries they share with neighbouring atoms. The boundary of the topological atom satisfies the quantum condition that defines an open system.[1,2] Thus all of the properties of an atom in a molecule or a crystal employed in the atomic classification of the properties of matter are predicted by quantum mechanics. The same topological properties of the charge density which define an atom as an open quantum subsystem, also lead to the definition of bonds, structure and structural stability[3] and the whole of the molecular structure hypothesis – that a molecule is a collection of atoms with definable properties, linked by a network of bonds – is given a basis in physics.

The topology of ρ yields a faithful mapping of the chemical concepts of atoms, bonds and structure. There is however, no indication of maxima in ρ corresponding to the electron pairs of the Lewis model, a model secondary only to the atomic hypothesis itself in our interpretation of chemical reactivity and molecular geometry. The physical basis of this most important model is one level of abstraction above the visible topology of the charge density and appears instead in the topology of the Laplacian of ρ. This function is the scalar derivative of the gradient vector field of the charge density, the quantity $\nabla^2\rho$, and it determines where electronic charge is locally concentrated and depleted. The local charge concentrations provide a mapping of the electron pairs of the Lewis model.[4,5] The Laplacian of the charge density plays a dominant role throughout the theory, appearing in the constraint determining the boundary of a quantum subsystem and relating the spatial properties of the charge density to the local contributions to the energy.

The experimental determination of the electronic charge density is therefore, of paramount importance. The charge density contains the information that enables one to define the atoms of chemistry and predict many of their static and reactive properties including the sites of electrophilic and nucleophilic attack. In addition, one can define and classify the bonds which link the atoms and determine the system's molecular structure. The manner in which this information can be obtained begins with the description and classification of the topological properties of the electronic charge density.

TOPOLOGICAL PROPERTIES OF THE CHARGE DENSITY

The Charge Density

The state function Ψ determines all of the information that can be known about a quantum system. If one desires a theory of molecular structure that is free of arbitrary or subjective assumptions, no information beyond that contained in Ψ should be used in its development. The state function for a molecular system is a function of the electronic and nuclear coordinates and of the time t, $\Psi(x, X, t)$, where x denotes the collection of electronic space and spin coordinates and X the collection of nuclear coordinates. While the general theory applies to the time dependent case, we are primarily interested in systems in stationary states whose properties do not change with time. Denoting a solution to Schrödinger's stationary state equation for a fixed arrangement of the nuclei by $\psi(x;X)$, the probability of finding each one of N electrons in a particular volume element $d\tau_i = dx_i dy_i dz_i$ with spin coordinate σ_i (equal to either the α or β spin coordinate) for a given configuration of the nuclei X is given by

$$\psi^*(x;X)\psi(x;X)dx_1 dx_2 \ldots dx_N \tag{1}$$

where $dx_i = d\tau_i \sigma_i$. The corresponding probability independent of spin is obtained by summing eqn (1) over the spin coordinates. If the summing over all spins is followed by an integration over the spatial coordinates of all the electrons but one (it matters not which electron is chosen, since ψ is an antisymmetrized function and hence all electrons are equivalent) the resulting expression gives the probability of finding one electron in some elemental volume, independent of the positions of the remaining electrons,

probability of one electron being in $d\tau_1$ =

$$\Sigma(\text{spins})\left\{\int d\tau_2 \int d\tau_3 .. \int d\tau_N \psi^*(x;X)\psi(x;X)\right\}d\tau_1 \tag{2}$$

Multiplication of this probability by the number of electrons N gives the probability of finding any one of the electrons in $d\tau_1$, that is, the total probability of finding electronic charge in $d\tau_1$. The corresponding probability density, the probability per unit volume, is called the electronic charge density and is denoted by the symbol $\rho(r;X)$, the space coordinate of a single electron being denoted by the position vector $r = ix + jy + kz$,

$$\rho(r;X) = N \Sigma(\text{spins})\left\{\int d\tau_2 \int d\tau_3 .. \int d\tau_N \psi^*(x;X)\psi(x;X)\right\} \tag{3}$$

The subscript 1 is suppressed in this equation since the result refers to all the electrons. The form of integration given in eqn (3) which yields the density of charge in real space, recurs throughout the theory and is designated in an abbreviated manner as

$$\rho(r;X) = N \int d\tau' \psi^*(x;X)\psi(x;X) \tag{4}$$

where $\int d\tau'$ denotes the summation over the spin coordinates of all electrons and integration over the cartesian coordinates of all electrons but one.

The Dominant Form in the Charge Density

The local maxima in a charge distribution are illustrated in Fig. 1. The charge density in B_2F_4, is displayed in three planes, as a projection in the third dimension above each geometric plane. The charge distribution exhibits a maximum at the position of each nucleus in Fig. 1a, the plane of the nuclei. Fig. 1b is for a plane perpendicular to that shown in 1a, obtained by a rotation about the boron–boron axis and containing these nuclei. The density again exhibits maxima at the positions of the boron nuclei. The observation of a maximum in $\rho(r)$ at a nuclear position is true when the distribution is viewed in any plane containing the nucleus. It is this property that is classified as a local maximum in $\rho(r)$.

This behaviour of $\rho(r)$ is to be contrasted with that displayed at the midpoint of the boron-boron axis. In the planes shown in 1a and 1b, $\rho(r)$ has the appearance of a saddle at this point, while in the plane perpendicular to and bisecting the boron–boron axis, $\rho(r)$ is a maximum at this same point. In this case $\rho(r)$ is a maximum in one particular plane. Knowledge of $\rho(r)$ in one or two dimensions is insufficient to characterize its three-dimensional form. What is needed is a method of summarizing in a precise manner the principal topological features of a charge distribution. This information is provided by the curvatures of $\rho(r)$ at its critical points[6,7].

Critical Points and their Classification

Each topological feature of $\rho(r)$, whether it be a maximum, a minimum or a saddle, has associated with it a point in space called a *critical point*, where the first derivatives of $\rho(r)$ vanish. Thus at such a point, denoted by the position vector r_c, $\nabla\rho(r_c) = 0$, where $\nabla\rho$ denotes the operation

$$\nabla\rho = i\partial\rho/\partial x + j\partial\rho/\partial y + k\partial\rho/\partial z \qquad (5)$$

Whether a function is a maximum or a minimum at an extremum is, of course, determined by the sign of its second derivative or curvature at this point. The second derivative of a function $f(x)$ at x is the limiting difference between its two first derivatives or tangent lines which bracket that point as expressed in eqn (6).
$$d_2 f(x)/dx^2 =$$

$$\lim\left[[\lim\{[f(x + \Delta x) - f(x)]/\Delta x\} - \lim\{[f(x) - f(x - \Delta x)]/\Delta x\}]/\Delta x\right] \qquad (6)$$

At a point where $f(x)$ is a minimum, the second derivative is the difference between a positive and a negative curvature and is, therefore, greater than zero while at a maximum in $f(x)$, the second derivative is the difference between a negative and a positive curvature and the result is a value less than zero. For values of x lying between such extrema, both slopes are either positive or negative and the curvature can be of either sign depending upon whether x is in the region of a maximum or a minimum. The second derivative undergoes a change in sign at the point of inflexion where its value is zero. It

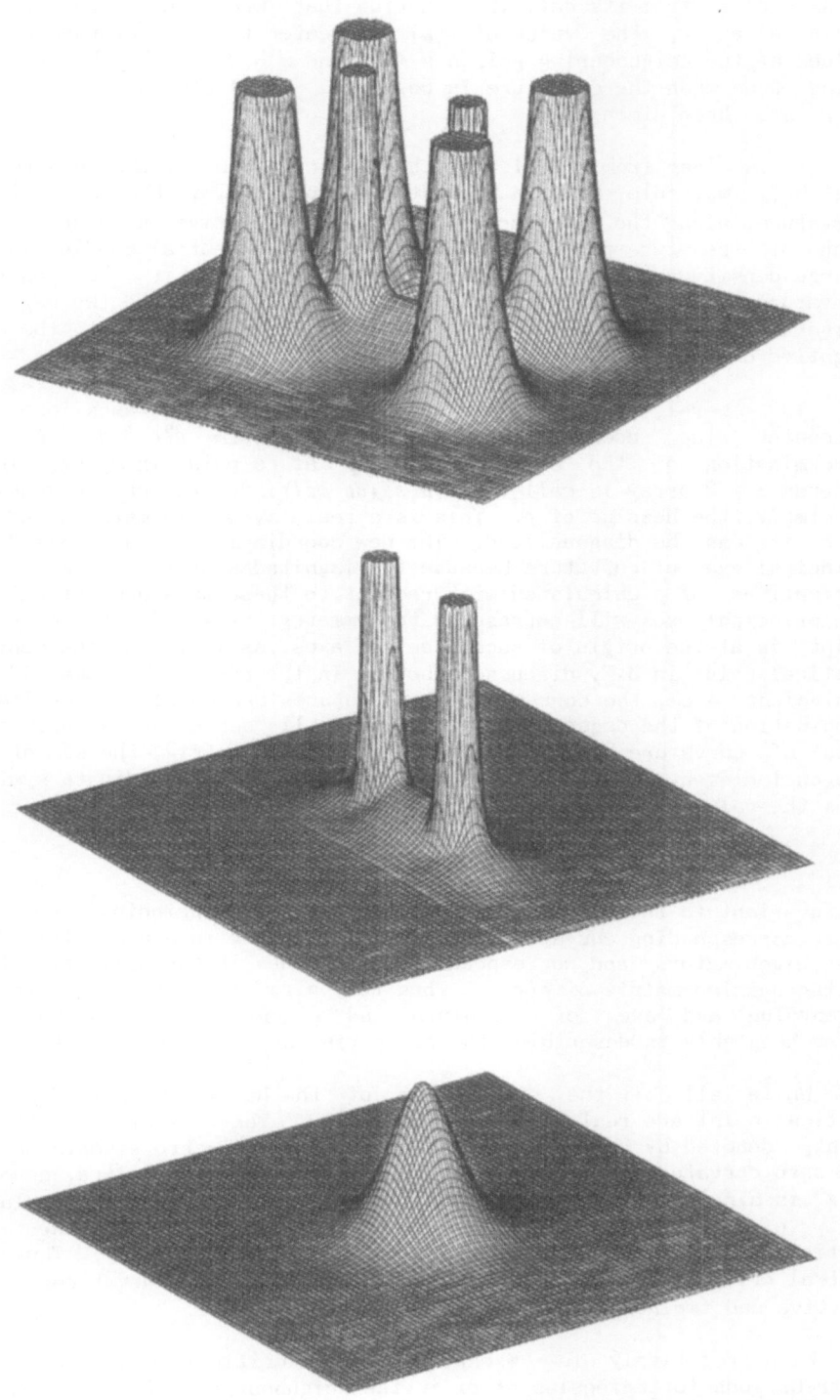

Fig. 1   Relief maps of the electronic charge density in B_2F_4 in three orthogonal planes containing the critical point midway between the boron nuclei.

is also clear from its defining equation that when the curvature is negative at x, the value of f(x) is greater than the average of its values at the neighbouring points x + dx and x − dx, with the reverse being true when the curvature is positive. These considerations carry over into three dimensions.

It is clear from Fig. 1 that the critical point at the centre of the B_2F_4 molecule between the two boron nuclei has one positive curvature, along the internuclear axis, and a negative curvature along each of the two perpendicular axes. Thus the central saddles in the charge density appearing in Figures 1a and 1b exhibit the positive curvature along the boron-boron axis and one each of the two negative curvatures, while the maximum shown in Fig.1c exhibits just the two negative curvatures.

In general, for an arbitrary choice of coordinate axes, one will encounter nine second derivatives of the form $\partial^2\rho/\partial x\partial y$ in the determination of the curvatures of ρ at a point in space. Their ordered 3 × 3 array is called the *Hessian matrix* of the charge density, or simply, the Hessian of ρ. This is a real, symmetric matrix and as such it can be diagonalized. The new coordinate axes are called the principal axes of curvature because the magnitudes of the three second derivatives of ρ calculated with respect to these axes are extremized. The principal axes will correspond to symmetry axes if the critical point is at the origin of such a set of axes, as it is for the central critical point in B_2F_4 discussed above. In the case of symmetrically equivalent axes, the corresponding curvatures are equal and any linear combination of the degenerate set of axes will serve as a principal axis of curvature. The trace of the Hessian matrix, the sum of its diagonal elements, is invariant to a rotation of the coordinate system. Thus the value of the quantity $\nabla^2\rho$, called the Laplacian of ρ,

$$\nabla^2\rho = \nabla\cdot\nabla\rho = \partial^2\rho/\partial x^2 + \partial^2\rho/\partial y^2 + \partial^2\rho/\partial z^2 \qquad (7)$$

is invariant to the choice of coordinate axes. The principal axes and their corresponding curvatures at a critical point in ρ are obtained as the eigenvectors and corresponding eigenvalues in the diagonalization of the Hessian matrix of $\rho(r_c)$. Thus the pairs of names 'curvature and eigenvalue' and 'axes of curvature and eigenvectors' can be used interchangeably in describing the properties of a critical point in ρ.

While all of the eigenvalues of the Hessian matrix of ρ at a critical point are real, they may equal zero. The *rank* of a critical point, denoted by ω, is equal to the number of non-zero eigenvalues or non-zero curvatures of ρ at the critical point. The *signature*, denoted by σ, is simply the algebraic sum of the signs of the eigenvalues, i.e., of the signs of the curvatures of ρ at the critical point. The critical point is labelled by giving the duo of values (ω,σ). Thus the central critical point in B_2F_4 with three non-zero curvatures, one positive and two negative, is a (3,−1) critical point.

With relatively few exceptions, the critical points of charge distributions for molecules at or in the neighbourhood of energetically stable geometrical configurations of the nuclei are all of rank three. The near ubiquitous occurrence of critical points with $\omega = 3$ in such cases is another general observation regarding the topological behaviour of molecular charge distributions. It is in terms of the properties of critical points with $\omega = 3$ that the elements of molecular structure are defined. A critical point with $\omega < 3$, i.e., with at least one zero curvature, is said to be degenerate. Such a critical point is unstable in the sense that a small change in the charge density, as caused by a displacement of the nuclei, causes it to either vanish or to bifurcate into a number of non-degenerate or stable ($\omega = 3$) critical points. Since structure is generic in the sense that a given structure or arrangement of bonds persists over a range of nuclear configurations, the observed limited occurrence of degenerate critical points is not surprising. One correctly anticipates that the appearance of a degenerate critical point in a molecular charge distribution denotes the onset of structural change.

There are just four possible signature values for critical points of rank three. They are:

(3,-3)   all curvatures are negative and ρ is a local maximum at r_c

(3,-1)   two curvatures are negative and ρ is a maximum at r_c in the plane defined by their corresponding axes. ρ is a minimum at r_c along the third axis which is perpendicular to this plane.

(3,+1)   two curvatures are positive and ρ is a minimum at r_c in the plane defined by their corresponding axes. ρ is a maximum at r_c along the third axis which is perpendicular to this plane.

(3,+3)   all curvatures are positive and ρ is a local minimum at r_c.

Critical Points of Molecular Charge Distributions

The coulombic potential becomes infinitely negative when an electron and a nucleus coalesce and, because of this, the state function for an atom or molecule must exhibit a cusp at a nuclear position. Thus, while the charge density is a maximum at the position of a nucleus, this point is not a true critical point because $\nabla\rho$, like $\nabla\psi$, is discontinuous there. However, this is not a problem of practical import and the nuclear positions behave topologically as do (3,-3) critical points in the charge distribution and hereafter they will be referred to as such.

All of the saddle points shown in Fig. 1a are (3,-1) critical points. A (3,-1) critical point is found between every pair of nuclei which are considered to be linked by a chemical bond in the B_2F_4 molecule. Fig. 2 shows representations of the charge distribution, similar to those shown in Fig. 1, for the three symmetry planes of the diborane molecule. While the topology of the charge distribution for the plane of the four terminal hydrogen nuclei and the two boron nuclei shown in Fig. 2a is indistinguishable from that shown in Fig. 1a for B_2F_4, the central critical point in the former molecule is not a (3,-1) critical point. This is made clear from Fig. 2b which shows the charge density in the plane of the bridging hydrogens obtained by a $90°$ rotation about the boron-boron axis. The charge density is a minimum

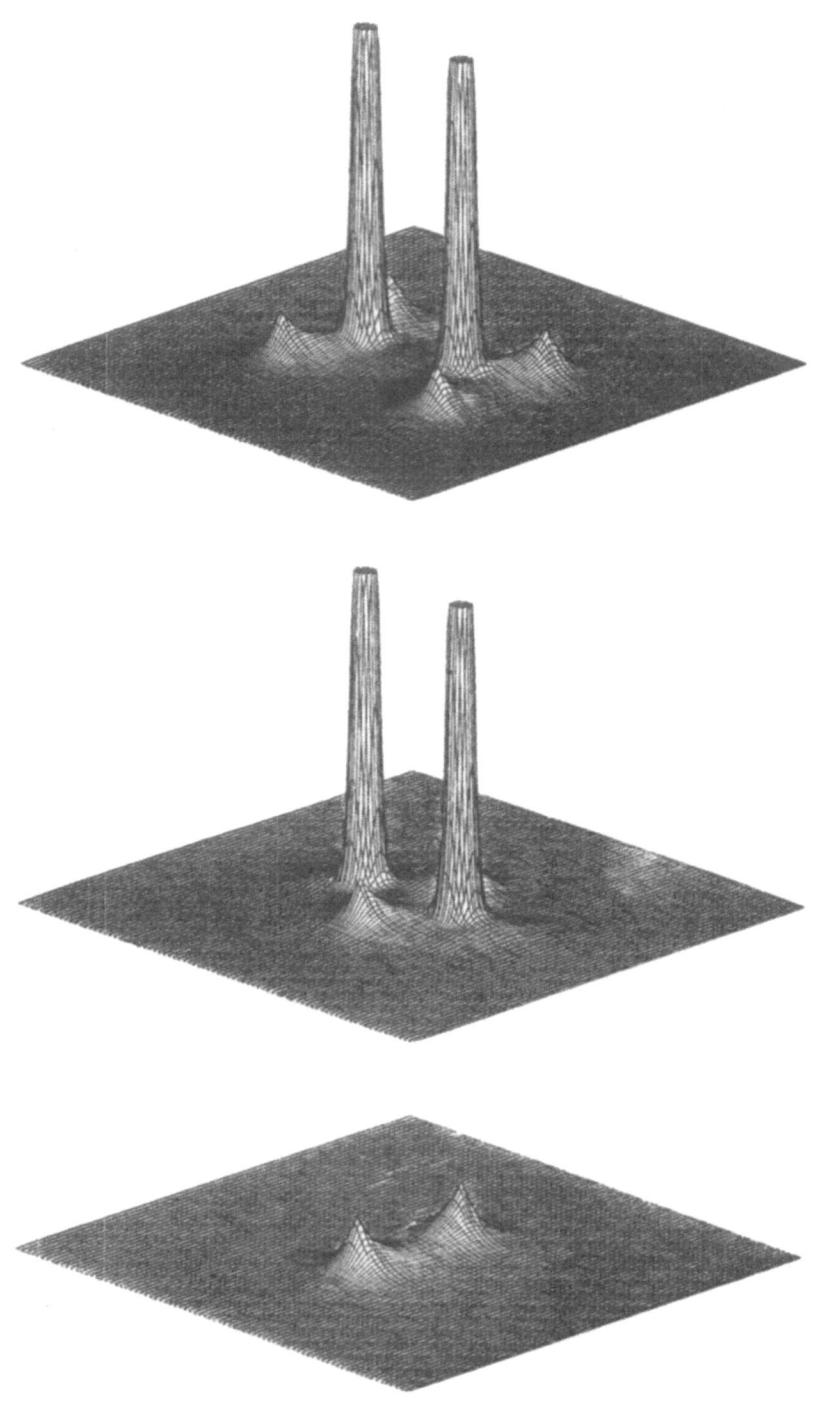

Fig. 2   Relief maps of the electronic charge density in B_2H_6 in three
orthogonal planes containing the critical point midway between
the boron nuclei.

at the central point in this plane showing that it is a (3,+1) critical point in the charge density. The final view of this critical point is shown in Fig. 2c, in the plane perpendicular to the boron-boron axis and containing the bridging protons, where it again appears as a saddle in ρ. The axis of the single negative curvature of this critical point is perpendicular to the plane shown in Fig. 2b. Thus Figures 2a and 2c exhibit this one negative and one each of the two positive curvatures and ρ has the appearance of a saddle point in these planes. While (3,-1) and (3,+1) critical points can both appear as (2,0) critical points when viewed in specific planes where every critical point is of rank two, their full three-dimensional behaviour is quite different. The other critical points appearing as saddles connecting neighbouring nuclei in Figures 2a and 2b are indeed (3,-1) critical points in agreement with the chemical structure usually assigned to this molecule, i.e., terminal BH_2 groups linked by two bridging protons. One notes for future reference that the minimum of the (3,+1) central critical point in Fig. 2b is bounded by a ring of four (3,-1) critical points which link the bridging protons to the boron nuclei.

The fourth and final kind of stable critical point is illustrated in Fig. 3 which gives representations of the charge distribution in $B_6H_6^{-2}$. The chemical structure assigned to this molecule is that of a cage. The critical point at the centre of the molecule appears as a minimum in the Figure and, because of the symmetry possessed by this molecule, this critical point will have the same appearance when viewed in any plane. It is a (3,+3) critical point and the charge density is a local minimum at the centre of the cage structure. Fig. 3a is interesting as it gives two-dimensional views of all four kinds of stable critical points. In terms of the structural properties associated with the critical points as illustrated above, the reader can satisfy herself or himself that the (2,0) saddles between each boron and its neighbouring hydrogen are (3,-1) critical points, that the two (2,-2) maximum are in the unique plane of the (3,-1) critical points between two pairs of out-of-plane boron nuclei and that the remaining four (2,0) saddles are (3,+1) critical points in the triangular faces formed by the boron nuclei joined by (3,-1) critical points.

We have demonstrated the existence of a connection between the number and kind of critical points appearing in a charge distribution and its conventional chemical structure. We next show that these qualitative associations of topological features of ρ with elements of molecular structure can be replaced with a complete theory, one which recovers all of the elements of structure in a manner that is totally independent of any information other than that contained within in the charge density. The underlying structure of the charge density is brought to the fore in its associated gradient vector field. The boundary condition of a quantum subsystem is also stated in terms of this field. It is this most remarkable coincidence of reasons for studying the gradient vector field, one from the quantum definition of a subsystem, the other from the quite independent demonstration that its form yields a mapping of the elements of molecular structure, that gives the theory of atoms in molecules its unified structure.

Gradient Vector Field of the Charge Density

The gradient vector field of the charge density is represented through a display of the trajectories traced out by the vector $\nabla\rho$. A trajectory of $\nabla\rho$, also called a gradient path, starting at some

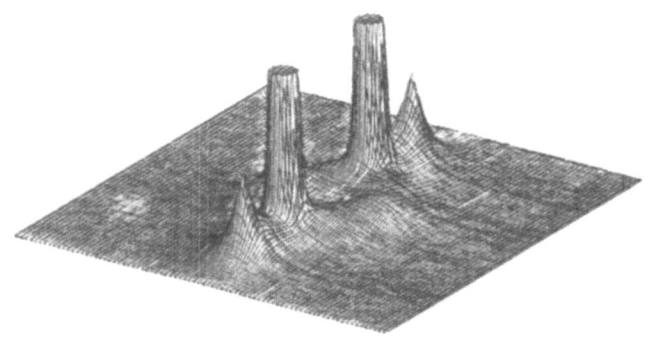

a

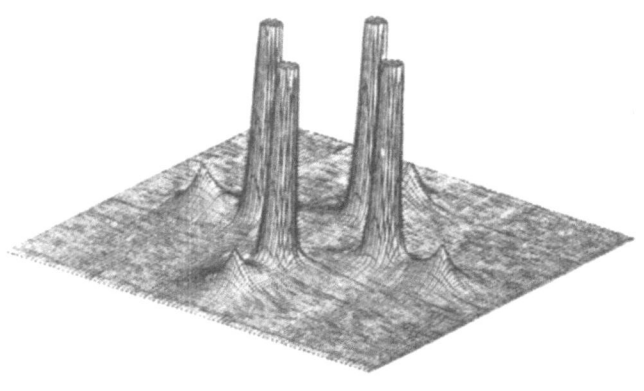

b

Fig. 3   Relief   maps   of the   electronic   charge   density   for   the
octahedral $B_6H_6^{-2}$ molecule in two orthogonal planes containing
the central cage critical point.   The third orthogonal plane
is symmetrically equivalent to the lower figure.

arbitrary point r_o is obtained by calculating $\nabla\rho(r_o)$, moving a distance Δr away from this point in the direction indicated by the vector $\nabla\rho(r_o)$ and repeating this procedure until the path so generated terminates. This operation is the three-dimensional analogue of approximating a function $f(x)$ in terms of its tangent line at x, $f(x+\Delta x) = f(x) + (df/dx)\Delta x$, an expression which, as a consequence of the definition of a derivative, becomes exact in the limit $\Delta x \to dx$.

Trajectories of $\nabla\rho$ possess the following properties:

a) Since the gradient vector of a scalar points in the direction of greatest increase in the scalar, the trajectories of $\nabla\rho$ are perpendicular to lines of constant density, i.e., to contour lines of ρ.

b) The vector $\nabla\rho(r)$ is tangent to its trajectory at each point r.

c) Every trajectory must originate or terminate at a point where $\nabla\rho(r)$ vanishes, i.e., at a critical point in ρ.

d) Trajectories cannot cross since $\nabla\rho(r)$ defines but one direction at each point r.

The differential equation for $\nabla\rho(r)$ is

$$dr(s)/ds = \nabla\rho(r(s)) \tag{8}$$

where the notation $r(s)$ implies that a point r on a given trajectory is dependent upon a path parameter s. Eqn (8) represents three first-order differential equations and it yields unique solutions only when particular values are assigned to three constants of integration. This corresponds to fixing some initial point on a trajectory, at $s = s_1$, for example. Then every other point on the trajectory that passes through the point $r(s_1)$ is obtained by integrating eqn (8) with the three constants of integration given by the components of $r(s_1)$,

$$r(s) = r(s_1) + \int_{s_1}^{s} \nabla\rho(r(t))dt \tag{9}$$

A trajectory of the gradient vector field of $\rho(r)$ is, therefore, a parameterized integral curve, a solution curve, of the differential equation for $\nabla\rho(r)$. By fixing a point on a given trajectory, all other points which lie on the same trajectory can be obtained by solving eqn (9).

Phase Portraits of the Gradient Vector Field

An eigenvalue and its associated eigenvector of the Hessian of ρ (a principal curvature and its associated axis) at a critical point define a one-dimensional system. If the eigenvalue or curvature is negative, then ρ is a maximum at the critical point on this axis and a gradient vector will approach and terminate at this point from both its left and right hand side as illustrated in Fig. 4 for the case $(1,-1)$, a system of rank 1 and signature -1. If the eigenvalue is positive then ρ is a minimum at the critical point on this axis and two gradient vectors will originate at this point as illustrated for the case $(1,+1)$. In two dimensions, if both eigenvalues are negative, then ρ is

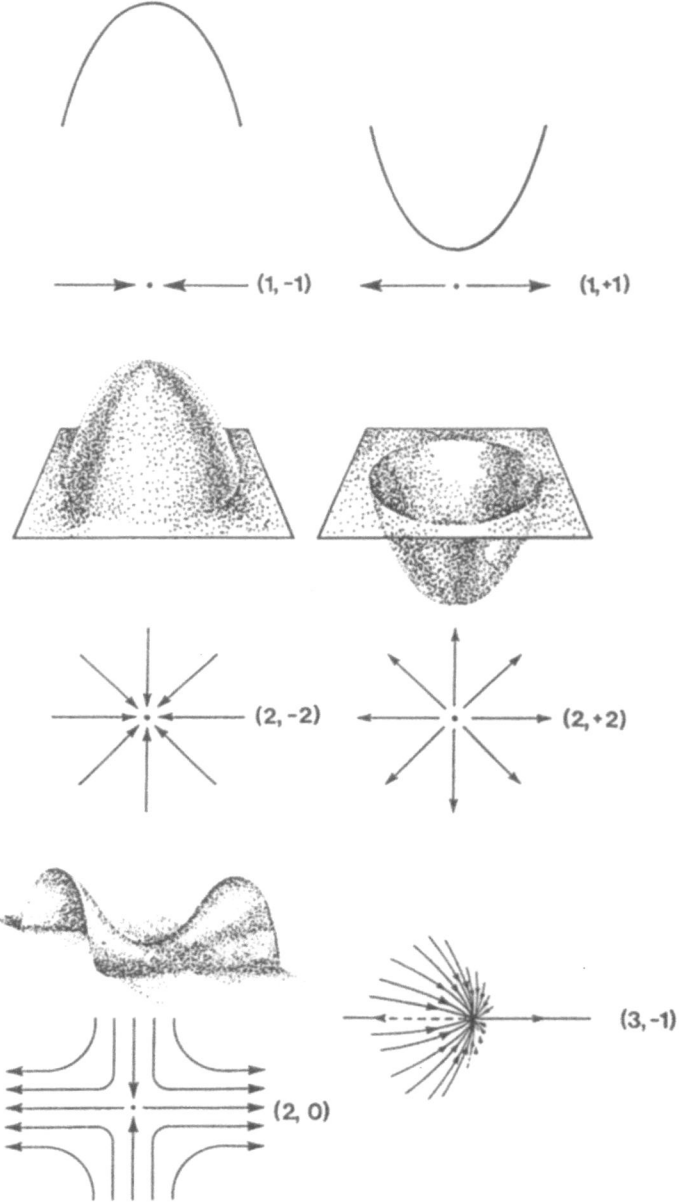

Fig. 4   Phase   portraits for one-, two- and three-dimensional critical
points.

a maximum at the critical point and all trajectories of $\nabla\rho$ will terminate at the critical point as illustrated for the case $(2,-2)$. This set of trajectories is defined by all possible linear combinations of the two associated eigenvectors which span a two-dimensional space. Similarly, if both eigenvalues are positive and ρ is a minimum at the critical point, all trajectories will originate at the critical point and again define a surface as illustrated for the case $(2,+2)$. A more interesting situation is obtained when the eigenvalues are of opposite sign (the signature is zero) and the charge density in a plane has the form of a saddle as illustrated for the case $(2,0)$. In this situation the two trajectories associated with the axis of the negative curvature terminate at the critical point while the two associated with the positive curvature originate there. The trajectories formed by linear combinations of the two associated eigenvectors neither terminate nor originate at the critical point but instead avoid this point as indicated for the case labelled $(2,0)$.

In three dimensions, the pairs of eigenvectors associated with the two negative or two positive eigenvalues of a $(3,-1)$ or $(3,+1)$ critical point, respectively, will again define a surface. Unlike the two-dimensional examples discussed above, these surfaces will be planar only if the critical point r_c lies in a symmetry plane. If not, the surface will be planar only in the immediate neighbourhood of r_c and will, in general, be curved beyond this region but still defined by the unique set of trajectories which terminate at a $(3,-1)$ critical point or originate at a $(3,+1)$ critical point. The final diagram in Fig. 4 is a display of the three-dimensional form of the phase portrait of a $(3,-1)$ critical point. The unique pair of trajectories associated with the single positive eigenvalue originate at the critical point, as in the $(1,+1)$ case illustrated above. The set of trajectories defined by linear combinations of the pair of eigenvectors associated with the two negative eigenvalues terminate at the critical point and define a surface. The charge density is a maximum in the surface at the critical point and a minimum at this same point along the perpendicular axis. The behaviour of the charge density at a $(3,+1)$ critical point is just the opposite of this and its phase portrait is obtained by reversing all of the arrows shown for the $(3,-1)$ case.

Elements of Molecular Structure

The gradient vector field of the charge density in the plane containing the nuclei of the B_2F_4 is illustrated in Fig. 5. A $(3,-3)$ critical point, such as occurs at each of the nuclear positions, serves as the terminus, the ω-limit set, of all the paths starting from and contained in some neighbourhood of the critical point. A $(3,-3)$ critical point exhibits the property which defines a point attractor in the gradient vector field of the charge distribution: there exists an open neighbourhood B of the attractor which is invariant to the flow of $\nabla\rho$ such that any trajectory originating in B terminates at the attractor. The largest neighbourhood satisfying these conditions is called the basin of the attractor.

Since $(3,-3)$ critical points in a many-electron charge distribution are generally found only at the positions of the nuclei, the nuclei act as the attractors of the gradient vector field of $\rho(r,X)$. The result of this identification is that the space of a molecular charge distribution, real space, is partitioned into disjoint regions, the basins, each of which contains one point attractor or

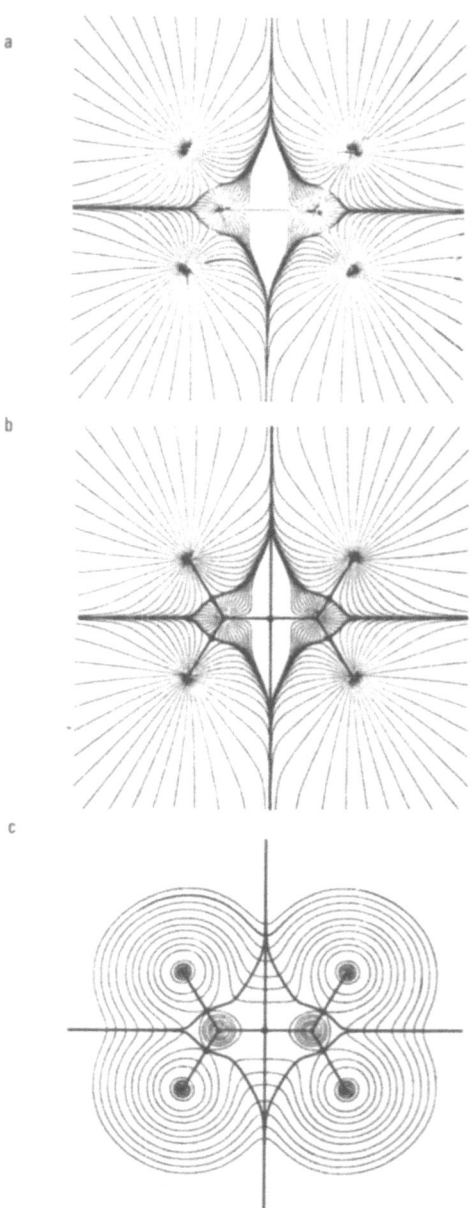

Fig. 5   a) and b) Gradient vector field of the charge density of B_2F_4
          in the plane containing the nuclei.   c) Contour map of $\rho(r)$
          in same plane showing bond paths and interatomic surfaces.
          A bond critical point is denoted by a dot.

nucleus. This fundamental topological property of a molecular charge distribution is illustrated in Fig. 5a which depicts only those gradient paths of the charge density which terminate at each of the nuclear attractors in the molecule. While Fig. 5a illustrates this property for only one plane, it is to be emphasized that because ρ is a local maximum at a nucleus, a $(3,-3)$ critical point, the basin of an attractor is a region of three-dimensional space and the partitioning so clearly indicated in the Figure extends throughout all of space. *An atom, free or bound, is defined as the union of an attractor and its associated basin.*

Alternatively, an atom can be defined in terms of its boundary. The basin of the single nuclear attractor in an isolated atom covers the entire three-dimensional space R^3. For an atom in a molecule the atomic basin is an open subset of R^3. It is separated from neighbouring atoms by interatomic surfaces. The existence of an interatomic surface S_{AB} denotes the presence of a $(3,-1)$ critical point between neighbouring nuclei A and B. The presence of such a critical point between certain pairs of nuclei was noted above as being a general topological property of molecular charge distributions. Their presence now appears as providing the boundaries between the basins of neighbouring atoms. As discussed above and illustrated in Fig. 4, the trajectories which terminate at a $(3,-1)$ critical point define a surface, the *interatomic surface* S_{AB}. In a sufficiently small neighbourhood of the critical point at r_c, the interatomic surface coincides with its tangent plane at r_c, which is linearlly spanned by v_1 and v_2, the eigenvectors associated with the negative eigenvalues of the Hessian of ρ at r_c. The entire interatomic surface can be obtained by solving the differential eqn (8), for initial conditions $r_o = r(0)$, such that each r_o belongs to the intersection of the surface with the above neighbourhood of r_c and is thus expressible as a linear combination of v_1 and v_2.

The reader will recall Fig. 1c showing that the charge density is a maximum at a $(3,-1)$ critical point in this surface. Fig. 5b shows in addition to the trajectories that terminate at each of the nuclei in B_2F_4, the trajectories associated with each of the $(3,-1)$ critical points. Two trajectories terminate at each such critical point in the plane of the diagram. They denote the intersection of the interatomic surfaces with this plane. It should be borne in mind that each such pair of trajectories are but two of an infinity of gradient paths all of which terminate at a $(3,-1)$ critical point and define a surface in three dimensions, as illustrated in Fig. 4. The atomic surface S_A of atom A is defined as the boundary of its basin. Generally this boundary comprises the union of a number of interatomic surfaces, separating two neighbouring basins, and some portions which may be infinitely distant from the attractor. The atomic surface of a boron atom in B_2F_4 as seen in Fig. 5 for example, consists of three interatomic surfaces, two with fluorines and another with a boron atom.

If the topological property which defines an atom is also one of physical significance, then it should be possible to obtain from quantum mechanics an equivalent mechanical definition. This can be accomplished through a generalization of the quantum action principle to obtain a statement of this principle which applies equally to the total system or to an atom within the system. The result is a *single* variational principle which defines the observables, their equations of motion and their average values for the total system or for an atom within the system.[1,2]

The generalization of the action principle to a subsystem of some total system is unique, as it applies only to a region that satisfies a particular constraint on the variation of its action integral. The constraint requires that the subsystem be bounded by a surface of zero flux in the gradient vector of the charge density as stated in eqn (10),

$$\nabla\rho(r)\cdot n(r) = 0 \qquad \text{for all points on the surface } S(r) \qquad (10)$$

In order for the scalar product of n, the vector normal to the surface with $\nabla\rho$ to vanish, it is necessary that the atomic surface not be crossed by any trajectories of $\nabla\rho$ and as such it is referred to as a *zero flux surface*. The state function ψ and $\nabla\psi\cdot n$, where the gradient is taken with respect to the coordinates of any one of the electrons, vanish on the boundary of a bound system at infinity. Thus ρ and $\nabla\rho$ vanish there as well and a total isolated system is also bounded by a surface satisfying eqn (10). Since the generalized statement of the action principle applies to any region bounded by such a surface, the zero flux surface condition places the description of the total system and the atoms which comprise it on an equal footing.

Because of the dominant topological property exhibited by a molecular charge distribution - that it exhibits local maxima only at the positions of the nuclei -- the imposition of the quantum boundary condition of zero flux leads directly to the topological definition of an atom. Indeed the interatomic surfaces, along with the surfaces found at infinity, are the only closed surfaces of R^3 which satisfy the zero flux surface condition of eqn (10). This is a natural result of the property of an atomic basin as shown in Fig. 5a – all the trajectories in the vicinity of a given nucleus terminate at that nucleus and no trajectories cross from one basin to another. Since trajectories of $\nabla\rho$ never cross, the zero flux surface condition follows directly from the definition of an interatomic surface in terms of the set of trajectories which terminate at a (3,-1) critical point. In terms of this same definition, it follows that the vector $\nabla\rho(r)$ will be tangent to the surface $S(r)$ of an atom at every point r.

Chemical Bonds and Molecular Graphs

Fig. 5b also shows the pairs of gradient paths which originate at each (3,-1) critical point and terminate at the neighbouring attractors. As previously discussed and illustrated in Fig. 4, each such pair of trajectories is defined by the eigenvector associated with the unique positive eigenvalue of a (3,-1) critical point. These two unique gradient paths define a line through the charge distribution linking the neighbouring nuclei along which $\rho(r)$ is a maximum with respect to any neighbouring line. Such a line is found between every pair of nuclei whose atomic basins share a common interatomic surface, and in the general case it is referred to as an *atomic interaction line*.[4,8,9]

The existence of a (3,-1) critical point and its associated atomic interaction line indicates that electronic charge density is accumulated between the nuclei that are so linked. This is made clear by reference to the displays of the charge density for such a critical point, as given in Fig. 1 for example, particularly Fig. 1c which shows that the charge density is a maximum in an interatomic surface at the position of the critical point. This is the point where the atomic interaction line intersects the interatomic surface and charge is so accumulated between the nuclei along the length of this line. Both theory and observation concur that the accumulation of electronic charge between a pair of nuclei is a necessary condition if two atoms are to be bonded to one another. This accumulation of charge is also a sufficient condition when the forces on the nuclei are balanced and the system possess a minimum energy equilibrium internuclear separation. Thus the presence of an atomic interaction line in such an equilibrium geometry satisfies both the necessary and sufficient conditions that the atoms be bonded to one another. In this case the line of maximum charge density linking the nuclei is called a *bond path* and the (3,-1) critical point referred to as a *bond critical point.*[4]

For a given configuration X of the nuclei, a *molecular graph* is defined as the union of the closures of the bond paths or atomic interaction lines. Pictorially the molecular graph is the network of bond paths linking pairs of neighbouring nuclear attractors. The molecular graph isolates the pair-wise interactions present in an assembly of atoms which dominate and characterize the properties of the system be it at equilibrium or in a state of change.

A molecular graph is the direct result of the principal topological properties of a system's charge distribution: that the only local maxima, (3,-3) critical points, occur at the positions of the nuclei thereby defining the atoms, and that (3,-1) critical points are found to link certain, but not all pairs of nuclei in a molecule. The network of bond paths thus obtained is found to coincide with the network generated by linking together those pairs of atoms which are assumed to be bonded to one another on the basis of chemical considerations. Molecular graphs for a sampling of molecules in equilibrium geometries with widely different bonding properties are illustrated in Fig. 6. The existence and position of a bond or (3,-1) critical point in this and other figures is indicated by a black dot.

The recovery of a chemical structure in terms of a property of the system's charge distribution is a most remarkable and important result. The representation of a chemical structure by an assumed network of lines has evolved through a synthesis of observations on elemental combination and models of how atoms combine, particularly models of chemical valency – a model which states that the ability of a given type of atom to form bonds, its valency, can be saturated and that valency is determined by the number of valence electrons. A great deal of chemical knowledge goes into the formulation of a chemical structure and, correspondingly, the same information is successfully and succinctly summarized by such structures. The demonstration that a molecular structure can be faithfully mapped onto a molecular graph imparts new information to them – that nuclei joined by a line in the structure are linked by a line through space along which electronic charge density, the glue of chemistry, is maximally accumulated. Finding the physical basis for a molecular structure also leads to a broadening of the concept – that the dominant interactions between atoms, be they attractive or repulsive, have a common physical representation. This is not an entirely surprising result since the

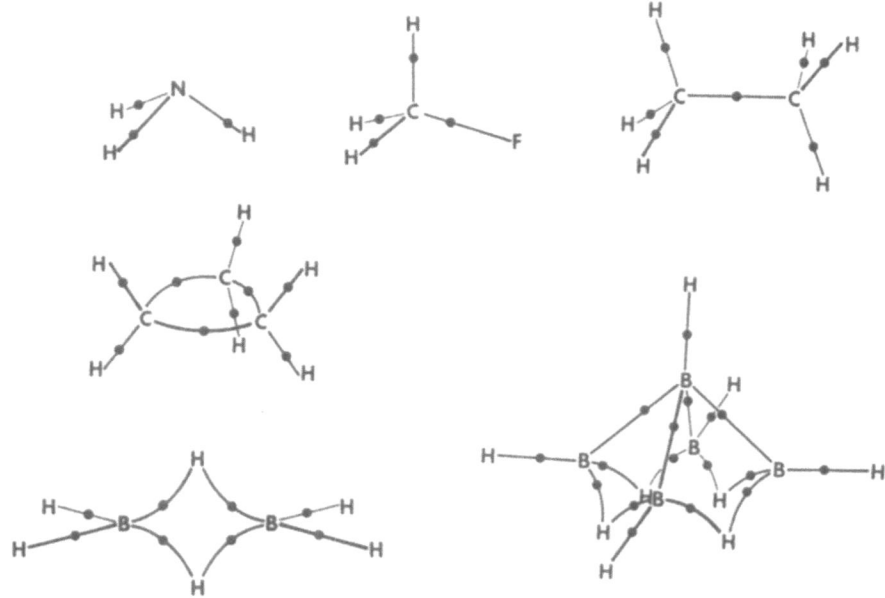

Fig. 6   a) Molecular graphs as determined by the distribution of electronic charge in each molecule.

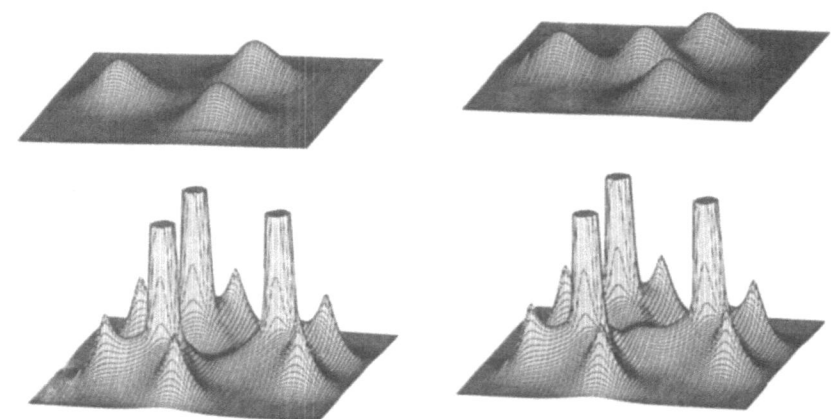

Fig. 6   b) Relief maps of $\rho(\mathbf{r})$ for bicyclooctane (LHS) and [2.2.2]propellane and bicyclopentane (LHS) and [1.1.1]propellane in symmetry plane bisecting the bridgehead C–C axis.

ever present nuclear excursions from an equilibrium separation between a pair of atoms force a sampling of these same portions of a potential surface even though the atoms are considered to be bonded to one another. It is in answer to the closely related questions of what is meant by the making and breaking of chemical bonds that leads one to consider the most important extension of the molecular structure concept. The dynamic behaviour of the molecular graphs as caused by the relative motions of the nuclei forms the basis for the definition of structural stability and the analytical description of the mechanisms of structural change.[3]

It is to be stressed that a bond path is not to be understood as representing a "bond". The presence of a bond path linking a pair of nuclei implies that the corresponding atoms are bonded to one another. As demonstrated later, the interaction can be characterized and classified in terms of the properties of the charge density at its associated (3,-1) critical point.

The agreement of structures predicted in this manner by quantum mechanics for a wide variety of systems with chemically accepted structures, shows that the many models which are used to rationalize the network of bonds in a molecule may be replaced with a single theory of molecular structure. The small ring propellanes such as [1.1.1]propellane possess unusual structures with "inverted" geometries at the bridgehead carbon atoms. A bond path links the bridgehead nuclei in the propellane molecules even though this results in structures that have four bond paths, all to one side of a plane, terminating at each of the bridgehead nuclei. While the model of hybridized orbitals cannot describe such a situation, even assuming bent bonds, the molecular graphs for the propellanes demonstrate that the charge distribution of a carbon atom can be so arranged as to yield bond paths -- lines of maximum charge density - which correspond to an inverted structure.

The structures of the propellanes consist of three rings, either three- or four-membered, sharing a common bond, the bridgehead bond. Addition of a molecule of hydrogen to a propellane breaks the bridgehead and yields the corresponding bicyclic molecule. These molecules possess cage structures, the interior of each molecule being bounded by three curved ring surfaces and containing a cage critical point. The propellanes and their bicyclic congeners exemplify the contrasting behaviour exhibited by the charge density between a pair of nuclei in situations where the atoms are and are not bonded to one another. Fig. 6b gives relief maps of the charge density in the plane which bisects and is perpendicular to the bridgehead internuclear axis for [2.2.2] and [1.1.1]propellane and its bicyclic analogue. Such a plane contains the critical point in the interatomic surface between the bridgehead atoms. It also contains the critical point in the interatomic surface between the methylenic carbon atoms linked by a peripheral bond and/or the C and H nuclei of a peripheral CH_2 group in the case of three-membered ring in a propellane or corresponding group in the bicyclic molecule. The charge density is a maximum at a bond critical point in the interatomic surface. Thus the diagrams for the [2.2.2] systems both exhibit three maxima in the charge density corresponding to the three peripheral bonds between the methylene groups. These maxima in bonded density are replaced by the nuclear maxima of a CH_2 group in [1.1.1]propellane. The purpose of the diagram is to contrast the behaviour of the charge density at the bridgehead critical point for the propellanes with that found for the bicyclic molecules. *In the propellanes which possess a bridgehead bond, the charge density is a maximum at this point, while in the bicyclic molecules which do not possess a bridgehead bond, the charge density is a minimum at this same point.* In the former molecules there is a line of maximum charge density linking the nuclei while in the latter such a line is absent and the charge density is instead a local minimum at the central critical point in the bicyclic molecules. There is, therefore, an essential, qualitative difference in the manner in which electronic charge is distributed along a line linking a pair of bonded nuclei (as between the bridgehead nuclei in the propellanes) and along a line linking two nuclei that are not bonded (as between the nuclei in the corresponding bicyclic structures).

In addition to the fundamental difference in the form of the bridgehead charge density between the two sets of molecules, there are also substantial quantitative differences in the values of ρ. Thus its value at the bond critical in [1.1.1]propellane is 0.203 au, four-fifths the value of a normal C-C bond while the value of ρ at the corresponding cage critical point in the bicyclic molecule is 0.098 au. It is also clear from the figure that the values of ρ at the bond critical point and the three adjacent ring critical points in [1,1.1]propellane are almost equal, giving rise to a very broad bonded maximum in ρ in the bridgehead interatomic surface and to a "fat" bridgehead bond.

The fundamental difference in the manner in which electronic charge is distributed in the bridgehead internuclear region, between the propellanes and the corresponding bicyclic molecules, is not made apparent in density difference maps. Such deformation maps show a region of charge depletion between these nuclei for the propellanes as well as for the bicyclic molecules. First, there is no physical basis for demanding that a density difference map *in a polyatomic molecule* show a charge buildup between a pair of nuclei if the nuclei are to be considered bonded to one another. Second, the reference density, in addition to being physically nonrealizable, is arbitrary in its construction. Different results are obtained and different conclusions are reached depending on whether one employs spherical atom densities or densities of atoms in prepared valence states in the construction of the promolecule density. Third, in performing a comparison between density difference maps for pairs of molecules, [1.1.1]propellane and various cyclic and bicyclic molecules for example, one is actually comparing four different charge distributions. It is clear that the spherical atom promolecule density for bicyclo[1.1.1]pentane will, because of the larger bridgehead separation of 1.87 Å, subtract considerably less charge from the bridgehead symmetry plane than will the corresponding promolecule density for the equilibrium geometry of [1.1.1]propellane where the corresponding separation is calculated to be 1.54 Å. Thus the observation of an "essential similarity" in the deformation density distributions for these two molecules, in particular the lack of a charge buildup between the bridgehead nuclei, is an artifact of the promolecule distributions and is not a reflection of the relative properties of the two charge distributions of interest. Qualitatively and of fundamental importance, the propellane molecule does accumulate charge at the bond midpoint and along the resultant bond path, while the bicyclic compound exhibits a local minimum in ρ in this same region. In addition, as demonstrated by a direct quantitative comparison of their charge distributions, the bicyclic molecule has significantly less charge in this region than does the propellane molecule. These essential observations are lost in a comparison of the deformation densities because the reference density for propellane removes more charge from the critical bridgehead region than does that for the bicyclic molecule. Why complicate a comparison of two distributions through the introduction of two more distributions which are arbitrary in their definition and are of no direct physical interest?

Rings and Cages

The remaining critical points of rank three occur as consequences of particular geometrical arrangements of bond paths and they define the remaining elements of molecular structure – rings and cages. If the bond paths are linked so as to form a ring of bonded atoms, as in the bridging plane of the diborane molecule for example, whose molecular graph is shown in Fig. 6, then a (3,+1) critical point is found in the interior of the ring. As discussed above and illustrated in Fig. 4, the eigenvectors associated with the two positive eigenvalues of the Hessian matrix of ρ at this critical point generate an infinite set of gradient paths which originate at the critical point and define a surface, called the ring surface. This behaviour is illustrated here by the gradient paths in the bridging plane of the diborane molecule as shown in Fig. 7. All of the trajectories which originate at the critical point at the centre of the ring of nuclei, the (3,+1) or ring critical point, terminate at the ring nuclei, but for the set of single trajectories each of which terminates at one of the bond critical points whose bond paths form the perimeter of the ring. These bond paths are noticeably inwardly curved away from the geometrical perimeter of the ring, a behaviour characteristic of systems which are electron deficient. The remaining eigenvector of a ring critical point, its single negative eigenvector, generates a pair of gradient paths which terminate at the critical point and define a unique axis perpendicular to the ring surface at the critical point. In diborane this axis is perpendicular to the plane shown in Fig. 7. It represents the intersection of the boundaries of the basins of the hydrogen and boron atoms forming the ring. *A ring*, as an element of structure, *is defined as a part of a molecular graph which bounds a ring surface.*

If the bond paths are so arranged as to enclose the interior of a molecule with ring surfaces then a (3,+3) or cage critical point is found in the interior of the resulting cage. The charge density is a local minimum at a cage critical point. The phase portrait in the vicinity of a cage critical point is shown in Fig. 8 for $B_6H_6^{-2}$. Trajectories only originate at such a critical point and terminate at nuclei, and at bond and ring critical points, thereby defining a bounded region of space. *A cage*, as the final element of molecular structure, *is a part of a molecular graph which contains at least two rings, such that the union of the ring surfaces bounds a region of R^3 which contains a (3,+3) critical point.* While it is mathematically possible for a cage to be bounded by only two ring surfaces, the minimum number found in an actual molecule so far is three, as in bicyclo[1.1.1]pentane, for example.

The number and type of critical points which can coexist in a system with a finite number of nuclei is governed by the Poincaré–Hopf relationship. With the above association of each type of critical point with an element of molecular structure, this relationship states

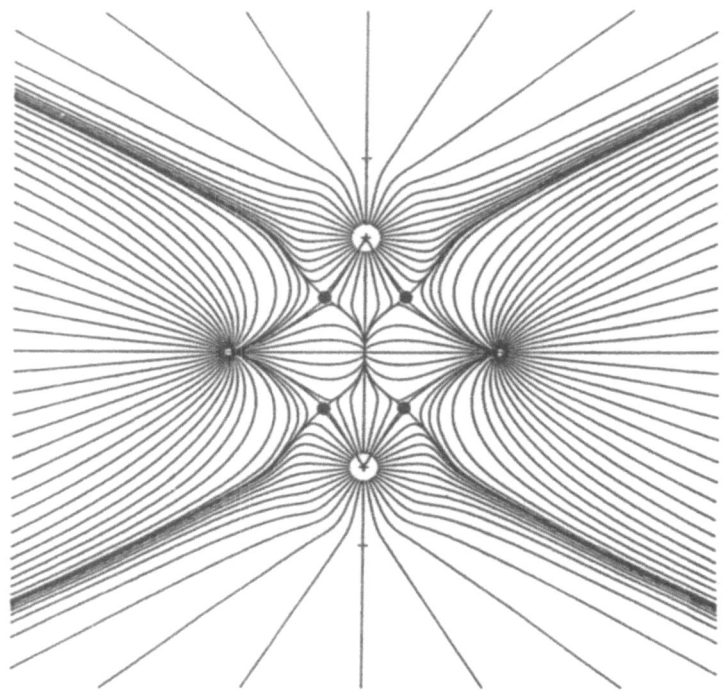

Fig. 7   Gradient vector field of $\rho(r)$ for B_2H_6 in same plane   as   that
shown in Fig. 2b and containing the ring surface.

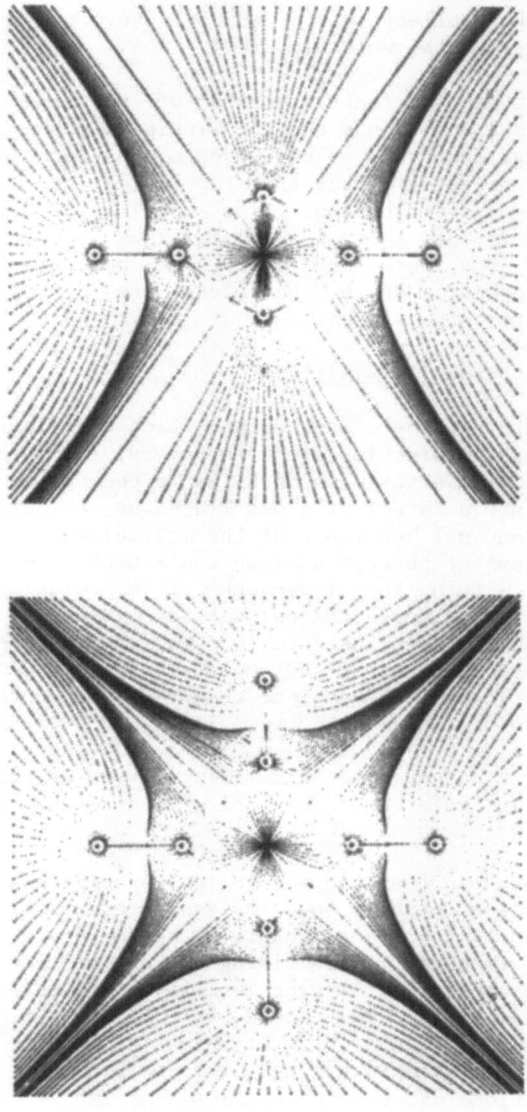

Fig. 8   Gradient vector field for $\rho(r)$ for $B_6H_6^{-2}$ in same planes as
those shown in Fig. 3 and containing the cage critical point.

that[6]

$$n - b + r - c = 1 \qquad (11)$$

where n is the number of nuclei, b is the number of bond paths (or atomic interaction lines), r is the number of rings and c is the number of cages. The collection of numbers (n,b,r,c) is called the *characteristic set* of the molecule.

It has been shown that elements of the molecular structure hypothesis, atoms and bonds and the linking together of the atoms to form chains, rings and cages as evidenced by the forms of the molecular graphs, follow directly from the topological properties of the electronic charge distribution. The experimental development of the structural aspects of this hypothesis thus appear as inevitable consequences of the form and properties possessed by the charge distribution, properties which themselves are a reflection of the forces acting within the system.

Structure and Structural Stability

The finding that the charge distribution contains the information required to define a molecular graph for every geometrical arrangement of the nuclei allows for much more than a pedestrian recovery of the notion of structure as a set of atoms linked by a network of bonds. Instead, the dynamical behaviour of the molecular graphs as induced by a motion of the system through nuclear configuration space is shown to be describable within the framework of a relatively new field of mathematics, one which demonstrates that the notions of structure and structural stability are inseparable. The application of these ideas to the dynamical properties of the molecular graphs defined by the charge density leads to a complete theory of structure and structural stability.

The reader is referred to an existing review of this part of the theory, given in reference 3. We give here only a summary of the principal findings.

It has been shown that the topological properties of a system's charge distribution enable one to assign a molecular graph to each point X in the nuclear configuration space of a system. This assignment corresponds to defining a unique network of atomic interaction lines to each molecular geometry. A gradient vector field of the charge density, the field $\nabla\rho(r;X)$ where X denotes a set of nuclear coordinates, exists for every geometry X. The definition of molecular structure makes use of the mathematical device of an equivalence relation of vector fields over R^3. The equivalence relation is defined as follows: two vector fields V and V' over R^3 are said to be equivalent if and only if there exists a homeomorphism which maps the trajectories of V into the trajectories of V'. By applying this definition to the gradient vector fields $\nabla\rho(r;X)$, $X\epsilon R^Q$, one obtains an equivalence relation operating in nuclear configuration space R^Q. An equivalence relation is then obtained for the molecular graph defined by each $\nabla\rho(r;X)$ and *a molecular structure is defined as an equivalence class of molecular graphs.*

The result of applying the equivalence relationship to the field

$\nabla\rho(r;X)$ is a partitioning of nuclear configuration space R^Q into a finite number of non-overlapping regions, each of which is characterized by a unique molecular structure. These structurally stable, open regions are separated by boundaries, hypersurfaces in the space R^Q. A point on a boundary possesses a structure which is different from but transitional to the structures characteristic of either of the regions it separates. Since a boundary is of dimension less than R^Q, arbitrary motions of the nuclei will carry a point on the boundary into neighbouring stable structural regions and its structure will undergo corresponding changes. The boundaries are the loci of the structurally unstable configurations of a system. In general, the trajectory representing the motion of a system point in R^Q will carry it from one stable structural region through a boundary to a neighbouring stable structural region. The result is an abrupt and discontinuous change in structure. A change in structure is catastrophic and for this reason the set of unstable structures is called the catastrophe set. A point in a structurally stable region of nuclear configuration space is termed a *regular point*, and a point on one of the structurally unstable boundaries is termed a *catastrophe point*. This definition of molecular structure associates a given structure with an open neighbourhood of the most probable nuclear geometry, and removes the need of invoking the Born-Oppenheimer approximation for the justification or rationalization of structure in a molecular system.

By appealing to the theorem of structural stability of Palis and Smale[10] one can show that only two kinds of structural instabilities or catastrophe points can arise and that there are therefore, only two basic mechanisms for structural change in a chemical system.

Palis and Smale's theorem of structural stability when used to describe structural changes in molecular system predicts a configuration $X \in R^Q$ to be structurally stable if $\rho(r,X)$ has a finite number of critical points such that:

(a)   each critical points is non-degenerate, and

(b)   the stable and unstable manifolds of any pair of critical points intersect transversely.

The immediate consequence of the theorem is that a structural instability can be established through one of two possible mechanisms which correspond to the bifurcation and conflict catastrophes. A change in molecular structure can only be caused by the formation of a degenerate critical point in the electronic charge distribution or by the attainment of an unstable intersection of the submanifolds of bond and ring critical points as illustrated in Fig. 9.

CLASSIFICATION OF CHEMICAL BONDS

Bond Order, Bond Path Angle, Bond Elliplicity and Structural Stability

The properties of a given structure are usefully characterized in terms of the properties of the charge density at the $(3,-1)$ or bond critical points. For bonds between a given pair of atoms one may

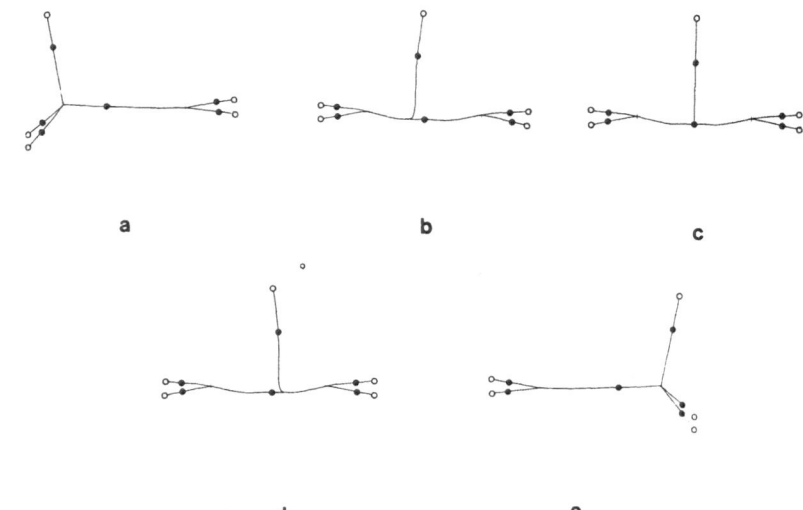

Fig. 9   Molecular graphs a to e represent the change in structure, $CH_3-CH_2^+ \rightarrow CH_2^+-CH_3$, by the conflict mechanism. Open circles denote protons and black dots are bond critical points. Structure c is unstable. Its structure, which is transitional between reactants and product structures, exists for but one geometry along the reaction path.

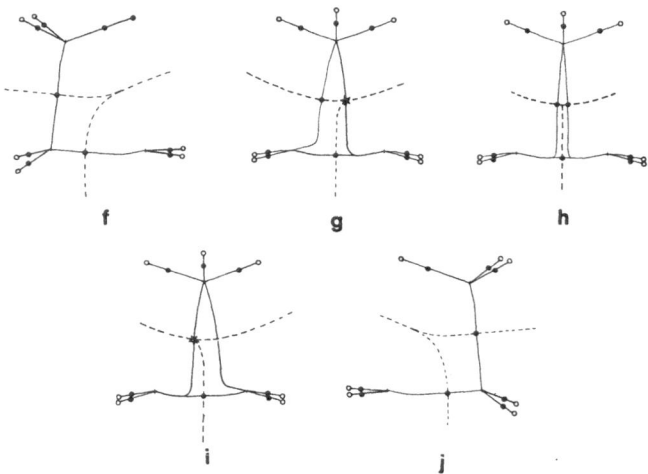

The graphs f to j represent the change in structure $CH_3-CH_2-CH_2^+ \rightarrow CH_2^+-CH_2-CH_3$, by the bifurcation mechanism. In g an unstable critical point is created which for further motion along the reaction coordinate bifurcates into a bond and a ring critical point and the open structure is transformed into a ring. For further motion, the ring critical point migrates towards the original $C-CH_3$ bond critical point, eventually coalescing with and annihilating it. At this point the ring is opened to yield the product structure.

define a bond order whose value is determined by ρ_b, the value of the charge density at the bond critical point.[11] The extent of charge accumulation in the interatomic surface and along the bond path increases with the assumed number of electron pair bonds and this increase is faithfully monitored by the value of ρ_b. When applied to the ρ_b values of the C-C bonds in hydrocarbons one obtains bond orders of 1.0, 1.6, 2.0 and 3.0 for ethane, benzene, ethylene and acetylene, respectively. In a hydrogen bond AH-BX, obtained when an acid AH binds to a base atom B of a base BX, the value of ρ_b is relatively small and only slightly greater than the sum of the unperturbed densities of the H and B atoms of the acid and base molecules at the degree of penetration found in the dimer. The strength of the hydrogen bond is found to parallel this degree of penetration of the van der Waals envelopes of the acid and base molecules and to again increase with an increase in ρ_b.[12,13] A related property is the bonded radius of an atom, the distance from the nucleus to the associated bond critical point, a quantity closely palleling the relative electronegativity of two bonded atoms.

A bond path, unless dictated to be so by symmetry, is not necessarily coincident with the internuclear axis and when it is not, the bond path length R_b is greater than the internuclear separation R_e. Such bent bond paths are present in those systems where classical structural arguments predict the presence of strain such as in the small ring hydrocarbons.[9,14] In these instances the bond paths in general, are outwardly bent from the geometrical perimeter of the ring. In molecules which are electron deficient, such as the boranes, the bond paths linking the bridging hydrogen atoms are strongly bent towards the interior of the ring so as to maximize the binding from a minimum amount of electron density, see Fig. 7. Wiberg and co-workers have shown that the presence of bent bond paths is more prevalent than anticipated and that the degree of bending is a useful parameter in understanding structural effects in molecules. This property of bond paths is however, more usefully catalogued using the idea of a bond path angle rather than of the bond path length. The bond path angle α_b, is the angle subtended at a nucleus by the pair of bond paths linking it to the two nuclei which define the corresponding geometrical bond angle α_e.[14] The difference $\Delta\alpha = \alpha_b - \alpha_e$ provides a measure of the degree of relaxation of the charge density away from the geometrical constraints imposed by the nuclear framework. In general, for a strained molecule $\Delta\alpha > 0$ and in these cases the bonds are less strained than the geometrical angles α_e would suggest. In cyclopropane for example, the bond path angle exceeds the $60°$ C-C-C bond angle by $18.8°$. Wiberg and Breneman[15] and Wiberg and Murcko[16] have found large negative values for $\Delta\alpha$ for the H-C-X angle in a variety of methyl derivatives CH_3X when X is more electronegative than C, and positive values when X is less electronegative. These effects are related to both steric and electronic effects. Wiberg and Laidig,[17] in a study of the origin of the rotational barriers adjacent to double bonds have used the bond path angle to determine the degree of p character in the bonds to a carbonyl carbon. They find the angle opposite the more electronegative atom to have the larger bond angle and the orbital directed to it has high p character. The geometrical angles do not reflect the anticipated changes in hybridization that are revealed through the study of the bond path angles.

The charge density along a bond path attains its minimum value at the bond critical point and the associated curvature or eigenvalue of the Hessian of ρ at r_c, λ_3, is thus positive. The charge density in an interatomic surface on the other hand, attains its maximum value at the bond critical point and the two associated curvatures of ρ at r_c, λ_1 and λ_2, those directed along axes perpendicular to the bond path, are thus negative. In a bond with cylindrical symmetry, these two negative curvatures of ρ at the bond critical point are of equal magnitude. However, if electronic charge is preferentially accumulated in a given plane along the bond path (as it is for a bond with π-character, for example), then the rate of fall-off in ρ is less along the axis lying in this plane than along the one perpendicular to it, and the magnitude of the corresponding curvature of ρ is smaller. If λ_2 is the curvature of smallest magnitude, then the quantity $\epsilon = [\lambda_1/\lambda_2 - 1]$, the ellipticity of the bond, provides a measure of the extent to which charge is preferentially accumulated in a given plane.[11] The axis of the curvature λ_2, the major axis, determines the relative orientation of this plane within the molecule. The ellipticities of the C-C bonds in ethane, benzene and ethylene are 0.0, 0.23 and 0.45 respectively, for densities calculated from basis sets containing proper polarizing functions and the major axis of the ellipticity in each of the latter two molecules is perpendicular to the plane of the nuclei. The bond ellipticities faithfully recover the anticipated consequences of the conjugation and hyperconjugation models of electron delocalization.[11]

The chemistry of a three-membered ring is very much a consequence of the high concentration of charge in the interior of the ring relative to that along its bond paths, a fact which is reflected in substantial bond ellipticities.[11,18] The values of ρ_r, the value of the charge density at a ring critical point, is generally only slightly less than, and in some cases almost equal to, the values of ρ_b for the peripheral bonds in the case of a three-membered ring of carbon atoms. In four-membered and larger rings of carbon atoms the values of ρ_r are considerably smaller, as the geometrical distance between the bond and ring critical points is greater than in a three-membered ring. Because electronic charge is concentrated to an appreciable extent over the entire surface of a three-membered ring, the rate of fall-off in the charge density from its maximum value along the bond path towaard the interior of the ring is much less than its rate of decline in directions perpendicular to the ring surface. Thus the C-C bonds have substantial ellipticities, and their major axes lie in the plane of the ring. The ellipticity of a C-C bond in cyclopropane is actually slightly greater than that for the "double bond" in ethylene, indicating that the extent to which charge is preferentially accumulated in the plane of the ring is greater than that accumulated in the π-plane of ethylene. This property accounts for the well-documented ability of three-membered rings to act as an unsaturated system with the charge distribution in the plane of the ring exhibiting properties characteristic of a π-like system, one that is able to conjugate with a neighbouring unsaturated system. Such conjugation is illustrated by the interaction of the cyclopropyl group with the (formally) vacant 2p orbital of CH_2^+ in cyclopropylcarbinyl cation, $C_3H_5CH_2^+$. The major axis of the ellipticity induced in the $C-CH_2$ bond has an overlap of 0.97 with the corresponding axes of the neighbouring C-C bonds of the cyclopropyl group. (The overlap is determined simply by taking the scalar product of the eigenvectors defining the major axes of the two bond critical points.) Such conjugating ability of the cyclopropyl group is rationalized using molecular orbital models through the choice of a particular set of orbitals, the so-called Walsh orbitals. Theory shows that the "π-like" nature of a three-membered ring is a property of its total charge distribution, one that results from the proximity of its ring and bond critical points. Understanding the physical basis of this effect enables one to predict its appearance and consequences in other systems.

The opening of a ring structure results from the coalescence of the ring and a bond critical point, the positive curvature of the ring point annihilating the in-plane negative curvature λ_2 of the bond point to yield a zero curvature characteristic of an unstable or degenerate critical point, Fig.9. The decrease in the magnitude of λ_2 and its eventual disappearance means the ellipticity of the bond which is to be broken increases dramatically and becomes infinite at the geometry of the bifurcation point. Thus a structure possessing a bond with an unusually large ellipticity is potentially unstable. The two equivalent ring bonds of the cyclopropylcarbinyl cation, $C_4H_7^+$ provide an example of this behaviour.[18] The two long bonds of the three-membered ring of this are of order 0.6 and exhibit ellipticities equal to 6.7. Their corresponding paths are very inwardly curved and their bond path length exceeds the internuclear separation by 0.20 Å. The structure verges on instability since either of the bond critical points of the long bonds can be annihilated by coalescence with the ring critical point. The curvature of ρ at the ring critical point which lies almost on the line joining the two bond critical points, is close to zero and correspondingly, the associated negative curvature, λ_2, of each of the neighbouring bond critical points is equally small in magnitude. As anticipated for long bonds, the positive curvature of ρ along their bond paths is relatively large, as is the second and parallel positive curvature of ρ at the ring critical point. The values of ρ at the bond and ring critical points differ by only 0.001 au. Thus there is a nearly flat-bottomed trough in the distribution of charge linking these three critical points and little energy is required to cause a migration of the ring point along the trough to coalesce with a bond point and yield a ring opened structure. It is a general observation that little energy is required for the nuclear motions which result in a migration of a critical point along an axis associated with vanishing by small curvature of the charge density. Thus the energy surface in the neighbourhood of this structure is very flat for such a motion of the nuclei and the open structure differs from it in energy by less than a kcal/mole.[18]

Further examples of potentially unstable structures being revealed through exceptionally high bond ellipticities are provided by the propellanes, particularly [2.1.1]propellane. The bridgehead bond critical point and each of the ring critical points of the two three-membered rings in this molecule are separated by only 0.07Å and the value of ρ_b = 0.197 au exceeds that of ρ_r by only 0.001 au. The close proximity and nearly equal values for the bond and ring critical points results in a zero value for the curvature of the density at the bond critical point in the direction of the three-membered ring critical points. The result is a very large ellipticity, equal to 7.21, for the bridgehead bond in this molecule. The bridgehead bonds in both [2.2.1] and [2.1.1]propellane are predicted to be the most susceptible to rupture by the bifurcation mechanism and both molecules readily undergo polymerization at 50°K. The bifurcation catastrophe undergone by the [1.1.1]propellane molecule has been used to illustrate the mathematical modelling of a structural instability [6].

These properties of a charge distribution have been applied to a study of the position of equilibrium between [10]-annulene and dinorcaradien as a function of the substituents R. While X-ray diffraction studies yield the geometries of the

relevant species, and in particular the C1-C6 internuclear separation, they do not enable one to determine whether or not carbons C1 and C6 are bonded to one another and hence to determine which of the above two structures is the correct one for a given set of substituents R and R'. Gatti et al[19] determined the topological properties of the theoretically determined charge distributions at the experimentally measured geometries for combinations of substituents R and R' with C1-C6 separations ranging from 1.543 Å for R = R' = CN to 2.269 Å for R = R' = F. The study demonstrated that the systems with a C1-C6 separation of 1.770 Å or less possessed a C1-C6 bond. The bond order of the C1-C6 bond and the separation between its critical point and the ring critical point of the three-membered ring undergo a continuous decrease and, correspondingly, the C1-C6 bond ellipticity exhibits a continuous increase through these molecules. A display of the gradient vector field of the next member in the plane of the three-membered ring, which possesses a C1-C6 separation only slightly greater at 1.783 Å, indicated that this separation is past the geometry of the bifurcation point created by the coalescence of the ring and bond critical point, as both critical points are absent from the display. It has also been shown[18] that the topological theory of molecular structure can also be used to treat in an unambiguous manner the corresponding problem of determining whether or not homoaromatic conjugation is present in a given system, a property which is also determined by the properties particular to a cyclopropyl ring.

PROPERTIES OF THE LAPLACIAN OF THE ELECTRONIC CHARGE DENSITY

The Laplacian of the electronic charge density is defined in eqn (7). Its use in models of molecular geometry and reactivity are the subject of a recent review.[20] The present discussion is limited to an overview of the role which this function plays throughout the theory and of the physical basis it provides for the models based on the electron pair concept of Lewis[5] and for the classification of chemical bonds.

The Role of the Laplacian of the Charge Density in the Quantum Theory of Molecular Structure

The Laplacian of the charge density appears as an energy density in the theory, that is as L(r), the quantity

$$L(r) = -(\hbar^2/4m)\nabla^2\rho(r) \tag{12}$$

The integral of L(r) over an atom Ω to yield L(Ω)

$$L(\Omega) = \int_\Omega L(r)d\tau = (-\hbar^2/4m)\int_\Omega \nabla^2\rho(r)d\tau$$

$$= (-\hbar^2/4m)\oint dS(\Omega,r)\nabla\rho(r)\cdot n(r) = 0 \tag{13}$$

vanishes because of the zero flux boundary condition, eqn (10), which defines an atom in a molecule. The demonstration that an atom is an open quantum subsystem is obtained by a variation of Schrödinger's energy functional $\mathcal{E}[\psi,\Omega]$ for a stationary state and by a variation of the action integral for a time dependent system.

$$W_{12}[\Psi,\Omega] = \int_{t_2}^{t_2} \mathcal{L}[\Psi,\Omega]\,dt \tag{14}$$

In each case the zero flux boundary condition is introduced by imposing the variational constraint that

$$\delta L(\Omega) = \delta\{\int_{\Omega} \nabla^2\rho(r)\,d\tau\} = 0 \tag{15}$$

at every stage of the variation. The possibility of introducing the constraint in this manner is a consequence of the property of the functional $\mathcal{E}[\psi,\Omega]$ and of the Lagrange integral $\mathcal{L}[\Psi,\Omega]$, that at the point of variation where the appropriate Schrödinger equation is satisfied, they both reduce to an integral of the density $L(r)$. The property given in eqn (13) is common for an atom and for the total system and it is this property which endows them with similar variational properties, thereby making possible the generalization of the principle of stationary action to an atom in a molecule.

It is because $L(\Omega)$ vanishes for an atom that $T(\Omega)$, the electronic kinetic energy of an atom is well defined. The density $L(r)$ appears in the local expression for the virial theorem, eqn (16). This is an important result, since it relates a property of the charge density to the local contributions to the total energy,

$$(\hbar^2/4m)\nabla^2\rho(r) = 2G(r) + \mathcal{V}(r) \tag{16}$$

The electronic potential energy density $\mathcal{V}(r)$, the virial of the forces exerted on the electrons and the electronic kinetic energy density $G(r)$ define the electronic energy density $E_e(r)$

$$E_e(r) = G(r) + \mathcal{V}(r) \tag{17}$$

Because $L(\Omega)$ vanishes for an atom, integration of eqn (16) over the basin of an atom yields the atomic virial theorem

$$2T(\Omega) = -\mathcal{V}(\Omega) \tag{18}$$

and as a consequence, the electronic energy of an atom in a molecule satisfies the following identities

$$E_e(\Omega) = \int_{\Omega} E_e(r)\,d\tau = -T(\Omega) = \tfrac{1}{2}\mathcal{V}(\Omega) \tag{19}$$

It is a property of the Laplacian of a scalar function such as the charge density $\rho(r)$, that it determines where the function is locally concentrated, where $\nabla^2\rho(r) > 0$, and locally depleted, where $\nabla^2\rho(r) < 0$. Electronic charge is concentrated in those regions of space where the Laplacian of the charge density is negative. The expressions 'local charge concentrations' and 'local charge depletions' will refer to maxima and minima in the function $-\nabla^2\rho(r)$, extrema which are to be distinguished from local maxima and minima in the charge density itself. This property of the Laplacian can be used to determine the dominant contributions to the local energy of the electronic charge distribution using the local expression for the virial theorem.

The potential energy density $V(r)$ is everywhere negative, while the kinetic energy density is everywhere positive. Thus the sign of the Laplacian of the charge density determines, via eqn (16), which of these two contributions to the total energy is in excess over their average virial ratio of 2:1. *In regions of space where the Laplacian is negative and electronic charge is concentrated, the potential energy dominates both the local total electronic energy $E_e(r)$ and the local virial relationship. Where the Laplacian is positive and electronic charge is locally depleted, the kinetic energy is in local excess.*

An energy density is dimensionally equivalent to a force per unit area or a pressure. Thus the Laplacian may alternatively be viewed as a measure of the pressure exerted on the electronic charge density relative to the value of zero required to satisfy a local statement of the virial theorem, i.e., $V(r) + 2G(r) = 0$. In regions where the Laplacian is negative, the charge density is tightly bound and compressed above its average distribution. In regions where the Laplacian is positive, the charge density is expanded relative to its average distribution, the pressure is positive and the kinetic energy of the electrons is dominant.

The Laplacian of the charge density and the Lewis electron pair model

Neither the electronic charge density nor the electronic pair density offer any evidence of the localized bonded and nonbonded pairs of electrons evoked in the Lewis model of electronic structure. The relatively simple topology exhibited by the charge density has already been described and while it accounts for elements of molecular structure, it does not offer any suggestion of the existence of spatially localized pairs of electrons.

The electron pairs of Lewis and the associated models of geometry and reactivity find physical expression in the topology of the Laplacian of the charge distribution. The Laplacian distribution recovers the electronic shell model of an atom by exhibiting a corresponding number of pairs of shells of charge concentration and charge depletion.[4,21,11] For a spherical free atom, the outer or valence shell of charge concentration, the VSCC, contains a sphere over whose surface electronic charge is maximally and uniformly concentrated. Upon entering into chemical combination, this valence shell of charge concentration is distorted and maxima, minima and saddles appear on the sphere of charge concentration. The maxima correspond in number, location and in size to the localized pairs of electrons assumed in the Lewis model. The VSEPR model of molecular geometry[23] is a direct extension of the Lewis model and it predicts the geometries of closed-shell molecules about some central atom which contains from two to seven pairs of electrons in its valence shell. All of the properties postulated in this model for bonded and nonbonded pairs of electrons are recovered by the maxima in the valence shell of charge concentration of the central atom and the Laplacian of the charge density provides a physical basis for this most successful of models of molecular geometry.[20,24] It has been shown that positioning the reference electron at the critical point corresponding to a maximum in the Laplacian distribution maximally localizes its Fermi hole.[24]

The discussion is so far has focused on the properties of the local charge concentrations of the Laplacian distribution and on how they recover the Lewis and VSEPR models of electron pairs. The Lewis model, however, encompasses chemical reactivity as well, through the concept of a generalized acid–base reaction. Complementary to the local maxima in the VSCC of an atom for the discussion of reactivity are its local minima. A local charge concentration is a Lewis base or a nucleophile, while a local charge depletion is a Lewis acid or an electrophile. A chemical reaction corresponds to the combination of a "lump" in the VSCC of the base combining with the "hole" in the VSCC of the acid. In terms of the local virial theorem, eqn (16), the reaction of a nucleophile with an electrophile is a reaction of a region with excess potential energy on the base atom with a region of excess kinetic energy on the acid atom. The accompanying rearrangement of the charge is such that at every stage of the reaction, $L(\Omega)$ remains equal to zero for each atom. Thus reductions in the magnitudes of the local concentrations or depletions of charge requires opposing changes in other parts of the atom to satisfy the constraint on its charge distribution as given in eqn (13).

The positions of the local charge concentration and depletion together with their magnitudes, are determined by the positions of the corresponding critical points in the VSCC's of the respective base and acid atoms. This information enables one to predict positions of attack within a molecule or the geometries of approach of the reactants. For example, a keto oxygen in the formamide molecule has two large nonbonded charge concentrations in the plane of the nuclei ($\nabla^2 \rho = -6.25$ and -6.30 au) while the nitrogen atom exhibits two such maxima of less magnitude ($\nabla^2 \rho = -2.14$ au) above and below this plane. On the basis of this information one correctly predicts that the formamide molecule will preferentially protonate at the keto oxygen, specifically at the position of the largest of the two charge concentrations and in the plane of the nuclei. There are holes in the VSCC of a carbonyl carbon and they determine the position of nucleophilic attack at this atom. These holes are above and below the plane of the nuclei of the keto grouping and the corresponding critical point for a number of ketones are positioned to form angles of $110° \pm 1°$ with respect to the C=O bond axis. This is the angle of attack predicted for the approach of a nucleophile to a carbonyl carbon.[25]

Similar predictions have been made for the Michael addition reaction, specifically for the nucleophilic attack of an unsaturated carbon in acrylic acid, $CH_2{=}CH{-}CO_2H$ and methyl acrylic acid.[26] The properties of the Laplacian distribution correctly predict that the attack occurs at the terminal carbon of the methylene group, the carbon of the unsubstituted acid being most reactive, and that the approach of the nucleophile will be from above or below the plane of the nuclei along a line forming an angle of $115°$ with the C=C bond axis, the latter prediction being in agreement with calculations of the potential energy surface for this reaction. Bader and Chang[27] have given a discussion of the use of the Laplacian distribution in the prediction of the sites of electrophilic attack in a series of substituted benzenes.

Electrostatic potential maps have been used to make predictions similar to these.[28] Such maps, however, do not in general reveal the location of the sites of nucleophilic attack, as the maps are determined by only the classical part of the potential. The local virial theorem eqn (16), which along with the kinetic energy density, determines the sign of the Laplacian of the charge density, involves the full quantum potential. The potential energy density $V(r)$ contains the virial of the Ehrenfest force, the force exerted on the electronic charge at a point in space. The classical electrostatic force is but one component of this total force.

The Laplacian distribution has been used to predict the structures of a large number of hydrogen bonded complexes by aligning the (3,+3) critical point, a local charge depletion on the nonbonded side of the proton in the acid HF, with the (3,-3) critical point of the base, a local concentration of charge, for which $-\nabla^2 \rho$ attains its largest value.[29] With only a few exceptions, the geometries of the complexes predicted in the SCF calculations (which agree with experiment where comparison is possible), are those predicted by the properties of the Laplacian as outlined above.

The Characterization of Atomic Interactions

The gradient vector field of the charge density identifies the set of atomic interactions within a molecule. These interactions, which define the molecular structure, can be characterized in terms of the properties of the Laplacian of the charge density. The local expression of the virial theorem eqn (16), relates the sign of the Laplacian of ρ to the relative magnitudes of the local contributions of the potential and kinetic energy densities to their virial theorem averages. By mapping those regions where $\nabla \rho^2 < 0$, the regions where electronic charge is concentrated, one is mapping those regions where the potential energy density makes its dominant contributions to the lowering of the total energy of the system.

As previously discussed, the interaction of two atoms leads to the formation of a critical point in the charge density at which the Hessian of ρ has one positive eigenvalue labelled λ_3 and two negative eigenvalues labelled λ_1 and λ_2, implying that ρ exhibits one positive and two negative curvatures at the point r_c. Since the two perpendicular curvatures of ρ, whose eigenvectors define the interatomic surface, are negative, the charge density is a maximum at r_c in the interatomic surface and charge is locally concentrated there with respect to points in the surface. The curvature of ρ along the interaction line is positive, charge density is locally depleted at r_c relative to neighbouring points along the line and ρ is a minimum at r_c along this line. *Thus the formation of a chemical bond and its associated interatomic surface is the result of a competition between the perpendicular contractions of ρ towards the bond path which lead to a concentration or compression of charge along this line, and the parallel expansion of ρ away from the surface which leads to its separate concentration in each of the atomic basins.* The sign of the Laplacian of ρ at the bond critical point, the quantity $\nabla^2 \rho(r_c)$, determines which of the two competing effects is dominant and because of the appearance of $\nabla^2 \rho(r_c)$ in the local expression for the virial theorem eqn (16), its sign also serves to summarize the essential mechanical characteristics of the interaction which creates the critical point. There is therefore, an intimate link between the topological properties of $\rho(r)$ and its Laplacian, the trace of the Hessian of ρ, and through the properties of the Laplacian one may begin to bridge the gap between the form of the charge distribution and the mechanics which govern it. The reader is referred to reference[4] for a full discussion and for numerical and pictorial illustrations.

When $\nabla^2 \rho(r_c) < 0$ and is large in magnitude, electronic charge is concentrated in the internuclear region as a result of the dominance of the perpendicular contractions of ρ towards the interaction line, or equivalently in these bound systems, towards the bond path. The result is a sharing of electronic charge by both nuclei, as is found for interactions usually characterized as covalent or polar and they shall

56

be referred to as *shared interactions*. In shared interactions, as exemplified for N_2 in Fig. 10, the region of space over which the Laplacian is negative and which contains the interatomic critical point, is contiguous over the valence regions of both atoms and the VSCC's of the two atoms form one continuous region of charge concentration. The interaction is dominated by the lowering of the potential energy associated with the formation of the $(3,-1)$ critical point. In a shared interaction, the nuclei are bound as a consequence of the lowering of the potential energy associated with the concentration of electronic charge between the nuclei, eqn (16).

This concentration of electronic charge in the interatomic surface is reflected in relatively large values of $\rho(r_c)$, the value of ρ at the $(3,-1)$ critical point, for molecules with shared interactions and the ratio of the perpendicular contractions of ρ to its parallel expansion, as measured by the ratio $|\lambda_1/\lambda_3|$, is greater than unity.[4] In cases of tight binding, as evidenced by the large negative values of $\nabla^2\rho(r_c)$, as found in N_2 for example, λ_3 as well as λ_1 and λ_2 is large in magnitude, but the ratio $|\lambda_1/\lambda_3|$ is still greater than unity. Occupation of the antibonding 2π orbital of AB or of the corresponding π_g orbital of A_2, causes an increase in λ_3.

The second limiting type of atomic interaction is that occurring between closed-shell systems, such as found in noble gas repulsive states, in ionic bonds, in hydrogen bonds and in van der Waals molecules. One anticipates that such interactions will be dominated by the requirements of the Pauli exclusion principle. Thus for *closed-shell interactions*, $\rho(r_c)$ is relatively low in value and the value of $\nabla^2\rho(r_c)$ is positive.[4] The sign of the Laplacian is determined by the positive curvature of ρ along the interaction line, as the exclusion principle leads to a relative depletion of charge in the interatomic surface. These interactions are dominated by the contraction of charge away from the interatomic surface towards each of the nuclei. The Laplacian of ρ is positive over the entire region of interaction and the kinetic energy contribution to the virial from this region is greater than the contribution from the potential energy. The spatial display of the Laplacian of ρ given in Fig. 10 for NaCl is atomic-like for this example of a closed-shell interaction. The regions where the Laplacian is negative are, aside from small polarization effects, identical in form the those of a free atom or ion. Thus the spatial regions where the potential energy dominates the kinetic energy are confined separately to each atom, reflecting the contraction of the charge towards each nucleus, away from the region of the interatomic surface. The ratio $|\lambda_1/\lambda_3| < 1$ in all the examples of closed-shell interactions.[4]

A hydrogen bond results from the interaction of two closed-shell systems and the properties of ρ at the associated bond critical point reflect all of the characteristics associated with such interactions; a low value for $\rho(r_c)$ and $\nabla^2\rho(r_c) > 0$. The same characteristics, with even smaller values of $\rho(r_c)$, are found for the bond to hydrogen formed in a van der Waals complex of an acid such as HF or HCl with an inert gas atom. *A hydrogen bond, which includes the van der Waals complexes, is defined to be one in which a hydrogen atom is bound to the acid fragment by a shared interaction, $\rho(r_c)$ large and $\nabla^2\rho(r_c) < 0$, and to the base by a closed-shell interaction, $\rho(r_c)$ small and $\nabla^2\rho(r_c) > 0$.* These are the very characteristics exhibited by the experimentally determined charge density at the bond critical points for the bond paths linking a proton in crystals of amino acids.

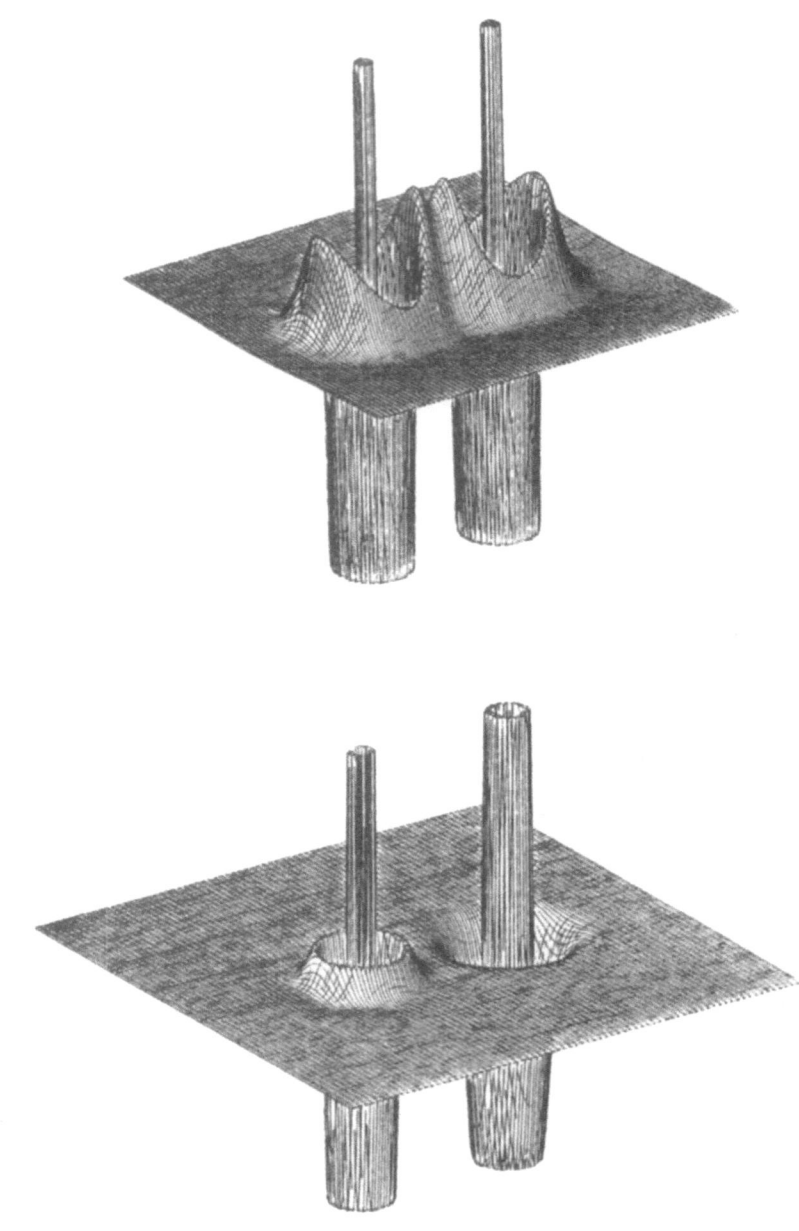

Cl   Na

Fig. 10 Relief maps of the Laplacian of $\rho(r)$ for the ground states of
N_2 and $NaC\ell$.

The examples so far considered have demonstrated the existence of two extremes of atomic interactions, one set being the opposite of the other in terms of the regions of charge concentrations and depletions and the associated mechanical consequences. The whole spectrum of possible interactions lying between these two limiting extremes are found to occur for interactions that do and do not involve interatomic charge transfer.

The electrostatic and virial theorems, coupled with the properties of the Laplacian of the charge density, also enable one to classify a given atomic interaction as belonging to a bound or unbound state. The regions of charge concentration as defined by the Laplacian can be used in conjunction with the electrostatic theorem to determine whether the forces on the nuclei will be attractive or repulsive. The electronic charge in these same regions makes the dominant stabilizing contributions to the potential and total energies of the system and in this way the Laplacian provides a link between the force and the energy. The reader is referred to reference 4 for this discussion.

QUANTUM MECHANICS OF AN OPEN SYSTEM

The generalized variation of the quantum action integral which leads to Schwinger's principle of stationary action, can be extended to obtain a quantum description of an open system and its properties.[1] The operational statement of the principle of stationary action for an open system in a stationary state is

$$\delta G[\psi,\Omega] = -\{(i/\hbar)\langle\psi[\hat{H},\hat{F}]\psi\rangle_\Omega + \text{complex conjugate}\} \qquad (20)$$

where $G[\psi,\Omega]$ is the energy functional for an open system, $\hat{H}$ is the Hamiltonian operator and the operator $\hat{F}$ is the generator of an infinitesimal unitary transformation in ψ causing the variation in $G[\psi,\Omega]$. This result is obtained only for a region of space satisfying the boundary condition of zero flux in the gradient vector field of the charge density, eqn (10), a condition satisfied by both the total system and the atoms which comprise it. Thus a single quantum description applies to the total system and to the atoms within it. Eqn (20) is a variational statement of the atomic hypervirial theorem for the observable $\hat{F}$. Its derivation defines the observables and their average values for an atom in a molecule. It yields the theorems, such as the Ehrenfest and virial theorems, which govern these average values.

All properties of an atom in a molecule are defined, including the atomic contributions to the electric polarizability and magnetic susceptibility.[30] The reader is referred to references 1, 2 and 30 for a full discussion. We give here the reasons why these atoms of theory are the atoms of chemistry. First, they are the most transferable pieces of a system that one can define in real space and which simultaneously exhaust the space of the system. Second, their properties, as well as being defined by quantum mechanics, are directly determined by their charge distribution. That is, if an atom has the same form in the real space of two systems, then it contributes the same amounts to the total properties in each system. Third, the atomic contributions $M(\Omega)$ to a property M are additive to yield the molecular value. Thus

$$\langle\hat{M}\rangle = \sum_\Omega M(\Omega) \qquad (21)$$

When an atom is transferable between molecules with little or no change, then its properties are transferable (point 2) as well as additive (point 3) and the theory thus predicts the existence of group additivity schemes, such as the additivity observed in the heats of formation of the hydrocarbons. Fourth and finally, the atoms of theory recover the experimentally measurable properties of atoms in molecules, such as the additive group contributions to the heats of formation.

TUTORIAL

The tutorial sessions will introduce the reader to the programs which implement the theory of atoms in molecules, as outlined in this chapter. The tutorials will use charge distributions obtained from theoretical calculations but it is to be emphasized that the same analysis can be applied to charge densities obtained from experiment. If possible, their application to an experimental density distribution will be illustrated.

The first program, called EXTREME RHO, locates and classifies by rank and signature the critical points in a charge distribution. This information enables one to assign a molecular structure to the system under investigation and to classify its atomic interactions. The program EXTREME DELSQURHO locates and classifies the critical points in the Laplacian distribution. This information identifies the sites of electrophilic and nucleophilic attack and also determines their relative propensity to attack.

The program PLOTDEN enables one to plot the charge density, the Laplacian and other properties of the density either in the form of a contour or a relief map. It also can be used to display a given envelope of the same functions. In the case of the Laplacian distribution, the envelope plot yields a display of a molecule's reactive surface. These graphical displays are frequently useful in implementing the other programs in the series.

The program SCHUSS plots the trajectories of the gradient vector field of ρ. It can be used to obtain a full gradient vector field map or the molecular graph which is obtained by the linking of the trajectories which originate at the bond critical points and terminate at the nuclei. The same program calculates the trajectories which lie in an interatomic surface and it can be used to overlay a contour map with the system's bond paths and the intersections made by the interatomic surfaces with the plane of the diagram.

The program PROAIM, PRoperties Of Atoms In Molecules, enables one to calculate the average properties of an atom in a molecule. These properties include the kinetic energy of the electrons, and from this via the virial theorem, the total energy of the atom, its potential energies of interaction, its monopole, dipole and quadrupole moments and the degree of physical localization of the electronic charge within the basin of the atom. The input for this program, which consists of all the critical points lying in the atom's surface, is obtained from EXTREME RHO. The program SURPROS calculates the average values of properties in an interatomic surface.

The theory of atoms in molecules, as applied through these programs, enables one to determine a system's molecular structure, to classify the bonds which make up the structure, to define the atoms and their properties within the system and to determine the sites of chemical reactivity. The electronic charge density, its gradient vector field, and its Laplacian distribution serve to define the conceptual models of chemistry and in this manner the molecular structure hypothesis and its associated classifications are transformed into a physical theory.

REFERENCES

1. R. F. W. Bader and T. T. Nguyen Dang, Ad. Quantum Chem. 14:63 (1981).

2   R. F. W. Bader, Pure Appl. Chem. 60:145 (1988).

3. R. F. W. Bader, T. T. Nguyen Dang and Y. Tal, Rep. Prog. Phys. 44:893 (1981).

4. R. F. W. Bader and H. Essén, J. Chem. Phys. 80:1943 (1984).

5. R. F. W. Bader, P. J. MacDougall and C. D. H. Lau, J. Am. Chem. Soc. 106:1594 (1984).

6. K. Collard and G. G. Hall, Int. J. Quantum Chem. 12:623 (1977).

7. V. H. Smith, J. F. Price and I. Absar, Israel J. Chem. 16:187 (1977).

8. R. F. W. Bader in "MTP International Review of Science. Physical Chemistry Ser. 2. vol 1 Theoretical Chemistry," A. D. Buckingham and C. A. Coulson eds., Butterworths, London (1975).

9. G. R. Runtz, R. F. W. Bader and R. R. Messer, Can. J. Chem. 55:3040 (1977).

10. J. Palis and S. Smale, Pure Math, 14:223 (1970).

11. R. F. W. Bader, T. S. Slee, D. Cremer and E. Kraka, J. Am.Chem. Soc. 105:5061 (1983).

12. M. T. Carroll and R. F. W. Bader, Mol. Phys. 65:695 (1988).

13. R. J. Boyd and S. C. Choi, Chem. Phys. Letters 129:62 (1986).

14. K. B. Wiberg, R. F. W. Bader and C. D. H. Lau, J. Am. Chem. Soc. 109:985,1001 (1987).

15. K. B. Wiberg and C. M. Brenneman, private communication.

16. K. B. Wiberg and M. A. Murcko, J. Mol. Struct. (Theochem) 169:355 (1988).

17. K. B. Wiberg and K. E. Laidig, J. Am. Chem. Soc. 109:5935 (1987).

18. D. Cremer, E. Kraka, T. S. Slee, R. F. W. Bader, C. D. H. Lau, T. T. Nguyen-Dang and P. J. MacDougall J. Am. Chem. Soc. 105:5069 (1983).

19. C. Gatti, M. Barzaghi and M. Simonetta J. Am. Chem. Soc. 107:878 (1985).

20. R. F. W. Bader, R. J. Gillespie and P. J. MacDougall, in "Molecular Structure and Energetics", J. F. Liebman and A. Greenberg eds., VCH Publishers, Fla. (1989).

21. R. P. Sagar, A. C. Ku, V. H. Smith and A. M. Sinas, J. Chem. Phys. 88:4367 (1988).

22. Z. Shi and R. J. Boyd, J. Chem. Phys. 88:4375 (1988).

23. R. J. Gillespie, "Molecular Geometry," Van Nostrand Reinhold, London (1972).

24. R. F. W. Bader, R. J. Gillespie and P. J. MacDougall, J. Am. Chem. Soc. 110:7329 (1988).

25. H. B. Burgi and J. D. Dunitz, Acc. Chem. Res. 16:153 (1983).

26. M. T. Carroll, J. R. Cheeseman, R. Osman and H. Weinstein, J. Phys. Chem. 93:5120 (1989).

27. R. F. W. Bader and C. Chang, J. Phys. Chem. 93:2946 (1989).

28. E. Scrocco and J. Tomasi, Adv. Quantum Chem. 11:116 (1978).

29. M. T. Carroll, C. Chang and R. F. W. Bader, Mol. Phys. 63:387 (1988).

30. R.F.W. Bader, J. Chem. Phys. (1989).

ELECTROSTATIC PROPERTIES OF MOLECULES FROM DIFFRACTION DATA

Robert F. Stewart

Department of Chemistry
Carnegie Mellon University
Pittsburgh, PA 15213 USA

INTRODUCTION

Single crystal x-ray diffraction analysis today has become an essential methodology for scientists in a wide range of disciplines that span solid-state physics, synthetic chemistry and molecular biology. Although x-ray diffraction was first exploited by the Braggs in 1913 for crystal structure determination,[1] it has only been the last three decades that the method has become routine for establishing the atomic structure in a crystallographic unit cell. The advent of digital computers has made it possible to implement direct methods, the evaluation of relations between measured diffraction intensities and phases, for the assignment of phases to observed x-ray structure factor moduli. Accurate measurements of the diffracted intensities are now made with diffractometers that operate in an automatic mode with supporting software to select out the photon counts which are mostly due to elastic scattering. For many crystals of organic compounds, which have rather low Debye temperatures, it is routine in many labs to cool the sample to about 100 K with a cold stream of nitrogen and thereby measure Bragg diffraction intensities at these reduced temperatures to larger scattering angles than is possible at room (300 K) temperature. Thus full atomic resolution is often achieved. When the phase estimates of the structure factors are adequate for locating the atoms within ten or so picometers of their mean thermal positions, least squares programs are then used to adjust both atomic positions and mean square amplitudes of vibration by fitting a structure factor model to the "observed" structure factor moduli or their squares. In this latter stage of a crystal structure anlaysis, the observation to parameter ratio is greater than ten to one and, for the case of high resolution data, the ratio may exceed fifty to one. The very large redundancy - the larger number of observations to number of atomic parameters - in x-ray crystallography is what makes the experiment a definitive method for structure determination. Upon completion of a crystal structure determination, it is routine to report atomic positions with estimated standard deviations in the range of tenths of picometers. In some cases a dynamical analysis of the structure, such as a rigid body motion model, is included in an attempt to deduce the equilibrium structure of the atoms in the crystal. The final objective, a complete determination of the atomic arrangement with precise metrics in a crystal unit cell, has been met.

In the present lecture, however, a far more ambitious enterprise in x-ray crystal structure analysis is addressed. With a standard crystal structure determination, the precise locations of local maxima in the mean thermal electron density distribution is the desired result; the actual values of the electron density at these sites are not a matter of concern. On

The Application of Charge Density Research to Chemistry and Drug Design
Edited by G. A. Jeffrey and J. F. Piniella, Plenum Press, New York, 1991

the other hand, an investigator seeking a determination of the charge density would like to establish the values of the electron density everywhere in the unit cell. To map the electron density everywhere would require a virtual infinity of measured (and phased) structure factors; the advantage of over-determination vanishes. With a finite data set, one is forced to compromise the objective of "everywhere" and to adopt various strategies that can lead to evaluation of the electron density within some level of tolerance.

A crystal may be thought of as a very large molecule. The number of electrons and nuclei in a crystal make the system macroscopic. A quantum chemical treatment may provide some guidance, but a statistical mechanical approach is needed to characterize x-ray data even at the most elementary level. A simple crystal structure may have a very elaborate charge density distribution. Evidence of this complexity may be found in the Fourier sum of suitably phased and accurately measured x-ray structure factors. The x-ray structure factors are the Fourier components of the thermally weighted vibrational average of the electron density distribution in the crystallographic unit cell. Neutron structure factors are the Fourier components of the thermally weighted vibrational average of the nuclear distribution. Neutron diffraction data may be analyzed for the nuclear distribution by least squares with a small Gram-Charlier expansion[2] about each nucleus; in principle this representation gives the mean thermal probability distribution of the nuclei, which is due solely to the lattice dynamics of the crystal. By using nuclear charges as the nuclear form factor and the cumulants from a neutron diffraction analysis, structure factors for the nuclear charge distribution may be combined with suitably phased x-ray structure factors so that the total charge density can be mapped within the resolution of the sphere of x-ray diffraction data. If high energy electrons were kinematically diffracted by crystals (they are not), then the total charge density, or more appropriately, the electrostatic potential could be directly mapped from the electron data with appropriate phase assignments.

Several strategies may be used to display charge density results. They all have inherent limitations. The appropriate level for comparison with theory is not obvious, particularly for a moderately complicated structure. The major task, however, is to explore the extent to which experimentally derived charge density distributions provide an understanding of chemical bonding in crystals. The most useful results that may evolve in charge density determinations from diffraction data of crystals are the sundry physical properties that depend on the distribution. Even this desired goal may not be sufficient to characterize the bulk properties of all crystalline materials, since defects can be crucial to the property of interest. Clearly, the burden is to use our modest results to help resolve disparity between theoretical models, or, better to complement the data and conclusions from another experiment on the same crystal.

Several approaches and a few results on the mapping of electrostatic properties from diffraction data are outlined below. Some fundamental physical theory of x-ray scattering is first covered; it is essential to understanding the validity of the form factor approximation and the nature of the canonical ensemble average of elastic x-ray scattering over the accessible vibrational states of a crystal. Relationships between structure factors and physical properties are reviewed. The use of generalized x-ray scattering factors in least squares fits to reduced x-ray diffraction intensities is discussed. The gsf (generalized x-ray scattering factor) model may be used to evaluate many physical properties and the limitations of this procedure are emphasized. Our methods for evaluating the variances of all physical properties are also covered.

ELASTIC X-RAY SCATTERING

X-ray diffraction is basically a Rayleigh light scattering experiment at 10^{18}-10^{19} Hz. The magnitude of the incident wavevector, $2\pi/\lambda$, is equal to the magnitude of the wavevector of the scattered x-ray; the difference vector of scattered from incident wavevectors, K, is 2π times the Bragg vector, H, which is normal to a diffraction plane in a crystal.

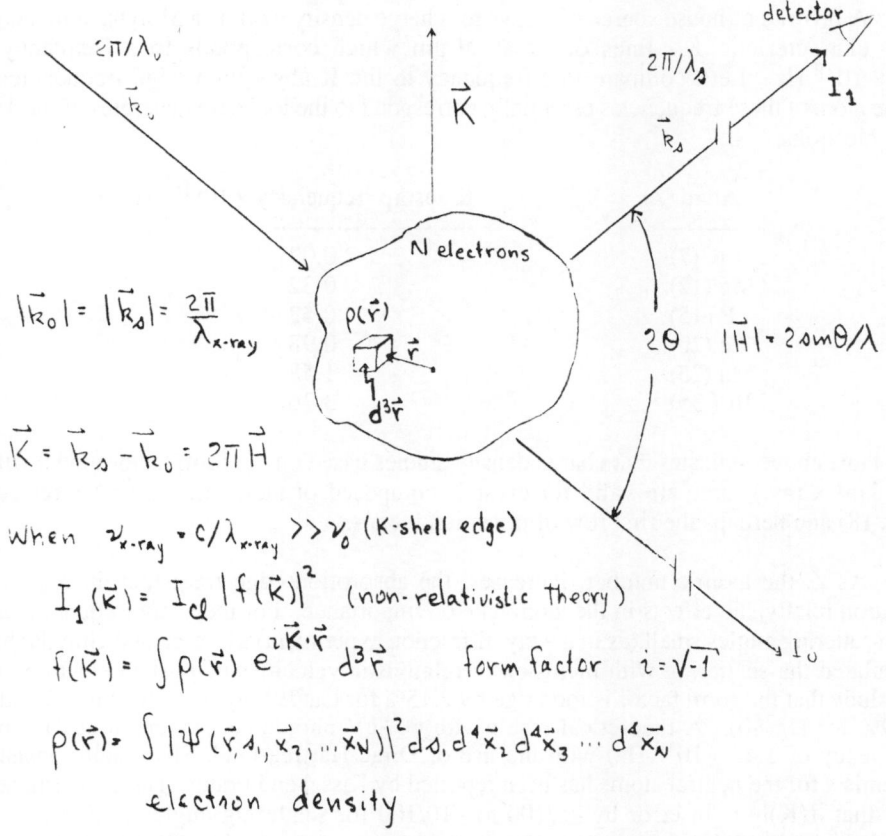

$$|\vec{k}_0| = |\vec{k}_s| = \frac{2\pi}{\lambda_{x\text{-}ray}}$$

$$\vec{K} = \vec{k}_s - \vec{k}_0 = 2\pi\vec{H}$$

$$|\vec{H}| = 2\sin\theta/\lambda$$

When $\nu_{x\text{-}ray} = c/\lambda_{x\text{-}ray} \gg \nu_0$ (K-shell edge)

$$I_1(\vec{K}) = I_{cl}\,|f(\vec{K})|^2 \quad \text{(non-relativistic theory)}$$

$$f(\vec{K}) = \int \rho(\vec{r})\, e^{i\vec{K}\cdot\vec{r}}\, d^3\vec{r} \qquad \text{form factor} \quad i = \sqrt{-1}$$

$$\rho(\vec{r}) = \int |\Psi(\vec{r},\sigma_1,\vec{x}_2,...,\vec{x}_N)|^2\, d\sigma_1\, d^4\vec{x}_2\, d^4\vec{x}_3 \cdots d^4\vec{x}_N$$

electron density

The diagram above illustrates the basic relations between the intensity of elastically scattered x-rays and the electron density of a many electron atom. The theory is applicable to the high frequency limit, whereby the x-ray frequency is greater than any natural resonance frequencies of electrons in the atom - but not so large that relativistic effects are important. The latter starts to occur when the energy of the x-ray photon exceeds 500 keV or has a wavelength less than 2.5 picometers. It is interesting that even in the early papers on the quantum theory of radiation scattering with wavelengths short compared to the size of an atom, relativistic effects were a natural part of the theory.[3] The elastically scattered intensity is proportional to the square of the atomic form factor in the non-relativistic theory; the proportionality is the Thomson cross-section for an electron $(e^2/mc^2)^2$, a polarization factor which is $[3 + \cos(4\theta)]/4$ for unpolarized x-rays, and $1/R^2$ where R is the distance from diffracting crystal to the detector. The atomic form factor is the Fourier transform of the one-electron density distribution of the atom. It is the probability of observing an electron, regardless of spin, in the average of all the other electrons at some point between r and $r + d^3r$. This function can be constructed from a many electron wavefunction by integrating the squared modulus of the wavefunction over all the spins and positions of the electrons but one and over the spin of the last electron. The spin independence of the scattering is due to the very small cross-section of magnetic scattering compared to the electrical scattering - the ratio is one part per million or less.[4] Because the nuclei have masses that are at least two thousand times greater than an electron mass, electrical scattering by the nucleus is even smaller than magnetic scattering by an electron. In this high frequency regime, the physical theory is telling us that all the electrons are scattering the x-ray at its incident frequency with no phase shift so that no electron is distinguishable from any other electron.

A favorite in-house source of x-rays for charge density work is a Molybdenum anode. The characteristic $K\alpha$ lines occur at 71 pm which corresponds to a frequency of 4.2×10^{18} Hz. Let's compare this frequency to the K absorption edge frequencies of some atoms - these frequencies essentially correspond to the ionization energies of the 1s or core electrons.

Atom (Z)	K absorp frequencey $\times 10^{-18}$ Hz
N (7)	0.097
Mg (12)	0.32
P (15)	0.52
Ca (20)	0.98
Mn (25)	1.58
Br (35)	3.26

The table above indicates that charge density studies based on the form factor model with a standard x-ray source are valid for crystals composed of atoms in the first three rows (Z < 18) and perhaps the first row of transition metals.

As Z, the atomic number, increases, the absorption edge frequency increases; in addition relativistic effects in the atom take on importance. For the x-ray frequency large and scattering angles small (as in a γ-ray diffraction experiment) Florescu and Gravila[5] have calculated the scattering with inclusion of relativistic velocities of K-shell electrons and conclude that the form factor is too large by 2.15% for Cu(29), by 6.8% for Sn(50) and by 13.9% for Hg(80). A theoretical result with 60 keV photons (wavelength at 21 pm or frequency of 1.45×10^{19} Hz) with the use of Dirac-Hartree-Fock-Slater self-consistent potentials for the neutral atoms has been reported by Kissel and Pratt.[6] These investigators find that $|f(K)|^2$ is in error by $-5/100$ to $-10/100$ for scattering angles in the range of $60 - 150°$ for Zn(30), but is too large by nearly a factor of two for Ta(73) with scattering in the angular range of 90 to 150°. Pratt gives a good review of his theoretical work on Rayleigh scattering cross-sections for x-ray and γ-ray photons by atoms in a 1984 paper.[7]

To put the present discussion in perspective we should note that bonding effects cause deviations from standard atomic form factors in the range of 5 to 15 parts per hundred. The theoretical results cited above either exceed or are in this range. The absolute accuracy of the relativistic calculations is not the major issue; the value is that they serve to define a useful range of atomic numbers for which a meaningful charge density analysis can be sensibly undertaken with the form factor approximation. My personal preference is to restrict charge density analyses to accurate diffraction measurements from either Mo or Ag $K\alpha$ radiation of crystals with atomic numbers less than eighteen or to use either synchrotron radiation or γ-ray diffraction data for studies of first-row transition metals. It should also be appreciated that with larger atomic numbers the absorption correction becomes progessively larger (absorption is due to the $\hat{p} \cdot \vec{A}$ term in scattering theory) and anomalous scattering corrections to $f(\vec{K})$ become appreciable (this is due to the other two terms of order $|\vec{A}|^2$ in scattering theory). [The jargon just used refers to the momentum operator of an electron, $\hat{p}$, and the vector potential of the electromagnetic field, $\hat{A}$.] When correction factors such as these exceed bonding effects by factors of three or more, then the reliability of charge density "results" can be severely compromised.

For molecules and crystals, the elastically scattered x-rays depend on the nuclear configuration $\vec{Q}$ (a 3N dimensional vector, with N the number of nuclei in the molecule or microcrystal). In the adiabatic approximation of molecular quantum mechanics, we can express the electron density as parametrically dependent on the nuclei and hence the elastic intensity as a function of $\vec{Q}$:

$$I_{el}(\vec{K};\vec{Q}) \propto \left| \int \rho(\vec{r};\vec{q}) \, e^{i\vec{K}\cdot\vec{r}} \, d^3\vec{r} \right|^2 = |F(\vec{K};\vec{q})|^2 \tag{1}$$

The detector counts photons of intensity $I_1(\vec{K})$, which is a canonical ensemble average over $\vec{Q}$,

$$I_1(\vec{K}) \propto \sum_v e^{-E_v/kT} \int \chi_v^*(\vec{Q}) \, I_{el}(\vec{K};\vec{Q}) \, \chi_v(\vec{Q}) \, d^{3N}\vec{Q} \tag{2}$$

The $X_V(\vec{Q})$ are vibrational wavefunctions with vibrational energy E_V. In adiabatic theory, these vibrational wavefunctions are solutions to the Schrodinger wave equation, where the electronic energy of the electrons in the field of fixed nuclei plus the expectation value of the nuclear kinetic energy serve as the potential for the nuclear vibrations. In formulating (2) note that initial and final vibrational states are the same in the sum over states, so that thermal diffuse scattering is not included. For real diffraction, TDS (thermal diffuse scattering) does occur and arises from phonon exchange with the x-ray - in the language of a molecular spectroscopist it is Raman scattering but with frequency shifts in the acoustal range as well as the more familiar gigahertz range. One-phonon exchange scattering does peak at Bragg diffraction angles[8] and can be a substantial fraction of the detected scattered radiation at higher scattering angles. This bothersome inelastic scattering can be corrected approximately with known elastic constants[9] or partially suppressed by reducing the temperature of the crystal during diffraction measurements. But the other point to note is that even if (2) were the only event measured, $I_1(\vec{K})$ does not seem to have a simple relation with a molecular (or crystal) form factor. The thermal vibrational average of the square of the molecular form factor is not the same as the square of the thermal vibrational mean of the molecular form factor:

$$I_1(\vec{K}) = I_{cl} \sum_v e^{-E_v/kT} \int \chi_v^*(\vec{Q}) \, |F(\vec{K}\cdot\vec{Q})|^2 \, \chi_v(\vec{Q}) \, d^{3N}(\vec{Q}) \tag{3'}$$

but,

$$I_B(\vec{K}) = I_{el} \left| \sum_v e^{-E_v/kT} \int \chi_v^*(\vec{Q}) \, F(\vec{K},\vec{Q}) \, \chi_v(\vec{Q}) \, d^{3N}(\vec{Q}) \right|^2 \tag{4}$$

Stewart and Feil[10] did show that for a large number of vibrational degrees of freedom, the relative difference between (3') and (4) is of order 1/N, so that even for a microcrystal with a million normal modes of vibration, (3'), which is ideally measured, does not differ appreciably from (4), which is used to interpret the event. This means that with a large number of normal modes of vibration, the atoms appear to vibrate independently and correlated motion averages out to a virtual null affect on the scattering intensity. With (3') measurably the same as (4),

$$I_1(\vec{K}) \simeq I_B(\vec{K}) = |F_B(\vec{K})|^2 \tag{5}$$

where, $\quad F_B(\vec{K}) = \int \bar{\rho}(\vec{r}) \, \exp(i\vec{K}\cdot\vec{r}) \, d^3\vec{r} \tag{6}$

and $\bar{\rho}(\vec{r})$, the mean thermal electron density distribution, is

$$\sum_v e^{-E_v/kT} \int \chi_v^*(\vec{Q}) \rho(\vec{r}; \vec{Q}) \chi_v(\vec{Q}) d^{3N}\vec{Q} \qquad (7)$$

For a crystal $\bar{\rho}$ has a long range order; $\bar{\rho}(\vec{r} + n\vec{L}) = \bar{\rho}(\vec{r})$. When $\lambda \sim |\vec{L}|$, diffraction occurs so that,

$$F_B(\vec{K}) \implies F_{\vec{H}} \qquad \{\vec{H};\ (h\vec{a}^*,\ k\vec{b}^*,\ l\vec{c}^*)\ ;\ h,k,l\ \text{integers}\}$$

$I_1(\vec{K})$ becomes $I_1(\vec{H})$ and is observed only at discrete angles. Much of this intensity is elastic scattering, but inelastic components due to the crystal vibrations can also occur under the Bragg reflections (vide supra). More importantly, (5) implicitly treats the scattering as a single event - kinematic scattering. If multiple scattering becomes dominant, then the intensities are subject to extinction effects and cannot be reduced to structure factors in any simple way. The last caveat: avoid crystals that suffer severe extinction in the diffraction process.

Before a charge density analysis is undertaken, one takes on the following responsibilities:

(i)   Measure $I_{\vec{H}}$ for near kinematic conditions.

(ii) Reduce $I_{\vec{H}}$ to $|F_{\vec{H}}|^n$ (n = 1 or 2).

(iii) Phase $|F_{\vec{H}}|$ with an appropriate model which in principle should contain most (if not all) aspects of $\bar{\rho}$.

PROPERTIES FROM FOURIER SUMS

The reduced data may serve as coefficients in Fourier sums that can be used to evaluate a variety of electrostatic properties within the crystallographic unit cell. Table 1 is a summary of these properties and their respective dependence on $|\vec{H}|$ ($|\vec{H}| = d^* = 2\sin\theta/\lambda$).

Table 1.   Electrostatic Properties from Fourier Sums
of Structure Factors

Property	Type	Dependence on $|\vec{H}|^n/n$
Electrostatic potential	scalar	−2
Diamagnetic shielding tensor	2nd rank	−2
Electrostatic energies	scalar	−2
Electric field	vector	−1
Electric field gradient	traceless 2nd rank tensor	0
Charge density	scalar	0
Diamagnetic current density	vector	0
Gradient of field gradient	3rd rank	1
Gradient of charge density	vector	1
Grad-grad of field gradient	4th rank	2
Hessian of charge density	2nd rank	2
Laplacian of charge density	scalar	2

Explicit formulae for the sums have been reported elsewhere[11] for most of the properties. The list above is quite extensive and is designed to keep an ambitious, but perhaps foolish crystallographer, out of the taverns and dance halls for years. The dependence of the sum on $|\vec{H}|$ is the key facet to note. For example, it is much easier to map the electrostatic potential than the Laplacian of the electron density at a desired level of resolution from a finite data set. In order to implement evaluation of the several sums it is essential that one first study the effects of termination ripple in the maps. A straightforward computation of a sum with $F_{\vec{H}}$ to map the electron density in a unit cell is not a practical method. (See reference 12 on a map by Cromer for melamine at $|\vec{H}|$max of 2.0 Å^{-1}.) We have explored three strategies for mapping out properties from Fourier sums; in all cases but one, these sums are combined with direct space lattice sums to arrive at the final result.

The first study employed combined electron and nuclear structure factors, F_N-$F_{\vec{H}}$, as Fourier coefficients in the sum. The reduced nuclear structure factor is constructed with the nuclear charge as the form factor and cumulants, as mentioned above; the electron structure factor is from the x-ray data, but scaled and phased with a gsf refinement model. For this case, the total mean thermal electrostatic property is mapped. Several maps of the potential and electric field in the unit cell of stishovite (SiO$_2$) with $|\vec{H}|$max of 2.4 Å^{-1} have been constructed. These maps display larger termination ripple for the electric fields than for the potential.[13] But the data proved to be too limited even for a faithful mapping of the potential.

Our second and most intensely studied type of Fourier coefficients are the difference structure factors, $F_{\vec{H}}$-$F_{\vec{H}}$(IAM), where $F_{\vec{H}}$(IAM) is a calculated structure factor with neutral atom, spherically symmetrical Hartree-Fock form factors. The independent atom model (IAM) has the great virtue that direct space lattice sums of all its electrostatic properties, whether it be potentials, densities, Laplacians, fields, or field gradients, are rapidly convergent. For some properties, such as the potential, the nuclear contribution is included as part of the IAM for the direct space lattice sum. The main question to pursue with these ΔF's, is the extent of data in reciprocal space that is necessary to map out the deformation property to some tolerance level. In this case it is desirable to first generate model crystal data, which are error free, and determine the convergence behavior for each property.

Molecular form factors from diatomic Hartree-Fock wavefunctions, spanned by Slater-type functions, served as a starting point in the resolution studies; these form factors had been evaluated out to $|\vec{H}|$max of 8/R, where R is the internuclear distance. Several selected molecules were assigned realistic mean square amplitudes of vibration and packed into cubic crystal lattices (space group Pa3 for N$_2$ and P2$_1$3 for CO and SiO), so that a set of thermal structure factors could be generated out to $|\vec{H}|$max of 5 or 6 Å^{-1}. A tolerance level of 1% or less was taken as the standard for resolving each property. A sphere of 2.0 to 2.2 Å^{-1} in $|\vec{H}|$max was adequate for a viable map of the deformation electrostatic potential. For the corresponding electric field, a sphere to 2.8-3.1 Å^{-1} was necessary, while for both the deformation density and the deformation electric field gradient, $|\vec{H}|$max must extend to at least 4.0 Å^{-1} before the termination effects are less than 1%. If we recall that the physical limit of MoKα data is 2.8 Å^{-1} in $|\vec{H}|$max, it's clear that only the electrostatic potential can be mapped to a 1% tolerance level from an accurate x-ray diffraction data set, which is obtained by standard methods. It is then a simple matter to add in the electrostatic potential of the IAM's with a direct space lattice sum. We thus learned that accurate and accessible x-ray structure factors can be used to determine the total electrostatic potential in a crystal with a simple model for correcting series termination effects, whereas a similar approach to the derivation of the total charge density eludes us. The mapping of electrostatic potentials in various crystals by the method outlined here has been reported previously.[14]

The third type of Fourier coefficients we have employed in reciprocal lattice sums do not directly involve x-ray structure factors. The coefficients are usually a special feature of a structure factor model, such as monopole terms, and have good convergence properties in $\vec{H}$ space, but are very slowly convergent in direct space. For example, when evaluating electric field gradients at some site in a lattice, the contribution from a "diffuse" or valence type monopole distribution function is most easily determined with a Fourier sum in $\vec{H}$ rather than $\vec{r}$ space. Some terms in the evaluation of electrostatic energies are also conveniently implemented with reciprocal space lattice sums.

GENERALIZED X-RAY SCATTERING FACTORS

A generalized x-ray scattering factor (gsf) is a form factor model for the local electron density centered on a nucleus. Each atomic center is assigned a small finite multipole expansion of the atomic form factor with a variable population coefficient for each surface harmonic and is weighted by the thermal function of the associated nucleus; the thermal function may include cumulants higher than the second order (harmonic) mean square amplitudes of vibration. This model corresponds to a rigid pseudoatom in that the local, non-spherical electron density rigidly follows the nucleus to which it is assigned. For most applications, the radial parts of these gsf's are Fourier-Bessel transforms of analytical, radial density basis functions. To date, we have used single exponentials (STF's), many exponentials from Hartree-Fock SCF AO products, expansions of Laguerre functions with degree $2l + 2$, where l is the order of the surface harmonic, and nodeless form factors that are transforms of modified spherical Bessel functions of the third kind. These radial functions all satisfy Poisson's equation for small r and approach $\exp(-cr)$ at large r. The major advantage to these functions is that transforms to $\vec{r}$-space from $\vec{H}$-space are free from termination effects. The gsf's are used as basis functions in least square fits to $|F_{\vec{H}}|$ or $|F_{\vec{H}}|^2$. With a finite data set, termination effects in a least squares projection onto these functions are restricted to the variable population and radial parameters. The electronic parameters determined by the least square fits to the reduced x-ray data afford an analytical representation of the electron density in the crystal and may be used to compute any property that depends on the electron or total charge density in the unit cell. It is also instructive to use the gsf's to construct a molecular moiety, remove it from the lattice and evaluate the electrostatic property. This procedure is akin to ORTEP plots of molecular structures as found in a crystal; in this case, isolated molecules are represented with atomic amplitudes of motion (thermal ellipsoids) that are peculiar to the crystal lattice. The electrostatic potential, for example, can be mapped for the "molecule" isolated from the lattice, but nonetheless will reflect the state in which it interacts with the crystal environment. The examples given in the present lecture all refer to systems (crystals or molecules) with stationary nuclei treated as point charges, but one could display mean thermal charge densities and its dependent properties as well.

The gsf model is rotationally invariant since the spherical, surface harmonics span all the irreducible representations of the rotation groups. This means the choice of a coordinate system is independent of the least squares result. In practice we use Cartesian representations of the surface harmonics with the X, Y, Z axes as some simple transformation of the reciprocal or direct space lattice vectors (eg., $X \parallel \vec{a}$, $Y \parallel \vec{c} \times \vec{a}$, and $Z \parallel \vec{c}^*$). X-ray diffraction intensities are usually measured without measuring the incident beam intensity. The elastic cross-section data are therefore on a relative scale. Although the reduced data are put on an approximate absolute scale by refinements with IAM's, the gsf model is on a relative scale and thus also the electron density distribution model. As a result, if extinction becomes large (exceeds 10%) or anomalous dispersion is an appreciable fraction of the monopole form factors, use of the gsf model becomes cumbersome. In extreme cases the "electronic" parameters, particularly radial scalars, no longer represent the electron density distribution but instead mimic defects in the kinematic scattering model by virtue of the fitting to non-kinematic diffraction data. The caveat here is that gsf's have the flexibility to accommodate systematic errors in the reduced data; the investigator must be on the constant lookout for deviations from kinematic scattering conditions. The gsf model is NOT constrained to give a positive-definite electron density everywhere in the unit cell. Our standard practice at the completion of a gsf analysis is to map the electron density from the least squares result in the unique part of the crystallographic unit cell. The values of ρ are then compared with the corresponding variances of the density model. If the density distribution satisfies the statistical tests, we proceed to evaluate other physical properties.

The gsf model is often used to scale and to phase the structure factor data. In addition the inverse least squares matrix elements are used explicitly in subsequent variance analyses. In our least squares programs, the correct Hessian of the mean square error function is evaluated (*i.e.*, explicit second derivatives are included). This allows us to determine if we are at a true minimum or on some type of saddle point at the completion of the study. If the Hessian is not positive definite, one moves in the direction of largest negative curvature in the Hilbert space of the parameters. Once a minimum (not necessarily global) has been found, the inverse Hessian elements are used to estimate standard deviations of A_c and B_c, as well as standard deviations for any property based on gsf's. Thus the variance of A_c is,

$$\sigma^2(A_c) = (gof)^2 \sum_{j \geq k} (2 - \delta_{jk}) \frac{\partial A_c}{\partial P_j} \, \varepsilon^{jk} \, \frac{\partial A_c}{\partial P_k} \tag{8}$$

with a corresponding expression for the variance of B_c. The P_j are variables (or dimensions) in the least squares space and the ε^{jk} are the inverse matrix elements. The gof is the square root of the weighted mean square error divided by the degrees of freedom.

$$gof = \left\{ \left(\sum_{H} w_H \Delta_H^2 \right) / (No - Nv) \right\}^{1/2} \tag{9}$$

The Fourier coefficients for a particular property are constructed from the components of $|F_{\vec{H}}|$ by the relations,

$$A_0 = k|F_{\vec{H}}|A_c/(y|Fc|^2)^{1/2}$$
$$B_0 = k|F_{\vec{H}}|B_c/(y|Fc|^2)^{1/2} \quad \text{and} \tag{10}$$

With (8) and (10), the variance of A_0 is,

$$\sigma^2(A_0) = A_0^2 \left\{ \left(\frac{\sigma_{(k)}}{k} \right)^2 + \left(\frac{\sigma_{|F_H|}}{|F_H|} \right)^2 + \left(\frac{\sigma_{(y)}}{2y} \right)^2 + \left(\frac{B_c}{4|Fc|} \right)^4 \left(\left(\frac{\sigma_{A_c}}{A_c} \right)^2 - \left(\frac{\sigma_{Bc}}{Bc} \right)^2 \right) \right\} \tag{11}$$

and a similar expression occurs for B_0. For properties, $G(\vec{r})$, derived from the gsf model directly, the corresponding variance is

$$\sigma_G^2 = (gof)^2 \sum_{j \geq k} (2 - \delta_{jn}) \frac{\partial G}{\partial P_j} \, \varepsilon^{jn} \frac{\partial G}{\partial P_k} \tag{12}$$

In practice the P_j in (12) are restricted to electronic parameters since in almost all cases, the errors in the atomic position parameters make a marginal or negligible contribution to the variance of $G(\vec{r})$. [This may not be the case for the Laplacian of ρ.] For some properties, such as a molecular dipole moment, a total variation such as (12) is not warranted so that undetermined Lagrange multipliers are introduced. The multipliers may be found by minimizing the variance of the property.

EXAMPLES OF RESULTS FROM DIFFRACTION DATA

The examples for this lecture will be restricted to organic molecular crystals. In all four cases, neutron diffraction results from measurements at the same temperature are used in the gsf analysis. The mean thermal nuclear positions of the aoms are fixed at the values from least squares fits to the neutron structure factors and for the H atoms, the mean square amplitudes of vibration for the protons or deuterons are also taken from the neutron diffraction results. The radial functions in the gsf model are based on shell localized SCF AO's,[15] which are expanded with orthogonal Laguerre functions,[16] and on single exponential functions or STF's. It is emphasized here, that although a rigid pseudoatom model is used to represent the several properties from the charge density distribution, the electrostatic property is not that for an equilibrium structure on a Born-Oppenheimer surface, but is a property based on the mean thermal nuclear distribution in a crystal lattice. The deconvolution model used in this work does not provide the investigator with a static charge denstiy distribution for the equilibrium structure in the lattice.

Urea

Urea forms a molecular crystal with a tetragonal lattice in space group $P\bar{4}2_1m$. X-ray diffraction intensities were measured out to $|H|$ of 2.30 Å^{-1} at 123 K with a MoK source.[17] The data were reduced to 318 unique structure factor amplitudes. Neutron diffraction measurements on a different crystal specimen at 123 K were used to determine the nuclear sites and mean square amplitudes of vibration.[18] The site symmetry of the C and O atoms is mm and that for N and H is m; there are two molecules or urea per unit cell. The O atom accepts four hydrogen bonds in this crystal.

For the gsf analysis of the x-ray data, the first and second cumulants (positions and thermal parameters) were fixed at the neutron values, so that only electronic parameters were varied in the projection work. The monopole functions on the C, N and O atoms consisted of two shells, which were constructed from the localized SCF s orbitals and the 2p orbital; the first monopole is an is×is product that consists of 21 STF's and the second is $(2 \times os \times os + n \times (2p)^2/(2+n)$ where n is 2, 3 and 4, for C, N, and O, respectively. These "outer" shell form factors consisted of 9 Jacobi functions. (The Fourier-Bessel transform of the Laguerre function is a Jacobi function.) The H atom monopole was a single shell, which is a ls STF $(\exp(-r))$. Each monopole was assigned a variable population parameter, with initial values of 2 for the "inner" shells and 4, 5, and 6 for the "outer" shells of C, N, and O; the H atoms started at a value of 1 for the monopole shell. The "inner" shell population variables were constrained to be the same for C, N, and O; the "outer" shell monopoles were shaped with a variable radial scaling parameter, κ,[19] and the exponential parameters, α_0, for the two H atoms were also varied, but constrained to be the same. For the higher multipole radial form factors on all the atoms, single exponential functions were used. The exponential parameters were taken from optimized fits to a set of theoretical structure factors for urea[17] and were not varied in the least squares projection of the gsf's onto the experimentally derived structure factors. Multipoles up to third order (octopoles) were included on C, N, and O and up to second order (quadrupoles) on H; variable population parameters, subject to the constraints imposed by site symmetry mm or m, started at zero. Altogether, 39 unique electronic parameters were varied in the least squares fit to the reduced data for 318 reflections. The residuals for the least squares error function were constructed with $|F_0|^2-|F_c|^2$ and weights were taken from the estimated standard deviations of the experimentally derived $|F_0|^2$. At completion of the least square variations, the criterion for convergence was $1.62 \times (10^{-12})$; the full Hessian (*i.e.* inclusion of second derivative terms) was constructed at this minimum and inverted for use in computing the variances of all properties based on the gsf expansion. $R_w(|F|^2)$ was 0.0227, $R(|F|^2)$ was 0.0213 and the gof was 1.183. Tabulation of the radial density basis functions for the gsf fits to urea is given below in Table 2. Values in ( ) are the estimated standard deviations and indicate a variation of a radial parameter.

Table 2. Radial Density Functions Used for Urea

Atom	Pole order	Type	κ/α (Å^{-1})
O	0	is×is	1.0
	0	(2×os×os+4×2p×2p)/6	0.9788(43)
	1	$(r^2) \times \exp(-\alpha_O r)$	9.58092
	2	$(r^2) \times \exp(-\alpha_O r)$	9.58092
	3	$(r^3) \times \exp(-\alpha_O r)$	9.58092
N	0	is×is	1.0
	0	(2×os×os+3×2p×2p)/5	0.9562(71)
	1	$(r^2) \times \exp(-\alpha_N r)$	7.97465
	2	$(r^2) \times \exp(-\alpha_N r)$	7.97465
	3	$(r^3) \times \exp(-\alpha_N r)$	7.97465
C	0	is×is	1.0
	0	(2×os×os+2×2p×2p)/4	0.0329(128)
	1	$(r^2) \times \exp(-\alpha_C r)$	4.42196
	2	$(r^2) \times \exp(-\alpha_C r)$	4.42196
	3	$(r^3) \times \exp(-\alpha_C r)$	4.42196
H	0	$\exp(-\alpha_0 r)$	5.946(287)
	1	$r \times \exp(-\alpha_H r)$	4.91329
	2	$(r^2) \times \exp(-\alpha_H r)$	4.91329

In order to display the electrostatic potential based on a gsf analysis, it is necessary to include the nuclear distribution with that of the electrons. Since the gsf model represents the electron density distribution on a relative scale some sort of scaling procedure must be adopted. For the results presented in this lecture, the gsf's are uniformly rescaled to give a unit cell that is electrically neutral. The nuclei are now included as three-dimensional δ functions with a charge of Z, so that the corresponding form factor is Z which is a constant of $|\vec{H}|$. The potential is the Fourier transform of the gsf's with inclusion of the nuclear form factor and a weight of $|\vec{H}|^{-2}$. This function is evaluated on a suitable grid and converted to lines of equipotentials with a contour plot. The plot for urea was constructed from the gsf's in the asymmetric unit plus the mirror related gsf's for N and the two H's. Units used here are in electrons/Å (1 e/Å is 14.40 volts or 1389 kJ/Faraday).

A contour plot of equipotential lines in the plane of urea is shown in Fig. 1a. The electropositive lines are terminated at 0.5 e/Å, after which the point charge potential of the nucleus dominates the contour lines in the plot. The double minima near the oxygen is about –486(92) kJ/mol and twice the depth of the single minimum value of –243 kJ/mol that was found from a theoretical calculation for an isolated urea molecule. The double minimum is a feature peculiar to the H bonding in the urea crystal; an annulus of either the –0.35 or –0.30 e/Å contour about the carbonyl oxygen is not observed. A second feature to note in Fig. 1a is the extension of the smallest electropositive contour (0.05 e/Å) from the H nuclei.

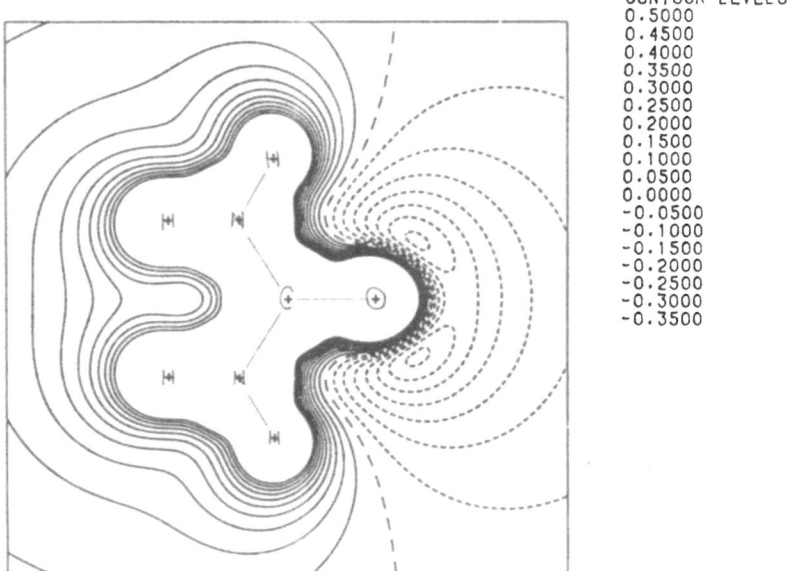

CONTOUR LEVELS
0.5000
0.4500
0.4000
0.3500
0.3000
0.2500
0.2000
0.1500
0.1000
0.0500
0.0000
-0.0500
-0.1000
-0.1500
-0.2000
-0.2500
-0.3000
-0.3500

Fig. 1a.   The electrostatic potential of urea from pseudoatom fragments removed from the crystal lattice.   The section is the crystallographic plane of the atoms. Contours are in intervals of 0.05 e/Å. Maximum is 0.50 e/Å and minimum is –0.35 e/Å. Long dashed line is the null.

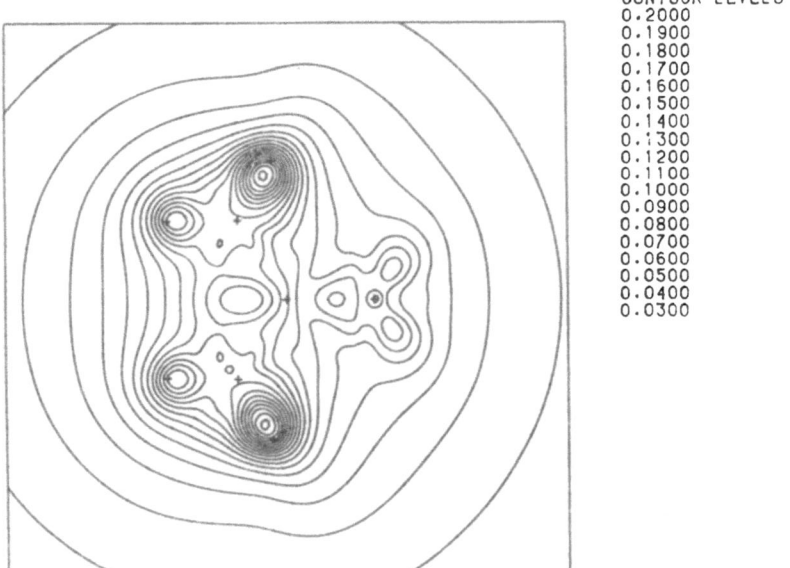

CONTOUR LEVELS
0.2000
0.1900
0.1800
0.1700
0.1600
0.1500
0.1400
0.1300
0.1200
0.1100
0.1000
0.0900
0.0800
0.0700
0.0600
0.0500
0.0400
0.0300

Fig. 1b.   The estimated standard deviation of the potential from urea gsf's. Contours are in intervals of 0.01 e/Å with the minimum value at 0.03 e/Å. The section is the same as in Fig. 1a.

It extends beyond 470 pm from the protons, but only to about 320 pm in the isolated molecule calculation. But in the urea crystal the four hydrogens are all donors to the O···H bonds. The exaggerated polarity compared to an isolated urea molecule probably reflects the donor-acceptor interactions in crystalline urea.

The variances of the potential from the urea gsf's were calculated as outlined in eqn (12) above. The root mean square of this function is plotted as contour lines in Fig. 1b and serve as the estimated standard deviations of the equipotential lines.

It can be seen that at the sites of the double minima in Fig. 1a, the esd from Fig. 1b is 0.065 e/Å. This corresponds to a relative error of nearly 20%. Most physical properties based on gsf fits to accurate x-ray diffraction data have estimated standard deviations that are between 15% to 35% of the value for the property. It is in this context that I refer to charge density results by these methods as modest.

The double minimum in the potential near the O atom in the plane of urea is associated with the H bond in the lattice where the proton is 206.7 pm away from the oxygen. In the plane normal to this one, two other N–H donors from two urea molecules form H bonds with the protons 200.9 pm from the oxygen. A study of small clusters of urea was undertaken to observe the change in potential near the acceptor oxygen. The potential due to two types of clusters of urea are displayed in Fig. 2a and 2b as contour plots. In Fig. 2a the molecule on the right is a double N–H donor. The double minimum is now reduced to –0.11 e/Å. The three molecule cluster, shown in Fig. 2b has a minimum of about –0.12 e/Å. When the four molecule cluster is formed as it occurs in the lattice, the potential about the oxygen becomes electropositive everywhere. This result is shown in Fig. 3.

For any cluster that becomes sufficiently large the potential in the interior will always become electropositive everywhere since the positive charge sources (nuclei) are much more localized than the electrons. What the study shows here is that the minimum number for an electropositive potential about the acceptor oxygen is a cluster of four urea molecules with the geometry of that in the crystal. This kind of study may serve as crude test for H bonding, although the results for 9-methyladenine (vide infra) do not bear up to an electropositive bridge as a criterion for H bonding interactions.

The next property we examine is the electron density distribution from the gsf model fit to the urea structure factor amplitudes. The first step is to evaluate the electron density with a direct space lattice sum and search for any large negative electron densities in the cell. A minimum value of –0.032 e/Å^3 was observed with an estimated standard deviation of 0.028 e/Å^3. The density does not significantly differ from zero, so that this model cannot be rejected as non-physical based on a statistical criterion. The gsf's were converted into an electron density distribution function for urea removed from the lattice in a manner analogous to the construction of the electrostatic potential. This is shown as a 3-D plot in Fig. 4. The maximum electron density has been truncated at 64 e/Å^3 in order to reveal the local maxima for the H atoms. The major feature to note is the sharp peaking of density on the nuclear sites and small differences in the curvatures of the density between the atoms. The electron density for the H bonding regions in the crystal has been evaluated by a direct space lattice sum of the Fourier transform of the gsf's. These results are shown in Fig. 5. The plane in Fig. 5a corresponds to that in Figs. 2a and 3a and displays the density for the "long" H bond of 207 pm. The local maxima on the O and two N atoms are truncated at 64 e/Å^3. Fig. 5b corresponds to Figs. 2b and 3b; the truncated maxima occur on the C and O atoms. The H bonds in this figure are 201 pm long. In both plots one can see that curvatures in the density away from the O···H directions are much less than those changes in the density normal to the approximate midpoints of the N–H bonds or of the C–O bond.

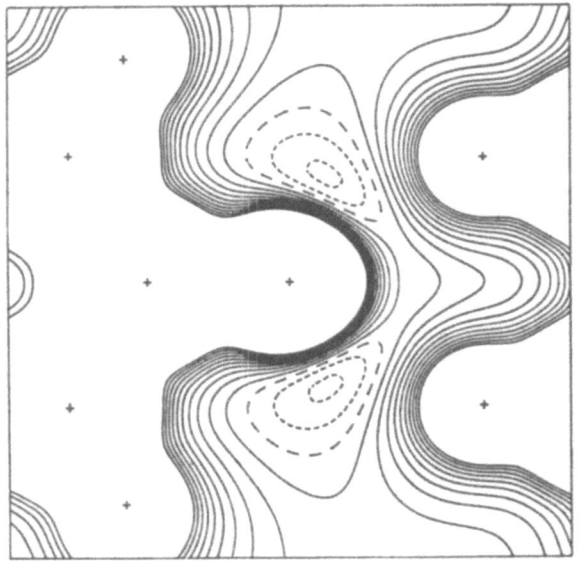

CONTOUR LEVELS
0.5000
0.4500
0.4000
0.3500
0.3000
0.2500
0.2000
0.1500
0.1000
0.0500
0.0000
-0.0500
-0.1000

Fig. 2a

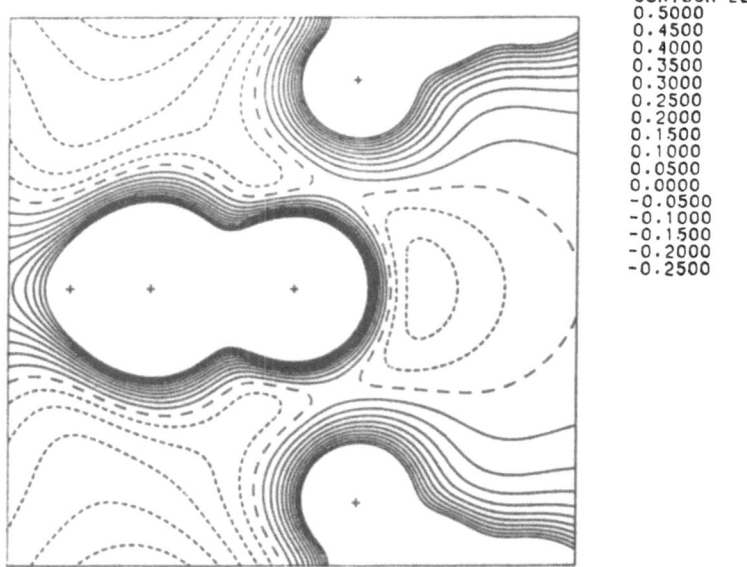

CONTOUR LEVELS
0.5000
0.4500
0.4000
0.3500
0.3000
0.2500
0.2000
0.1500
0.1000
0.0500
0.0000
-0.0500
-0.1000
-0.1500
-0.2000
-0.2500

Fig. 2b

Electrostatic potential for small clusters of urea. Contours as in Fig. 1a. Fig. 2a is for a two molecule cluster; Fig. 2b, in a plane normal to 2a, is a three molecule cluster..

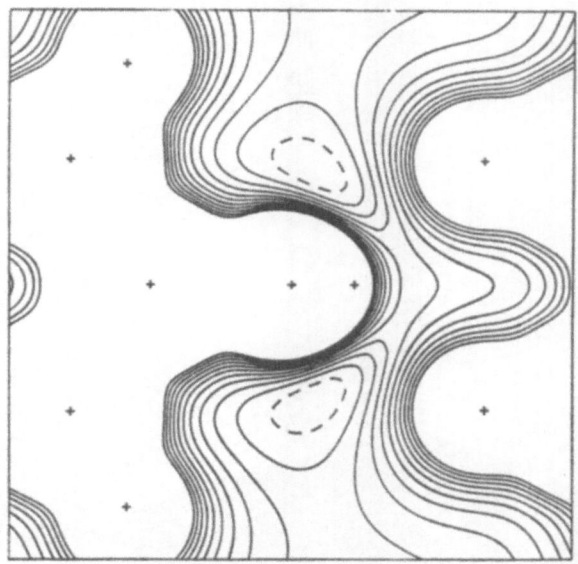

```
CONTOUR LEVELS
0.5000
0.4500
0.4000
0.3500
0.3000
0.2500
0.2000
0.1500
0.1000
0.0500
0.0000
```

Fig. 3a

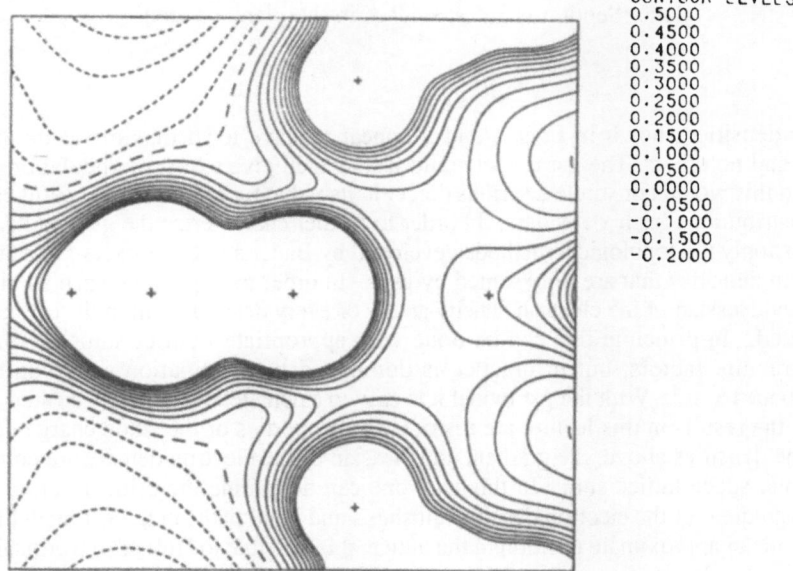

```
CONTOUR LEVELS
0.5000
0.4500
0.4000
0.3500
0.3000
0.2500
0.2000
0.1500
0.1000
0.0500
0.0000
-0.0500
-0.1000
-0.1500
-0.2000
```

Fig. 3b

Potential for cluster of four ureas. Plane 3a same as 2a and
plane 3b same as 2b.

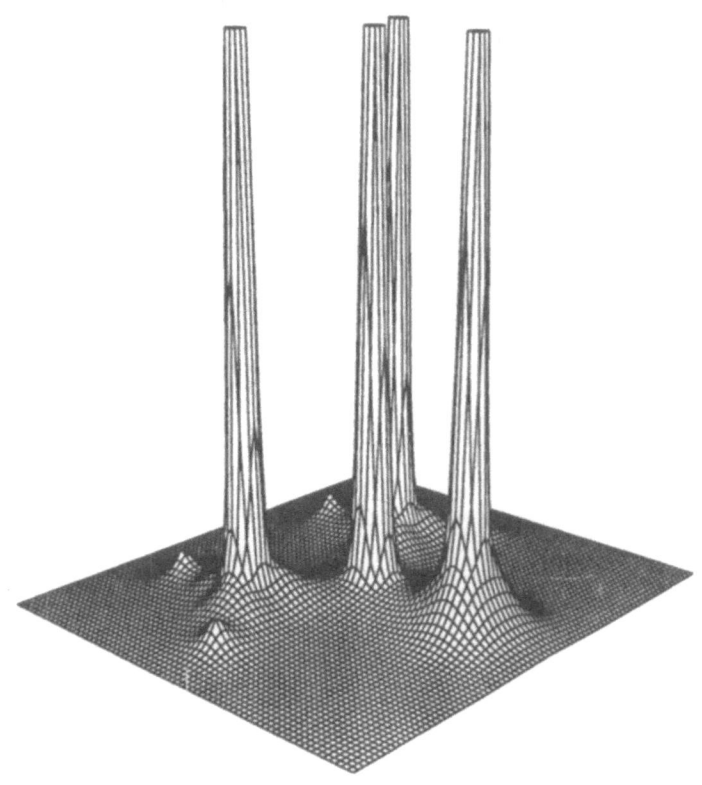

Fig. 4.   The electron density distribution from the gsf fit to the urea data.   Section is 5.4 Å × 4.7 Å in the plane of urea.

The densities shown in Figs. 4 and 5 appear to have local maxima at the nuclear positions and no others. The gsf model could just as well give a host of undulations in the crystal and this would constitute a serious defect in its value for extraction of useful electron density distributions from x-ray data. In order to further characterize the gsf results, it was decided to apply the topological methods developed by Bader and co-workers[19] to a study of the electron densities that are represented by gsf's. In order to implement such studies, the gradient and Hessian of the electron density at any or every desired point in the crystal must be evaluated. In principle this can be done with appropriate Fourier sums of correctly phased structure factors, but in practice is dominated by termination ripple unless $|\vec{H}|$ exceeds about 16 Å^{-1}. With the gsf model it is easy to calculate F_c out to any desired sphere in $|\vec{H}|$, but the results in this lecture are restricted to properties of the static charge density. As with the densities above, the gradient and Hessian of the electron density are computed with a direct space lattice sum. In this way one can determine the critical points - sites where the gradient of the electron density vanishes - and bond paths in the gsf model for the crystal. With an approximate position in the lattice, it is possible to "refine" the critical point by inverting the Hessian at the initial value and shifting to the site where the vector for the gradient of the density diminishes. As long as the Hessian has three non-zero principal values, the Newton method will give convergence. The refinement procedure employed in this work is somewhat akin to differential synthesis of atomic positions - (3,–3) critical points - used by crystallographers over twenty-five years ago.

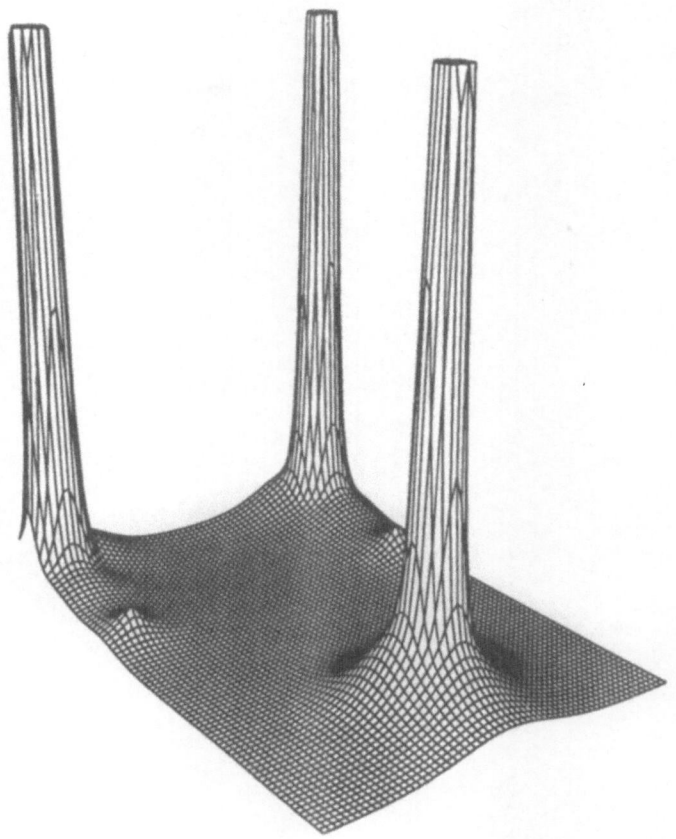

Fig. 5a. The (110) plane with 2.7 Å × 3.6 Å area in the electron density
plots for the H bonding regions in the urea crystal.

For the present gsf model of the urea crystal, the only local maxima found in the density occur at the nuclear sites determined from the neutron diffraction data. At the next level, six (3,–1) critical points in the asymmetric part of the unit cell were detected and refined. Each one of these can be associated with a chemical bond path in the crystal. A summary of the results is given in an abbreviated form in Table 3. For comparison, results from the density distribution of IAM's are also included.

We see that the (3,–1) critical points from the IAM density are rather close to those of the gsf density distribution. Larger differences occur in comparing the density, Laplacian of the density, and ellipticities from the two models. The negative traces of the Hessian of the electron density at the (3,–1) critical points, $i.e.$ the Laplacian of ρ, is rather more negative for the gsf model and presumably reflects a feature of covalent bonding inherent in the measured structure factors that is not as well represented with IAM's. The Laplacians at the H bond critical points are positive in both models but the ellipticities for the gsf model seem quite erratic. For the $O \cdots H1$ bond, one eigenvalue for the Hessian was -0.06 $(e/Å^5)$ and this critical point is probably not statistically different from a (3,1) or (2,0) point. In order to give a more critical account of the gsf values in Table 3, a determination of the variances of these properties must be undertaken. Programs to evaluate these, as outlined in eqn (12) above, are under development.

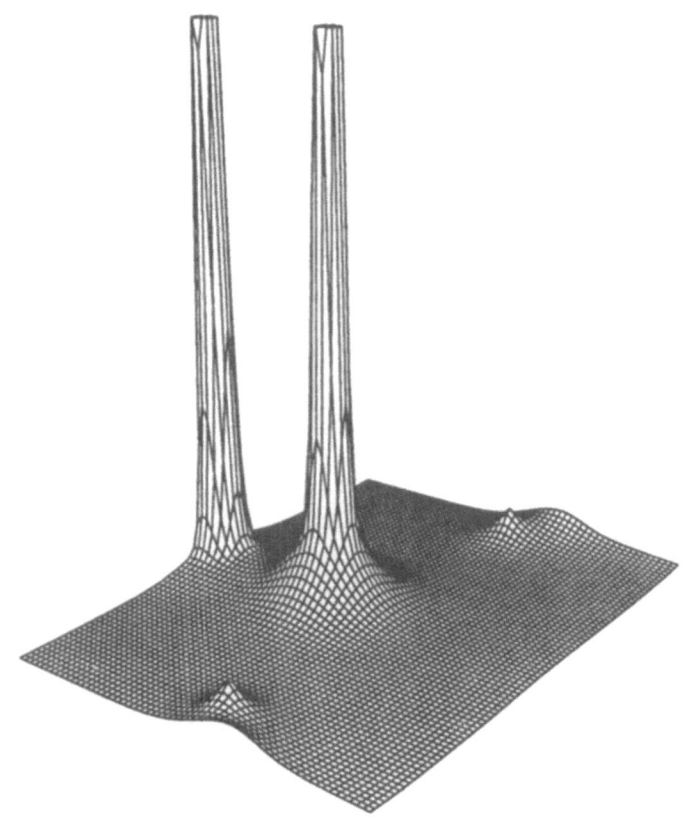

Fig. 5b. The (110) plane with 4.6 Å × 3.45 Å area in the electron density
plots for the H bonding regions in the urea crystal.

Table 3. (3,−1) Critical points in the urea crystal for gsf
and IAM density distributions

	Unit cell fractions			ρ $(e/\text{Å}^3)$	$\nabla^2\rho$ $(e/\text{Å}^5)$	ε	Bond path
	x	y	z				Bond length (pm)
gsf	0.0625	0.5625	−0.1398	0.08	1.54	0.61	O· · ·H2
IAM	0.0815	0.5815	−0.1834	0.17	1.88	0.06	206.7
gsf	0.3308	0.8308	0.3598	0.06	2.32	8.83	O· · ·H1
IAM	0.3540	0.8540	0.3339	0.19	2.04	0.07	200.9
gsf	0.0000	0.5000	0.4206	2.51	−10.42	0.05	C–O
IAM	0.0000	0.5000	0.4250	0.71	0.71	0.03	125.8
gsf	0.0507	0.5507	0.2811	2.30	−14.63	0.19	N–C
IAM	0.0608	0.5608	0.2650	1.79	−3.49	0.03	134.1
gsf	0.2273	0.7273	0.2582	2.38	−26.20	0.21	N–H1
IAM	0.2287	0.7287	0.2585	1.57	−8.89	0.01	100.7
gsf	0.1435	0.6435	0.0238	2.23	−20.95	0.05	N–H2
IMA	0.1434	0.6434	0.0166	1.59	−9.45	0.01	100.0

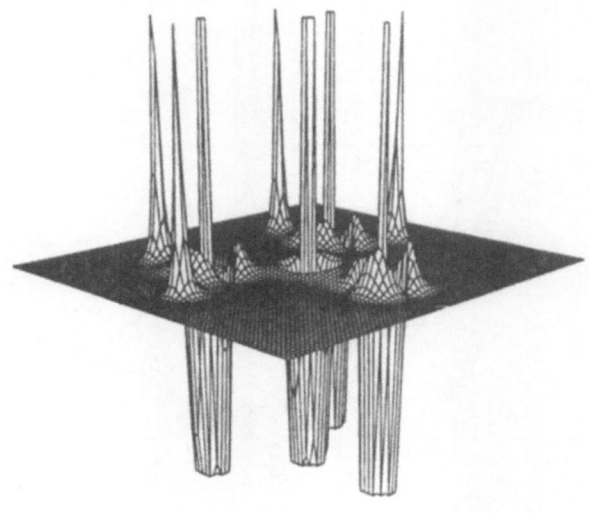

Fig. 6a

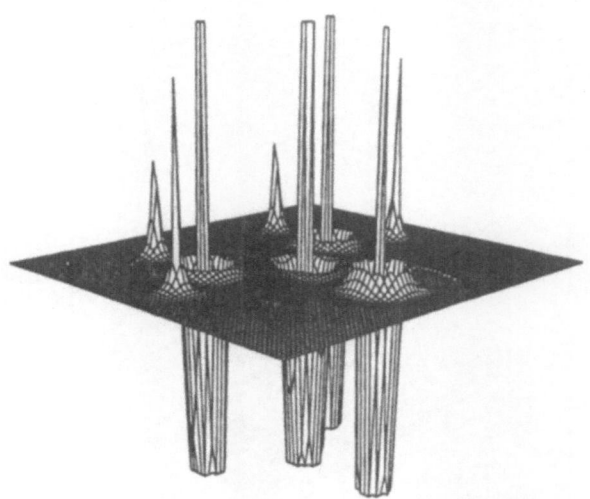

Fig. 6b

Negative Laplacian of ρ for model densities of urea. Maxima
are truncated at 1024 e/Å^5 and minima at -896 e/Å^5. Fig. 6a is
the gsf model; Fig. 6b is IAM.

One may gain a more global view of the gsf model by plotting the negative Laplacian
of its density in the crystal or some subset removed from the crystal. As discussed by
Bader *et al*,[20] these scalar maps reveal the local concentration of the electron density with the
associated formation of chemical bonds, the shell structure of atoms in the molecule and
possible "lone pair" structure. A 3-D plot, similar in construction to the density shown in
Fig. 4, of $-\nabla^2\rho$ for the gsf model of urea removed from the lattice is shown in Fig. 6a. As
a point of reference, the $-\nabla^2\rho$ for the IAM's in the same geometry is given in Fig. 6b.

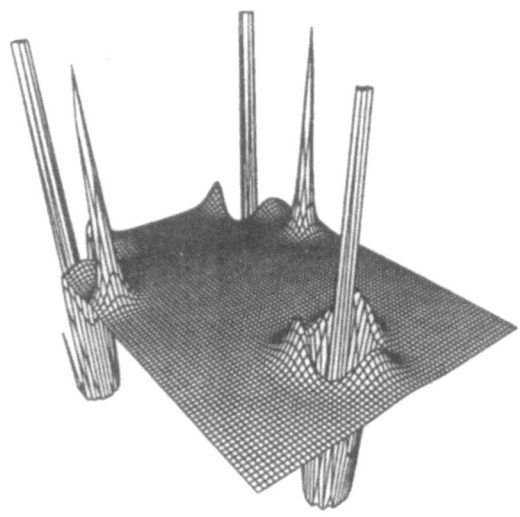

Fig. 7a

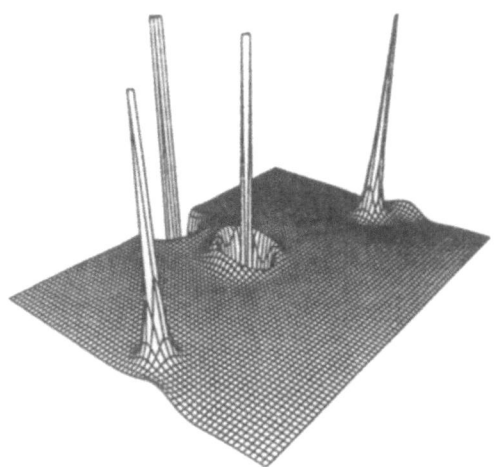

Fig. 7b

Negative Laplacian of the gsf density in the H bonding regions
in the urea crystal. Maxima truncated at 1024 e/Å^5 with minima
at −896 e/Å^5. Section 7a same as 5a with 207 pm H bond.
Section 7b same as 5b with 201 pm bond.

Both maps in Figs. 6a and 6b clearly reveal the shell structure of the C, N, and O
atoms and only the single shell of the H atoms. For the IAM plot the L shells are perfectly
circular and increase in their local maxima (*i.e.*, local concentration of electron density) as
the atomic number increases from C to O. The smaller maximum for H2 in Fig. 6b is an
artifact in the plot that depends on the choice of grid; H1 and H2 are equivalent IAM's. The
L shell features in the gsf model, in contrast to Fig. 6b, are highly structured. The local

maxima about the N atoms are directed towards the C and H atoms, to which they are covalently bonded and the O atom has two distinct lone pairs as well as a local maximum towards the C atom to which it forms a covalent bond. The critical points all lie in positive regions of the plot in Fig. 6a (see Table 3 above). The "bond" critical point for C–O in the IAM resides in a small negative region in the plot shown in Fig. 6b.

For the H bonding regions in urea the Laplacian of the gsf density was evaluated with a lattice sum. The results are shown as 3-D plots in Figs. 7a and 7b. The corresponding densities are given above in Figs. 5a and 5b.

We see that the lone pair structure near the O atom is localized in the $(1\bar{1}0)$ plane and is absent in the $(\bar{1}\bar{1}0)$. This feature was also revealed in the electrostatic potential of the urea gsf's removed from the lattice. The critical points for the $O\cdots H$ bonds lie in very small negative regions in both plots. The surfaces, particularly in Fig. 7b, are very flat and expose the difficulty in characterizing these $(3,-1)$ critical points in a quantitative way. (The eigenvalues from the Hessian at the $(3,-1)$ point in the 200 pm H bond were -0.59, -0.06 and 2.96 $(e/\text{Å}^5)$, respectively, with the eigenvector of the least negative curvature normal to the $(\bar{1}\bar{1}0)$ plane in Fig. 7b.)

Imidazole

The nitrogen heterocycle, imidazole, $C_3H_4N_2$ crystallizes in a monoclinic lattice with space group $P2_1/c$ and four molecules per unit cell. MoKα x-ray data were collected at 103 K out to $|H|$ of 2.6 Å^{-1}.[21] Neutron diffraction measurements on a different crystal specimen, but at a temperature of 103 K, have also been reported.[22] The x-ray intensities were reduced to 3279 unique structure factor amplitudes. A gsf model, similar to that for urea, was used for a projection analysis of the experimental structure factors, all of which were assumed to be on the same relative scale. The positions of all atoms were fixed at the neutron values as were the second cumulants for the protons. The U_{ij}'s for C and N were treated as variables, however. The inner shell monopoles for N and C were constructed from shell localized SCF AO's of the quartet S N atom and triplet P C atom; the populations

Table 4. Radial Density Functions Used for Imidazole

Atom	Pole order	Type	κ/α (Å^{-1})
N	0	is×is	1.0
	0	(2×os×os+3×2p×2p)/5	0.9995(58)
	1	$(r^2)\times\exp(-\alpha_N r)$	5.523(187)
	2	$(r^2)\times\exp(-\alpha_N r)$	5.523(187)
	3	$(r^3)\times\exp(-\alpha_N r)$	5.523(187)
C	0	is×is	1.0
	0	(2×os×os+2×2p×2p)/4	1.0148(80)
	1	$(r^2)\times\exp(-\alpha_C r)$	6.114(179)
	2	$(r^2)\times\exp(-\alpha_C r)$	6.114(179)
	3	$(r^3)\times\exp(-\alpha_C r)$	6.114(179)
H	0	$\exp(-\alpha_0 r)$	4.641(136)
	1	$r\times\exp(-\alpha_H r)$	4.881(257)
	2	$(r^2)\times\exp(-\alpha_H r)$	4.881(257)

were constrained to be the same. The outer shell monopoles, as for urea, were scaled radially ("κ refined"), but constrained to be the same for the atom type, and assigned separate population parameters. The H atom monopoles were separately populated, but constrained to be the same in radial variation. All atoms have site symmetry 1 in this structure so that for higher multipoles, three, five and seven populations were varied for each C and N gsf dipoles, quadrupoles and octopoles, respectively; the H atoms were spanned with dipole and quadrupole gsf's. Radial parameters for single exponential functions were varied, but constrained to be the same among the poles of an atom type.

The total number of variable parameters was 152 of which 122 were for gsf's and the other 30 were U_{ij}'s, which in a nominal way are considered nuclear parameters. The residuals for the mean square error function were $|F_o|-|F_c|$ and weights were based on experimental data reduction. Upon completion of the least squares variations, the criterion for convergence was $4.65 \times (10^{-18})$; the full Hessian of the mean square function was inverted and saved for variance calculations of gsf properties. The final $R_w(|F|)$ was 0.0382, R($|F|$) was 0.0848, and gof was 1.237.

An electrostatic potential from the imidazole gsf's and the nuclear form factors as point charges was computed in the least squares plane of the N, C, and H nuclei. A contour plot of this potential is shown in Fig. 8a with a corresponding contour plot of the estimated standard deviations of the potential in Fig. 8b. A minimum of -375 kJ/mol occurs at about 1.1 Å from one N atom with an esd of 114 kJ/mol. Note that the estimated error is about 30% at this site. An overall feature of the map in Fig. 8a is the large polarity conveyed on imidazole with the gsf model. In contrast to the closeness of the $+0.05$ e/Å contour to the H atoms bonded to the C atoms, notice the large extension of this electropositive contour away from the proton bonded to N. In the lattice this H atom is a donor to the non-protonated N atom in a glide related imidazole molecule with formation of a hydrogen bond.

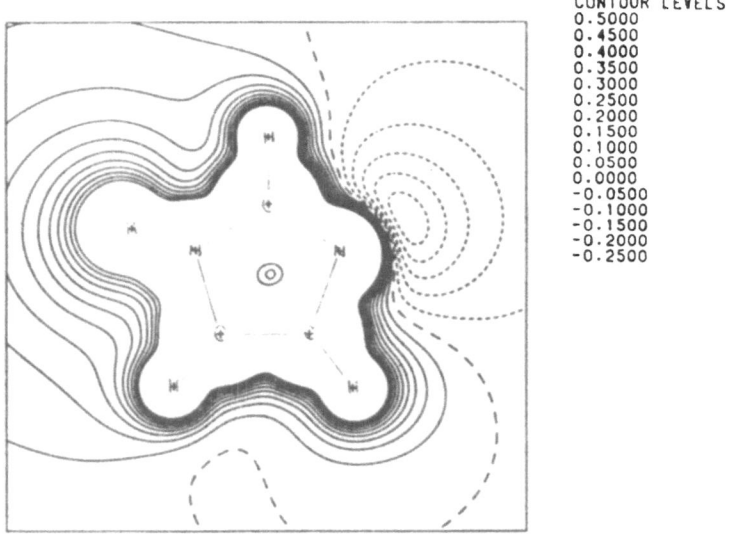

CONTOUR LEVELS
0.5000
0.4500
0.4000
0.3500
0.3000
0.2500
0.2000
0.1500
0.1000
0.0500
0.0000
-0.0500
-0.1000
-0.1500
-0.2000
-0.2500

Fig. 8a

Fig. 8. The electrostatic potential and esd of the potential for imidazole gsf's removed from the lattice. The section is 7.5 Å × 7.5 Å in the least square plane of all the atoms. Fig. 8a has maximum contour of 0.5 e/Å, increments of 0.05 e/Å and minimum at -0.25 e/Å. Null is long dashed line. Fig. 8b is 0.04 e/Å at the periphery and increases in increments of 0.01 e/Å to 0.15 e/Å.

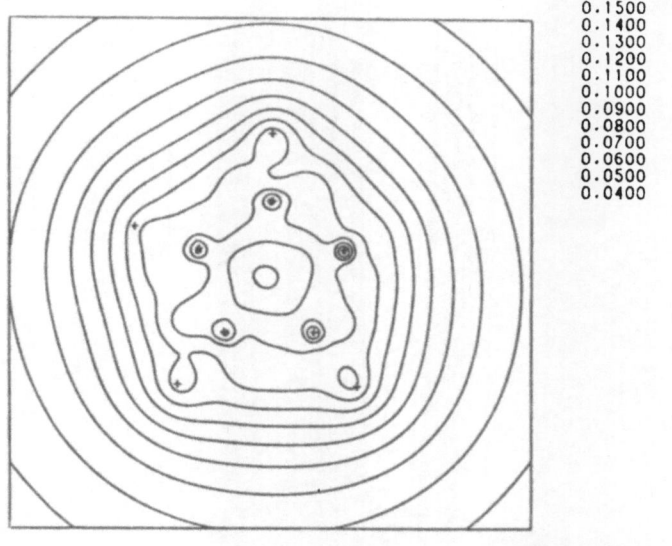

CONTOUR LEVELS
0.1500
0.1400
0.1300
0.1200
0.1100
0.1000
0.0900
0.0800
0.0700
0.0600
0.0500
0.0400

Fig. 8b

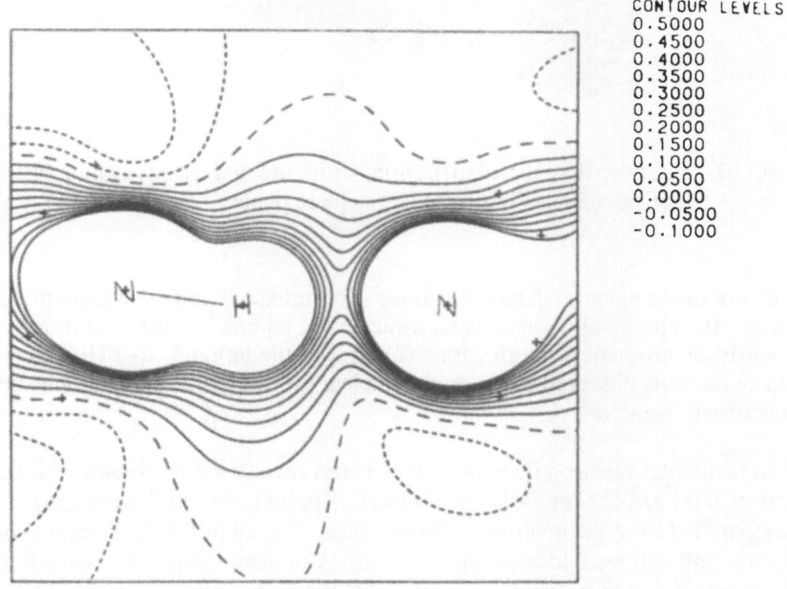

CONTOUR LEVELS
0.5000
0.4500
0.4000
0.3500
0.3000
0.2500
0.2000
0.1500
0.1000
0.0500
0.0000
-0.0500
-0.1000

Fig. 9. Electrostatic potential for a cluster of two molecules of
imidazole. Contours as in Fig. 8a.

The imidazole molecules form a single hydrogen bond in the lattice. A cluster of two
molecules was formed and the potential evaluated in a plane that bisects the dihedral angle
between the least squares plane of imidazole and its glide related plane. It is clear that the
cluster of two molecules is sufficient to produce an electropositive bridge of 0.32 e/Å
compared to the electronegative value of –0.27 e/Å for the isolated molecule.

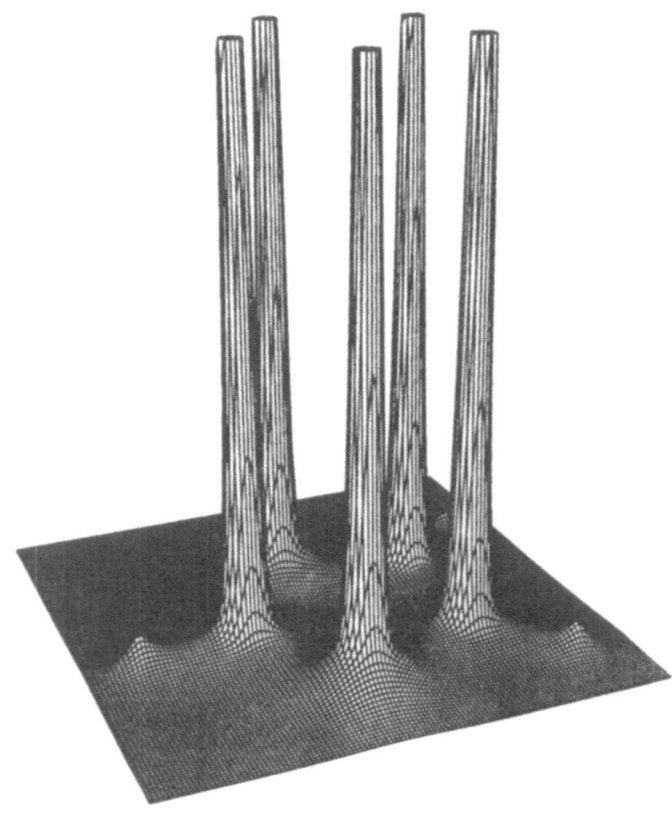

Fig. 10. Electron density distribution from the gsf fit to imidazole.
Section is 5 Å × 5 Å in the least square plane.

A 3-D plot of the electron density from gsf's for imidazole removed from the lattice is shown in Fig. 10. The maxima have been truncated at 64 e/Å^3 as for urea in Fig. 4. The view is towards the aromatic N atom with C(2)-H(2) to the left and C(4)-H(4) to the right. The only local maxima observed occur on the nuclear positions and saddle points appear to be present between the atoms.

A scan of the gsf electron density in the crystal revealed a minimum of –0.03 e/Å^3 with an esd of 0.04 e/Å^3. The only (3,–3) critical points observed were at the nuclear positions determined from the neutron diffraction data. A total of 10 (3,–1) critical points in the asymmetric unit cell were located and "refined". Correspondingly, the same number of saddle points were determined from the IAM density distribution. The IAM (3,–1) critical points occur at about the same places as do the corresponding gsf values. This was also the case for urea. The major differences in the two models are found in the ellipticities and the Laplacian of the density. The gsf model is representing covalency to the chemical bonding in contrast to the IAM, which is merely an overlapping of locally spherical density functions. The single H bond in the crystal has a positive Laplacian at its (3,–1) critical point; the large ellipticity, if it be significant, would suggest that this H bonding feature is not simply one of "closed shell ions". It is clearly desirable that a variance analysis of these gsf properties at the (3,–1) critical points be carried out.

Table 5. (3,–1) Critical points in the imidazole crystal
for gsf and IAM density distributions

	Unit cell fractions			ρ (e/Å^3)	$\nabla^2\rho$ (e/Å^5)	ε	Bond path Bond length (pm)
	x	y	z				
gsf	0.2046	0.2522	0.4282	0.17	2.69	1.07	N(3)· · ·H(1)
IAM	0.2054	0.2757	0.4158	0.31	2.51	0.01	181.2
gsf	0.1845	0.2638	0.1402	2.20	−19.83	0.31	N(1)–C(2)
IAM	0.1835	0.2677	0.1408	1.78	−3.05	0.05	134.8
gsf	0.1818	0.2769	0.2367	2.27	−15.04	0.32	C(2)–N(3)
IAM	0.1787	0.2723	0.2327	1.84	−4.43	0.05	132.5
gsf	0.2659	0.4585	0.3031	2.16	−14.51	0.20	N(3)–C(4)
IAM	0.2655	0.4627	0.3023	1.69	−1.40	0.05	137.5
gsf	0.3171	0.5530	0.2367	2.08	−17.62	0.53	C(4)–C(5)
IAM	0.3122	0.5477	0.2302	1.55	−3.29	0.05	137.0
gsf	0.2863	0.4600	0.1363	2.13	−23.67	0.64	C(5)–N(1)
IAM	0.2745	0.4540	0.1307	1.73	−1.98	0.06	136.5
gsf	0.2090	0.2826	0.0102	2.09	−34.79	0.08	N(1)–H(1)
IAM	0.2070	0.2838	0.0082	1.46	−6.09	0.00	104.3
gsf	0.1043	0.1113	0.1593	2.15	−24.46	0.22	C(2)–H(2)
IAM	0.1029	0.0999	0.1589	1.27	−5.22	0.01	106.4
gsf	0.3429	0.6454	0.3566	1.88	−17.61	0.17	C(4)–H(4)
IAM	0.3505	0.6464	0.3609	1.26	−5.04	0.01	106.9
gsf	0.3619	0.6338	0.1269	1.84	−15.94	0.08	C(5)–H(5)
IAM	0.3616	0.6319	0.1288	1.24	−4.17	0.01	107.7

A 3-D plot of the negative Laplacian of the electron density from the gsf model fit to imidazole is shown in Fig. 11. The plane and view of the molecule is the same as in Fig. 10. The anisotropy of local maxima in the L shell about the aromatic N atom is rather less than for the other atoms, all of which include bonds to H atoms, in the five-membered ring. As can be seen from Table 5, the bond critical points all lie in positve regions of the $-\nabla^2\rho$ plot in Fig. 11. The eigenvectors for the least negative curvatures at the (3,–1) critical points for the ring atoms are virtually parallel to the least squares plane vector except for C(5)–N(1) with a 21.5 tilt and N(1)–C(2) with a 17.6 tilt. Thus the plot in the plane of Fig. 11 approximately displays the sum of the positive and most negative curvatures in the vicinity of the critical points.

A lattice sum of the negative Laplacian of the gsf model in the N(3)· · ·H(1) region of the crystal was calculated and represented with 3-D plots. These are shown in Fig. 12 for two different planes. The bond critical point, which is about 0.7 of the distance from the N to the H atom, lies in a slightly negative region in these plots. The eigenvalues of the Hessian at the H bond critical point were −1.49, −0.72 and 4.90 e/Å^5, respectively. The plane in Fig. 12a is at an angle of 114° with respect to the least squares plane with the N(3) acceptor and for Fig. 12b the plane is at a corresponding angle of 29°. The eigenvector for the positive curvature has an angle of 105° to the plane with the acceptor atom N(3). A vector map of the gradient of ρ in this H bonding region of the crystal might prove to be illuminating.

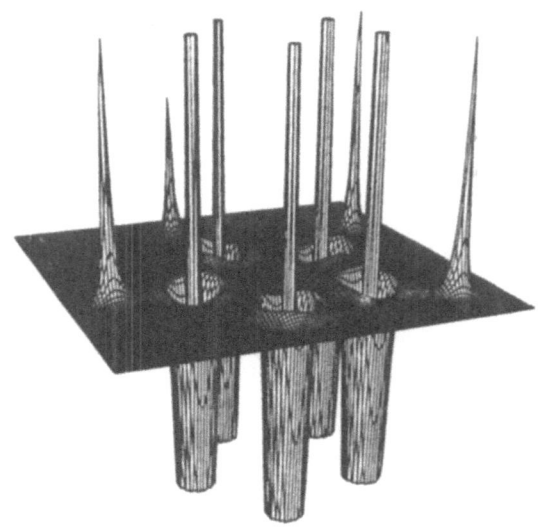

Fig. 11. Negative Laplacian of ρ for gsf density of imidazole in its least squares plane. Maxima are truncated at 1024 e/Å^5 and minima at −896 e/Å^5.

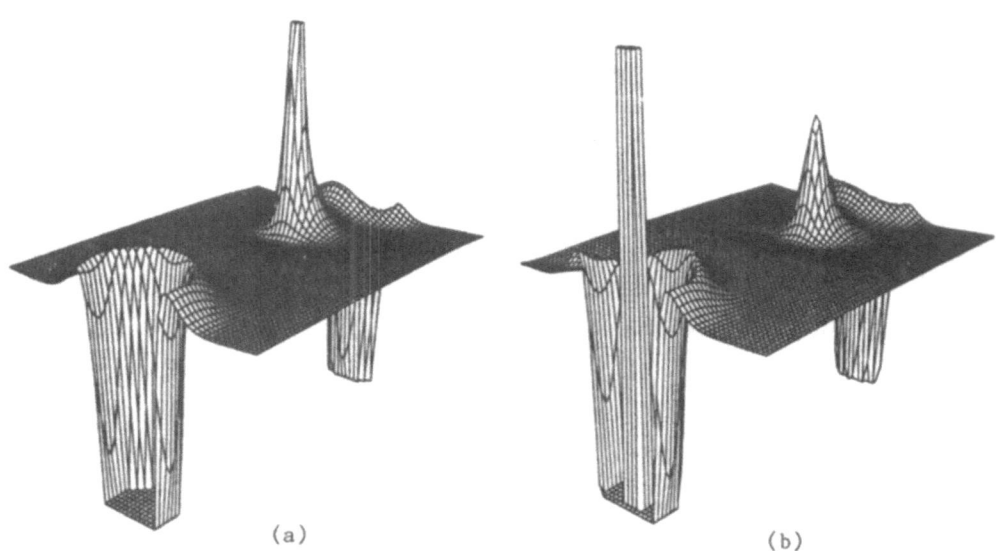

(a)   (b)

Fig. 12. -$\nabla^2\rho$ of the gsf density in the H bonding of imidazole. Max at 256 e/Å^5 and min at −256 e/Å^5. (a) 2.8 × 2.2 Å in plane of positive and least negative curvature; (b) same size as (a) but plane of positive and most negative curvature.

9-Methyladenine

The molecular crystal of 9-methyladenine is monoclinic in space group $P2_1/c$ with four molecules per unit cell. Both neutron[23] and Mokα x-ray diffraction[24] measurements at 126 K have been reported. The integrated x-ray intensities, which extended to |H| of 1.99 Å^{-1}, were reduced to a unique set of 5277 structure factor amplitudes. The gsf model was the same as adopted for analysis of imidazole. Nuclear positions were fixed at the neutron values as well as H thermal parameters. In addition to variation of U_{ij}'s for N and C, anisotropic secondary extinction parameter was included with the Becker-Coppens model.[25] The residuals in the mean square error function were $|F_o|^2 - |F_c|^2$; weights were based on experimental data reduction. The error function was spanned with 313 variables. At completion of the fitting procedure, the criterion for convergence was $1.95 \times (10^{-12})$. The full Hessian of the mean square error function was positive definite and inverted for variance evaluations of the gsf physical properties. The final radial parameters and functions are $R_w(|F|^2) = 0.0791$, $R(|F|^2) = 0.0421$ and gof = 1.329.

Table 6. Radial Density Functions Used for 9-MeAd

Atom	Pole order	Type	κ/α (Å^{-1})
N	0	is×is	1.0
	0	(2×os×os+3×2p×2p)/5	1.0090(57)
	1	$(r^2) \times \exp(-\alpha_N r)$	6.808(250)
	2	$(r^2) \times \exp(-\alpha_N r)$	6.808(250)
	3	$(r^3) \times \exp(-\alpha_N r)$	6.808(250)
C	0	is×is	1.0
	0	(2×os×os+2×2p×2p)/4	1.0260(84)
	1	$(r^2) \times \exp(-\alpha_C r)$	5.602(162)
	2	$(r^2) \times \exp(-\alpha_C r)$	5.602(162)
	3	$(r^3) \times \exp(-\alpha_C r)$	5.602(162)
H	0	$\exp(-\alpha_0 r)$	4.196(103
	1	$r \times \exp(-\alpha_H r)$	3.709(243)
	2	$(r^2) \times \exp(-\alpha_H r)$	3.709(243)

The gsf fits to 9-MeAd were used to evaluate an electrostatic potential of the molecule removed from the lattice. A contour plot of this result in the least squares plane of the non-hydrogen atoms is shown in Fig. 13a and the corresponding estimated standard deviations of the potential are in Fig. 13b. The acceptor N atoms N(1) and N(7) both show electronegative extrema of about –275 kJ/mol, but the N(3) atom gives a similar result as well. In the lattice, N(3) does not appear to participate in a hydrogen bond. If the esd map is consulted one finds large random error for the equipotential lines in Fig. 13a. For example, at the local minima, the esd is 0.15 e/Å or 210 kJ/mol. The plot in Fig. 13b indicates that indeed the potential for 9-MeAd is a modest result and probably at best is only suggestive of the donor-acceptor facets of H bonding by this molecule.

A cluster calculation, similar to imidazole, was undertaken. For the two molecule case, 9-MeAd and one glide-related 9-MeAd, the potentials reduce to about –135 kJ/mol near N(1) and N(7), but do not form an electropositive bridge. A contour plot of this potential in the least squares plane of N(1), N(7), H(O1) and H(O2) is shown in Fig. 14. A rather larger cluster will be necessary before these electronegative regions become positive. It should be noted, however, that the esd of the potential from this calculation is nearly 0.3 e/Å in the H bonding region.

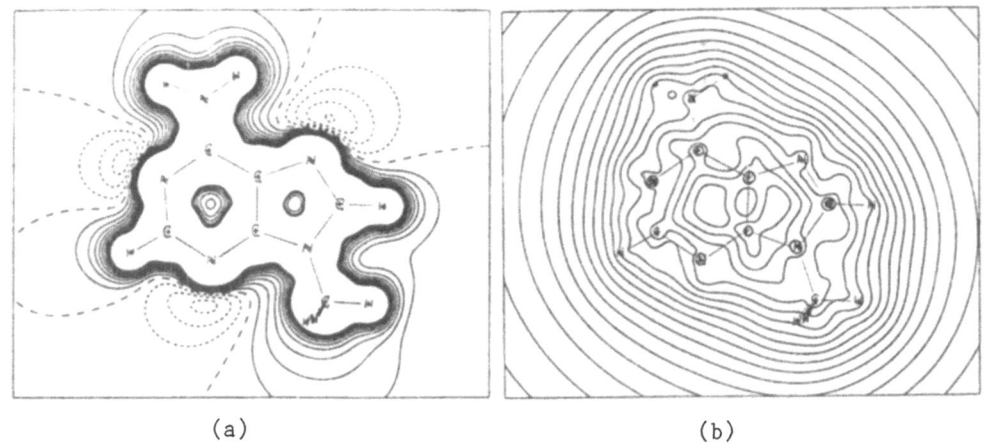

(a)   (b)

Fig. 13. Electrostatic potential and esd contour plots for 9-MeAd. 12 Å ×
10 Å sections. (a) Max contour at 0.5 e/Å; min contour at –0.2
e/Å; increments of 0.05 e/Å. (b) Esd min contour at edges start at
0.06 e/Å and increase in increments of 0.01 e/Å to a max of
0.23 e/Å.

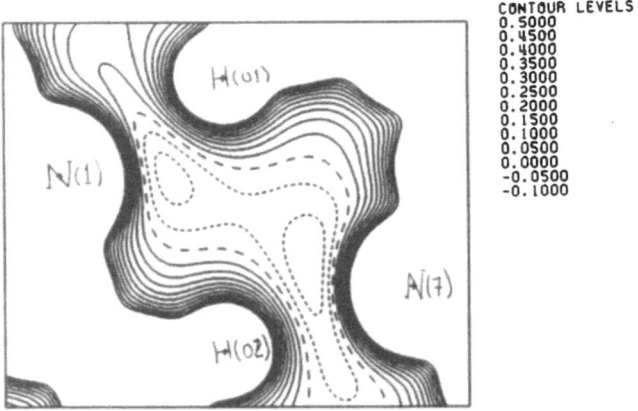

CONTOUR LEVELS
0.5000
0.4500
0.4000
0.3500
0.3000
0.2500
0.2000
0.1500
0.1000
0.0500
0.0000
-0.0500
-0.1000

Fig. 14. Electrostatic potential for a cluster of two 9-MeAd molecules.
Section is 4 Å × 4.7 Å. Contours as in Fig. 13a.

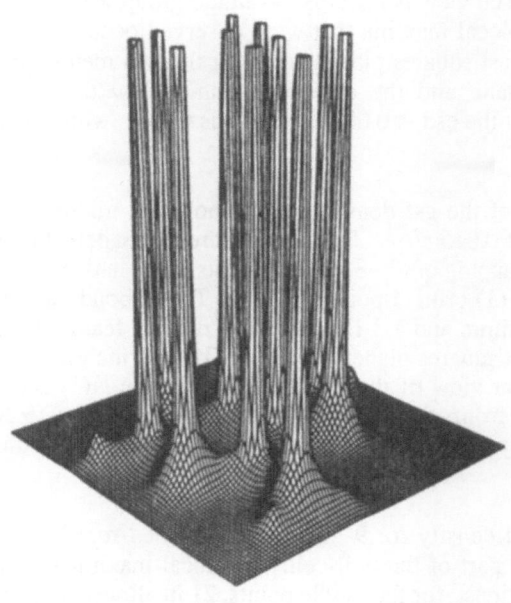

Fig. 15. Electron density distribution from 9-MeAd gsf fit. Local maxima truncated at 32 e/Å³. Section is 7.5 Å × 7.5 Å.

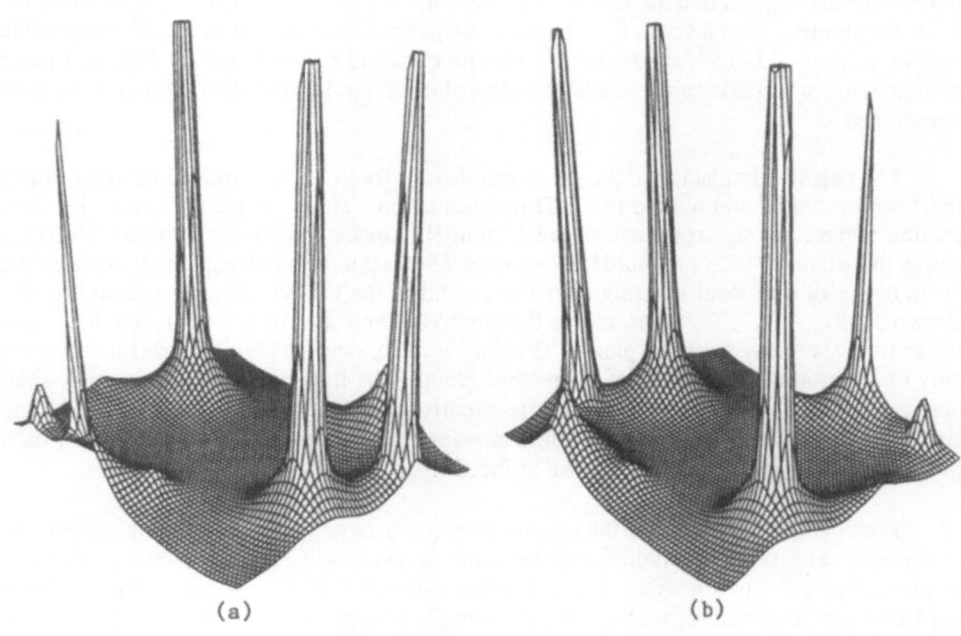

(a)   (b)

Fig. 16. Density from the gsf model in the lattice for the H bonding region in 9-MeAd. Sections are 4.0 Å × 4.7 Å. Maximum density truncated at 32 e/Å³. (a) H(O1)···C(6) in front. (b) H(O2)···N(7) in front.

The electron density distribution for 9-MeAd gsf's removed from the lattice was evaluated in the same plane as for the potential shown in Fig. 13a. A 3-D plot of this result is given in Fig. 15. The view is towards the amino group with N(1) to the left and N(7) to the right. The only local maxima that were observed occurred at the sites of the nuclei projected onto the least squares plane except for the two methyl protons, one of which is 93 pm above the plane and the other is 83 pm below the plane. The lowest density was -0.009 e/Å^3, but the esd is 0.007 e/Å^3 in this region, which is near N(1) and towards the C(2)–H(2) bond.

A lattice sum of the gsf density model showed a minimum of -0.061 e/Å^3 in the crystal with an esd of 0.015 e/Å^3. This 4 σ result reveals a defect in the gsf model that does not afford reliable mapping of (3,+3) critical points in the lattice; this may also hold true for determination of (3,+1) critical points as well. The H bonding regions of the gsf model were all positive definite and a 3-D plot of this density feature is shown in Fig. 16. The section is in the least squares plane of N(7), H(O1) and the glide related N(1) and H(O2). Fig. 16a gives a front view of the "Watson-Crick" type N(1) acceptance to H bonding; Fig. 16b shows the front view of "Hoogsteen" type H bond to N(7). The esd of the density in the vicinity of H bond critical points is 0.02 e/Å^3 with a density value about 0.2 e/Å^3.

The gsf crystal density for 9-MeAd was scanned for all (3,–3) and (3,–1) critical points in the unique part of the unit cell. The local maxima occurred only at the mean thermal nuclear positions; for the saddle points, 21 in all were detected and refined. These results are summarized in Table 7. It is of some interest to compare the ellipticities and Laplacians of the (3,–1) critical points in the five-membered ring of 9-MeAd with the imidazole results. For 9-MeAd the ellipticities in the C(4)–C(5)–N(7)–C(8)–N(9) portion of the moelcule vary from 0.14 to 0.28 with an average of 0.22. On the other hand, imidazole varies from 0.20 to 0.64 with an average of 0.41. The gsf projection results from the imidazole data suggest that the C(4)–C(5) is very much like a bond in ethylene, whereas the 9-MeAd numbers are closer to C–C bonds in benzene. The Laplacians are all comparable and vary from -13 e/Å^5 to -24.5 e/Å^5 with no easily identifiable trend. Before a more quantitative comparison can be made, the esd's of the $\nabla^2\rho(\vec{r}_c)$ and of the ellipticities must be determined.

The negative Laplacian of the gsf model density for 9-MeAd removed from the lattice has been evaluated and plotted in a 3-D representation. These are shown in Fig. 17. The section is the least squares plane of the 11 non-H atoms. The amino hydrogen H(O1) is above this plane by 24.3 pm, but H(O2) is only 9.9 pm out of the plane, which explains the prominence of one local maxima over the second in the region around the amino group shown in Fig. 17a. The protons on the methyl group are 10.8, –93.1 and 83.3 pm, respectively, removed the the plane. Thus in Fig. 17b, one sees a local maximum due to only one proton in the vicinity of the methyl group. The lone-pair structure in the L-shell maxima about N(1), N(3) and N(7) are clearly evident in Figs. 17a and 17b. The anisotropy about N(7) in 9-MeAd is rather more pronounced than the corresponding L shell ring about N(3) for imidazole as shown in Fig. 11.

The negative Laplacian of the crystal, gsf density model in the H bonding region was evaluated. The sections studied are the same as discussed above for the density and displayed as 3-D plots in Figs. 16a and 16b. 3-D plots of the corresponding negative Laplacian are shown in Figs. 18a and 18b. The (3,–1) critical points for both H bonds are 0.64 of the bond lengths away from the N acceptor atoms. For N(1)· · ·H(O1), best viewed in Fig. 18a, the critical point is at a slight negative value of -1.3 e/Å^5. The bond path vector (eigenvector of the positive eigenvalue) forms an angle of about 2° with the interatomic vector of N(1) to H(O1). The corresponding critical point values for N(7)· · ·H(O2), best viewed in Fig. 18b, are -0.9 e/Å^5 and an angle of 3°. The bond paths for the two hydrogen bonds are essentially linear with this gsf model.

Table 7. (3,–1) Critical points in the 9-MeAd crystal for gsf and IAM density distributions

	Unit cell fractions			ρ (e/Å^3)	$\nabla^2\rho$ (e/Å^5)	ε	Bond path
	x	y	z				Bond length (pm)
gsf	0.1873	0.1386	0.1577	0.21	0.91	0.65	N(7)· · ·H(O2)
IAM	0.1932	0.1361	0.1685	0.19	1.80	0.01	203.7
gsf	0.2158	0.3096	0.3305	0.23	1.26	0.56	H(O1)· · ·N(1)
IAM	0.2126	0.3205	0.3288	0.23	2.08	0.04	195.4
gsf	0.2637	0.0719	0.7560	2.48	−24.65	0.16	N(1)–C(2)
IAM	0.2626	0.0710	0.7561	1.78	−3.51	0.03	134.2
gsf	0.2804	−0.0119	0.7379	2.46	−18.05	0.16	C(2)–N(3)
IAM	0.2842	−0.0087	0.7440	1.82	−4.31	0.03	133.0
gsf	0.2767	−0.0430	0.6078	2.16	−11.27	0.20	N(3)–C(4)
IAM	0.2748	−0.0428	0.5980	1.78	−3.50	0.03	134.2
gsf	0.2427	0.0210	0.4909	2.29	−22.24	0.25	C(4)–C(5)
IAM	0.2415	0.0203	0.4880	1.50	−2.28	0.07	139.4
gsf	0.2183	0.1161	0.4999	2.16	−19.44	0.17	C(5)–C(6)
IAM	0.2185	0.1150	0.5036	1.44	−1.70	0.04	141.5
gsf	0.2295	0.1439	0.6312	2.23	−16.73	0.22	C(6)–N(1)
IAM	0.2266	0.1465	0.6280	1.75	−2.68	0.04	135.3
gsf	0.2005	0.2049	0.5336	2.55	−24.69	0.22	C(6)–N(10)
IAM	0.2017	0.2020	0.5355	1.80	−3.75	0.03	133.8
gsf	0.2083	0.0718	0.3735	1.98	−12.98	0.22	C(5)–N(7)
IAM	0.2094	0.0697	0.3731	1.66	−0.86	0.05	138.6
gsf	0.2036	0.0122	0.2598	2.73	−24.48	0.14	N(7)–C(8)
IAM	0.2064	0.0067	0.2610	1.85	−4.73	0.04	132.1
gsf	0.2396	−0.0636	0.3153	2.22	−16.95	0.21	C(8)–N(9)
IAM	0.2342	−0.0630	0.3146	1.72	−1.99	0.05	136.6
gsf	0.2564	−0.0597	0.4771	2.09	−16.84	0.28	N(9)–C(4)
IAM	0.2602	−0.0604	0.4718	1.71	−1.64	0.06	137.0
gsf	0.2724	−0.1649	0.4088	1.91	−12.09	0.02	N(9)–C(11)
IAM	0.2773	−0.1644	0.4108	1.47	2.05	0.01	145.3
gsf	0.3003	−0.2388	0.3434	1.84	−18.46	0.15	C(11)–H(11)
IAM	0.2967	−0.2393	0.3419	1.25	−4.71	0.00	107.6
gsf	0.2042	−0.2432	0.4159	1.88	−21.25	0.05	C(11)–H(12)
IAM	0.2029	−0.2436	0.4198	1.23	−4.34	0.00	108.6
gsf	0.3894	−0.2285	0.5051	1.98	−21.47	0.17	C(11)–H(13)
IAM	0.3934	−0.2293	0.5099	1.23	−4.33	0.00	108.7
gsf	0.1905	0.2820	0.4215	2.04	−28.98	0.14	N(10)–H(O1)
IAM	0.1905	0.2816	0.4224	1.52	−7.56	0.01	102.3
gsf	0.1897	0.3057	0.5770	2.13	−33.14	0.06	N(10)–H(O2)
IAM	0.1910	0.3045	0.5760	1.53	−7.85	0.01	101.9
gsf	0.2085	−0.0584	0.1732	1.85	−18.40	0.12	C(8)–H(8)
IAM	0.2070	−0.0599	0.1665	1.23	−4.56	0.01	108.1
gsf	0.2901	0.0174	0.8677	1.88	−20.64	0.13	C(2)–H(2)
IAM	0.2912	0.0172	0.8728	1.21	−4.10	0.01	109.2

Benzene

Single crystal x-ray and neutron diffraction measurements have been reported for fully deuterated benzene at 123 K.[26,27] The molecule crystallizes in the orthorhombic space group Pbca with four molecules per unit cell. In this crystal, the benzene lies on a center of symmetry with the three unique C and D atoms in general positions. The x-ray data, reduced to relative structure factors, consisted of 2335 unique reflections, which extended to an |H| of 2.36 Å^{-1}. The C and D positions and the D thermal parameters were held fixed at the neutron values for the gsf analyses of the x-ray structure factors. The C atom monopoles were constructed with shell localized SCF AO's for the triplet P state; higher poles up to the octopole level were represented with single exponentials. The L shell monopole was scaled with a variable radial parameter, κ, and constrained to be the same for the three C atoms. The exponent radial parameters for the higher poles on the C atoms were varied, but constrained to be the same for all poles and all atoms. The D atoms included harmonics up to the quadrupole level with single exponentials that had variable radial parameters, all of which were constrained to be the same. The population parameters of the C atom K shell monopoles were also constrained to be equal. The radial functions and parameters are listed in Table 8. The total number of variable electronic parameters in the gsf model was 79. In addition the U_{ij}'s of the C atoms were treated as least squares variables so the total number of variables in the mean square error function was 97 with residuals |F_0|-|F_c| and weights based on the reduced experimental data. The final figures of merit were R_w(|F|) of 0.0313, R(|F|) of 0.1026 and a gof of 1.178 at completion of the minimization process. A value of $2.04 \times (10^{-12})$ for the criterion for convergence was achieved.

The gsf properties evaluated to date have been restricted to the unique pseudoatoms and the centrosymmetrically related ones removed from the lattice. Crystal properties have not been directly evaluated by lattice sums. All sections for the various plots below are in planes parallel to the least squares plane of the three C and three D nuclei and the centrosymmetrically related ones. The average deviation of the nuclei from the plane is 3.8 pm with D(3) at a maximum of 8.1 pm.

Table 8. Radial Density Functions Used for Benzene

Atom	Pole order	Type	κ/α (Å^{-1})
C	0	is×is	1.0
	0	(2×os×os+2×2p×2p)/4	0.9923(62
	1	$(r^2)\times\exp(-\alpha_C r)$	6.442(159
	2	$(r^2)\times\exp(-\alpha_C r)$	6.442(159
	3	$(r^3)\times\exp(-\alpha_C r)$	6.442(159
H	0	$\exp(-\alpha_H r)$	4.490(71)
	1	$r\times\exp(-\alpha_H r)$	4.490(71)
	2	$(r^2)\times\exp(-\alpha_H r)$	4.490(71)

The lines of equipotential for the gsf model of benzene are shown in Figs. 19a (in the plane of the molecule) and 19b (1 Å above the plane). As expected for this type of hydrocarbon, the potential about the molecule *in its plane is* electropositive everywhere. Note that the potential contours in Fig. 19a reflect only a center of symmetry, which is necessarily the case in the crystal. In a plane 100 pm above the molecule (Fig. 19b) the potential is most electronegative above the middle of the ring and somewhat resembles the

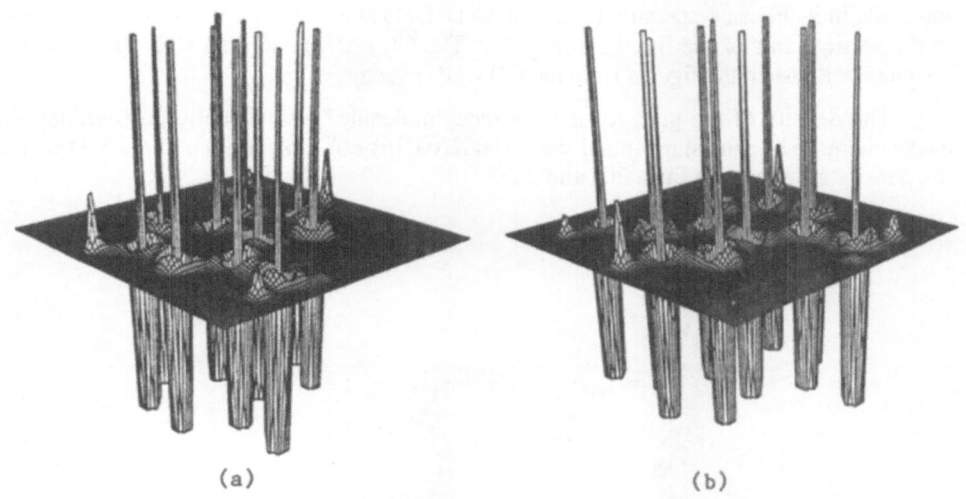

(a)   (b)

Fig. 17. Negative Laplacian of ρ for the gsf density of 9-MeAd. Maxima
truncated at 1024 e/Å^5 and minima at –896 e/Å^5. Sections are
7.5 Å × 7.5 Å. (a) reveals N(1), N(7) and amino group as in
Fig. 15. (b) is view of N(3), C(2)–H(2), N(9) and the methyl
group in front.

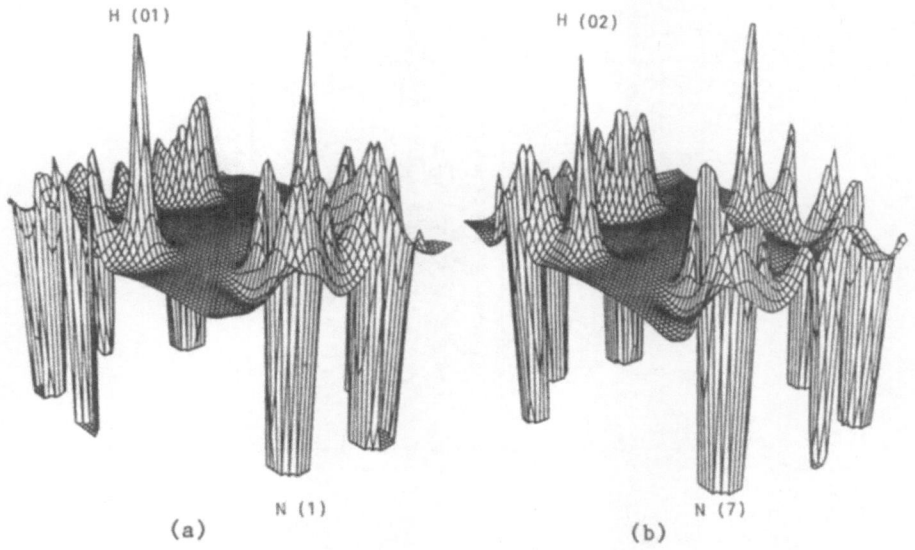

(a)   (b)

Fig. 18. $-\nabla^2\rho$ from the gsf model in the lattice for the H bonding region in
9-MeAd. Sections are 4.0 Å × 4.7 Å. Extrema truncated at 256 and
–256 e/Å^5. (a) H(O1)$\cdots$N(1) in foreground. (b) H(O2)$\cdots$N(7) in
front view with N(1)$\cdots$H(O1) in the back.

potential in crystalline graphite.[14] The symmetry of the equipotential lines in Fig. 19b is not centric and presumably reflects, in the peripheral regions, the packing interactions of the molecule in its Pbca crystal structure. The C(1)–D(1) atomic vector is upward and parallel to the vertical axis of the figures in Fig. 19. The C(2)–D(2) and then C(3)–D(3) proceed counterclockwise in the figures from the C(1)–D(1) vector.

The density of the gsf's for the benzene "molecule" was virtually positive definite everywhere. The minimum found was –0.002(0.015) e/Å^3. A contour and a 3-D plot of the density are shown in Figs. 20a and 20b.

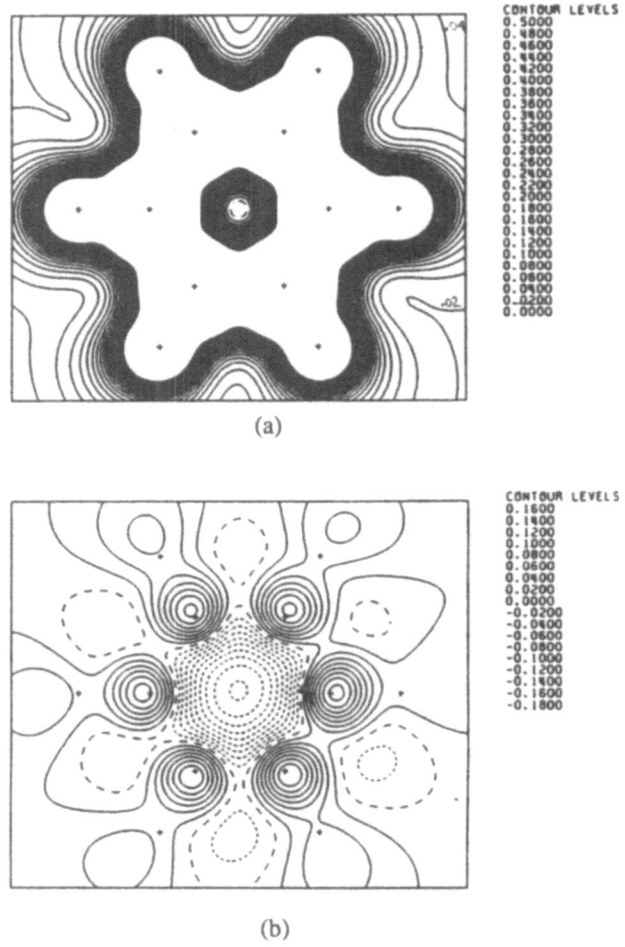

(a)

(b)

Fig. 19. Electrostatic potential contour plot for benzene from gsf's. Sections are 7 Å × 6 Å. Contour intervals at 0.02 e/Å. (a) Plot in the least squares plane. Max contour is 0.5 e/Å; min is 0.0 e/Å. (b) 1 Å above the plane. Max contour is 0.16 e/Å; min is –0.18 e/Å. Long dash is null.

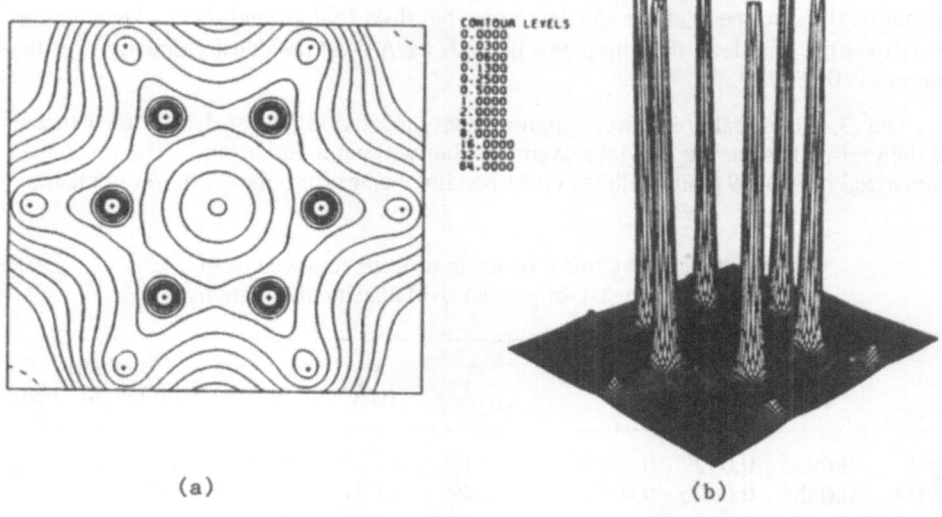

(a)   (b)

Fig. 20. The electron density from a benzene gsf model. Sections are 5.5 Å × 4.9 Å. (a) Contours start at 0.025 e/Å^3 and increase in multiples of two to 64 e/Å^3. Null is long dash. (b) 3-D plot with maximum densities truncated at 64 e/Å^3.

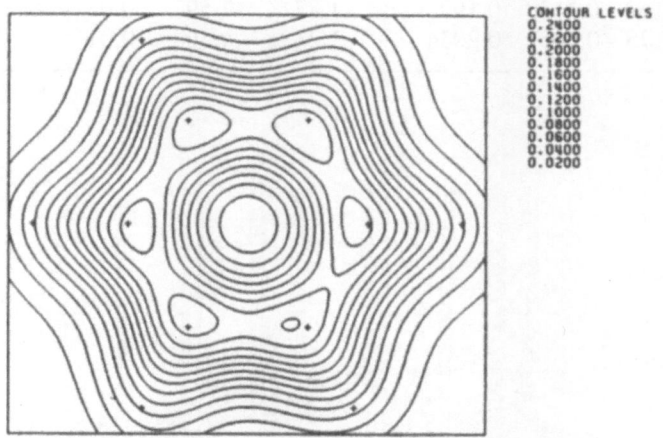

Fig. 21. Electron density distribution for benzene gsf's at 1 Å above the ring. Contour increments of 0.02 e/Å^3 with minimum at 0.02 e/Å^3 and maximum at 0.24 e/Å^3.

Both representations of the electron density show local maxima only around the nuclear sites. Saddles in the density near the bond critical points are more easily seen in Fig. 20b than 20a. The electron density in the same section as shown for the potential in Fig. 19b was evaluated and plotted. This result is shown in Fig. 21 as a contour map. Although the density has local maxima above the C nuclei, it has a more uniform distribution around the ring in the plane 100 pm above the nuclei. Note that the density distribution does not peak above the deuterons, but does fold around them. The electron density over the middle of the ring is just under 0.1 e/Å³, yet the density directly over the deuterons is 0.08 e/Å³.

The (3,–1) critical points were approximately located in the gsf density for benzene and then refined as for the previous examples, but without a lattice sum. The results are summarized in Table 9 along with the corresponding values from the IAM. As in the three

Table 9. (3,–1) Critical points in benzene removed from
the crystal for gsf and IAM density distributions

| | Unit cell fractions | | | ρ (e/Å³) | $\nabla^2\rho$ (e/Å⁵) | ε | Bond path |
	x	y	z				Bond length (pm)
gsf	–0.0975	0.0929	0.0588	2.12	–16.27	0.28	C(1)–C(2)
IAM	–0.0986	0.0920	0.0601	1.49	–2.51	0.03	139.2
gsf	–0.1086	–0.0226	0.1288	2.13	–16.71	0.19	C(2)–C(3)
IAM	–0.1071	–0.0253	0.1290	1.49	–2.45	0.03	139.4
gsf	–0.0085	–0.1179	0.0698	2.19	–17.59	0.20	C(3)–C(1')
IAM	–0.0085	–0.1173	0.0689	1.50	–2.58	0.04	139.0
gsf	–0.0910	0.2133	–0.0095	1.83	–15.19	0.06	C(1)–D(1)
IAM	–0.0919	0.2144	–0.0092	1.24	–4.50	0.01	108.2
gsf	–0.2093	0.0671	0.1915	1.81	–15.16	0.06	C(2)–D(2)
IAM	–0.2114	0.0674	0.1936	1.23	–4.45	0.01	108.4
gsf	–0.1160	–0.1430	0.1991	1.87	–17.59	0.05	C(3)–D(3)
IAM	–0.1178	–0.1457	0.2034	1.24	–4.56	0.01	108.1

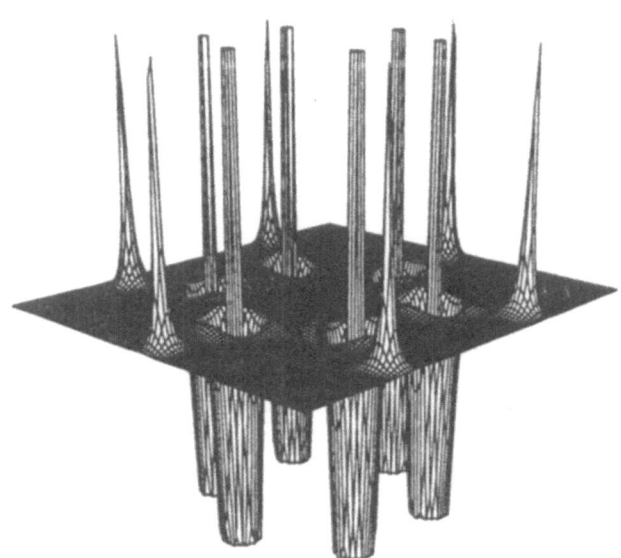

Fig. 22. Negative Laplacian of ρ for gsf density of benzene in its least squares plane. Maxima are truncated at 1024 e/Å⁵ and minima at –896 e/Å⁵. Section is the same as for Fig. 20b.

previous examples, the IAM atomic densities have critical points at positions which are nearly the same as the gsf result. On the other hand, the ellipticities are dramatically different for the C–C bonds spanned with gsf's and have an average value of 0.22. A theoretical value for the benzene molecule is 0.25. Both C–C and C–D have traces of the Hessian at the critical points with much greater negative values than the IAM. The average Laplacian of the gsf density at the $(3,-1)$ critical points for C–C is -16.9 e/Å^5 and for C–D it is -16.0 e/Å^5.

A 3-D plot of the negative Laplacian of ρ for the gsf fit to the x-ray data from deuterated benzene is shown in Fig. 22. All $(3,-1)$ critical points lie in positive regions which are near the midpoints for the C–C bonds and about two thirds of the C–D bonds from C to D. The gsf model apparently promotes a local concentration of electron charge density in the directions of the covalent bonds formed among the C and D atoms. The eigenvectors of the least negative curvatures at the critical points were all approximately normal to the plane shown in Fig. 22 with the exception of C(3)–C(1') which forms an angle of $-34°$ with respect to the plane normal. Before a serious critique of this peculiar result can be undertaken, however, the esd's of the Hessians of the gsf density must first be evaluated.

DISCUSSION

The tenor of this lecture has focused on a few physical properties that are derived from the gsf fits to real diffraction data from four different organic molecular crystals, for which neutron diffraction results at the same temperature are available. The nuclear distribution densities, therefore, have not been significantly altered from the neutron refinements in the present study. The gsf results depend parametrically on the nuclear distribution; the electrostatic potentials, an extranuclear property, have a much weaker dependence on the probability distribution function for the nuclei than the Laplacian of the electron density, which is an internuclear property. In particular, the treatment of H atoms with gsf's is greatly simplified with the use of neutron data and makes it possible to seriously explore the gsf electron density as well as its gradient and Hessian in regions of H bonding in the crystal.

The gsf parameters have been determined by least squares projections onto experimentally derived structure factor amplitudes, or their squares, with weighting functions that are derived from the inferred random error in the intensity measurements. The quality of the four x-ray data sets has not been addressed in this lecture. The approach has been to take what has been given, carry out the projection work, and see what emerges for the electron density distribution and a few of its properties. The final figures of merit, $R_w(|F|)$, varied from 0.011 to about 0.04 (essentially twice these figures for $R_w(|F|^2)$) with overdetermination factors between 8.2 for urea to 24.1 for benzene. For all data sets the maximum $|\vec{H}|$ extended to 2 Å^{-1} or greater. The particular values of the gsf parameters have also not been discussed, but rather the total representation has been emphasized as the object for your evaluation, be it derisive or skeptical. A comment on the radial parameters listed in Table 4 (imidazole) compared to those in Table 6 (9-methyladenine) is in order, however. The N radial parameters for the higher multipoles in the two crystals differ markedly compared to the esd's. For imidazole, the N pseudoatom radial functions are actually more diffuse than for the C pseudoatoms, yet the lowest mean square error value was found in this region of the "alpha" space that spans the mean square error function. It is indeed possible to explore other regions of the space (reverse the magnitudes for C and N) and seek a second minimum, which may be higher or lower than the present value. That has not been done in the present study. The Fourier-Bessel transform of these radial functions are essentially the same on an infinite interval in r as one truncated at about 0.6 of the cube-root of the unit cell volume. The functions have not become so diffuse that they can be rejected on the basis of the infinite volume approximation in atomic form factor evaluations. It may be that the radial parameters in Table 4 reflect systematic error in the reduced experimental data for imidazole, but at the present stage I think a definitive answer is not evident. In general, one should always include a final density error map, with a Fourier sum of F_O-F_C(gsf), as an illustration of the quality of fit. These have been published for gsf models similar to the ones given here for urea,[17] imidazole,[21] and 9-MeAd[24] and are not included among the sundry plots above. The benzene work is not yet complete since other radial type multipoles are under investigation.

Almost all of the gsf properties covered above are scalars, but it is possible to map out vector fields (electric fields and gradients of the electron density) and second rank tensors, such as electric field gradients and Hessians, with the gsf model. The electrostatic potentials clearly reveal functionality around the periphery of the molecules and reflect the crystal environment in contrast to a vacuum around an isolated molecule. For a viable construction of these potentials with the gsf model it is essential that scaling procedures treat the experimental structure factor amplitudes on a relative scale. To constrain the gsf model to a neutral unit cell during the least squares minimization process can be unproductive. Such efforts bias the model towards the IAM scale factor and may result in an exaggerated electropositive contribution from the monopoles and undue weight from the higher multipoles. The electrostatic potential from a superposition of IAM's is electropositive everywhere. The esd maps, based on a variance calculation, of the potentials shown in this work reveal random errors that range from 15% to 50% of the values in electronegative regions of key interest. That the esd's are relatively large is not only due to the random error in the reduced data, but also reflect the "flatness" of the mean square error sum with respect to variation of the gsf electronic parameters. We saw that the densities from the gsf's for all four examples are non-positive definite in some regions of space, but, with the exception of 9-MeAd, none of these appear to be statistically significant. By contrast an IAM density is positive definite everywhere. The gsf densities in these studies of organic molecular crystals had local maxima only at the nuclear sites and no others. (A gsf result from an anlaysis of Be x-ray data revealed a local maximum in the density that was not at a nuclear site.[28]) Application of Bader's topological methods to the gsf results have proved to be informative and valuable. It was shown, for example, in Fig. 6b for urea that although the IAM density displays the shell structure of atoms, it does not reveal any effects of chemical bonding. The negative Laplacian of the gsf density, on the other hand, revealed several facets of chemical bond formation as shown in Fig. 6a. The local concentration of electron density is clearly oriented towards atoms participating in the bonds and a lone pair structure in the L shell of the oxygen atom emerges as well. We emphasize that the gsf model is not constrained to any particular orientation; the shaping of the local electron density in the pseudoatom is due solely to projections of the gsf multipoles onto the reduced experimental data. For all four examples, plots of the negative Laplacian of the gsf density suggest that the results are qualitatively correct and that projection methods with gsf's can retrieve chemical bonding information from present day x-ray diffraction data. The quantitative features have been partially summarized in tables of $(3,-1)$ critical points that were located in the several crystals. The traces of the Hessians and ellipticities at the critical points are in accord with quantum chemical values that Bader *et al* have reported but depart markedly from corresponding values for IAM densities. A critical evaluation of the exact numerical results awaits variance computations of the gradients and Hessians of the gsf density.

The last point to note in this work is that the gsf density CANNOT in principle be a pure quantum state result. Even if we employed a gsf model that was constrained to be positive definite and derivable from an N-representable first order density matrix, it is only the mean thermal form factors which are projected onto the reduced data. Although the convolution approximation has been used to represent the real world of a "perfectly" measured x-ray structure factor amplitude, the nuclear vibrations mix in electronic states, however small they may be. In the limit of absolute zero temperature, the zero point vibrations of the nuclei will dominate the distortions of the electron density from that of stationary nuclei; the latter configuration is not truely an observable with elastic x-ray scattering. At best, a gsf density is a ghost of a quantum object. Any effort to extract "virial" fragments from the results presented above is emphatically discouraged by this investigator.

ACKNOWLEDGEMENT

I am grateful to Mark Spackman for some of his collaborative work on electrostatic potentials. Much of this research has been supported by a grant from the University of Pittsburgh.

REFERENCES

1. W. H. Bragg and W. L. Bragg, *Proc. Roy. Soc. London*, **88**, 428 (1913); **89**, 246 (1913).
2. C. K. Johnson and H. A. Levy, *International Tables for X-Ray Crystallography*, Vol. IV, pp. 311-336 (1974).
3. I. Waller, *Phil. Mag.*, **4**, 1228 (1927).
4. P. M. Platzman and N. Tzoar, *Phys. Rev.*, **B2**, 3556 (1970).
5. V. Florescu and M. Gavrila, *Phys. Rev.*, **A14**, 211 (1976).
6. L. Kissel and R. H. Pratt, *Phys. Rev. Lett.*, **40**, 387 (1978).
7. R. H. Pratt, *Indian J. Phys.*, **58A** (Suppl.), 1-11 (1984).
8. M. Born, *Rep. Prog. Phys.*, **9**, 294 (1942-1943).
9. B. T. M. Willis and A. W. Pryor, *Thermal Vibrations in Crystallography*, Cambridge University Press, Chapter 9 (1975).
10. R. F. Stewart and D. Feil, *Acta Cryst.*, **A36**, 503 (1980).
11. R. F. Stewart, *Chem. Phys. Lett.*, **65**, 335 (1979).
12. R. F. Stewart, in *Critical Evaluation of Chemical and Physical Structural Information*, eds. D. R. Lide, Jr. and M. A. Paul, National Academy of Sciences, Washington DC, p.540 (1974).
13. M. A. Spackman and R. F. Stewart, in *Methods and Applications in Crystallographic Computing*, eds. S. R. Hall and T. Ashida, Oxford University Press, Oxford, UK (1984).
14. R. F. Stewart, *God. Jugosl. Cent. Kristalogr.*, **17**, 1 (1982).
15. R. F. Stewart, in *Electron and Magnetization Densities in Molecules and Crystals*, ed. P. Becker, Plenum Press, NY, NASIS Series B: Physics, Vol. 48, 427 (1980).
16. R. J. van der Wal and R. F. Stewart, *Acta Cryst.*, **A40**, 587 (1984).
17. S. Swaminathan, B. M. Craven, M. A. Spackman and R. F. Stewart, *Acta Cryst.*, **B40**, 398 (1984).
18. S. Swaminathan, B. M. Craven and R. K. McMullan, *Acta Cryst.*, **B40**, 300 (1984).
19. R. F. W. Bader, T. T. Nguyen Dang and Y. Tal, *Rep. Prog. Phys.*, **44**, 893 (1981).
20. R. F. W. Bader and H. Essen, *J. Chem. Phys.*, **80**, 1943 (1984).
21. J. Epstein, J. R. Ruble and B. M. Craven, *Acta Cryst.*, **B38**, 140 (1982).
22. R. K. McMullan, J. Epstein, J. R. Ruble and B. M. Craven, *Acta Cryst.*, **B35**, 688 (1979).
23. R. K. McMullan, P. Benci and B. M. Craven, *Acta Cryst.*, **B36**, 1424 (1980).
24. B. M. Craven and P. Benzi, *Acta Cryst.*, **B37**, 1584 (1981).
25. P. J. Becker and P. Coppens, *Acta Cryst.*, **A30**, 129 (1974).
26. J. R. Ruble, private communication (1989) (x-ray data for C_6D_6).
27. G. A. Jeffrey, J. R. Ruble, R. K. McMullan and J. A. Pople, *Proc. Roy. Soc.*, **A414**, 47 (1987).
28. R. F. Stewart, *Acta Cryst.*, **A33**, 33 (1977).

X-RAY DIFFRACTION AND CHARGE DISTRIBUTION:

Application to the electron density distribution in the Hydrogen Bond

Dirk Feil

Chemical Physics Laboratory
University of Twente
POB 217
7500 AE Enschede
The Netherlands

The electron density distribution (EDD) in molecules closely resembles the joint EDD's of the atoms that constitute the molecule. In fact, the deviations are so small that for more than fifty years the scattering by X-rays could be accounted for by using atomic scattering factors. Better recording techniques and improved methods for analysis of the measured intensities allow a more accurate determination of the EDD. The last twenty years has seen many studies of the effect of covalent bonding on the EDD of molecules in crystals. The results show in general good qualitative agreement with the outcome of Hartree-Fock calculations. Although the number of quantitative studies, in which ab-initio structure factors are compared with experimental ones, is still small, we assume that a careful diffraction experiment will yield quantitative information on the charge distribution in molecules. The reliability of the results strongly depend on the way the experiment is carried out and the type of information that is wanted. To illustrate the possibilities to obtain useful information by diffraction experiments we study a number of simple systems: atoms and molecules.

The X-rays scattered by different parts of the atomic EDD interfere. Consequently the structure factors depend on the spatial distribution of the electron density. Fig.1a shows the EDD's of the Na-atom and the corresponding ion.

The electron density diminishes rapidly with increasing distance to the nucleus. Since the volume of spherical shells about the nucleus increases, the amount of charge in a certain spherical shell does not decrease nearly as rapidly, indicating a considerable amount of diffuse charge in the

The Application of Charge Density Research to Chemistry and Drug Design
Edited by G. A. Jeffrey and J. F. Piniella, Plenum Press, New York, 1991

103

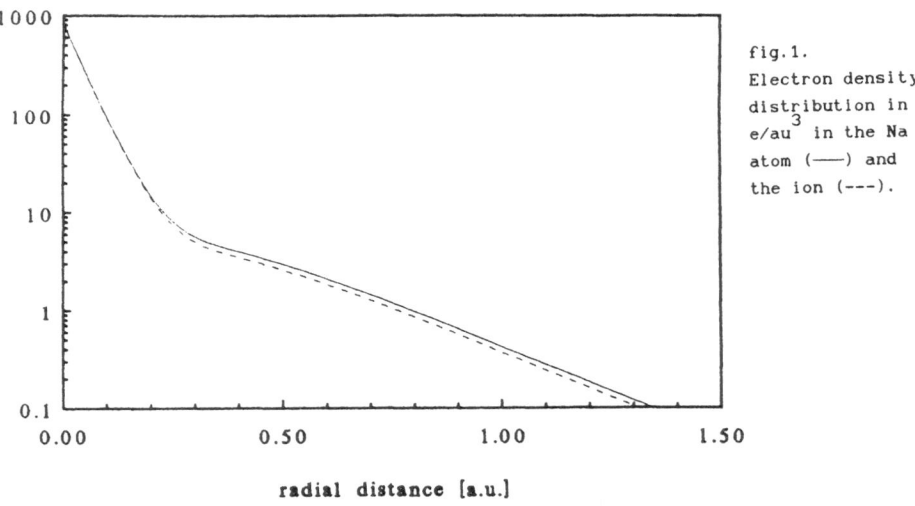

fig.1.
Electron density
distribution in
e/au^3 in the Na
atom (——) and
the ion (---).

radial distance [a.u.]

atom. The fact that the difference between an ion and an atom is situated
at the outer regions of the atom has serious consequences for the observa-
bility by X-ray diffraction. Fig.2 shows the Fourier transforms of the EDD
corresponding with the various atomic orbitals of the atom.

-- 1s   ···· 2s   —— 3s

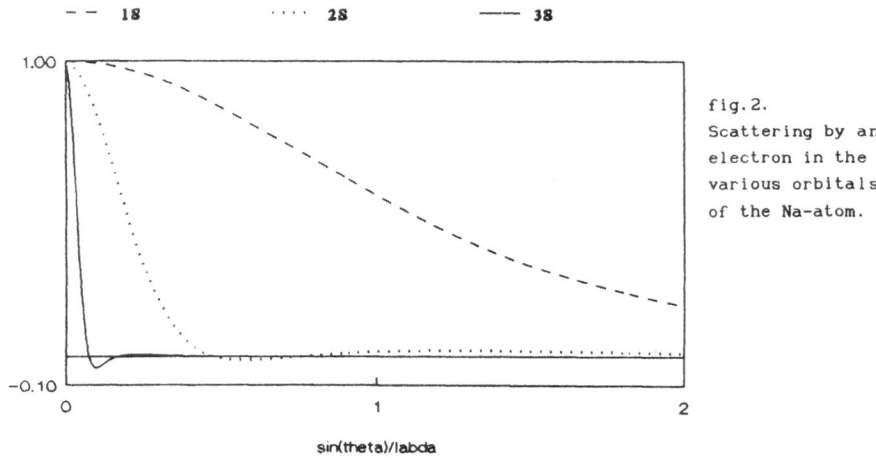

fig.2.
Scattering by an
electron in the
various orbitals
of the Na-atom.

sin(theta)/labda

It is clear that the diffuse charge is poorly represented in the form factor. The consequence for our objective, the derivation of charges from X-
ray data, is dramatic. As a result the form factors of neutral atoms and
the related ions differ only in the small-k region as witnessed by fig.3.
Consequently the possibility of deriving ionic charges and charge transfer
from X-ray data is a subject of hot debate[1]. To complicate the problem,
the charge distribution of the ion is modified by the electrostatic field
in the crystal, resulting in an ionic scattering factor that differs from
the one found in the International Tables for X-ray Crystallography.
The situation is quite different in molecules determined by covalent bon-

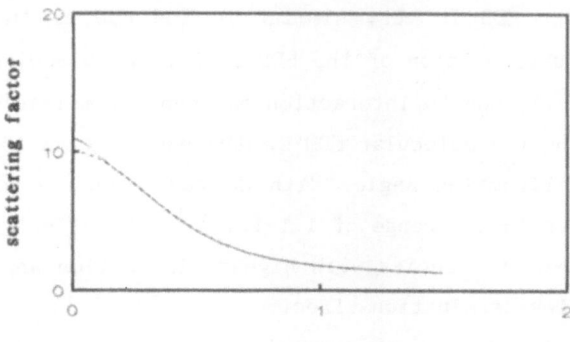

fig.3.
The scattering factor
of the Na-atom (——)
and the Na-ion (---).

sin(theta)/labda

ding. Now intra-atomic transfer of electrons is more important than the inter-atomic one and often, but not always, the bonding regions are involved. The EDD in a molecule closely resembles the EDD of the promolecule, defined as the linear combination of the atomic EDD's with each nucleus at the same position as in the molecule. Qualitative understanding of the difference between the two, the difference density distribution or DDD, is obtained by considering two consecutive changes of the EDD. The first one is related to the transformation of free or spherical averaged atoms into atoms in the 'prepared state ', the second one being the well known transfer of electron density from the outer atomic regions into the bonding region. Fig 4 shows the EDD of the H₂O molecule.

The bonding effects come out clearly when we plot DDD, the difference density distribution, (fig 5).

fig.4. The electron density distribution of the water molecule. Contours at 0.5, 1, 2, 4, 8, 16 e/Å^3.

fig.5. Difference density distribution of the water molecule. Contours at intervals of 0.1 e/Å^3.

In a molecular crystal the EDD is very similar to the EDD of the pro crystal, defined as the superposition of the EDD's of free molecules. The difference between the two is due to interaction between the molecules and to thermal motion smearing the molecular EDD's. Moreover observations are limited by the maximum diffraction angle. With Mo radiation the maximum $\sin(\Theta)/\lambda$ value usually lies in the range of 1.1-1.4 Å^{-1}. The effect of the omission of terms of the Fourier series with higher diffraction angles on the EDD is called the series termination effect.

Figures 6, 7 and 8 show the effects of thermal motion and series termination.

$$\sin(\Theta)/\lambda_{max} = 0.7 \qquad \sin(\Theta)/\lambda_{max} = 1.0 \qquad \sin(\Theta)/\lambda_{max} = 2.0$$

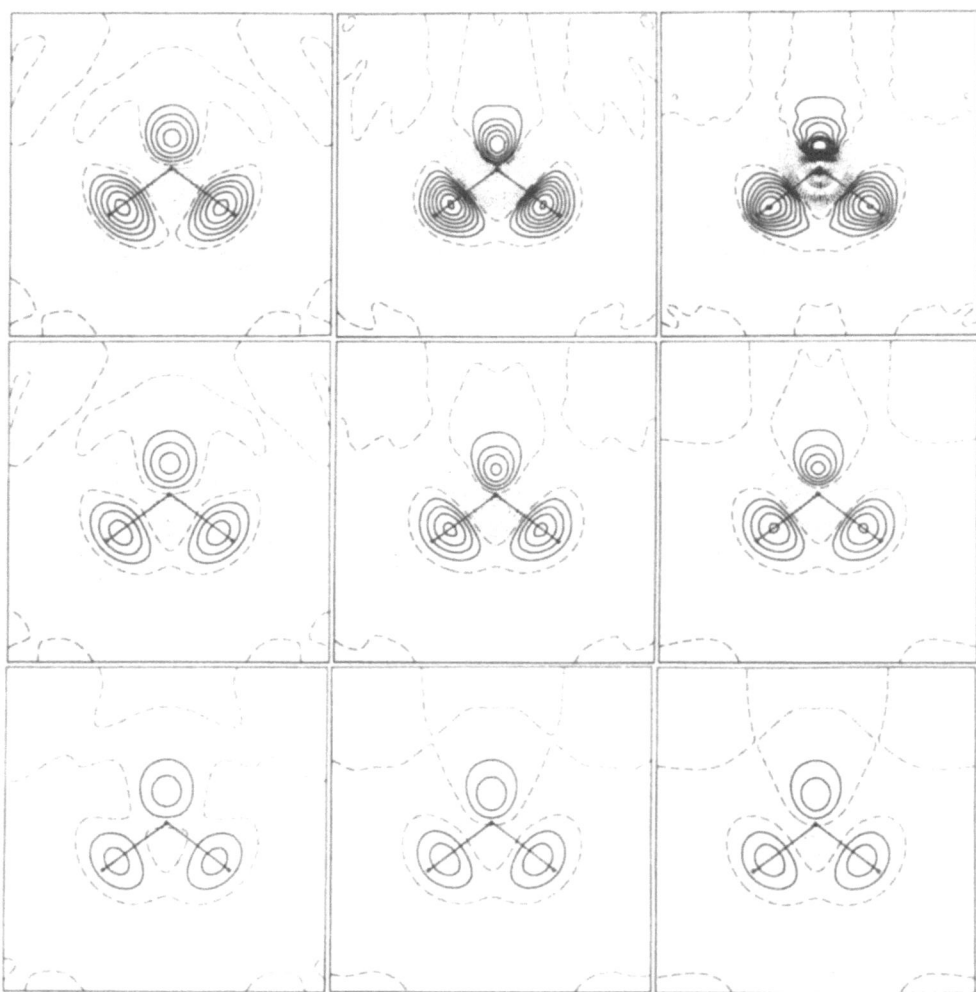

fig.6. The difference density of water at various temperatures and resolution. The columns differ in resolution, while the top, middle and bottom row give the results for a static molecule and molecules vibrating isotropically at 100K and 300K.

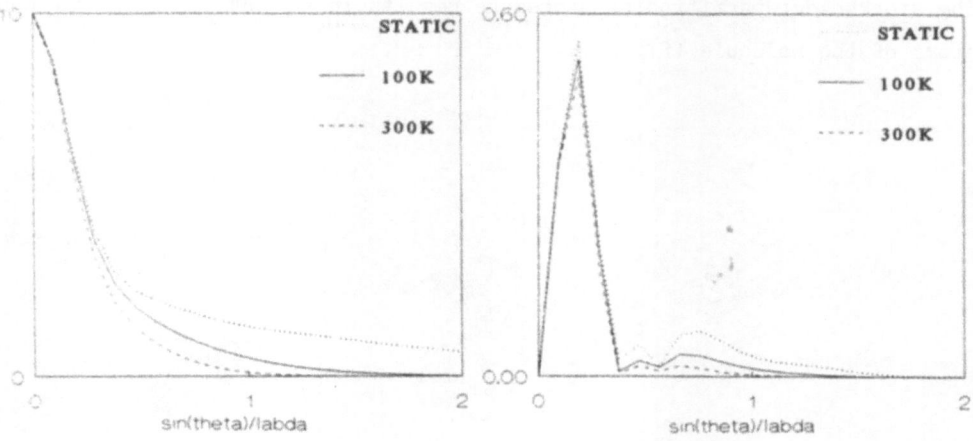

fig.7. Molecular scattering factor
(in units of e) along the 0k0 axis
of a water molecule in the xy-plane
with its bisectrix along the x-axis,
at various temperatures.

fig.8. Modulus of the molecular
scattering factor of the difference
density of water along the 0k0-axis
expressed in electrons.

It is clear that thermal motion washes out all details of the EDD. This
makes it superfluous to go to very high values of $(\sin\Theta)/\lambda$ by using small
wavelength radiation like Ag radiation.

In most accurate electron density studies either the EDD or the DDD is
developed in terms of functions centred on the various atoms (see the contributions of Robert Stewart and of Claude Lecomte). The functions centred
on a particular nucleus define an atom. It is clear that these atoms
strongly depend on the choice of functions, a problem closely related to
the basis set dependance of quantumchemically determined charges. Hirsh-
felds stockholder method[2] offers the possibility to partition both the EDD
and the DDD over the various atoms, circumventing the basis set effect.
Fig.9 shows the DDD for the urea molecule.

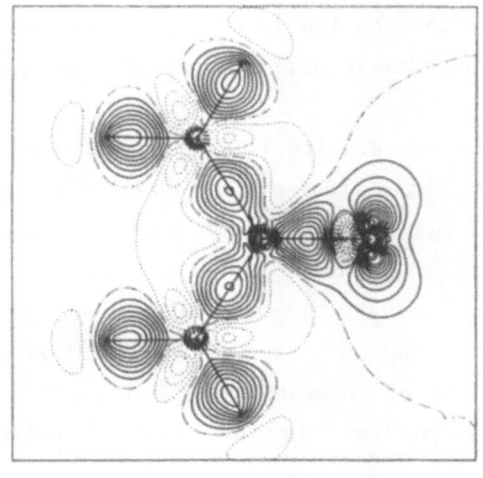

fig.9. The difference density of
the urea molecule. Contours at
intervals of 0.1 $e/\text{Å}^3$.

The stockholder partitioning procedure results in the DDD's of the various atoms of the molecule (fig.10).

O-atom   C-atom   N-atom

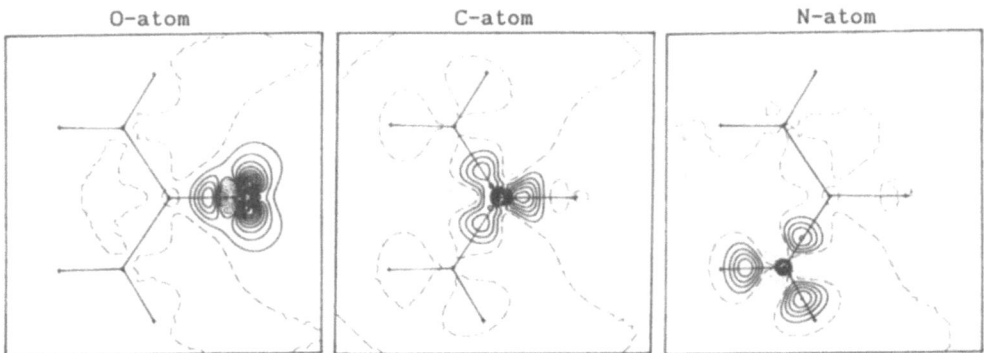

fig.10. The difference density of the various atoms of the urea molecule.
Contours as in fig.9.

The atomic EDD as obtained from the partitioning of the observed DDD is the convolution of the static EDD, corresponding with the outcome of a good quantumchemical calculation, and the thermal distribution function:

$$\rho^{obs}(\mathbf{r}) = \rho^{static}(\mathbf{r}) \otimes p(\mathbf{r})$$

The fourier transform is the product of the fourier transforms of the separate terms:

$$f(H) = f^{static}(H) * T(H)$$

The multiplication by $T(H)$ makes the high order terms vanish. Consequently it is impossible to determine $f^{static}(H)$ at high H-values from comparison with experimental data. Thus the models to describe the atomic EDD do not have parameters to fit the detailed EDD near the nuclei, and a rather featureless distribution results. It is concluded that a high resolution study of the core EDD is not feasible with X-ray diffraction and that the results of quantumchemical calculations on the inner regions of the atom should not be compared with the outcome of a refinement of experimental data.

The major features of a difference map are the EDD's corresponding with covalent bonding between the atoms. Many studies show qualitative agreement between the results of X-ray diffraction on molecular crystals and the outcome of quantum chemical calculations. The effect of molecular interaction in the crystal is much smaller than the effect of covalent bonding and for a long time it was not certain that X-ray diffraction could yield information on this type of interaction. The following will show that X-ray diffraction can contribute to the study of molecular interaction.

Molecular Interaction

When two molecules approach one another electrostatic interaction occurs. Assuming no mutual polarization of the molecules, the electrostatic energy can be obtained by summing the interaction of the various atomic multipoles on one molecule with those on the other one. When the intermolecular distance get smaller, mutual polarization of the molecules occurs. Since most molecules have a closed shell configuration overlap of the charge clouds will result in exchange repulsion and a redistribution of the electron density.

At short intermolecular distances appreciable correlation of electronic motion in the two molecules takes place. The corresponding lowering of the electronic energy is known as the London or dispersion energy. The changes in the EDD due to correlation are extremely small. The effects rapidly decrease with distance.

In principle electron density studies can contribute to the study of molecular interaction. The EDD of the individual molecules renders the multipole moments necessary to calculate the electrostatic interaction and the fields that cause the polarization of the partner in the intermolecular bond. Spackman *et al.* have applied this method to calculate the electrostatic interaction from Bragg diffraction data[3]. The various effects will be illustrated with a study of the hydrogen bond.

Hydrogen Bond

Whereas the qualitative aspects of the electron density distribution in the hydrogen bond are well understood[4], a quantitative treatment encounters large difficulties. The Hartree-Fock method is known to yield in principle good electron density distributions, but the method requires large basis sets to live up to its reputation. In particular the outer regions of the molecules with diffuse electron density distributions are difficult to handle. The highly regarded 6-31G** basis set does not satisfy the requirements[5]. Unfortunately hydrogen bonded dimer systems in general contain so many atoms that larger basis sets seem out of question.

The Hartree-Fock-Slater method seems to offer a solution. This method can be regarded as a variant of the density functional method, in which the local density approximation is applied. Owing to this approximation computing time does not increase so drastically with the number of basis functions and the method has shown to yield good electrostatic properties[6]. We shall use the method to calculate the electron density distribution in the bond and to analyze it in terms of polarization and exchange repulsion.

To be able to understand the various features of the electron density distribution in the hydrogen bond the important effect of polarization of a water molecule in a homogeneous field will be studied first. The difference between the electron density distribution in the water dimer, which will be studied subsequently, and the distribution in the monomers can be explained with mutual polarization. Owing to the large intermolecular distance interpenetration of the charge clouds plays a minor role.

In principle the electron density distribution is an observable quantity and the results of the calculations can be compared with the outcome of experiments. In practice only X-ray diffraction is available for this purpose and the system under study has to be a part of a crystal. Consequently the effects of the environment have to be taken into account. Recent extensive X-ray studies of the electron density distribution in the α-oxalic acid dihydrate complex makes it possible to compare the outcome of the quantum chemical calculations with experiment[7]. Ab-initio Hartree-Fock calculations on the oxalic acid molecule revealed rather large differences between theory and experiment[8]. These differences were qualitatively in agreement with the expected effect of hydrogen bonding. Similar calculations with a 4-31G basis set, augmented with bond polarization functions, on the doubly hydrated oxalic acid molecule suggested that hydrogen bonding had little influence on the electron density distribution in the oxalic acid molecule, being the hydrogen bond donor, but a more pronounced effect on the accepting water molecules[9]. This outcome is in disagreement with the result of our model study on the water dimer, where we noticed the strongest effect in the O-H bond of the donor. To resolve the problem, Krijn & Feil [10] [11] carried out extensive Hartree-Fock-Slater calculations in which both the effect of hydrogen bonding and crystalline environment were taken into account. It was shown that the agreement with experiment is excellent and that the quantum chemical model has to include not only the interaction between the directly involved monomers, but also the effect of the charge distribution of the molecules in the environment on the dimer to account for the observed distribution.

Details of the potentials, the computations and the results are reported elsewhere[10].

The polarized water molecule

The effect of the electric field on the electron density distribution of the water molecule is shown by displaying the polarization density, defi-

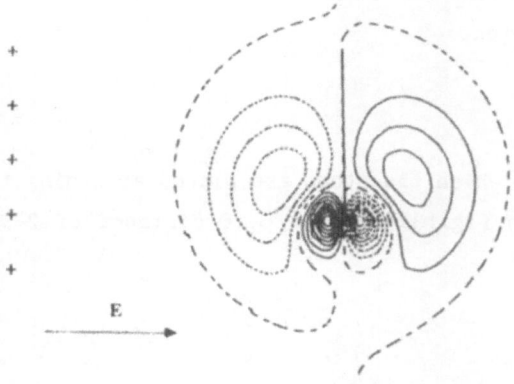

E

ned as the difference of the density distributions of the polarized and the unpolarized molecule. Fig. 11 shows the polarization density of H_2O in which the electric field is directed perpendicular to the plane of the molecule.

The contours show a distribution that is much more complex than one that corresponds with a mere shift of the electron density cloud in the direction of the field. The topography of the changes can be understood with elementary perturbation theory. Assuming that the outer orbitals contribute most to the polarization density, we consider the $1b_1$ orbital of the water molecule. It closely resembles the p_z-orbital of the O-atom. Under influence of the electric field some $4a_1$-orbital is mixed in:

$$\phi = 1b_1 + \lambda.(4a_1). \tag{1}$$

The contribution of ϕ to the polarization density is mainly given by the difference of ϕ^2 and $(1b_1)^2$. Assuming λ to be small, the polarization density is proportional to the product $(1b_1)*(4a_1)$. The two orbitals involved are shown in fig.12.

The outer electron density is shown to be so diffuse that contours are

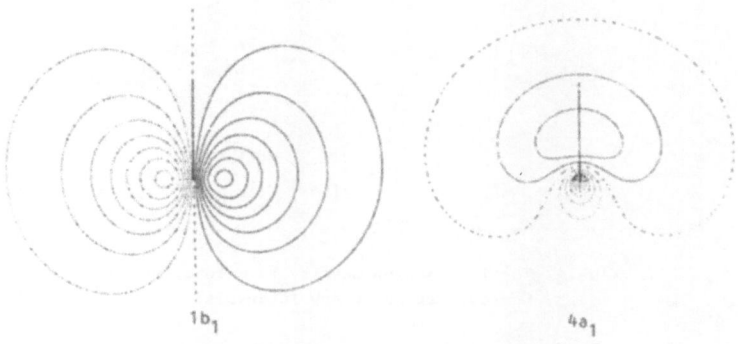

$1b_1$ $4a_1$

fig.12. The orbitals of the water molecule
that are mainly involved in polarization.
Contours at arbitrary intervals.

absent. Nevertheless it is this diffuse electron density distribution that accounts for the induced dipole moment.

The water dimer

We now turn to the water dimer, in which the atoms are placed according to the equilibrium geometry, shown in fig.13, with an O..O distance of 2.98 Å.

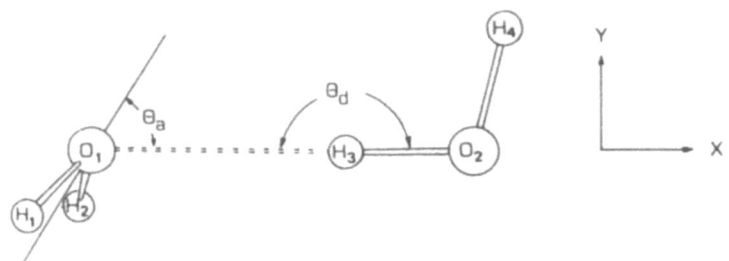

fig.13. Geometry of the linear water dimer.

The effect of bonding is reflected in the interaction density, defined as the difference between the density distributions of the dimer and the superimposed monomers. One of the problems in calculating differences is the fact that in calculating the dimer the system has a larger basis set available than in the case of the monomers. To prevent the resulting 'basis set superposition error' both monomers were calculated in the basis of the dimer[12]. Details of the geometry and the calculations are given by Krijn & Feil[5]. The interaction density is given in fig.14.

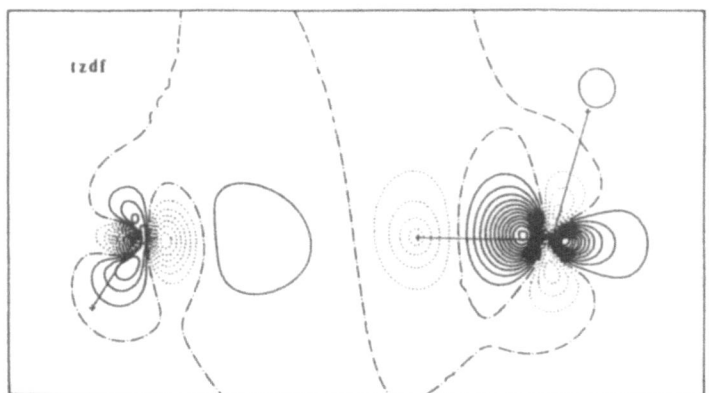

fig.14. The interaction density in a long H-bond.
Contours at arbitrary intervals.

The oxygen atom of the acceptor molecule shows the same pattern as the oxygen atom in the polarized water molecule, indicating the importance of polarization in the hydrogen bond. In fact when each water molecule is placed in the electric field of the unperturbed partner and the polarization densities were added, the result closely resembles the interaction density of fig.14.

In α-oxalic acid dihydrate, to be discussed shortly, the O..O distance is 2.48 Å, much shorter than the equilibrium distance in the water dimer. To show the effect of the reduced distance, we calculated the interaction density in a water dimer in which the angles of fig 11 are retained, but the O..O distance is reduced to 2.48 A. The result is shown in fig 15. We notice a considerable increase in the interaction density. The interpenetration of the charge clouds of the monomers makes exchange repulsion to play a large role.

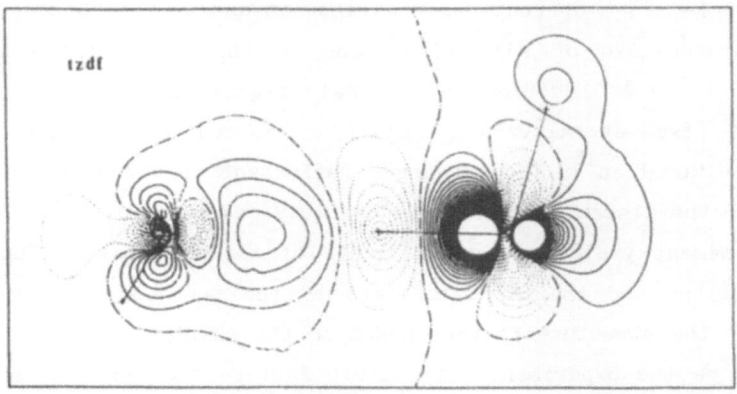

fig.15. The interaction density in a short H-bond.
Contours at arbitrary intervals.

Recent work by Bruning[13] allows the calculation of the effect of exchange repulsion. The product of the wavefunctions of the polarized molecules is antisymmetrized. The resulting wavefunction is a good approximation to the wavefunction of the dimer.

Experimental verification: Oxalic acid dihydrate

The crystal consists of layers of hydrogen bonded oxalic acid and water molecules. Fig.16 shows the oxalic acid molecule and its nearest neighbours.

It reveals the presence of both long and short hydrogen bonds. From the model studies on the water dimer we learned that calculation of the electron density in the short bonds requires supermolecule calculations, while the effect of the long bonds can be accounted for by including the charge distribution of the environment in the Hamiltonian.

To obtain the experimental density distribution a full-angle multipole

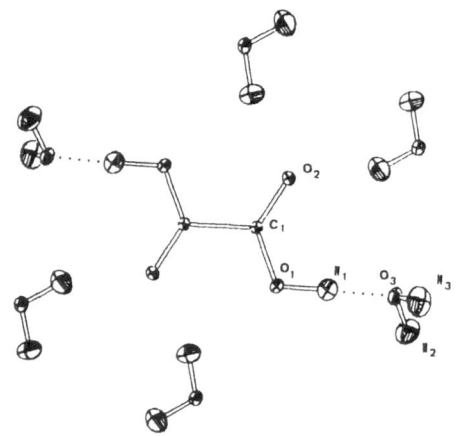

fig.16. Oxalic acid and the
surrounding water molecules in
oxalic acid dihydrate.

refinement was performed, using the cell-parameters and X-ray data of Dam, Harkema & Feil[14], corrected for extinction and anomalous dispersion. The data set consists of all reflections in the reciprocal sphere up to $(\sin\Theta)/\lambda = 1.3$ Å. Position and thermal parameters of the hydrogen atoms were kept fixed at the values taken from the neutron diffraction study of Koetzle & McMullan[15]. The converged refinement resulted in the following values of the discrepancy factors: R= 0.019 and R_w = 0.014.

The refinement yields the coefficients of the deformation functions and the positions and thermal parameters of the non-hydrogen atoms. In the following the structure factors based on the multipole density are considered to be the experimental structure factors and the Fourier summation based on the temperature corrected structure factors yields the experimental dynamic deformation density distribution shown in fig.17. This distribution is used as reference to test the various quantum chemical models.

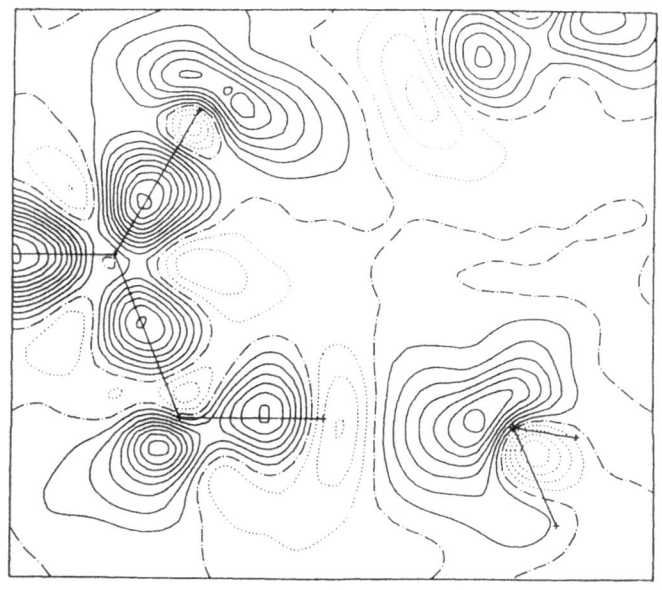

fig.17. The difference
between experimental and
promolecule density in
oxalic acid dihydrate.
Contours at intervals of
0.1 e/Å^3.

Since the number of observations greatly exceeds the number of parameters in the refinement model, the discrepancy between the resulting density and the real electron density is assumed to be much smaller than indicated by the R-factors that result from the refinement. The latter reflect the discrepancy between the individual measured structure factors and the ones based on the refined model. They are an indication of the accuracy of the individual measurements.

The first quantum chemical model of the electron density distribution in oxalic acid dihydrate consists of the summation of free atom electron density distributions. The difference between the experimental and the model results is the map of fig.17.

Since the omission of covalent bonding in the model is seen to cause the largest discrepancies, the second model consists of the superposition of independent molecules. The resulting difference with the experimental electron density distribution is shown in fig.18.

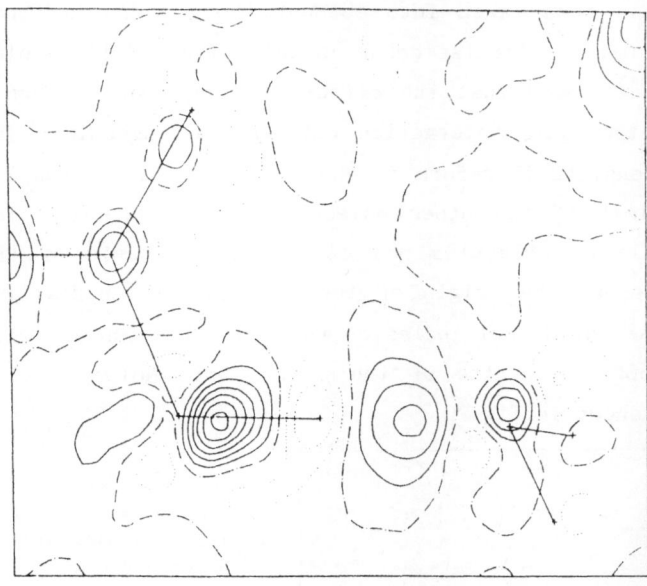

fig.18. The difference between experimental and molecular density in oxalic acid dihydrate. Contours at intervals of 0.05 e/$\overset{3}{\text{Å}}$.

The model does not take hydrogen bonding between the molecules into account. Assuming no errors in either the experimental map or in the quantum chemical method fig.18 shows the effect of hydrogen bonding. We recognize the same features as in the water dimer. The next improvement in the model consists of including the molecular interaction between the oxalic acid molecule and the nearest water molecules. A supermolecule calculation was carried out and the resulting density distribution was subtracted from the experimental distribution: fig.19.

The main features of the distribution in the hydrogen bond region have disappeared, confirming the assumption made above.

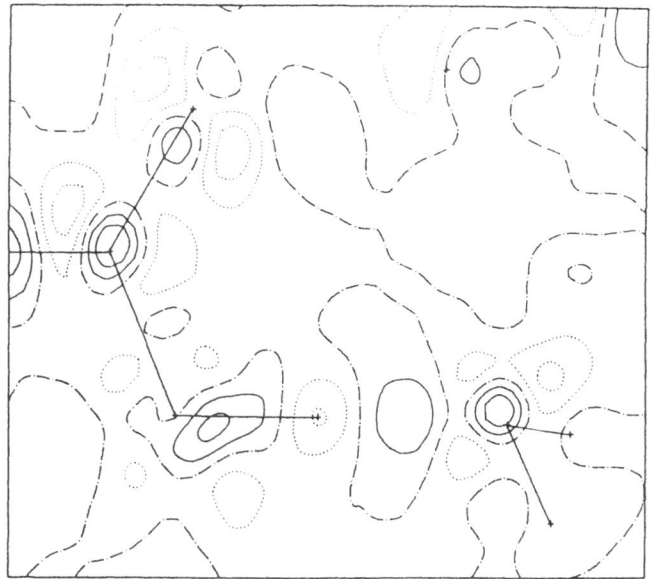

fig.19. The difference between experimental and trimer density in oxalic acid dihydrate. Contours at intervals of $0.05 \ e/\text{Å}^3$.

Although the shortest hydrogen bonds are included in the model, not all molecular interaction has been taken into account. Long hydrogen bonds between oxalic acid and water molecules occur as well. The model calculations on the water dimer showed that the effect of long hydrogen bonds consists mainly of electrostatic interaction between the molecules. The fourth and last model consists therefore of the oxalic dihydrate complex in the electrostatic field of the other molecules in the crystal. The field of the first shell of molecules was calculated on basis of the charge distribution, whereas the origin of the rest of the crystalline field was taken to be the set of multipoles on the atoms in a sphere with a radius of 18.5 A about the centre of the oxalic acid molecule. The resulting difference is shown in fig.20.

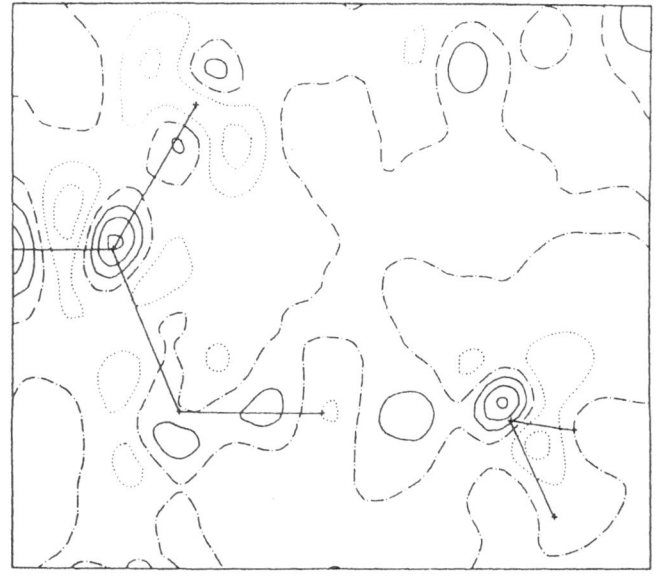

fig.20. The difference between the experimental density and a model in which covalent and hydrogen bonding and the effect of the crystal field have been taken into account. Contours at intervals of $0.1 \ e/\text{Å}^3$.

The differences between theory and experiment in the hydrogen bond region are of the order of 0.05 e/Å^3, the experimental error, expressing excellent agreement between theory and experiment. The considerable remaining differences occur at positions close to the nuclei where the experimental uncertainty is much larger due to the high electron density .The pattern of the distributions near the carboxylic oxygen and near the water oxygen atoms suggests that the thermal vibrations are not fully accounted for by the model: the vibrations may have some anharmonic character. Full details of the refinement are given by Krijn et al.[11].

Partitioning of the interaction density of the oxalic acid dihydrate complex with the Hirshfeld recipe results in the data reported in table 1.

Table 1. Partitioning of the interaction density in oxalic acid dihydrate. The charges, the dipole moments and the quadrupole moments are all given in atomic units.

Atom	δq	$\delta\mu_x$	$\delta\mu_y$	$\delta\mu_z$	$\delta\theta_{xx}$	$\delta\theta_{xy}$	$\delta\theta_{xz}$	$\delta\theta_{yy}$	$\delta\theta_{yz}$	$\delta\theta_{zz}$
C(1)	-18	13	10	-3	-19	-8	-5	-77	-1	-75
O(1)	-45	26	21	-1	-42	-19	-2	-59	-5	-54
O(2)	-35	-8	-36	-6	-50	-7	-2	-67	-6	-69
O(3)	14	65	24	-8	71	9	-11	107	-13	84
H(1)	10	0	7	-1	-7	1	0	21	-2	20
H(2)	36	22	-6	-8	70	11	1	64	-10	55
H(3)	38	21	3	4	68	10	-14	66	-7	53

The effect of the crystalline field on the atomic charges and higher momenta is shown in table 2.

Table 2. The effect of the crystalline field on the electron density distribution in the oxalic acid dihydrate trimer. All values in au.

Atom	δq	$\delta\mu_x$	$\delta\mu_y$	$\delta\mu_z$	$\delta\theta_{xx}$	$\delta\theta_{xy}$	$\delta\theta_{xz}$	$\delta\theta_{yy}$	$\delta\theta_{yz}$	$\delta\theta_{zz}$
C(1)	12	15	-2	12	-2	-21	17	-97	19	31
O(1)	13	51	5	2	-28	-37	0	-3	1	39
O(2)	-86	-23	-99	19	-153	16	21	-159	-4	-114
O(3)	-21	29	-68	16	8	-13	9	-59	19	-18
H(1)	12	-12	-8	2	-8	-10	3	3	6	19
H(2)	32	14	-17	4	30	0	3	5	2	20
H(3)	37	32	-23	26	54	-5	10	26	8	43

The present experience with charges and other momenta of the atomic charge distributions is very limited and the results do occasionally differ in an uncomfortable way. The values of these two tables should not be taken too literally.

The small amount of charge transfer contradicts the theory proposed by Reed[16].

Conclusions

The excellent agreement between the experimental, X-ray electron density distribution and the theoretical distribution calculated on basis of the density functional method, with the application of the local density approximation, shows the following:

-a careful measured electron density distribution of a hydrogen- bonded molecular crystal reflects the weak changes in the distribution due to exchange repulsion and polarization. Even the very weak effects of the crystalline field are noticeable.

-the Hartree-Fock-Slater method, alternatively known as the local density variant of the density functional method, yields reliable electron density distributions.

The electron density distribution in the hydrogen bond has the following characteristics:

-The effects of polarization and exchange repulsion balance to a large extent in the acceptor region, whereas they add up to a considerable effect in the donor region.

-the hydrogen bond reinforces the polarity of the molecules that participate in the bond. The acceptor becomes a better donor (for other bonds) and the donor becomes a better acceptor.

-The charge transfer between molecules is very small. In agreement with the fact that oxalic acid is a stronger acid than water, a small fraction of an electronic charge is transferred from water to (the carboxylic oxygen of) oxalic acid.

The effect of the crystalline field in oxalic acid dihydrate is to reinforce the effects of the strong hydrogen bond on the electron density distribution.

Acknowledgements

The contributions of J.Dam, who measured the X-ray data, of S. Harkema, who contributed to the X-ray analysis, of M.Krijn, who did most of the quantumchemical calculations, of H.Graafsma, who contributed to the comparison of theory and experiment, of H.Bruning, who carried out many calculations, and to R.de Vries, who calculated and plotted numerous figures and helped in putting the manuscript in final form, are gratefully acknowledged.

References

[1] F.L.Hirshfeld, *Cryst.Rev.* (1990) in press

[2] F.L.Hirshfeld, *Theor.Chim.Acta* **44**, 129 (1977).

[3] M.A.Spackman, H.P.Weber, and B.M.Craven, *J.Am.Chem.Soc.* **110**, 775 (1988).

[4] I.Olovsson *Electron and Magnetization Densities in Molecules and Crystals,* ed.P.Becker, New York: Plenum (1980) pp 831-894.
E.Morokuma & K.Kitaura *Molecular Interactions vol 1*, ed.H.Ratajczak & W.J.Orville-Thomas, New York: Wiley (1980), chapter 2.

[5] M.P.C.M.Krijn and D.Feil, *J.Chem.Phys.* **89** 5787 (1988)

[6] M.P.C.M.Krijn and D.Feil *J.Chem.Phys.* **85** 319 (1986)

[7] P.Coppens, J.Dam, S.Harkema, D.Feil, R.Feld, M.S.Lehman, R.Goddard, C.Krüger, E.Hellner, H.Johansen, F.K.Larsen, T.F.Koetzle, R.K.McMullan, R.K.Maslen & E.D.Stevens *Acta Cryst.* **A40** 184 (1984)

[8] E.D.Stevens *Acta Cryst* **B36** 1876 (1980)

[9] M.Breitenstein, H.Dannöhl, H.Meyer, A.Schweig, U.Seeger & W.Zittlau, *Int.Rev.Phys.Chem.* **3** 335 (1983)

[10] M.P.C.M.Krijn and D.Feil *J.Chem.Phys.* **89** 4199 (1988).

[11] M.P.C.M.Krijn, H.Graafsma & D.Feil *Acta Cryst.* **B44** 609 (1988)

[12] S.F.Boys and F.Bernardi, *Mol.Phys.* **19**, 553 (1970)
see also
M.Gutowski, J.H.van Lenthe, J.Verbeek, F.B.van Duijneveldt and G.Chałasinski *Chem.Phys.Lett.* **124**, 370 (1986).

[13] Bruning and Feil, to be published.

[14] J.Dam, S.Harkema & D.Feil *Acta Cryst* B39 760 (1983).

[15] T.F.Koetzle & R.K.McMullan (1980), reported by Coppens *et al.* (6) as set 4.

[16] A.E.Reed, F.Weinhold, L.A.Curtiss, and D.J.Pochatko *J.Chem Phys.* **84** 5687 (1986)

ELECTRON DENSITY MODELS : DESCRIPTION AND COMPARISON

Claude Lecomte

Laboratoire de Minéralogie-Cristallographie
U.R.A. C.N.R.S. 809
Université de Nancy I, Faculté des Sciences
B.P. 239, 54506 VANDOEUVRE CEDEX, FRANCE

I - INTRODUCTION

The following chapter deals with electron density models which may be applied to organic or organo-metallic compounds. In the usual least squares refinements, the continuous electron density of the crystal (or the molecule) is divided into atomic charge densities where the atoms are neutral and of spherical shape

$$\rho^{pro}(r) = \sum_{j=1}^{Na} \rho_j^{at}(r)$$

where $\rho_j^{at}(r)$ is the electron density of atom j considered as a free atom in vacuum and $\rho^{pro}(r)$ is called the promolecule density summed over all atoms Na.

When we have an accurate, absorption corrected low temperature data set (see Blessing and Lecomte in this book) we can map the experimental deformation density $\delta\rho^{exp}(r)$ (1) for centrosymmetric crystals.

$$(1) \qquad \delta\rho^{exp}(r) = \rho^{obs}(r) - \rho^{pro}(r)$$

This reveals the asphericity of the valence electron density due to chemical bonding. These $\delta\rho^{exp}(r)$ maps can be computed after a high order (HO) X, Y, Z, U^{ij} refinement which relies on the assumption that the high order data ($\sin\theta/\lambda > 0.8$ Å^{-1} or 0.9 Å^{-1}) are mainly core-scattering and therefore insensitive to chemical bonding (frozen-core approximation).

The Application of Charge Density Research to Chemistry and Drug Design
Edited by G. A. Jeffrey and J. F. Piniella, Plenum Press, New York, 1991

121

These maps show the main features of electron density asphericity and reveal the quality of the data. Accurate data, processed with care should give maps with a very low noise and almost equal density peaks on equivalent chemical bonds.

However these maps do not readily lead to numbers describing charges, expansion-contraction coefficients of the atomic density, d orbital populations and electrostatic properties.

Alternative and much more elegant methods are those using aspherical or multipole refinement.

In these refinements the continuous molecular density is also subdivided into atomic densities but the atomic density is written

(2) $$\rho^{at}(r) = \rho_1^{at}(r) + \delta\rho^{at}(r)$$

where $\rho_1^{at}(r)$ and $\delta\rho^{at}(r)$ are either the core density and the perturbed refinable non spherical valence density or the free atom total density and the deviations from this density.

In each case $\delta\rho^{at}(r)$ (3) is described as a static density given by the product of radial functions $R_n(r)$ with a set of orientation dependant functions (angular functions $A_n(\theta, \phi)$) defined on local axis centered on the atoms (Fig. 1)

(3) $$\delta\rho^{at}(r) = \sum_n C_n R_n(r) A_n(\theta, \phi)$$

Usually the population parameters C_n and an expansion-contraction κ coefficient are the parameters that are refined together with the xyz, U^{ij} coefficients.

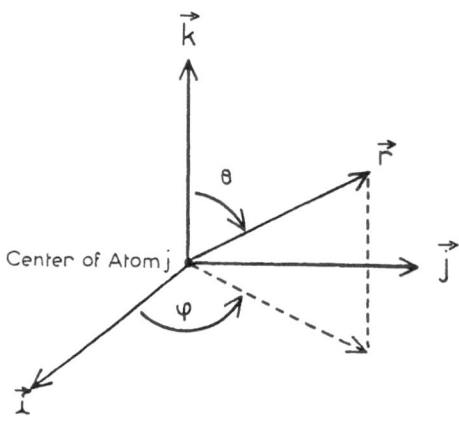

Fig. 1. Local axis and variable definitions.

These multipole models must be sufficiently flexible to describe the asphericity of the density without including so many parameters to prevent a stable least squares refinement. The ratio of number of parameters (Np) to the number of observations (NO) ratio (maximum $\sin\theta/\lambda$ at least equal to 1.1 Å^{-1}) should be greater than 7, at least

The multipole model has three advantages :

First it allows us to calculate structure factors phases as closely as possible to the true phases in the case of non-centrosymmetric crystals, in order to map the deformation density, $\delta\rho^{exp}(r)$, and some electrostatic properties (electrostatic potential, laplacian of rho, .gradient of rho, multipole moments).

$$\delta\rho^{exp}(r) = \frac{1}{V} \sum_{\text{all measured } H \text{ up to } \sin\theta/\lambda \text{ max}} \left[K^{-1}|F_o| e^{i\phi_{Mul}} - |F_{Sph}| e^{i\phi_{Sph}} \right] e^{-2\pi i \, H.r}$$

Secondly it allows the calculation of both dynamic $\delta\rho^{dyn}(r)$ (4) and static $\delta\rho^{stat}(r)$ (5) model deformation maps which are defined as follows :

(4)
$$\delta\rho^{dyn}(r) = \frac{1}{V} \sum_{\text{all } H \text{ up to } \sin\theta/\lambda \text{ max}} \left[|F_{mul}| e^{i\phi_{Mul}} - |F_{Sph}| e^{i\phi_{Sph}} \right] e^{-2\pi i \, H.r}$$

where $|F_{mul}|$, ϕ_{mul} are structure factor amplitudes and phases estimated from the multipole model and $|F_{sph}|$, ϕ_{sph} are calculated from the spherical atoms. In both terms the Debye Waller factors are included.

(5)
$$\delta\rho^{stat}(r) = \sum_{j=1}^{N_a} \delta\rho^{at}(r)$$

where N_a is the number of atoms of the molecule and $\delta\rho^{at}(r)$ is defined above (formula 3).

These latter maps are free of experimental noise but may be slightly model dependant and it is fundamental to test how the model fits the experimental density. A residual density map (6) must be calculated at each stage of the refinement to check that all significant density features of the experimental data are included in the model.

(6)
$$\delta\rho^{res}(r) = \frac{1}{V} \sum_{\text{all measured } H} \left[K^{-1}|F_o(H)| - |F_{Mul}(H)| \right] e^{i\phi_{Mul}} e^{-2\pi i \, H.r}$$

where F_o and K are respectively the observed structure factor and the scale factor.

As an example figure 2 shows the experimental deformation map (2a) and the residual map (2b) computed on a pyrrole ring of difluoro germanium octaethylporphyrin, OEPGeF$_2$[1], after a Hansen-Coppens[2] refinement. The residual map is without any

123

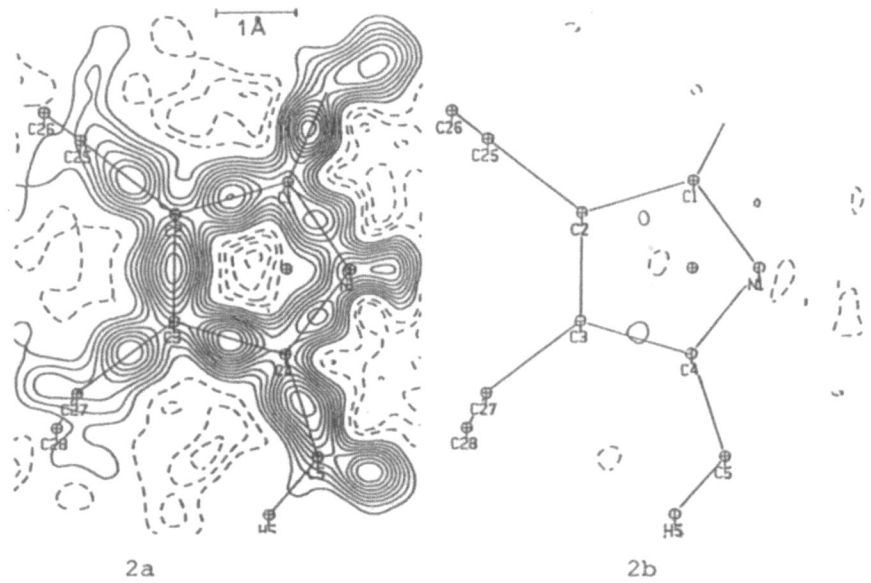

Fig. 2. Experimental deformation map (2a) of a pyrrole ring of OEPGeF$_2$[1] and residual map (2b). Contours interval 0.05 e Å^{-3} positive contours solid, negative dashed, zero contour omitted.

significant features showing that all the deformation density has been modelled.

- The third advantage is to provide parameters that can be used to calculate electrostatic properties (dipole moments, electrostatic potential, $\nabla^2(\rho)$, $\nabla\rho$...).

To illustrate this, the following is divided in three parts

I - Description of three multipole models : two of them are based on spherical harmonic functions, as proposed first by Stewart[3] (1969), but they use different radial functions and different local axis definitions : Molly[2] (Hansen-Coppens 1978), Pop[4] (Craven Weber 1980). The third is based on a set of cosine functions LSEXP[5] (Hirshfeld 1971, 1977).

II - The effect of phases derived from multipolar refinement on the density maps.
This effect is discussed for non-centrosymmetrical crystals with different space groups[6].

III - A comparison between the three models is made in the case of a centrosymmetric and of a non-centrosymmetric space group. It is shown that the density maps are slightly model dependant.

II - THE MULTIPOLE MODELS

A) MODELS BASED ON SPHERICAL HARMONICS EXPANSION AS PROPOSED BY STEWART[3]

1) Real spherical harmonic functions y_1^m

Mathematical definitions and properties of real spherical harmonics can be found in references [7] and [8].

Spherical harmonics in real form are a set of orthogonal functions i.e :

$$(7) \qquad \int_{\phi=0}^{2\pi} \int_{\theta=0}^{\pi} y_1^m (\theta,\phi) \, y_{1'}^{m'}(\theta,\phi) \, d\Omega = 0$$

if $m \neq m'$ and $1 \neq 1'$

where $d\Omega$ is the element of solid angle $d\Omega = \sin\theta \; d\theta \; d\phi$.

These functions can be used as angular basis functions to describe the asphericity of $\rho(r)$.

They can be subdivided in even (+) and odd (−) functions :

$$(8) \qquad \begin{cases} y_1^{m^+} = N_1^m \, P_1^m (\cos \theta) \cos m \, \phi \\[2mm] y_1^{m^-} = N_1^m \, P_1^m (\cos \theta) \sin m \, \phi \end{cases}$$

where P_1^m is the associated Legendre polynomial :

$$(9) \qquad P_1^m(z) = \frac{(1 - z^2)^{m/2}}{2^1 \; 1!} \; \frac{d^{m+1} (z^2 - 1)^1}{dz^{m+1}}$$

and N_1^m a normalization factor.

The normalization used in the multipolar electron density models is that of Hansen and Coppens[2] :

$$(10) \qquad \int |y_1^m| \; d\Omega = \varepsilon_1$$

where $\varepsilon_1 = 2 \quad 1 \neq 0$
$\varepsilon_1 = 1 \quad 1 = 0$

This normalization is different from that of the wave function : $\psi(r) = R'_{nl}(r)\, Y_1^m(\theta,\phi)$

(11) $$\int \psi^2(r)\, d\tau = 1 \;\Rightarrow\; \int_0^\infty R'^2_{nl}\, r^2\, dr \int_0^{4\pi} |Y_1^m|^2\, d\Omega = 1$$

where $R'_{nl}(r)$ is the radial function of the wave function ψ and Y_1^m are the complex spherical harmonics functions.

$$Y_1^m = N'\, P_1^m(\theta)\, e^{im\phi}$$

If $R'_{nl}(r) = r^{n'}\, e^{-\xi' r}$ (Slater type function) then the radial function of the density should be proportionnel to $\left(R'_{nl}(r)\right)^2$

$$\Rightarrow\quad R_{nl}(r) = r^n\, e^{-\xi r}$$

with $n = 2n'$ and $\xi = 2\xi'$.

Examples of calculation of these functions in spherical and cartesian coordinates are given in Ghermani, Habbou, Boumaïda and Lecomte. Table 1 gives the cartesian and spherical coordinates of the y_1^m functions ($1 \le 4$, $-1 \le m < 1$) without their normalization coefficients.

Table 1. Spherical Harmonic functions in real form.

Harmonic		Associated Polynomial*	
Order	Y_1^m	Cartesian coordinates	Spherical coordinates
0	Y_0^0	1	1
1	Y_1^0	z	$\cos\theta$
	Y_1^1	x	$\sin\theta\,\cos\phi$
	Y_1^{-1}	y	$\sin\theta\,\sin\phi$

Table 1.

2	Y_2^0	$(3z^2-1)$	$3\cos^2\theta-1$
	Y_2^1	$x\ z$	$\sin\theta\ \cos\theta\ \cos\phi$
	Y_2^{-1}	$y\ z$	$\sin\theta\ \cos\theta\ \sin\phi$
	Y_2^2	x^2-y^2	$\sin^2\theta\ \cos2\phi$
	Y_2^{-2}	$2\ x\ y$	$\sin^2\theta\ \sin2\phi$
3	Y_3^0	$(5z^2-3)z$	$\cos\theta(5\cos^2\theta-3)$
	Y_3^1	$(5z^2-1)x$	$\sin\theta(5\cos^2\theta-1)\ \cos\phi$
	Y_3^{-1}	$(5z^2-1)y$	$\sin\theta(5\cos^2\theta-1)\ \sin\phi$
	Y_3^2	$(x^2-y^2)z$	$\sin^2\theta\ \cos\theta\ \cos2\phi$
	Y_3^{-2}	$x\ y\ z$	$\frac{1}{2}\ \sin^2\theta\ \cos\theta\ \sin2\phi$
	Y_3^3	$(x^2-3y^2)x$	$\sin^3\theta\ \cos3\phi$
	Y_3^{-3}	$(3x^2-y^2)y$	$\sin^3\theta\ \sin3\phi$
4	Y_4^0	$7z^4-6z^2+3/5$	$7\cos^4\theta - 6\cos^2\theta + 3/5$
	Y_4^1	$(7z^2-3)\ xz$	$\frac{1}{2}(7\cos^2\theta-3)\sin\theta\ \cos\phi\ \cos\theta$
	Y_4^{-1}	$(7z^2-3)\ yz$	$\frac{1}{2}(7\cos^2\theta-3)\sin\theta\ \cos\theta\ \sin\phi$
	Y_4^2	$(7z^2-1)\ (x^2-y^2)$	$\frac{1}{2}(7\cos^2\theta-1)\ \sin^2\theta\ \cos2\phi$
	Y_4^{-2}	$(7z^2-1)\ xy$	$\frac{1}{2}(7\cos^2\theta-1)\ \sin^2\theta\ \sin2\phi$
	Y_4^3	$(x^2-3y^2)\ xz$	$\sin^3\theta\ \cos\theta\ \cos3\phi$
	Y_4^{-3}	$(3x^2-y^2)\ yz$	$\sin^3\theta\ \cos\theta\ \sin3\phi$
	Y_4^4	$x4 - 6x^2y^2 + y^4$	$\sin^4\theta\ \cos4\phi$
	Y_4^{-4}	$xy\ (x^2-y^2)$	$\sin^4\theta\ \sin4\phi$

* The expressions given are unnormalized functions ; see the original papers ([2], [4]) for normalization.

Figure 3 shows some of the multipole functions in polar coordinates (from N.K. Hansen).

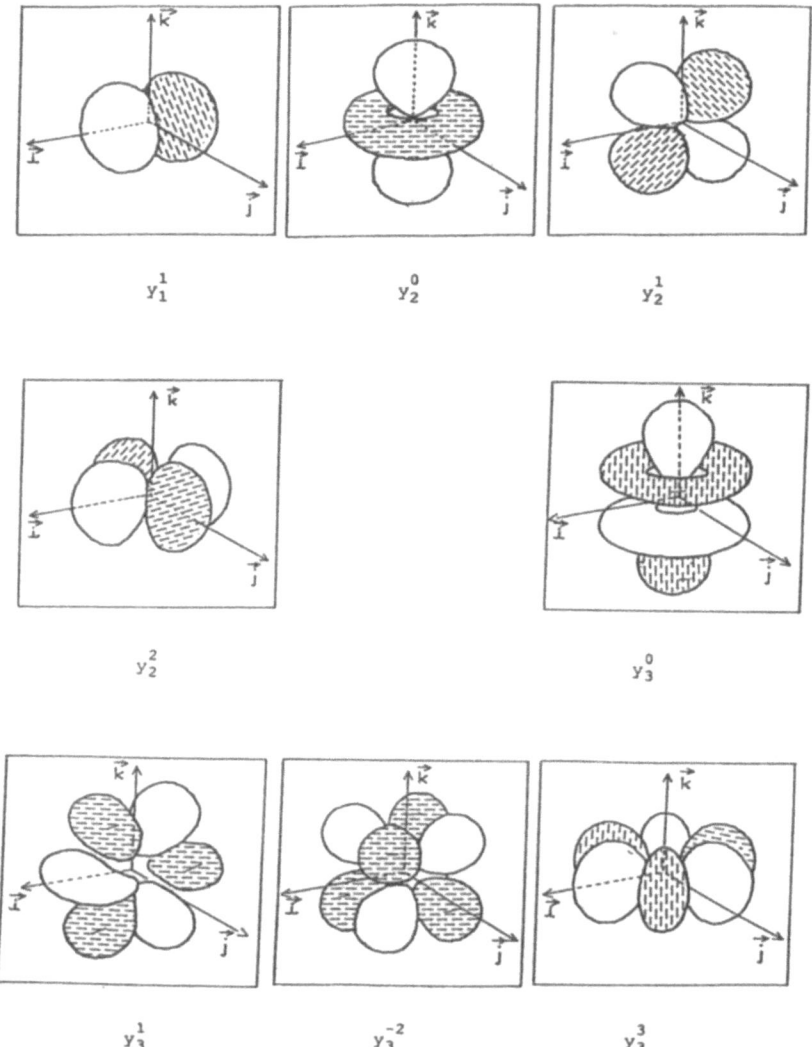

Fig. 3. Some spherical harmonic functions.

2) The structure factor calculation

In the local axis frame (figure 1), one term of the deformation density is written

(12) $\delta\rho_{nlm}(r) = R_{nl}(r)\, y_l^m(\theta,\phi)$

where $R_{nl}(r)$ is a Slater type function

(13) $R_{nl}(r) = N_{nl}\, r^{n_l}\, e^{-\xi r}$

($n \geq 1$ to give rise to a Coulombic potential).

with $\displaystyle\int_0^\infty R_{nl}(r)\, d\tau = 1$ * $\Rightarrow\quad N_{nl}\displaystyle\int_0^\infty r^{nl}\, e^{-\xi r}\, 4\pi r^2\, dr = 1$

$$\Rightarrow\quad N_{nl} = \frac{1}{4\pi}\, \frac{\xi^{n_l+3}}{(n_l+2)!}$$

The form factor expression for one term $f_{nlm}(s)$ is :

(14) $f_{nlm}(s) = \displaystyle\iiint \delta\rho_{nlm}(r)\, \exp(i\, s.r)\, d\tau$

where $d\tau = r^2 \sin\theta\, dr\, d\theta\, d\phi$ and $s = 2\pi H$.

To obtain the form factor expression we use the following expansion of $e^{i\, s.r}$ [9]

(15) $e^{i s.r} = \displaystyle\sum_{\lambda=0}^\infty \sum_{\mu=0}^\lambda i^\lambda \varepsilon_\mu (2\lambda+1) \frac{(\lambda-\mu)!}{(\lambda+\mu)!}\, P_\lambda^\mu(\cos\theta)\, P_\lambda^\mu(\cos u)\, j_\lambda(sr)\, \cos(\mu(\phi-v))$

$\varepsilon_\mu = 2\quad \mu \neq 0$

$\varepsilon_\mu = 1\quad \mu = 0$

θ, ϕ, u, and v are respectively the spherical coordinates of r and s defined on figure 4.
$j_\lambda(sr)$ is the spherical Bessel function of order λ.

---

* In the Molly program the normalization factor is

(15) $N_{n1} = \dfrac{\xi^{n_1+3}}{(n_1+2)!}$ i.e $\displaystyle\int_0^\infty R_{n1}(r)\, r^2\, dr = 1$

129

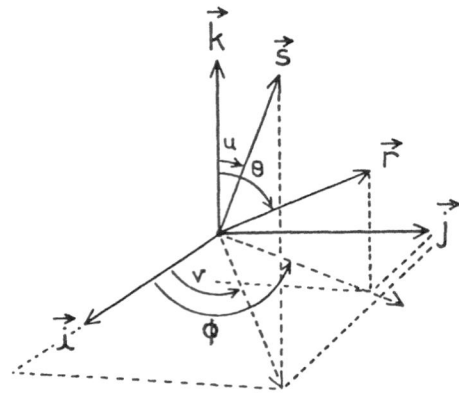

Fig. 4. Spherical coordinates of **r** and **s**.

The form factor expression becomes (in the case of even harmonic functions and omitting the normalization factor) :

$$f_{nlm}(s) = \iiint R_{nl}(r)\, P_l^m(\cos\theta)\, \cos m\phi \sum_\lambda \sum_\mu i^\lambda \varepsilon_\mu (2\lambda+1) \frac{(\lambda-\mu)!}{(\lambda+\mu)!} P_\lambda^\mu(\cos\theta)\, P_\lambda^\mu(\cos u)\, j_\lambda(sr) \times$$

$$\cos\mu(\phi-v)\, r^2 \sin\theta\, d\theta\, d\phi\, dr$$

Due to orthogonality of the associated Legendre polynomials only the terms with $\lambda = 1$ and $\mu = m$ do not vanish

$$\Rightarrow\ f_{nlm}(s) = K \int_0^\infty r^2 R_{nl}(r)\, j_l(sr)dr \int_0^\pi [P_l^m(\cos\theta)]^2 \sin\theta\, d\theta \int_0^{2\pi} \cos(m(\phi-v))\cos m\phi\, d\phi$$

$$= K \times \Phi_{nl}(s) \times I_2 \times I_3\, (mv)$$

where $\quad K = \varepsilon_m(2l+1) \dfrac{(l-m)!}{(l+m)!}\, i^l P_l^m(\cos u)$

The expression of these three integrals (Φ, I_2 and I_3) are given in appendix 1 and finally the form factor has the following form :

$$(19) \qquad f_{nlm}(s) = i^l P_l^m(\cos u) \cos(mv)\, \Phi_{nl}(s) = i^l \Phi_{nl}(s)\, y_l^m(u,v)$$

where $\Phi_{nl}(S)$ is the Fourier Bessel transform of the radial function and can be calculated according to Stewart[10]*.

Remarks a) the radial scattering factors $\Phi_l(S)$ are small finite polynominals as soon as $n > 1$ [10].

b) the form factor can be real or imaginary depending of the parity of l and we can expect correlation between dipole, octopole terms (l = 3) and, the positional parameters

* Analytical expressions of the radial integral were published very recently by SU and COPPENS (J. Appl. Cryst., 23, 71-73 (1990).

and between quadrupole (l = 2), hexadecapole terms (l = 4) and the thermal motion parameters.

c) As soon as the scattering vector $\vec{s}$ is expressed in the local axis system the calculation of the form factor is straightforward.

The structure factor expression becomes :

$$(20) \quad F(H) = \sum_{j=1}^{Na} \left[f_j^{Sph} + \sum_{l=0}^{l_{max}} \sum_{m=-l}^{l} i^l P_{jlm} \, \Phi_{jl}(2\pi H) \, y_l^m(u,v) \right] \times T(H) \times \exp{(2\pi i \, H.R_j)}$$

where f^{sph} is the free atom (or the core) scattering factor, Na is the number of atoms of the unit cell, P_{jlm} are population refinable parameters of atom j ; P_{00} is the monopole population giving an estimate of the charge of the atom and P_{1m}, P_{2m}, P_{3m}, P_{4m} are the dipole, quadrupole, octopole and hexadecopolar populations respectively. With the normalization used (formula 10) a population coefficient of +1 means that one electron has been removed from the negative lobe to the positive one of the y_l^m function (l ≠ 0).

T(H) is the thermal parameter including harmonic U^{ij} and anharmonic C^{ijk} and D^{ijkl} terms if necessary*.

R_j is the vector from the unit cell origin to atom j.

3) The Pop procedure[4] (Craven and Weber 1980)

Craven et Weber have implemented these calculations in a least squares program Called Pop: in this program the local axis system have the same orientation for each atom (Figure 5).

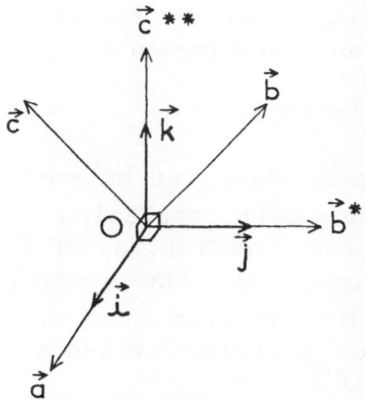

Fig. 5. Local axis system. O : center of atom j.

---

* C^{ijk} and D^{ijkl} are Gram Charlier expansion coefficients and are strongly correlated with population parameters as shown by Coppens et al.[11]. They may only be refined on high order data or can be provided by an independant neutron diffraction experiment.

The $\vec{i}$, $\vec{j}$, $\vec{k}$ vectors are unit vectors along the axis $\vec{a}$, $\vec{b}*$ and $\vec{c}** = \vec{a} \wedge \vec{b}*$

In this program the density is expressed as in relation (2) where ρ_1 ($\mathbf{r}$) can be chosen either as core or as free atom density. The radial function is defined as in (13) and the ξ parameter can be refined. No electroneutrality constraint is included in the least squares process. This program has the advantage that it gives all the population parameters in the same local axis system which can be used to compute the electrostatic properties.

In this refinement each atom may be described by a set of 35 parameters if the expansion is made up to the hexadecapolar level :

9 parameters	x, y, z	U^{ij}
1 parameter	$l = 0$	P_{00}
3 parameters	$l = 1$	dipoles
5 parameters	$l = 2$	quadrupoles
7 parameters	$l = 3$	octopoles
9 parameters	$l = 4$	hexadecapoles
1 parameter	ξ	radial exponent

For organic molecules the expansion can be truncated at the octopolar level whereas transition metal require an expansion up to $l = 4$. Then this type of refinement ($l = 4$) requires about 250 – 300 reflections per atom. The n and ξ parameters can be initially chosen according to Hehre, Stewart and Pople[12].

4) The Molly procedure[2] (Hansen and Coppens 1978)

a) The Kappa refinement[13] (Coppens, Guru Row, Leung, Stevens, Becker, and Wang 1979)

The Kappa formalism[12] allows for the expansion contraction of the perturbed valence density ($\rho'^{val}(r)$)

$$(21) \qquad \rho'^{val}(r) = P^{val} K^3 \rho^{val} (\kappa r)$$

where ρ^{val} is the unperturbed free valence ground state density and P^{val} the valence shell population ; the κ parameter is related to the expansion, contraction of the valence shell : if κ is smaller than one, then the density at the distance r equals the density of the free atom at the distance $\kappa r < r$ which means that the perturbed valence density is expanded relative to the free atom.

As shown by Coppens et al.[13], the structure factor becomes

$$(22) \qquad F(\mathbf{H}) = \sum_{j=1}^{Na} [f_j^{core}(\mathbf{H}) + P_j^{val} f_j^{val}(\mathbf{H}/\kappa)] \, T(\mathbf{H}) \exp (2\pi i \, \mathbf{H}.\mathbf{R}_j)$$

This formalism has been implemented in a program called RADIEL which contains an electro-neutrality constraint and certainly give a better estimate of net atomic charges than multipole models.

b) *The Molly procedure*[2]

This formalism is different from the Stewart[3]* or from the Pop procedures[4] : Each atomic density is described as :

$$(23) \qquad \rho^{at}(r) = \rho^{core}(r) + P^{val} \kappa^3 \rho^{val}(\kappa'r) + \sum_{l=0}^{4} \sum_{m=-l}^{l} \kappa''^3 R_{nl}(\kappa''r) P_l^m y_l^m (\theta, \phi)$$

Kappa refinement   +   Multipolar refinement

As written above the radial part $R_{nl}(r)$ is given by :

$$(24) \qquad R_{nl}(r) = \frac{\xi_l^{nl+3} r^{nl}}{(nl+2)!} \exp(-\xi_l r)$$

where the n and ξ coefficients can be chosen according to Hansen and Coppens or in such a way that the maximum of the radial function ($r_{max} = n_l/\xi_l$) must be at the density peak position.

P^{val} and P_{00} are two monopoles populations and can be usefull for separating (4s) (4p) contribution from the (3d) electrons in transition metal compounds. Furthermore the two Kappa parameters allow different expansions.

The local axis on each atom are defined by the operator. this flexibility has an advantage for big symmetrical molecules or/and for molecules containing chemically equivalent atoms : let us consider for example a iron porphin molecule $FeC_{20}N_4H_{12}$ (FeP) (figure 6).

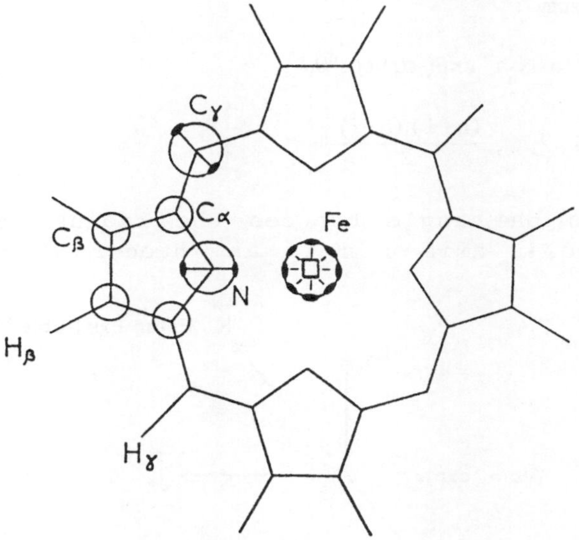

Fig. 6. FeP molecule.

---

* The Stewart Model is incorporated in a general program system called VALRAY, which includes multipole refinement as well as electrostatic properties calculations.

FeP contains 37 atoms, of which 25 non hydrogen atoms. If symmetry and chemical constraints are not in the refinement program, the number of parameters would be : 37 for iron (l_{max} = 4), 27 x 24 = 648 (l_{max} = 3, P_{00} = 0) for the C et N atoms and 12 x 9 for the hydrogen atoms (x, y, z, $<u^2>$, p^{val}, dipoles and Kappa) i.e 794 parameters including the scale factor (without extinction refinement) of which 524 are density parameters. In fact, they are seven chemical types of atoms : Fe, $C\alpha$, $C\beta$, $C\gamma$, $H\beta$, $H\gamma$. Furthermore we can assume a non crystallographic 4/m mm symmetry for the iron atom, a mm2 symmetry for the N and $C\gamma$ atoms and a mirror symmetry for $C\alpha$, $C\beta$ (Figure 5). If we choose carefully the local axis (see Ghermani et al. in this book) the number of refinable density parameters decreases from 524 to less than 100.

These constraints are equivalent to making a symmetry average deformation map and may be used if the data allow this average calculation*.

B) THE HIRSHFELD DEFORMATION FUNCTIONS[5] (HIRSHFELD 1971, 1977)

In the LSEXP program[5], the electron density is described by

$$(25) \qquad \rho_j(r) = \rho_j^{at}(r) + \sum_l C_{jl}\, \rho_{jl}(r)$$

where $\rho_j^{at}(r)$ is the free atom electron density and $\rho_{jl}(r)$ the atomic deformation functions. The atomic deformation, up to 35 non-orthogonal functions on each atomic center, is of the following form :

$$(26) \qquad \rho_{jl}(r) = N_n\, r^n\, \exp(-\alpha r)\, \cos^n \theta_K$$

$$K = 1, 2, ...,\ \frac{(n+1)\,(n+2)}{2}$$

where θ_K is the angle between the radius vector r and a specified polar axis ; n is an integer ($0 \leq n \leq 4$). The

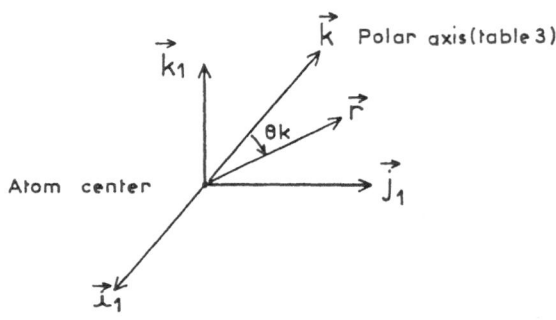

---

* For an application see references [14].

Table 2. Explicit expressions of the deformation functions

n	ρ_{j1}
0	$\dfrac{\alpha^3}{8\pi}\, e^{-\alpha r}$
1	$\dfrac{\alpha^4}{12\pi}\, r\, e^{-\alpha r}\cos\theta_K$
2	$\dfrac{\alpha^5}{32\pi}\, r^2\, e^{-\alpha r}\cos^2\theta_K$
3	$\dfrac{\alpha^6}{120\pi}\, r^3\, e^{-\alpha r}\cos^3\theta_K$

(27) $N_n = (n+1)\,\alpha^{n+3}/4\pi\,(n+2)!$ (Table 2).

normalization N_n (27) is chosen to make the volume integral of each even function equal to one electron. The functions with n odd have vanishing integrals

Then summing over all the even order coefficient gives the net increase of atomic charge arising from the deformation.

For each value of the n exponent there are $\dfrac{(n+1)\,(n+2)}{2}$ basis functions of the same shape on each atom and θ_K is measured from one of the $\dfrac{(n+1)\,(n+2)}{2}$ polar axis. These polar axis are chosen by defining a local coordinate $(\vec{i}_1,\ \vec{j}_1,\ \vec{k}_1)$ system on each atom according to the local symmetry. The polar axis have directions related to cubic symmetry $[1,0,0]$, $[1,1,0]$, $[1,1,1]$, $[\sqrt{2}-1,1,1]$ and are given in table 3.

Table 3. Directions of the local polar axis

n	Number	Direction
0	1	Spherically Symmetric
1	3	$[100]\,[010]\,[001]$
2	6	$[110]\,[1\bar{1}0]\,[101]\,[10\bar{1}]\,[011]\,[01\bar{1}]$
3	10	$[110]\,[1\bar{1}0]\,[101]\,[10\bar{1}]\,[011]\,[01\bar{1}]\,[111]\,[1\bar{1}\bar{1}]\,[\bar{1}1\bar{1}]\,[\bar{1}\bar{1}1]$
4	15	$[100]\,[010]\,[001]\,[\alpha11]\,[1\alpha1]\,[11\alpha]\,[\bar{\alpha}11]\,[1\bar{\alpha}1]\,[11\bar{\alpha}]\,[\alpha\bar{1}1]$ $[1\alpha\bar{1}]\,[\bar{1}1\alpha]\,[\alpha1\bar{1}]\,[\bar{1}\alpha1]\,[1\bar{1}\alpha]$ $(\alpha = \sqrt{2}-1)$

135

These 35 functions can be expressed as a combination of spherical harmonics, the ϕ dependance being implicitly contained in the choice of the polar axis (three monopole terms-for n = 0, 2, 4, two dipoles terms (n = 1, 3), two quadrupoles terms-for n = 2, 4, one octopole (n = 3) and one hexadecapole term (n = 4). Then this expansion is a more flexible expansion than the expansion described in part II.A because of the different radial functions.

III - EFFECT OF PHASES DERIVED FROM MULTIPOLAR REFINEMENT ON THE DENSITY MAPS [6]

The crystallographic model deformation density maps are calculated according to (28) :

$$(28) \qquad \Delta\rho(r) = V^{-1} \sum_{H} [|F_m(H)| \exp(i\,\phi_m(H)) - |F_s(H)| \exp(i\,\phi_s(H))] \exp(-2\pi\,i\,H.r)$$

where the subscripts m designate the atom centered multipolar density model, and s refers to the spherically averaged free atom superposition model.

Equation (28) shows that, with a non-centrosymmetric structure, reliable deformation density maps require two sets of amplitudes and two sets of phases.

A conventional $\Delta\,|F|$ synthesis

$$\Delta\rho = V^{-1} \sum_{H} (K^{-1}\,|F_o| - |F_c|)\,\exp i\,\phi_c\,\exp(-2\pi i\,H.r)$$

cannot be properly phased by a spherical atom model alone even if almost unbiased structural parameters are derived from a high order X-Ray or from a neutron refinement. The effect of phase difference $\Delta\phi = \phi_m - \phi_s$ can be calculated in the following way :

Let $\Delta|F| = |F_m| - |F_s|$ then equation (28) becomes

$$(29) \qquad \Delta\rho(r) = V^{-1} \sum [\,(|F_s| + \Delta|F|)\,\exp(i\,\phi_m) - |F_s|\,\exp(i\,\phi_s)]\,\exp(-2\pi\,i\,H.r)$$

$$= V^{-1} \sum \Delta|F|\,\exp(i\,\phi_m)\,\exp -2\pi\,i\,H.r +$$

$$V^{-1} \sum |F_s|\,[\exp(i\,\phi_m) - \exp(i\,\phi_s)]\,\exp -2\pi\,i\,H.r$$

$$(30) \qquad \Delta\rho(r) = \Delta\rho\,(\Delta\,|F|) + \Delta\rho\,(\Delta\phi)$$

where (appendix 2)

$$(31) \qquad \Delta\rho(\Delta\phi) = 2V^{-1} \sum_{H} |F_s|\,\sin(\Delta\phi/2)\,\exp[i\,(\phi_s + \phi_m + \pi)/2]\,\exp(-2\pi\,i\,H.r)$$

The amplitudes of the Fourier components of $\Delta\rho\,(\Delta\phi)$ are $2|F_s|\,\sin(\Delta\phi/2)$ which, as $\Delta\phi$ is usually small, reduces to

$|F_s|\Delta\phi$ and the phases are the average of ϕ_s and ϕ_m plus a phase advance of $\pi/2$.

For centrosymmetric structures the phases are restricted to 0 or π and there will be at most a few reflections with very small $|F|$ for which ϕ_m and ϕ_s are different ; thus, phases differences should make a negligible contribution to the total $\Delta\rho$ density.

In non-centrosymmetric structures, other things being equal, the higher the point group symmetry, the smaller the contribution from the phase differences. For example, the contribution will be smaller in point group 222 with centric h0l 0k0 and hk0 reflections than in point group 2 with only h0l centric reflections. On the other hand glide plane space group symmetry might extinguish half the reflections in a centric zone and thereby increase the relative contribution to the phase differences. This comment is illustrated by the following two peptide density studies[6][15]. We have calculated the deformation electron density of the peptide (1) and the pseudopeptide (2).

1 N-Acetyl-L-Tryptophan-Methylamide (figure 7a) Space group $P2_12_12_1$ (AcTr).

2 N-Acetyl-α,β-dehydrophenylalanine methylamide (figure 7b) Space group Cc (AcΔ).

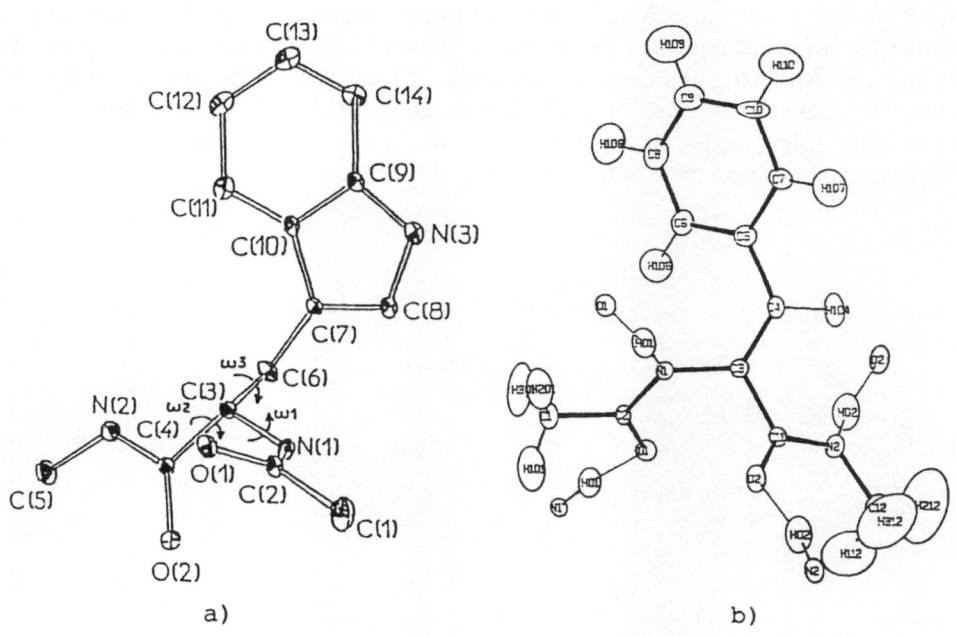

a)   b)

Fig. 7. ORTEP drawings of AcTr (a) and AcΔ (b)

Figure 8 shows maps of $\Delta\rho$ and $\Delta\rho(\Delta\phi)$ for the indole ring (C$_7$...N$_3$...C$_{10}$) (sinθ/λ max = 0.9 Å^{-1}), of AcTr.

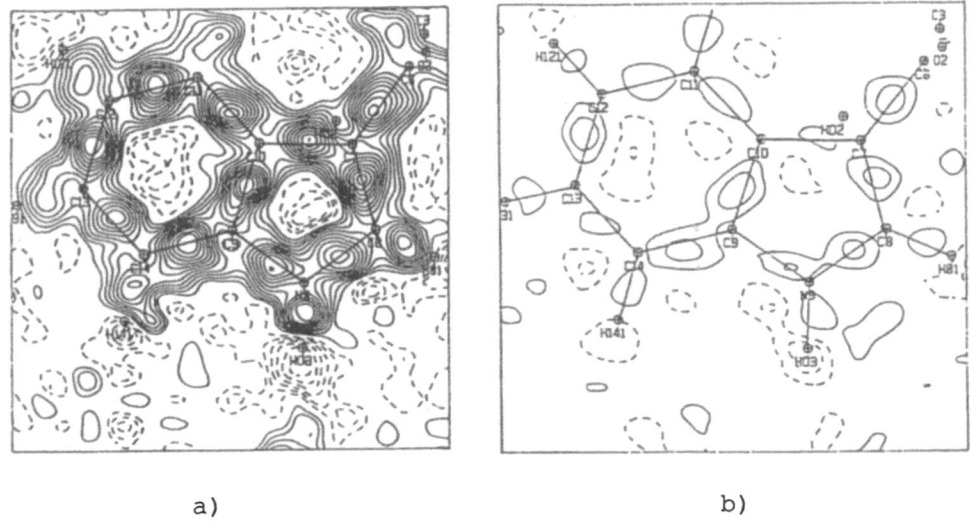

a)   b)

Fig. 8. Maps of (a) the total experimental deformation density
        $\Delta\rho(r)$ and (b) $\Delta\rho(\Delta\phi)$ in the plane of the indole ring of
        AcTr; contours interval 0.05 e Å^{-3}, positive contours
        solid, negative dashed, zero contour omitted.

The maximum $\Delta\rho(\Delta\phi)$ is 0.14 e Å^{-3} and the average $\Delta\rho(\Delta\phi)$ is
0.08 e Å^{-3}. It is obvious that the pattern of $\Delta\rho(\Delta\phi)$ density is
clearly significant. Because the hk0, h0l and 0kl zones are
centric $\Delta\rho(\Delta\phi)$ is much smaller than in AcΔ with only h0l
centric reflections : As shown on figure 9 the maximum and
average $\Delta\rho(\Delta\phi)$ density are 0.23 e Å^{-3} and 0.19 e Å^{-3}
respectively for the phenyl ring of AcΔ.

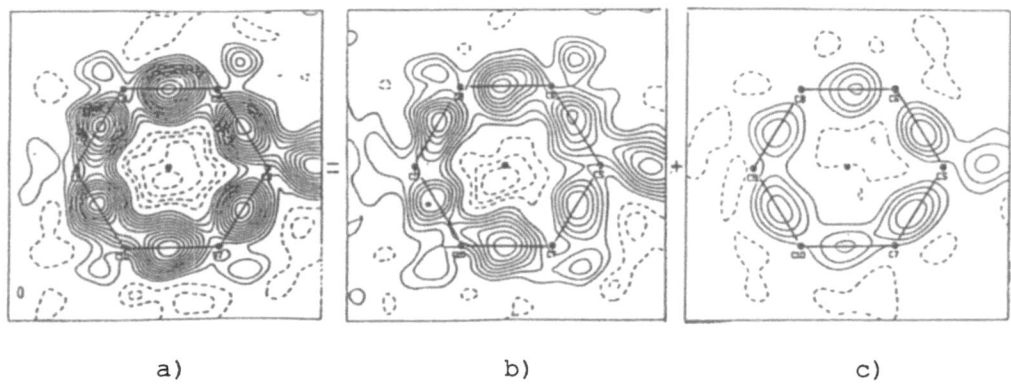

a)   b)   c)

Fig. 9. Maps of a) the total experimental deformation density
        b) $\Delta\rho|\Delta F|$ and c) $\Delta\rho(\Delta\phi)$ in the plane of the phenyl ring
        of AcΔ. Contours as in figure 8.

Furthermore, the addition of $\Delta\rho(\Delta\phi)$ clearly contributes to making the C-C bonding density equivalent.

Figure 10 shows the distribution of $F|\Delta\phi|$ a function of $\sin\theta/\lambda$ (a) and of $|F|$ (b) for AcΔ.

Predictably, $|F|\Delta\phi$ tends to increase with decreasing $(\sin\theta)/\lambda$ due to the diffuseness of the valence electron distribution, (figure 8-a). It is also apparent (8b) that the weak and medium weak reflections have a important influence. Consequently the measurement and the processing of the low angle, weak reflections require special care.

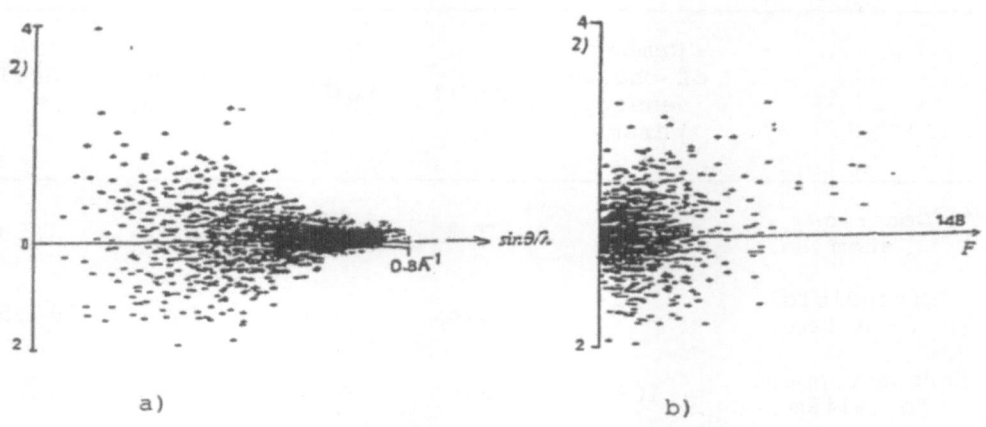

a)   b)

Fig. 10. Distribution of $|F|\Delta\phi$ as a function of (a) $\sin\theta/\lambda$ and of (b) $|F|$.

IV - COMPARAISON BETWEEN THE MOLLY, POP AND LSEXP REFINEMENTS

1). A centrosymmetric structure

Baert, Coppens, Stevens and Devos[16] applied the Molly and LSEXP refinements to low temperature data on pyridinium-1-dicyanomethylide $C_8N_3D_5$, space group $P2_1/m$, $\sin\theta/\lambda_{max}$ 1.1 Å^{-1}, 2390 very good quality reflections.

In both models the number of parameters was reduced by assumption of local symmetry. Table 4 gives a summary of multipole refinements reliability indices.

Ortep view of $C_8N_3D_5$ (Devos 1982)

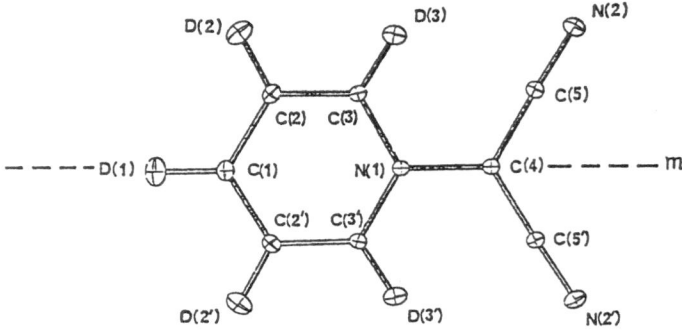

Table 4. Summary of refinements indices.

	Number of charge density params	R(F) %	R_w(F) %*	Goodness of fit	Scale Factor
Spherical atom	0	7.3	9.8	5.80	15.97
Hirshfield formalism	136	2.8	3.5	1.71	15.56
Hansen–Coppens formalism	108	2.9	3.9	1.85	15.58

* $R(F) = \dfrac{\sum||F_o|/K-|F_c||}{\sum|F_o|/K}$; $R_w(F) = \dfrac{[\sum w^2(|F_o|/K-|F_c|)^2]^{1/2}}{[\sum w^2(|F_o|/K)^2]^{1/2}}$; $G.O.F. = \dfrac{[\sum w\Delta^2]^{1/2}}{n_o - n_v}$

As expected (table 4) the agreement factors decrease dramatically when aspherical refinements are used. The smaller weighted indices (R_w(F) and G.O.F.) are obtained with the LSEXP refinement presumably due the larger number of variables in LSEXP. The much greater than one goodness of fit value is certainly due to an underestimation of the standard deviations of the structure factors because both models were able to reproduce all the deformation density. The residual maps are almost flat. The scale factors are in very good aggreement for both aspherical refinements, 3-4% smaller than that given by a spherical refinement. As shown in reference[16] the differences between positional and thermal parameters are small (4-5 σ).

The dynamic model maps obtained from the two refinements are almost identical for the peak heights but slightly different for the shapes, whereas static deformation model maps

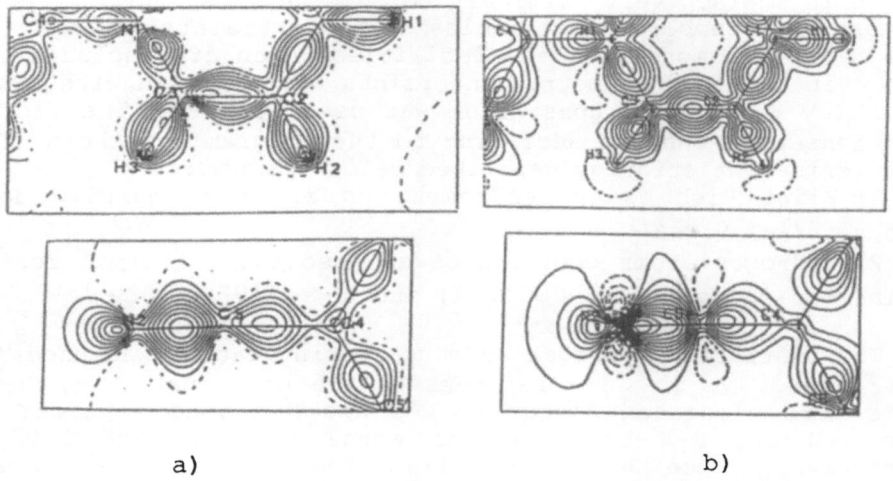

a)   b)

Fig. 11. Static model maps of $C_8N_3D_5$ (from reference 16) a)
Hirshfeld functions b) Multipole refinement.

are, as expected, much more sensitive to the fitting functions
used (figure 11).

When one compares net charges and dipole moments obtained
from these two refinements it appears that the Hirshfeld
charges are unrealistically large because of some diffuse
monopoles used leading also to an overestimated molecular
dipole moment. Anyhow both models are not really adequate to
yield net charges. As previously described a kappa monopole
refinement or a direct integration of the charge would give the
best estimate of the charges. In fact the concept of net charge
in an organic or organo-metallic is not a measurable physical
property. It means that charges are defined by the model used
and that only the properties calculated from this set of
charges can be compared. This is certainly the biggest problem
for the use of experimental net charges in molecular modelling.

2) A noncentrosymmetric structure

Souhassou, Lecomte, Blessing and Ghermani[15c,6b] have
applied three alternative refinements to the low temperature
data of N acetyl$-\alpha,\beta$ - dehydrophenylalanine methylamide (figure
7b), space group Cc. After a high order refinement on F(H) the
three different non spherical models were applied to the data.
The 18000 CAD4 data (MoKα radiation) were reduced according to
Blessing[17] and averaged to 7237 unique reflections (I > 0
$\forall\sigma(I)$, $(\sin\theta)/\lambda < 1.35$ Å^{-1}, $R_i(F^2)^* = 1.6\%$[6b].

$$^* R_i(F^2) = \frac{\sum\limits_{i=1}^{N_{obs}} |I_i - <I>|}{\sum\limits_{i=1}^{N_{obs}} I_i}$$

It is in fact very difficult to compare these three models, because the number of variables and constraints is not the same. For each model all the facilities which are included are used : in that way symmetry constraints were applied with LSEXP and Molly whereas no constraint was used with Pop*. The radial functions were choosen according to the program's authors. The same refinement strategy described below was used :

1) First High order refinement on all the positive data with $\sin\theta/\lambda > 0.9$ Å^{-1}.

2) Hydrogen atoms coordinates and isotropic thermal motion refinement on data with $I > 3\sigma(I)$ and $0 < \sin\theta/\lambda < 0.9$ Å^{-1}.

The contracted hydrogen atom scattering factor was used[18]. After the x, y, z, $<u^2>$ refinement, the coordinates of the H atoms were elongated along the C-H and N-H bond in order to make C-H and N-H bond lengths equal to 1.07 and 1.03 Å respectively. Then the hydrogen free atom scattering factor was used.

3) Refinement of monopole population (P_v) and scale factor then $P_v + \xi$ (or α) refinement together with the scale factor. There were no correlations greater than 0.8.

4) Multipole refinement and scale factor.

5) Thermal, positional and density parameters together.

In the steps 3, 4, 5, all the data with $0 < \sin\theta/\lambda < 1.35$ Å^{-1} and $I > 0$ were used (7732 reflections)

For all the three refinements the multipole expansion was truncated at the octopole level** for the non-hydrogen atoms, whereas monopole and dipole functions were used for the hydrogen atoms. The deformation was defined by (23) for Molly and by (2) for Pop and LSEXP where $\rho_1^{at}(r)$ is the free atom density.

In the Molly and LSEXP procedure, several symmetry and chemical constraints were imposed on the multipole density parameters to reduce the number of variables : The atoms of the phenyl ring had mm2 symmetry (m_1 in the ring plane and m_2 perpendicular to the ring plane passing through the center of the ring). Furthermore, all the C_6, C_7, C_8, C_9, C_{10} atoms were imposed to have the same density parameters. Two types of hydrogen atoms were defined H-N and H-C with a monopole and a single dipole along the H-C or H-N bond.

Figure 12 shows the local axis system used for the Molly and LSEXP refinements : Due to the imposed constraints the x axis of the atoms of the phenyl ring are all parallel to the local 2 axis and z is chosen perpendicular the phenyl ring.

---

* We are now refining without constraints with the Molly and LSEXP models.
** The l = 4 functions did not improve the refinement[6b].

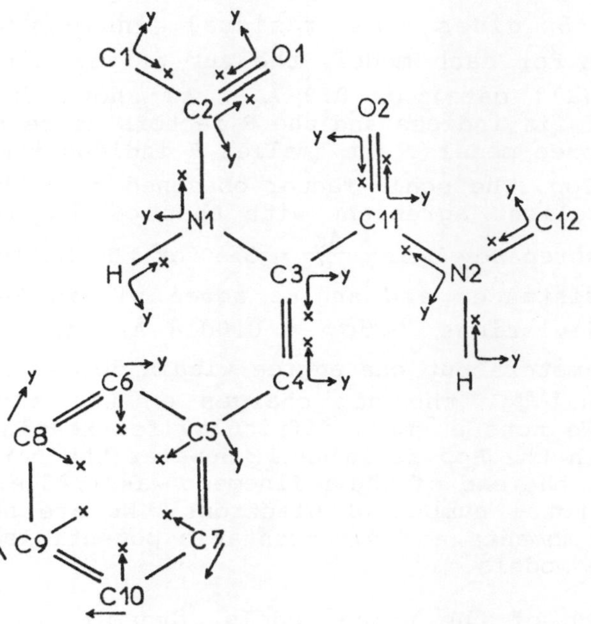

C1 O1
C2
O2
N1 C11 C12
H C3 N2
C6 C4 H
C8 C5
C9 C7
C10

AcΔ

Fig. 12. Local axis system.

Table 5. Residual indices.

Model	Number of density variables	R(F) %	wR(F) %	G.O.F.	Scale Factor
HO Spherical	0	7.89	5.30	0.85	0.6849
All Data I > 0					
Spherical	0	4.78	4.37	1.71	0.6881
Molly	201	3.81	2.35	0.97	0.6838
LSEXP	235	4.08	2.13	0.89	0.6886
Pop	317	3.60	2.00	0.86	0.6867
$\sin\theta/\lambda < 0.9$ Å^{-1} I > 0					
Spherical	0	3.25	3.94	2.27	0.6949
Molly	201	1.68	1.45	0.87	0.6919
LSEXP	235	2.05	1.70	0.99	0.6893
Pop	318	1.7	1.4	0.87	0.6867

Table 5 gives the residual indices for all the refinements. For each model, the cut off in $\sin\theta/\lambda$ was either 1.35 Å^{-1} (all data) or 0.9 Å^{-1}. As shown on table 5, the goodness of fit indices and the R factors decrease considerably with the three models. The smaller $\underline{R}$ indices are obtained with Molly and Pop. The scale factor obtained with the three models are in excellent agreement with that of H.O refinement. The highest discrepancy being $\frac{\Delta K}{\langle K \rangle} = 5‰$ for the Molly scale factor. All bond distances and angles agree within less than three standard deviations ($\langle 3\sigma \rangle = 0.0018$ Å) and all the thermal motion parameters but one aggree within 3σ. As shown before by Baert et al[16a]. the net charges on the atoms completely desagree. We note a small difficulty for keeping the molecule neutral with the Pop refinement. However the net charge of the molecule at the end of the refinement was 1.26 e; i.e less than 1% of the total number of electrons[*]. We are now calculating the dipole moments and electrostatic potentials obtained with these three models.

For each of the three models, dynamic model deformation maps according to formula (28) were calculated and are compared to the theoretical, calculated by Rohmer, Benard and Wiest[15b] from ab initio LCF calculations (double zeta basis for the valence orbitals). Table 6 gives the peak heights obtained. All experimental model peak heigths except for the oxygen lone pair are 10-15% higher than those obtained from ab initio calculations. Furthermore all experimental peaks agree within less than 0.1 e/Å^3 (2 standard deviations) and are in very good aggrement with the bond lengths differences.

Figures 13, 14 and 15 show the dynamic deformation maps on the $C_3=C_4$ bond and on the two peptide bonds together with the theoretical one. We note an excellent agreement on the $C_3=C_4$ bond, the size, the shape and the height of the double bond peak are strictly equivalent (figure 13).

$$O_1 \atop \|$$

The $C_1 - C_2 - N_1 - C_3$ peptide link deformation density is very well reproduced by the three models (figure 14). The biggest discrepancy being 0.1 e Å^{-3} on the $C_2=O_1$ bond. The position of the N_1-C_2 maximum appears to be different for the LSEXP map in which the bonding density is shifted to the nitrogen atom in accordance with a more electronegative character of the nitrogen atom. We also note an significant hole close to the oxygen atom with the Hirshfeld model.

---

[*] In Pop no electroneutrality constraint exists, whereas this constraint exists in the Molly system. For the LSEXP program the electroneutrality is forced by imposing that F(000) equals the number of electrons in the unit cell. This requirement is made by giving a sufficient statiscal weight to F(000).

Table 6. Comparison of the bonding peak heights ($\bar{e}$ Å^{-3}) : The experimental $\Delta\rho$ are from the dynamic model deformation map calculated from formula [4].

Bond	Distance (Å)	$\Delta\rho$ Theo 15b	$\Delta\rho$ Pop	$\Delta\rho$ Molly	$\Delta\rho$ LSEXP
$C_1 - C_2$	1.5036	0.45	0.40	0.40	0.40
$C_3 - C_{11}$	1.5026	0.45	0.50	0.50	0.50
$C_2 - O_1$	1.2383	0.45	0.60	0.50	0.60
$C_{11} - O_2$	1.2387	0.45	0.60	0.60	0.55
$C_2 - N_1$	1.3472	0.40	0.50	0.50	0.50
$C_{11} - N_2$	1.3356	0.40	0.50	0.50	0.50
$C_3 - N_1$	1.4125	0.30	0.35	0.35	0.35
$C_{12} - N_2$	1.4487	0.25	0.25	0.30	0.35
$C_3 - C_4$	1.3472	0.60	0.70	0.70	0.70
$<C_\phi - C_\phi>$	1.396	0.50	0.55	0.55	0.55

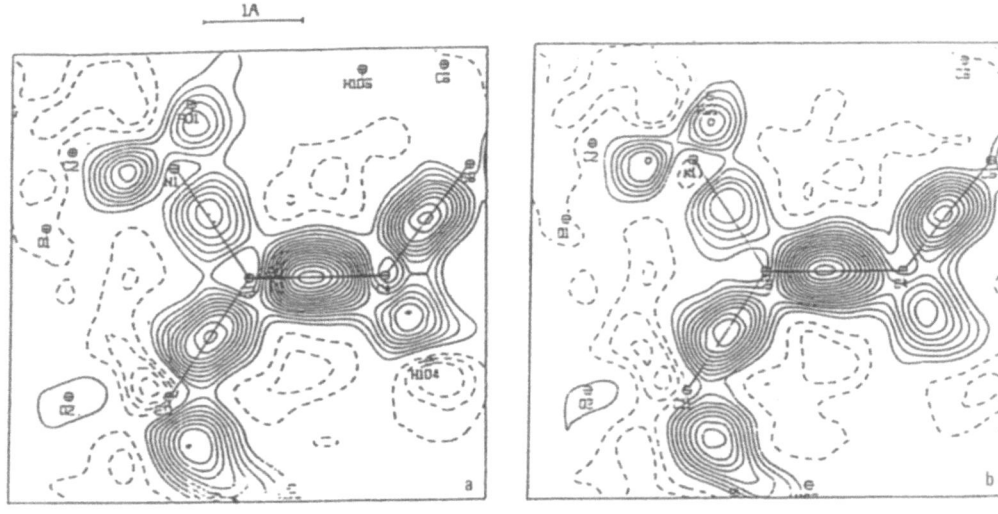

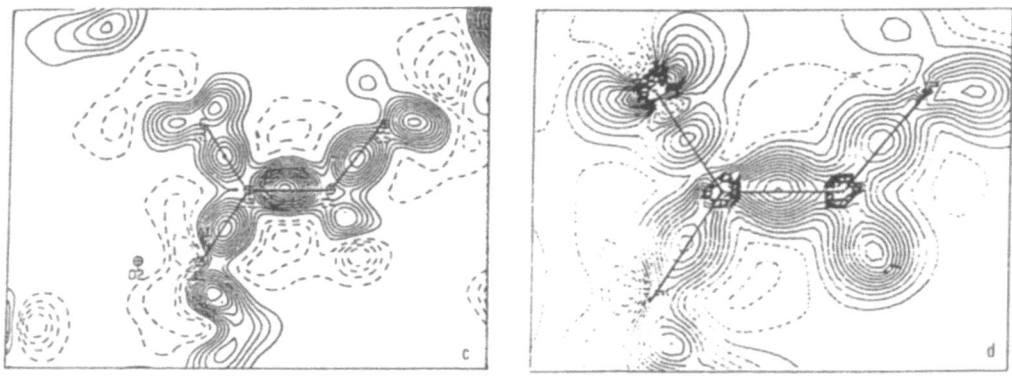

Fig. 13. Dynamic deformation map of the $C_3=C_4$ double bond
obtained from the Molly a), LSEXP b), Pop c), and ab
initio calculations d). Contours as in figure 2.

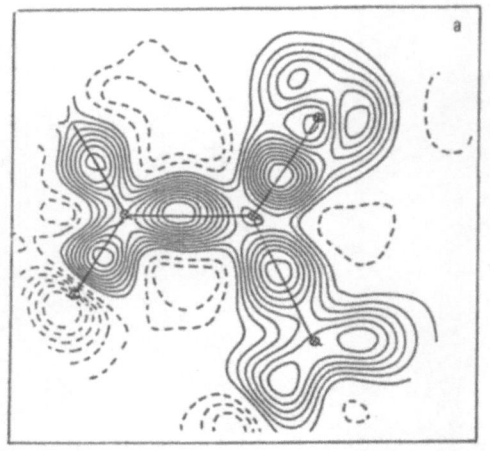
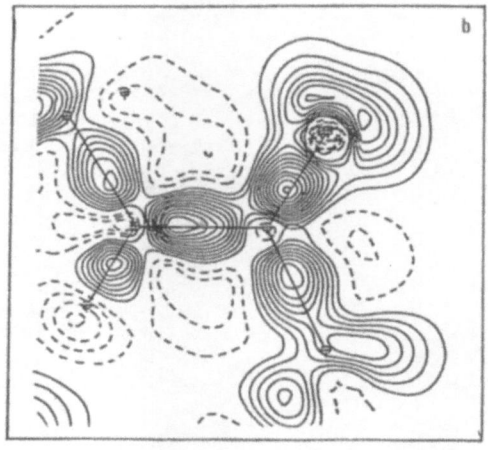
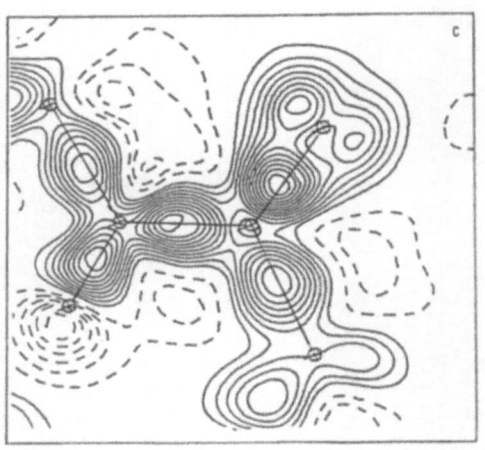
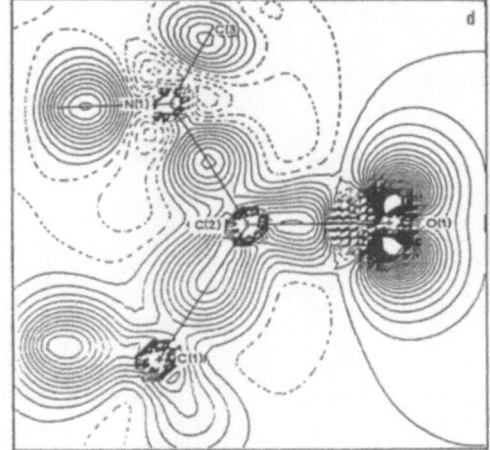

Fig. 14. Dynamic deformation map of the $\begin{bmatrix} C_3 \\ \diagdown \\ H_1 \end{bmatrix} - N_1 - C_2 \begin{matrix} \diagup O_1 \\ \diagdown C_1 \end{matrix}$

peptide link obtained from the Molly a), LSEXP b), Pop c), and ab initio calculations d). Contours as in figure 2.

A slightly poorer agreement appears on figure 15 close to the N_2 nitrogen atom where a trough of -0.2 e/$Å^3$ appears on the LSEXP dynamic model map. The maximum of this hole is 0.15 Å from the nucleus where the density error is bigger. The oxygen lone pair seems also to be very sensitive to the model. At this stage we should remember in all models that no symmetry constraint was imposed to the peptide link atoms.

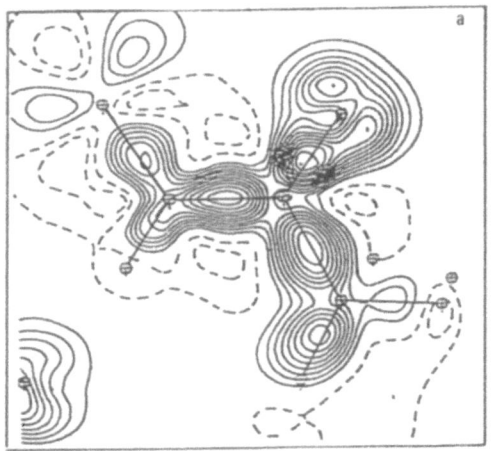 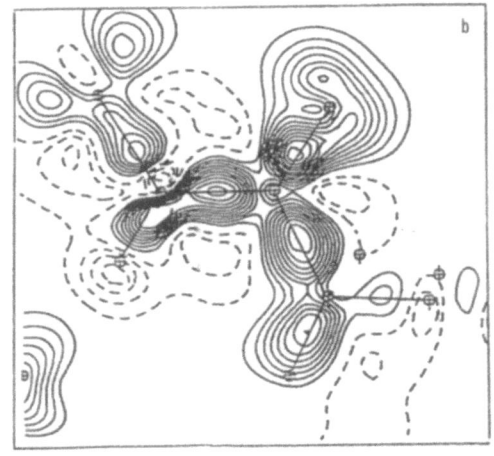

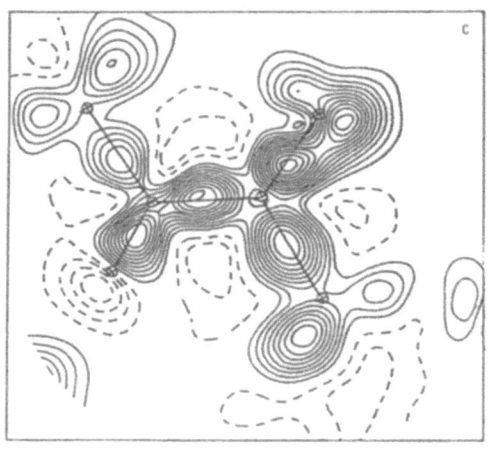 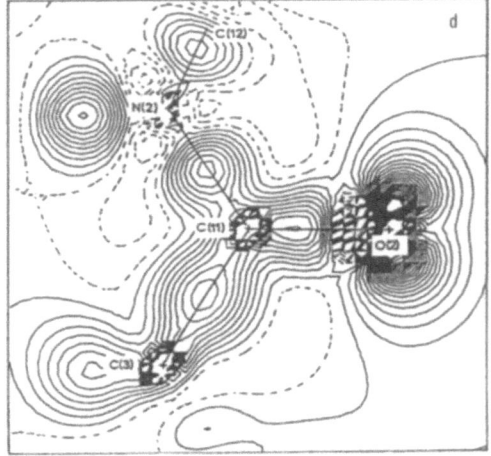

Fig. 15. Dynamic deformation map of the
$$\left[\begin{array}{c} C_{12} \\ H_2 \end{array} \!\!\!\! {\Large >} N_2 - C_{11} {\LARGE <} \!\!\!\! \begin{array}{c} O_2 \\ C_3 \end{array} \right]$$
peptide link obtained from the Molly a), LSEXP b), Pop c), and ab initio calculations d). Contours as in figure 2.

Figure 16 shows the experimental deformation maps at $0.9\,\text{\AA}^{-1}$ resolution in $\sin\theta/\lambda$ of the two peptide links calculated with the multipolar phases obtained from the Molly (16a,d) and from the Hirshfeld (16b,c) programs together with the "difference density" obtained between the two models (16c,f)

$$\Delta\left(\delta\rho^{exp}\right) = \delta\rho_{Molly}^{exp} - \delta\rho_{Hirshfeld}^{exp}$$

It is clear from figures 16c and 16f that the bonding density close to the center of the bond is equally reproduced by the two models within $0.05\ e\ \text{\AA}^{-3}$.

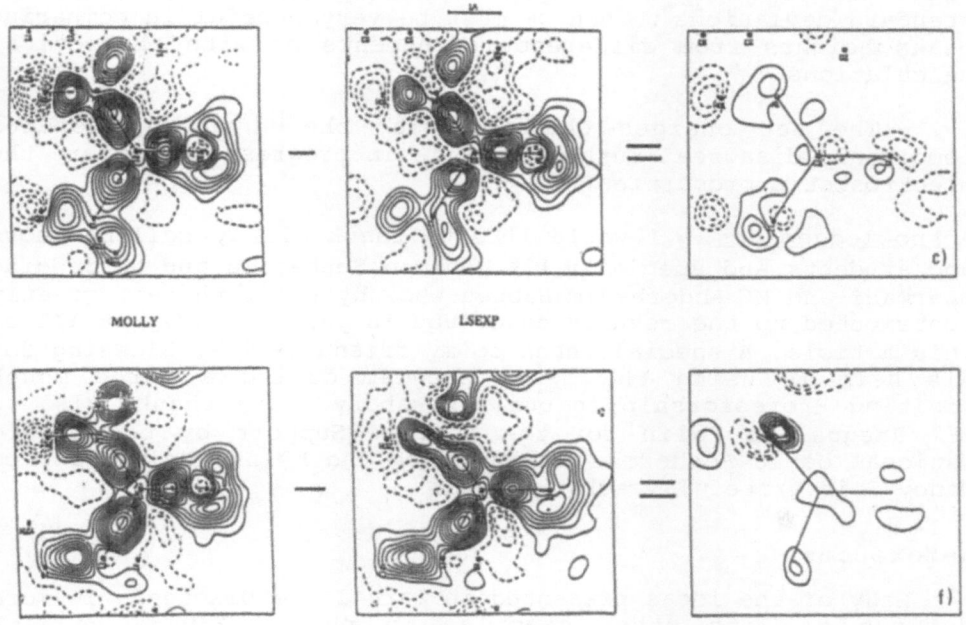

MOLLY   LSEXP

Fig. 16. Experimental deformation of the two peptide links and the difference between the two deformations (13c, 13f) calculated directly from $\Delta \left(\delta \rho^{exp} \right) = \delta^{exp}_{Molly} - \delta^{exp}_{LSEXP}$. Contours as in figure 2.

As discussed below, the discrepancy between LSEXP and Molly appears close to the N_2 nucleus and has a dipolar shape. This small but significant discrepancy comes from the phase value assigned to the observed structure factor and shows that the phase problem discussed in Part II is not completely solved.

In conclusion it appears :

- That the density in the bonding region is clearly and equally reproduced by the three models. The main difference occurs close to the nuclei and on the lone pairs. We therefore must be very careful in interpreting the role of the lone pair in hydrogen bond.

- It also means that, even with very accurate data, the error due to the model appears to be almost equal to two standard deviations[*]. Then we must be very careful in comparing peaks heights from different refinements or with theoretical calculations.

- The net charges obtained with the three refinements completely disagree. Further work is in progress to compare the electrostatic properties.

Acknowledgements : I would like to thank all my collaborators and students and specially Drs Mohamed Souhassou and Noureddine Ghermani and Mr Abderhahim Habbou who, by their theses, greatly contributed to the results described in part II and part III of this article. A special thank to my friend Dr R.H. Blessing for his help in using the LSEXP program during a three month visiting Professorship in our Laboratory. Many thanks also to Mme Jacqueline Palin for the typing. Support by the Centre National de la Recherche Scientifique and by the University of Nancy I is gratefully acknowledged.

References

Many of the ideas presented in Part I are discussed in more details or with other examples in the following general bibliography.

A - **General bibliography**

1. F.L. Hirshfeld Ed. : Electron Density Mapping in Molecules and Crystals, Isr. J. Chem. 16:87-229 (1977).
2. P. Becker Ed. : Electron and Magnetization Densities in Molecules and Crystals, N.A.T.O. Adv. Stud. Ser. B48, Plenum Press (1980).
3. P. Coppens and M.B. Hall Eds : Electron Distributions and the Chemical Bond, Plenum Press (1982).

---

[*] The problem of calculating the standard deviation in a non centric space group is still open ; the given sigmas are calculated from Cruickshank (Acta Cryst. **9**, 754 (1969). For centrosymmetric crystals an usefull expression has been derived by Rees (Acta Cryst. **A34**, 254 (1978).

B - References

1. A. Habbou, C. Lecomte, F. Benabicha, and M.M. Rohmer, to be published.
2. N.K. Hansen and P. Coppens, Acta Cryst. A34:909-921 (1978).
3. a) R.F. Stewart, J. Chem. Phys. 51:4569-4577 (1969).
   b) R.F. Stewart, J. Chem. Phys. 58:1668-1676 (1973).
   c) R.F. Stewart, Acta Cryst. A32:565-574 (1976).
4. a) B.M. Craven, H.P. Weber and X. He, Tech. Report TR-87-2, Department of Crystallography, University of Pittsburgh, PA 15260 (1987).
   b) J. Epstein, J.R. Ruble, and B.M. Craven, Acta Cryst. B38:140-149 (1982).
5. a) F.L. Hirshfeld, Acta Cryst. B27:769-781 (1971).
   b) F.L. Hirshfeld, Isr J. Chem. 16:198-201 (1977).
6. a) M. Souhassou, C. Lecomte, R.H. Blessing, A. Aubry, M.M. Rohmer, R. Wiest, M. Benard, and M. Marraud, Acta Cryst. B (1990) 000.
   b) M. Souhassou, Thèse de l'Université de Nancy I (1988).
   c) C. Lecomte, M. Souhassou, and N. Ghermani, ACA Meeting Seattle, p.37.
7. A. Angot, Compléments de Mathématiques, Editions de la Revue d'Optique, Paris (1957).
8. G. Arfken, Mathematical Methods for Physicits 2nd Ed. Academic Press, N.Y. (1970).
9. C. Flammer, "Spherical wavefunctions", Stanford University Press (1957).
10. R.F. Stewart, in : "Electron and Magnetization Densities in Molecules and Crystals", P. Becker, ed., Nato Adv. Stud. Ser. B48:439-442, Plenum (1980).
11. P.R. Mallinson, T. Koritzanski, E. Elkaim, N. Li, and P. Coppens, Acta Cryst. A44:336-342 (1988).
12. W.J. Hehre, R.F. Stewart, and J.A. Pople, J. Chem. Phys. 51:2657-2664 (1969).
13. P. Coppens, T.N. Guru Row, P. Leung, E.D. Stevens, P. Becker, and Y.W.Yang, Acta Cryst. A35:63-72 (1979).
14. a) E.D. Stevens, J. Am. Chem. Soc. 103:5087 (1981).
    b) C. Lecomte, D.L. Chadwick, P. Coppens, and E.D. Stevens, Inorg. Chem. 22:2982-2992 (1983).
    c) C. Lecomte, R.H. Blessing, P. Coppens, and A. Tabard, J. Am. Chem. Soc. 108:6942-6950 (1986).
    d) N. Li, P. Coppens, and J. Landrum, Inorg. Chem. 27:482-488 (1988).
15. a) M. Souhassou, C. Lecomte, and A. Aubry, Colloque Inserm 174:359-362 (1988).
    b) M. Souhassou, C. Lecomte, A. Aubry, R. Wiest, M.M. Rohmer, and M. Benard, in preparation.
    c) M. Souhassou, C. Lecomte, N. Ghermani, and R. Blessing, Acta Cryst. A, in preparation.
16. a) F. Baert, P. Coppens , E.D. Stevens, and L. Devos, Acta Cryst. A38:143-151 (1982).
    b) L. Devos, Thèse de Doctorat d'Etat, Université des Sciences et Techniques, Lille (F) (1982).
17. a) R.H. Blessing, Crystal Rev. 1:3-58 (1987).
    b) R.H. Blessing, and C. Lecomte, in this book.
18. R.F. Stewart, E.R. Davidson, and W.T. Simpson, J. Chem. Phys. 43,3:175-187 (1965).

Appendix 1

Expressions of the integrals arriving in the structure factor calculation.

I – $$I_2 = \int_0^\pi P_l^m(\cos\theta)^2 \sin\theta \; d\theta \qquad \text{if } z = \cos\theta \qquad dz = -\sin\theta \; d\theta$$

$$\Rightarrow I_2 = \int_{-1}^{+1} P_l^m(z)^2 \; dz = \frac{2}{2l+1} \frac{(l+m)!}{(l-m)!}$$

II – $$I_3 = \int_0^{2\pi} \cos m\phi \, \cos m(\phi-v) \; d\phi = \pi \cos mv \quad \text{if } m \neq 0$$

$$\text{if } m = 0 \quad I_3 = \int_0^{2\pi} d\phi = 2\pi$$

$$\Rightarrow \quad I_3 = \varepsilon_m \, \pi \cos mv$$

III – $$f_{nlms} = \varepsilon_m \pi i^l \cos mv \, (2l+1) \frac{(l-m)!}{(l+m)!} \; P_l^m(\cos u) \frac{2}{2l+1} \frac{(l+m)!}{(l-m)!} \; \phi_{nl}$$

(s)

$$= i^l \, 2 \, \varepsilon_m \, \pi \; \phi_{nl}(s) \; Y_l^m(u,v)$$

In conclusion :

$$\int_r R(r) \, Y_l^m(\theta, \phi) \, e^{2\pi i \, H.r} \; d\tau = 2\pi i^l \, Y_l^m(u, v) \int R_l(r) \, j_l(2\pi H.r) \, r^2 \; dr$$

The Fourier transform of a spherical harmonic $Y_l^m(\theta, \phi)$ is a also spherical harmonic function in reciprocal space :

$$F\left(Y_l^m(\theta, \phi)\right) = 2\pi i^l \, Y_l^m(u,v)$$

Appendix 2

Calculation of $A = \exp(i\,\phi_1) - \exp(i\,\phi_2)$

Let $\Delta\phi = \phi_1 - \phi_2$

Then
$$A = \exp i(\phi_2 + \Delta\phi) - \exp i(\phi_2)$$
$$= \exp(i\,\phi_2)\,[\exp(i\,\Delta\phi) - 1]$$
$$= \exp(i\,\phi_2)\,\exp i(\Delta\phi/2)\,[\exp i(\Delta\phi/2) - \exp -i(\Delta\phi/2)]$$
$$= \exp\,[i(\phi_2 + \Delta\phi/2)]\,[2i\,\sin(\Delta\phi/2)$$
$$= \exp[i(\phi_2 + \Delta\phi/2)]\,[2\,\exp(i\,\pi/2)\,\sin(\Delta\phi/2)$$
$$= 2\,\sin(\Delta\phi/2)\,\exp[i(\phi_2 + \phi_1 + \pi)/2]$$

Then
$$\Delta\rho(\Delta\phi) = V^{-1}\sum 2\,|F_s|\,\sin(\Delta\phi/2)\,\exp[i(\phi_s + \phi_m + \pi)/2]\,\exp(-2\pi i\,\mathbf{H}.\mathbf{r})$$

153

EXPERIMENTAL REQUIREMENTS FOR CHARGE DENSITY ANALYSIS

Robert H. Blessing

Medical Foundation of Buffalo
73 High Street
Buffalo, New York 14203, USA

Claude Lecomte

Laboratoire de Minéralogie et Cristallographie
U.R.A. C.N.R.S. 809
Université de Nancy I, Faculté des Sciences
Boîte Postale n° 239
54506 Vandœuvre-lès-Nancy Cedex, France

INTRODUCTION

Experimental charge density analysis by crystallographic methods requires accurate low-temperature X-ray and neutron diffraction measurements.

"Low temperature" means low relative to the Debye temperature of the crystal of interest, so that thermal vibrational smearing of the scattering density distribution is small, and the Bragg reflection intensities, especially those at high scattering angles, are large. Low temperature can mean room temperature for very hard crystals, such as minerals, metals, or ionic or covalent crystals of simple composition. But for the many cases of biochemical or organic molecular crystals, or crystals of transition metal complexes, low temperature means cryostatic temperatures achieved with liquid nitrogen or liquid helium devices. These devices and their use are discussed by Finn Larsen in another chapter of this book.

Neutrons are scattered by the atomic nuclei and not by the extra-nuclear electronic charge distribution; X-rays are scattered by the electrons and not by the nuclei. Thus, X-ray and neutron data measured at the same temperature are complementary. The X-ray data contain the crystallographically accessible electronic charge density information; the neutron data provide nuclear mean positions and mean-square thermal vibrational displacements, unbiased by the effects of non-spherical atomic electron density redistribution due to chemical bonding. The neutron results are especially valuable for

The Application of Charge Density Research to Chemistry and Drug Design
Edited by G. A. Jeffrey and J. F. Piniella, Plenum Press, New York, 1991

155

hydrogen atoms; they are also very valuable for other terminal atoms, such as carbonyl, carboxyl, amide, or hydroxyl oxygen atoms. The special technique of polarized neutron diffraction allows analysis of the unpaired electronic spin density distribution in crystals containing stable free radicals or magnetic metal atoms. The use of neutron diffraction in charge density studies is discussed by Hartmut Fuess in another chapter of this book.

This chapter then is concerned with the principles and practice of accurate single-crystal X-ray diffraction measurements. Our attention is further restricted to measurements made with conventional diffractometers and conventional sealed-tube sources. The increasing availability of greatly increased (~10×) flux from rotating anode sources and enormously increased (>100×) flux from synchrotron sources permits use of very small crystals and very short wavelengths to obtain data that are virtually absorption- and extinction-free. The special problems and great advantages of these sources are discussed by Philip Coppens in another chapter of this book.

GENERAL BACKGROUND AND FORMULARY

The mean thermal electron density in the unit cell can be calculated by Fourier summation of the X-ray crystal structure factors,

$$\rho(\mathbf{r}) = V^{-1} \Sigma_{\mathbf{h}} F(\mathbf{h}) \exp(-2\pi i \mathbf{h} \bullet \mathbf{r}) , \tag{1}$$

where the summation is, in principle, an infinite summation over all possible reciprocal lattice vectors $\mathbf{h}$. In practice, experimental resolution is always finite, and, depending on the resolution limit, the experimental total density $\rho(\mathbf{r})$ will always be affected by series termination errors or ripples. This one reason that one or another kind of difference density is often calculated,

$$\Delta\rho(\mathbf{r}) = V^{-1} \Sigma_{\mathbf{h}} [F_0(\mathbf{h}) - F_c(\mathbf{h})] \exp(-2\pi i \mathbf{h} \bullet \mathbf{r}) , \tag{2}$$

where the summation is made over the measured $\mathbf{h}$, and the series termination errors sum to zero.

The structure factor is in general complex,

$$F(\mathbf{h}) = |F(\mathbf{h})| \exp i\varphi(\mathbf{h}) = |F| \cos \varphi + i |F| \sin \varphi = A + iB , \tag{3}$$

and it can be calculated from the atomic form factors, mean positions, and mean-square displacements,

$$F(\mathbf{h}) = \Sigma_a f_a(\mathbf{h}) \exp(2\pi i \mathbf{h} \bullet \mathbf{r}_a) T_a(\mathbf{h}) . \tag{4}$$

Just as the unit cell electron density distribution is the (discrete) Fourier transform of the set of X-ray crystal structure factors, the atomic form factor for X-rays is the (continuous) Fourier transform of the atomic electron density,

$$f_a(\mathbf{h}) = \int_V \rho_a(\mathbf{r}) \exp(2\pi i \mathbf{h} \bullet \mathbf{r}) d^3\mathbf{r} . \tag{5}$$

Spherically averaged free-atom electron density distributions,

$$\rho_a^0(r) = (4\pi)^{-1} \int_0^{2\pi} d\varphi \int_0^{\pi} \sin \theta \, d\theta \, \rho_a(r,\theta,\varphi) , \tag{6}$$

are an approximation adequate for structural chemical crystallography, but parametrically multipolar atom-centered electron density models are central to accurate charge density analysis, as described in detail by Bob Stewart, Philip Coppens, Claude Lecomte, and Ed Stevens in other chapters of this book.

Usually, for the Debye-Waller thermal vibration factor, T in equation (4), an independent-atom, anisotropic, harmonic (rectilinear) vibration model is assumed, with six mean-square displacement parameters U_a^{jk}; the structure factor is then

$$F(\mathbf{h}) = \Sigma_a \, f_a(\mathbf{h}) \, \exp\,(2\pi i \, \Sigma_j \, h_j \, x_a^{\,j} - 2\pi^2 \, \Sigma_j \Sigma_k \, h_j h_k \, a^{*j} \, a^{*k} \, U_a^{\,jk}) \qquad (7)$$

$$[j,k = 1,2,3; \ \mathbf{h} = \Sigma_j \, h_j \mathbf{a}^{*j} = h\mathbf{a}^* + k\mathbf{b}^* + l\mathbf{c}^*; \ |\mathbf{h}| = d^* = 2(\sin\theta)/\lambda\,] \, .$$

Anharmonic and curvilinear (librational) vibrations are sometimes modeled via a finite Gram-Charlier higher moment expansion

$$F(\mathbf{h}) = \Sigma_a \, f_a(\mathbf{h}) \, \exp\,\{2\pi i \, h_j \, x_a^{\,j} - (2\pi^2 \, h_j h_k \, a^{*j} \, a^{*k} \, U_a^{\,jk}) \times$$

$$\{1 + (i^3/3!) \, c_a^{\,jkl} \, h_j h_k h_l + (i^4/4!) \, d_a^{\,jklm} \, h_j h_k h_l h_m\}\} \qquad (8)$$

$$(j,k,l,m = 1,2,3; \ \text{and here repeated indices imply summation}) \, .$$

The parameters c^{jkl} and d^{jklm} account for departures of the atomic displacement probability distribution from Gaussian normal form. Ten odd-function $\underline{c}$ coefficients account for skewness or acentricity of the distribution, and 15 even-function $\underline{d}$ coefficients account for kurtosis or non-normal peaking or flattening of the distribution.

Note that calculation of the structure factor implies calculation of both its amplitude $|F|$ and its phase φ, and calculation of a difference density, such as (2), requires two sets of amplitudes and two sets of phases. As described below, amplitudes, but not phases, can be derived from the diffraction intensity measurements. The needed phases must be derived from a computational model, and this phase modelling is discussed by Claude Lecomte in another chapter of this book.

In the kinematic scattering approximation, the intensity of the Bragg reflection is proportional to the squared structure factor amplitude,

$$I_{Bragg} = E\omega/I_0 = \lambda^3 \, r_e^{\,2} \, (v/V^2) \, (1/\sin 2\theta) \, [(1 + \cos^2 2\theta)/2] \, |F|^2 \qquad (9)$$

The proportionality is through factors of λ, the X-ray wavelength; $r_e = |e^2/(4\pi\varepsilon_0 m_e c^2)| = 2.818$ fm, the classical Thomson electron radius; $\underline{v}$, the crystal volume; $\underline{V}$, the unit cell volume; the Lorentz factor, $L = 1/\sin 2\theta$ for equatorial diffraction geometry; and the polarization factor, $P = (1 + \cos^2 2\theta)/2$ for an unpolarized (not crystal monochromated) incident beam, which is assumed to be homogeneous and bathe the crystal uniformly.

The measured reflection intensity includes the Bragg intensity along with other contributions for which appropriate corrections are required,

$$I_{meas}(\mathbf{h}) = k^{-2} \, \{I_{Bragg}(\mathbf{h}) \, A(\mathbf{h}) \, [1 + \alpha(\mathbf{h})] \, y(\mathbf{h})$$

$$+ \, \Sigma_m \, p_m \, I_{Bragg}(\mathbf{h}_m) + I_{bkg}\} \, . \qquad (10)$$

The Bragg intensity is due to the coherent, elastic part of the total scattering. The background includes: the incoherent, inelastic Compton scattering; most of the coherent, inelastic thermal (and disorder) diffuse scattering; scattering from the crystal mount and adhesive, or possibly a glass capillary crystal enclosure; possibly scattering from a cryostat vacuum enclosure; scattering from the direct beam stop; air scattering; and possibly X-ray florescence. Sometimes the measured intensity is noticeably affected by multiple reflection effects, which arise when one or more secondary reciprocal lattice points h_m are in reflecting position simultaneously with the primary point h. Identification and elimination of multiple reflection conditions is mainly a matter of controllable diffraction geometry, and after background subtraction a measured estimate of the squared structure factor magnitude is given by

$$|F|^2 = k^{-2} \, y \, I_{Bragg} / [L \, P \, A \, (1 + \alpha)] . \tag{11}$$

The scale factor $\underline{k}$ adjusts the measured relative intensity to the absolute scale. The extinction factor $0 < y < 1$ corrects for dynamical scattering effects that are responsible for departures from the kinematical approximation. The absorption factor $A < 1$ (actually a transmission factor) corrects for intensity losses by photoelectric absorption and Compton and Rayleigh scattering of the incident and reflected X-ray beams as they pass through the crystal. The thermal diffuse scattering term $\alpha > 0$ corrects for the part of the first-order or one-phonon thermal diffuse scattering intensity that is not removed by the background subtraction.

Practical aspects of the several intensity corrections are described in the following sections, and then summaries of experimental and computational procedures designed to achieve accuracy in intensity measurements are presented.

INTENSITY CORRECTIONS

Multiple Reflection

If for some crystal orientation, two reciprocal lattice points h_1 and h_2 intersect the Ewald sphere simultaneously, then the incident beam s_0 is scattered in the directions s_1 and s_2 simultaneously. The Ewald construction gives

$$(s_1 - s_0)/\lambda = h_1 \qquad \text{and} \qquad (s_2 - s_0)/\lambda = h_2 ,$$

and subtraction gives

$$(s_1 - s_2)/\lambda = h_1 - h_2 \qquad \text{or} \qquad (s_2 - s_1)/\lambda = h_2 - h_1 = -(h_1 - h_2) .$$

This means that the beam measured in the direction s_1 can be strengthened [$0 < p_m < +1$ for $h_m = h_2$ in equation (10)] by reflection of some of the beam s_2 into s_1 by the planes $(h_1 - h_2)$, or it can be weakened ($-1 < p_m < 0$) by the Friedel anti-reflection of s_1 into s_2 by the planes $-(h_1 - h_2)$. The sign of p_2 will depend on the relative reflectivities of the planes h_1 and h_2: $p_2 > 0$ if $I_{Bragg}(h_2) > I_{Bragg}(h_1)$, and $p_2 < 0$ if $I_{Bragg}(h_2) < I_{Bragg}(h_1)$. The magnitude of p_2 depends on the reflectivity of the planes $\pm(h_1 - h_2)$, and it will be large only if both $I_{Bragg}(\mathbf{h}_2)$ and $I_{Bragg}(\mathbf{h}_1 - \mathbf{h}_2)$ are large. Usually, $|p_m| <$ 0.1, and typical values are of the order 0.01.

In general, multiple reflection tends to weaken strong reflections, and strengthen weak ones. If the reflection $\mathbf{h}_1$ is a systematically absent, symmetry forbidden reflection, multiple reflection can produce a "Renninger reflection" at $\mathbf{h}_1$, which will usually be recognizably sharper than the normal single reflections, because the doubly reflected beam is effectively monochromated by its first reflection.

Multiple reflections will necessarily occur if the crystal is oriented with a crystallographic symmetry direction along the instrumental spindle axis, so the crystal mount should always be adjusted to a random, unsymmetrical orientation in order to reduce the number of multiple reflections to accidental occurrences. The number of these will be larger for higher symmetry crystals; for larger unit cells and shorter wavelengths, which imply a higher density of reciprocal lattice points; and for larger crystals and greater spectral width or wavelength spread of the incident beam, which imply coarser instrumental resolution, a larger volume for each reciprocal lattice "point", and a greater effective "thickness" of the Ewald sphere. In the absence of anisotropic absorption or extinction effects, multiple reflection is probably responsible for the largest part of the typically one or two percent excess scatter over and above the crystal positioning and counting statistical scatter among multiple equivalent reflection measurements.

Programs for detection and correction of multiple reflection intensity errors have been described by Coppens (1968), Tanaka and Saito (1975), and Le Page and Gabe (1979). Although multiple reflection effects can be troublesome for charge density studies, analysis of the effects using n-beam dynamical diffraction theory (Post, 1979) provides an avenue to experimental phase measurement for three-phase structure invariants $\Phi = \varphi(\mathbf{h}_1) + \varphi(\mathbf{h}_2) + \varphi(-\mathbf{h}_1 - \mathbf{h}_2)$.

Absorption

The transmission factor $\underline{A}$ is given by volume integration of the absorption function,

$$I_1/I_0 = \exp(-\mu t) ,$$

for the path lengths of the incident and reflected beams through the crystal,

$$A = v^{-1} \int_v \exp[-\mu(t_0 + t_1)]\, dv_t , \qquad (12)$$

where μ is the linear absorption coefficient. Computing the path lengths t_0 and t_1 requires that the crystal size and shape be known accurately. The crystal faces must be correctly indexed on axes with the same absolute orientation as the axes that define the reflection indexing, and the perpendicular distances of the faces from a common origin in or on the crystal must be measured, or calculated from other measurements of the crystal dimensions. Irregular fracture surfaces or hindered growth surfaces that are not cleavage planes or natural faces need to be approximated as pseudo-faces with large indices. The computational procedures for carrying out the diffraction geometry calculations and crystal volume integrations for absorption corrections have been reviewed by Chidambaram (1980), and a useful program, which allows for a crystal enclosed in a capillary, has been written by DeTitta (1985).

Figure 1 shows that the transmission factor for spherical crystals increases very nearly linearly with $(\sin \theta)^2$. Since the isotropic Debye-Waller factor,

$$T = \exp[-B(\sin \theta)^2/\lambda^2] \quad \text{with} \quad B = 8\pi^2 \langle u^2 \rangle, \quad (13)$$

decreases exponentially with $(\sin \theta)^2$, uncorrected absorption errors lead to underestimates of the mean-square atomic displacements $\langle u^2 \rangle$. For non-spherical crystals, uncorrected anisotropy of absorption biases the anisotropic mean-square displacement parameters U^{jk}, and produces false quadrupolar difference electron density features, analogous to the difference densities seen when an anisotropically vibrating atom is modeled as vibrating isotropically (see, e.g., Stout and Jensen, 1968, pp. 379-381). Such effects will be most noticeable for the heavy atoms in a structure, because the heavy atom scattering makes a relatively larger contribution at the higher scattering angles. For this reason, studies aimed at analysis of the thermal vibrations or electron density of transition metal or other strongly scattering, strongly absorbing atoms impose strict requirements for accuracy in the absorption correction.

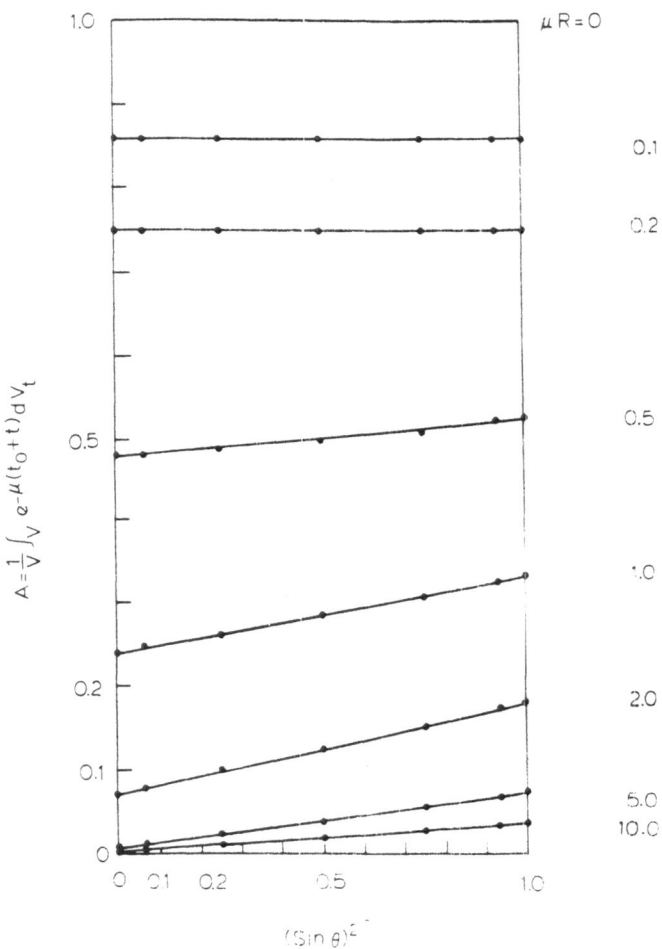

Figure 1. Spherical crystal transmission factors as computed and tabulated by Bond (1967).

A propagation of error calculation shows that an absorption corrected intensity $I_{corr} = I/A$ will have an estimated uncertainty

$$\sigma(I_{corr}) = I_{corr} \left([\sigma^2(I)/I^2] + (\mu t)^2 \left\{[\sigma^2(\mu)/\mu^2] + [\sigma^2(t)/t^2]\right\}\right)^{1/2} ,$$

where t is the absorption-weighted mean path length through the crystal,

$$t = \int_v (t_0 + t_1) \exp[-\mu(t_0 + t_1)] \, dv_t \Big/ \int_v \exp[-\mu(t_0 + t_1)] \, dv_t. \tag{14}$$

For crystals of the size ordinarily used for X-ray diffraction, $\sigma(t)/t$ will seldom be smaller than ~1%, and, due to uncertainties in the values of the mass absorption coefficients for X-rays (Hubbell, McMaster, Kerr Del Grande, and Mallett, 1974), $\sigma(\mu)/\mu$ will seldom be smaller than ~2%. Thus, intensity uncertainties due to the absorption correction will typically amount to at least several percent of μt.

The mean path lengths t for each reflection measurement are needed for a proper extinction correction, and the direction vectors of the incident and reflected beam paths t_0 and t_1, referred to crystal-fixed axes, are needed if the extinction is anisotropic. Thus even if absorption itself is negligible, an absorption calculation might be necessary as a preliminary to an extinction correction. For a crystal of anisotropic dimensions or a spherical crystal that exhibits anisotropic extinction, repeated measurements, but not symmetry or ψ-rotation equivalent measurements, should be averaged before the least-squares model fitting. Then, equivalent extinction-corrected data should be averaged before Fourier mapping of the electron density.

An absorption calculation using a Gaussian quadrature grid might also be necessary, even in the absence of significant absorption, as a means to correct for non-uniform illumination of the crystal by a beam of non-homogeneous intensity from an incident beam crystal monochromator, as described below.

For the difficult case of absorption by crystals of very ill-defined shape, for example, irregular single crystal fragments, adhesive coated or encapsulated crystals, or crystals ground or solvent-shaped to an ellipsoidal rather than spherical shape, an empirical absorption correction program based on fitting real spherical harmonic functions to the empirical transmission surface as sampled by multiple symmetry or ψ-rotation equivalent reflection measurements is under development (Blessing, 1990).

Extinction

Extinction phenomena represent the departure of diffraction by real crystals from the "ideally imperfect mosaic crystal" approximation of single-beam kinematical diffraction theory. If not too large, the effects can be treated as kind of perturbation within the kinematical framework. Extreme cases might require a more rigorous treatment in terms of n-beam dynamical diffraction theory. The extreme cases occur when the effective thickness of the perfect crystal domains or mosaic blocks exceeds the so called extinction length Λ,

$$t > \Lambda = V / (r_e \lambda P \, |F|) . \tag{15}$$

Examples of crystals with this high degree of perfection include some specimens of mineral and metal crystals; covalent crystals such as C, Si, Ge, BN, or GaAs, with the diamond structure; and simple ionic crystals such as LiF, NaCl, CaF_2, or BeO. Reviews of the extensive theoretical work on extinction have been presented by Becker (1980a) and Kato (1980), and these authors continue at work on the dynamical analysis at present.

Examples of extreme extinction in molecular crystals are rare. Most molecular crystals conform more-or-less closely to the ideally imperfect mosaic crystal model for the small crystals used for X-ray diffraction, but for the large crystals needed for neutron diffraction extinction can sometimes be quite severe, especially if the lattice forces are strong as they are in strongly hydrogen bonded or molecular salt crystals, such as α-oxalic acid dihydrate $(COOH)_2.2H_2O$, urea-phosphoric acid $(H_2N)_2CO.H_3PO_4$, or hydrazinium sulfate $N_2H_6^{2+} SO_4^{2-}$. Still, corrections derived within the kinematical framework are usually quite adequate.

For the quasi-kinematical treatment, two extinction processes are distinguished phenomenologically, as indicated in Figure 2.

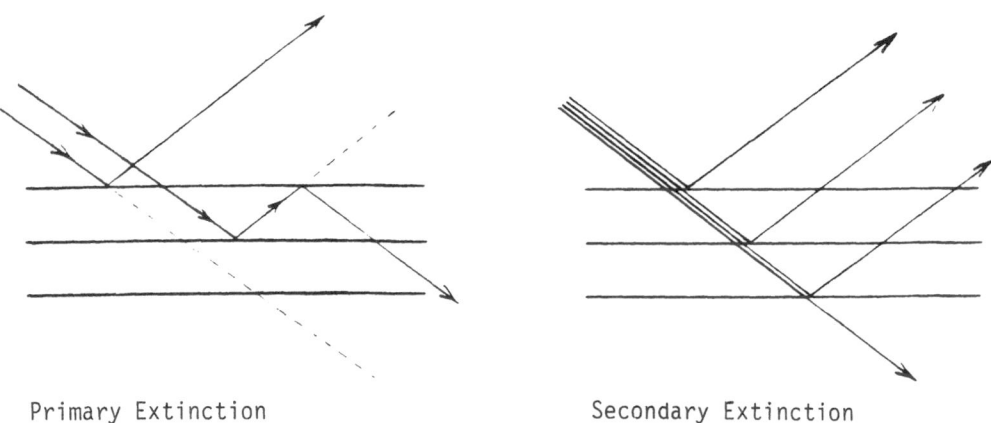

Primary Extinction   Secondary Extinction

Figure 2. Schematic Bragg reflection illustrations of primary and secondary extinction within an idealized mosaic block or perfect crystal domain.

Primary extinction is due to re-scattering of a scattered beam back into the incident beam. There is a phase change of $\pi/2$ with each reflection, so that beams reflected $\underline{n}$ times are π out of phase with beams reflected $\underline{n}$ - 2 times, and interfere destructively. This is in fact a perfect crystal, n-beam dynamical scattering effect for which

$$I \propto |F|^x \quad \text{with} \quad 1 < x < 2 .$$

Secondary extinction is due to the diminution of the intensity of the beam incident on planes in succeeding mosaic blocks due to single scattering by planes in preceding blocks. This is an imperfect crystal, kinematical scattering effect, for which, to a first approximation,

$$I \propto |F|^2 \exp [-(\mu + g'I)t] .$$

A comprehensive treatment of extinction in the quasi-kinematical approximation has been provided by Becker and Coppens (1974a, 1974b, 1975). Their analysis (see also Becker, 1980b) proceeds from the Darwin transfer equations,

$$\partial I_0/\partial x_0 = -\sigma I_0 + \sigma\, I_1$$

$$\partial I_1/\partial x_1 = +\sigma I_0 - \sigma\, I_1$$

$$\partial I_0/\partial x_0 + \partial I_1/\partial x_1 = 0\,,$$

where σ is the diffraction cross section per unit volume and unit intensity.

Integral solutions (integrated over the beam divergence, $d\varepsilon_1$, and over the incident and diffracted beam paths through the perfect crystal domains, dv_t, and the mosaic crystal volume, dv_T) yield the extinction factor,

$$y = I_{meas}/I_{Bragg} = y_p\, y_s(y_p)\,, \qquad (16)$$

where

$$y_p = (vQ)^{-1} \int_{-\infty}^{+\infty} d\varepsilon_1\ \sigma(\varepsilon_1) \times$$
$$\int_v dv_t\ J_0[2i\,\sigma(\varepsilon_1)\,\sqrt{(t_0 t_1)}]\, \exp\,[-\sigma(\varepsilon_1)(t_0 + t_1)]\,,$$

$$y_s = [vQA(\mu)]^{-1} \int_{-\infty}^{+\infty} d\varepsilon_1\ \overline{\sigma}(\varepsilon_1) \times$$
$$\int_v dv_T\ J_0[2i\,\overline{\sigma}(\varepsilon_1)\,\sqrt{(T_0 T_1)}]\, \exp\,\{[\,\overline{\sigma}(\varepsilon_1) + \mu](T_0 + T_1)\}\,,$$

$$Q = (a^2\lambda^3/V_{cell}{}^2)K^2|F|^2/\sin 2\theta = \int_{-\infty}^{+\infty}\ \sigma(\varepsilon_1)\, d\varepsilon_1\,,$$

and

$$\overline{\sigma} = \sigma * W(\eta)$$

is the convolution product of the unit scattering function with the mosaic orientation distribution function $W(\eta)$, which is modelled as either Lorentzian or Gaussian. The constant $\underline{a}$ equals 10 fm for neutrons or $r_e = 2.818$ fm for X-rays, and K is a polarization factor equal to 1 for neutrons and either 1 or $\cos 2\theta$ for X-rays. Analytical integration over the beam divergence, and Gaussian quadrature over spherical or ellipsoidal mosaic blocks and spherical or ellipsoidal crystal volumes ($T_{max}/T_{min} < 2$) gives

$$y_i = \{1 + C_i\,\xi_i + A_i(\theta)\,\xi_i{}^2/[1 + B_i(\theta)\,\xi_i]\}^{-1/2}\,, \qquad i = p \text{ or } s\,, \qquad (17)$$

where, as shown below, ξ_i is approximately proportional to $|F|^2/\sin 2\theta$. Least-squares fit against $\cos 2\theta$ gives numerical formulas for the coefficients A, B, and C. Orders of magnitude are 0.1 to 10 for ξ_i, -0.1 to +1 for A and B, and $C \approx 2$.

Note the approximate similarity of (17) to earlier approximations to the extinction factor (Zacharaisen, 1967)

$$y = (1 + 2x)^{-1/2} = 1 - x + (3/2)x^2 - (5/2)x^3 + ...,$$

$$x \approx g'p_Z|F_c|^2/\sin 2\theta\,,$$

$$p_Z = (1 + \cos^4 2\theta)/(1 + \cos^2 2\theta) \qquad \text{for X-rays,}$$
$$p_Z = 1 \qquad \text{for neutrons;}$$

or (Darwin, 1922, as cited by James, 1982)

$$y = \exp(-g'I_c) \approx 1 - g'I_c = 1 - g'|F_c|^2/(Lp),$$

where typical values of g' are of the order of 10^{-5}.

The practical formulas derived by Becker and Coppens for calculating extinction corrections in the least-squares refinement can be summarized as follows. The scaled, extinction corrected, calculated absolute structure factor magnitudes for least squares minimization of

$$\chi_n^2 = \Sigma\, w(|F_o|^n - |F|^n)^2, \qquad w = 1/(\sigma^2|F_o|^n),$$

are

$$|F|^2 = s^2\, y\, |F_c|^2 \qquad \text{or} \qquad |F| = s\, y^{1/2}\, |F_c|,$$

where here $s = k^{-1}$ is the inverse absolute scaling factor.

The general cases are:

for neutrons
$$y = y_p(\xi_p)\, y_s[y_p(\xi_p)\, \xi_s],$$

and for X-Rays
$$y = (\nu_0 y_0 + \nu_1 y_1)/(\nu_0 + \nu_1)$$

$$y = (\nu_0 + \nu_1)^{-1}\{\nu_0\, y_p(\nu_0\, \xi_p)\, y_s[y_p(\nu_0\, \xi_p)\, \nu_0\, \xi_s]$$
$$+\, \nu_1\, y_p(\nu_1\, \xi_p)\, y_s[y_p(\nu_1\, \xi_p)\, \nu_1\, \xi_s]\}$$

$$\nu_0 = (\cos 2\theta_m)^2, \qquad \nu_1 = (\cos 2\theta)^2$$

$$\nu_0 = 1 \text{ for an unpolarized incident beam,}$$

$$y_{p,s} = \{1 + C_{p,s}\, \xi_{p,s} + A_{p,s}(\theta)\, \xi_{p,s}^2/[1 + B_{p,s}(\theta)\, \xi_{p,s}]\}^{-1/2}.$$

The primary extinction variables are:

$$\xi_p = \gamma_p\, \rho^2\, |F_c|^2$$

$$\gamma_p = (3/2)\, a^2\, \lambda^4/V^2, \qquad\qquad a = 10 \text{ fm for neutrons}$$
$$a = r_e = 2.818 \text{ fm for X-rays}$$

$$\rho = r/\lambda$$

$$r = (\mathbf{u}^T\mathbf{E}\mathbf{u})^{-1/2} \text{ for anisotropic mosaic block size}$$

$$A_p = 0.20 + 0.45 \cos 2\theta \qquad\qquad -0.25 < A < 0.65$$
$$B_p = 0.22 - 0.12\,(0.5 - \cos 2\theta)^2 \qquad\qquad -0.05 < B < 0.19$$
$$C_p = 2$$

The secondary extinction variables are:

$$\xi_s = \gamma_s \, \psi \, |F_c|^2$$

$$\gamma_s = (a^2 \, \lambda^3/V^2) \, \overline{T}, \qquad \overline{T} = \int_v (T_0 + T_1) \exp\left[-\mu(T_0 + T_1)\right] dv_T \, / $$
$$\int_v \exp\left[-\mu(T_0 + T_1)\right] dv_T$$
$$\overline{T} = (vA)^{-1} \int_v (T_0 + T_1) \exp\left[-\mu(T_0 + T_1)\right] dv_T$$

Lorentzian $\qquad \psi_L = \rho/[1 + \rho(\sin 2\theta)/g]$, $\qquad\qquad \eta_L = (2\pi g)^{-1}$

Gaussian $\qquad \psi_G = \rho/[1 + \rho^2(\sin 2\theta)^2/g^2]^{1/2}$, $\qquad \eta_G = (2\pi^{1/2} g)^{-1}$

$g = (\mathbf{D}^T \mathbf{Y} \mathbf{D})^{-1/2}$ for anisotropic mosaic block misorientation

$A_{s,L} = 0.025 + 0.285 \cos 2\theta$		$-0.260 < A < 0.310$
$B_{s,L} = 0.15 - 0.2 \, (0.75 - \cos 2\theta)^2$	if $\cos 2\theta > 0$	$0.0375 < B < 0.45$
$B_{s,L} = -0.45 \cos 2\theta$	if $\cos 2\theta < 0$	
$C_{s,L} = 2$		

$A_{s,G} = 0.58 + 0.48 \cos 2\theta + 0.24 (\cos 2\theta)^2$	$0.34 < A < 1.30$
$B_{s,G} = 0.02 - 0.025 \cos 2\theta$	$-0.005 < B < 0.045$
$C_{s,G} = 2.12$	

The block size and orientation anisotropies are defined by unit direction vectors referred to crystal-fixed axes for the incident beam $\mathbf{u}_0$, the diffracted beam $\mathbf{u}_1$, and the normal to the equatorial plane $\mathbf{D} = \mathbf{u}_0 \times \mathbf{u}_1 / |\mathbf{u}_0 \times \mathbf{u}_1|$. The formulas given for the A and B values are valid for weakly absorbing crystals with values of μR or $\mu \, \overline{T}/2$ less than ~0.5. For more strongly absorbing crystals, A and B values must be obtained from Tables 2, 5, 6, 7, and 8 of Becker and Coppens (1974a).

Further practical approximations are: First, neglect primary extinction, that is, set $\xi_p = 0$ and $y_p = 1$. Second, distinguish two limiting cases of secondary extinction:

Type I. Dominated by mosaic block orientation,

$$\rho = r/\lambda \gg g \propto 1/\eta,$$

$$\rho \gg g, \quad \rho(\sin 2\theta)/g \gg 1, \quad \psi_L, \psi_G \to g/\sin 2\theta.$$

In this case the hypothetical blocks are large or the moasic misorientation is large, i.e., the crystal is composed of badly misaligned, large blocks, and the mosaic spread determines the rocking curve width.

Type II. Dominated by mosaic block size,

$$\rho \ll g, \quad \rho(\sin 2\theta)/g \ll 1, \quad \psi_L, \psi_G \to \rho.$$

In this case the blocks are small or the moasic misorientation is small; therefore, many blocks are parallel along the incident beam direction. For the closely aligned, small blocks, particle size broadening determines rocking curve width.

Severe extinction is usually of type I (usually with a Lorentzian mosaicity distribution), but in very severe cases it might be necessary to introduce a primary extinction component, $\xi_p \neq 0$ and $y_p < 1$, and consider

Type III. A general case,

$$\rho \approx g, \quad y = y_p(\xi_p)\, y_s(y_p\, \xi_s) \,.$$

In any case, because of the complicated dependence of y on g and ρ, it is expedient to evaluate the derivatives $\partial y/\partial g$ and $\partial y/\partial \rho$ numerically for building the least-squares normal equations. The needed formulas are:

For isotropic extinction,

$$(\partial |F|/\partial g) = (1/2)\, s\, y^{-1/2} |F_c|\, (\partial y/\partial g)$$

$$(\partial |F|^2/\partial g) = s^2 |F_c|^2\, (\partial y/\partial g) = 2|F|\, (\partial |F|/\partial g)$$

$$(\partial y/\partial g) = [y(g + \Delta g) - y(g)]/\Delta g \,, \qquad \begin{array}{ll} \Delta g = 10^{-3} g & \text{if } g > 0 \\ \Delta g = 1 & \text{if } g = 0 \end{array}$$

and exactly analogous expressions with ρ replacing g. In practice, typical values of the dimensionless parameters g and ρ are of the order of 10^4, corresponding to values of η of the order of 10 seconds of arc and values of r of the order of 1 μm,

$$g_L = (2\pi\eta_L)^{-1} \approx [2\pi \times 10" \times (1°/3600") \times (\pi/180°)]^{-1} = 0.33 \times 10^4 \,,$$

$$\rho = r/\lambda = 1 \times 10^{-6}\ \text{m}/1 \times 10^{-10}\ \text{m} = 1\ \mu\text{m}/10^{-4}\ \mu\text{m} = 10^4\ \text{Å}/1\ \text{Å} = 1 \times 10^4 \,.$$

For anisotropic extinction, with anisotropic mosaic block orientation,

$$g = (\mathbf{D}^T \mathbf{Y} \mathbf{D})^{-1/2}, \quad \mathbf{D} = \mathbf{u}_0 \times \mathbf{u}_1/|\mathbf{u}_0 \times \mathbf{u}_1| \,, \quad |\mathbf{u}_0 \times \mathbf{u}_1| = \sin 2\theta$$

$$(\partial |F|/\partial Y_{ij}) = (1/2)\, s\, y^{-1/2} |F_c|\, (\partial y/\partial g)\, (\partial g/\partial Y_{ij})$$

$$(\partial g/\partial Y_{ij}) = (-1/2)\, g^3\, (2 - \delta_{ij})\, D_i\, D_j \qquad i \leq j = 1, 2, 3 \qquad \begin{array}{l} \delta_{ij} = 0 \text{ if } i \neq j \\ \delta_{ij} = 1 \text{ if } i = j \end{array}$$

and, with anisotropic mosaic block size,

$$\rho = r/\lambda \,, \quad r = (\mathbf{u}^T \mathbf{E} \mathbf{u})^{-1/2}$$

$$(\partial |F|/\partial E_{ij}) = (1/2)\, s\, y^{-1/2} |F_c|\, (\partial y/\partial g)\, (\partial g/\partial E_{ij})$$

$$(\partial g/\partial E_{ij}) = (-1/2)\, (r^3/\lambda)\, (1 - \delta_{ij})\, u_i\, u_j$$

Analytical expressions for the derivatives for the least-squares normal equations have been given by Spackman (1987). The several possibilities for the extinction modelling are summarized in Table 1.

Table 1. Becker-Coppens-Hamilton-Zacharaisen schemes for extinction models

Type I.   Secondary extinction dominated by
large mosaic block misorientation
- Lorentzian orientation distribution
  - Isotropic
  - Anisotropic
- Gaussian orientation distribution
  - Isotropic
  - Anisotropic

Type II.   Secondary extinction dominated by
small mosaic block size
- Isotropic block size
- Anisotropic block size

Type III.   Primary extinction modelled by type II factored into
secondary extinction modelled by type I

Thermal Diffuse Scattering

The total X-ray scattering is independent of temperature, so that, as the elastic Bragg scattering intensity falls off due to atomic thermal vibrations, the inelastic thermal diffuse scattering (TDS) intensity builds up. Conservation of energy operates through exchanges of energy between the scattered radiation and the external lattice vibrational modes (and the very soft, lowest-energy internal molecular vibration modes in molecular crystals). The vibrational energy quanta are called phonons, and the TDS can be analyzed in terms of phonon-photon scattering, as well as in terms of thermal excitation of longitudinal and transverse travelling lattice waves of atomic vibrational displacements. Frequencies are in the ~10 to ~100 cm^{-1} range.

The study of lattice vibrations and TDS is a large field of crystallographic research interest in its own right (see, e.g., Willis and Pryor, 1975), but the main point of interest for our present purposes is that, like the Bragg intensity, the TDS intensity peaks at reciprocal lattice points, as indicated in Figure 3. The TDS is inelastic because there are both absorption and emission interactions with the Bragg scattering, and, therefore, the TDS peak has a broader wavelength distribution and broader intensity profile than the Bragg reflection peak. As Figure 3 indicates, most of the broad TDS peak is subtracted via a linear background correction, but a portion is not. Lattice dynamical analysis shows that the TDS intensity is proportional to the Bragg intensity, and the background-corrected measured net intensity is therefore

$$I_{meas} = I_{Bragg} + I_{TDS} = I_{Bragg}(1 + \alpha), \quad \alpha = I_{TDS}/I_{Bragg}. \tag{18}$$

In addition, the theory shows that, in an isotropic approximation,

$$\alpha \propto T\, [(\sin\theta)/\lambda]^2\, v_s^{-2}, \tag{19}$$

where here T is the absolute temperature, and v_s is the mean velocity of sound in the crystal, which is a measure of the "hardness" of the crystal or the mean force constant for the restoring forces experienced by the atoms they undergo their thermal vibrations. Equation (19) illustrates the practical consequences of TDS for the present context. The proportional dependence of α on $(\sin \theta)^2/\lambda^2$ means that uncorrected TDS is analogous to uncorrected absorption, and biases the mean-square atomic displacement parameters to values that are too small. More importantly, to the extent that the overall, unit-cell-averaged mean-square atomic displacement is anisotropic, α will be anisotropic; and

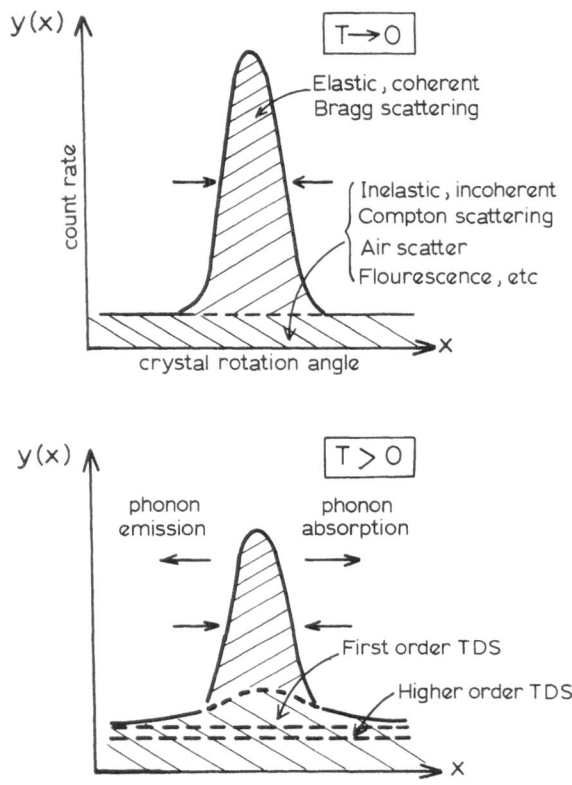

Figure 3. Schematic illustration of Bragg + TDS intensity profiles.

uncorrected TDS anisotropy can lead to the same kind of false quadrupolar difference density features as does uncorrected absorption anisotropy.

Programs for calculating the TDS correction have been developed by Stevens (1974); Sakata, Stevenson, and Harada (1983); and Helmholdt, Braam, and Vos (1983). In typical cases room temperature values of α at $(\sin \theta)/\lambda = 1 \text{Å}^{-1}$ are in the range ~0.05 to

~0.25. The obstacle to calculating the TDS correction is the need to know the elastic constants of the crystal in order to calculate v_s values. Elastic constants are difficult to measure, and values are available for only relatively few crystals. In most cases, the only practical possibility is to measure the diffraction data at the lowest possible temperature and at the highest possible instrumental resolution in order to minimize the TDS intensity included under the integrated Bragg peak. The long tails of the TDS peak fall off as $1/q^2$, where q is the distance in reciprocal space from the reciprocal lattice point. As Figure 3 indicates, it is essential to establish integration limits as close as possible to the true Bragg peak limits in order to minimize the TDS intensity included in the peak integration, without truncating the tails of the Bragg peak. It is a fortunate circumstance that the practically unavoidable TDS and scan profile truncation errors tend to cancel. (Note, incidentally, that the attenuation of the Bragg scattering due to the atomic thermal vibrations does not cause the Bragg peak to be broadened. The full-widths at half-height are the same in both parts of Figure 3.)

For cases of data from crystals with unknown elastic constants, for which an at least approximate TDS correction is believed necessary, Blessing (1984, 1986a, 1986b) has developed a program for deriving an approximate empirical TDS correction by a least-squares analysis, over the three-dimensional data set, of the slopes of the background profile close to the Bragg peak limits. The analysis has been shown (Blessing, 1986a) to give α values that have essentially the correct proportional temperature dependence for multiple temperature data sets from cubic ZnS, the correct inverse-square dependence on the Debye temperature ($\Theta \propto v_s$) for data sets from a series of alkali halides at room temperature, and qualitatively correct anisotropy for data sets from several simple molecular crystals at room temperature. In all of these test cases, elastic constants were known, so that the empirical correction could be compared with an analytical one. The comparison showed that the empirical α values tended to be under-estimates, sometimes by as much as about a third, but presumably an under-correction is less dangerous than an over-correction. The often observed significant differences between U^{jk} values from corresponding X-ray and neutron analyses are probably due at least in part to different TDS errors in the two sets of intensity data.

EXPERIMENTAL PROCEDURES

Crystal Specimen Selection

Crystal quality is the single most important factor determining the outcome of a charge density analysis. A specimen crystal of good quality is the sine qua non for high accuracy, and the best test of crystal quality is examination of a representative sample of reflection peak profiles, as described below.

Choice of sample will of course be dictated by considerations based on the biology, chemistry, or physics of the experimenter's research, but simple model compounds without extraneous substituent groups, such as bulky, hydrogen-rich alkyl or aryl groups, are to be preferred. Hydrogen-rich compounds are problematic, because hydrogen atoms scatter X-rays weakly and are therefore impossible to model very accurately from X-ray data alone. Bulky hydrocarbon-rich substituent groups usually undergo large amplitude librational vibrations, which are not fitted well by the harmonic, rectilinear vibration model.

The chosen compound should yield well-formed, approximately equi-dimensional crystals. Typically, a crystal of dimensions approximately $0.3 \times 0.3 \times 0.3 \text{ mm}^3$ is optimal. A good preliminary room temperature crystal structure analysis against data measured to a resolution corresponding to $(\sin \theta)/\lambda \sim 0.6 \text{ Å}^{-1}$ should converge to a value of

$$R = [\Sigma w(|F_o|^2 - k^{-2} |F_c|^2)^2 / \Sigma w(|F_o|^2)^2]^{1/2}, \quad w = 1/\sigma^2(|F_o|^2),$$

of ~0.05, or less for heavy atom crystals. The U^{jk} should be physically sensible, as judged by thermal ellipsoid drawings (Johnson, 1970), the rigid-bond test (Hirshfeld, 1976), and $TLS+\Phi_i$ molecular tensor analysis of the independent-atom U tensors (Dunitz, Schomaker, and Trueblood, 1988).

Crystals with as many as ~50 non-hydrogen atoms in the crystal chemical unit can be studied, if two to three months of diffractometer time are available, but such large problems will yield less accurate electron densities than the more typical 10-20 atom problems. Recent studies show that the electrostatic potential is a property much less sensitive to experimental errors and resolution limits than the electron density, and studies of only moderate accuracy can yield chemically meaningful molecular electrostatic potential maps. Although noncentrosymmetric space groups lead to substantial phase errors in X-ray-minus-neutron or X-ray-minus-X-ray (high order) difference density maps, multipole modelling of the electron density overcomes the phase problem for X-ray multipole-minus-spherical difference maps. Thus, multipole modelling opens the field of charge density analysis to the many interesting cases of chiral biological molecules.

An important restriction on charge density studies is imposed by the atomic number of the crystal's constituent atoms. Heavy atom crystals require large absorption corrections, and for many of the heavier atoms the mass absorption corrections are quite uncertain. There may also be uncertainties in heavy atom X-ray form factors due to sizeable relativistic and electron correlation effects. More importantly, from period to period in the periodic table, the ratio of valence to core electrons decreases, and the valence electrons are responsible for an ever smaller fraction of the X-ray scattering, as indicated in Table 2.

Furthermore, tables of X-ray form factors partitioned into core and valence shell components (Clementi, 1965; Cromer and Waber, 1974) show that with increasing atomic

Table 2. Valence/core electron ratios

Atom	H	C	N	O	P	S	Fe	Mo
Z	1	6	7	8	15	16	26	42
Core	0	2	2	2	8	8	18	36
Valence	1	4	5	6	5	6	8	6
Ratio	∞	2	2.5	3	0.625	0.753	0.444	0.167

radius, the outer-shell valence scattering becomes confined to ever smaller scattering angles. For example, significant scattering by Fe 3d electrons extends to a reciprocal space radius ($\sin \theta$)/λ of ~0.6 Å^{-1}, but ~90% of the scattering by Mo 4d electrons is confined within a radius of ~0.3 Å^{-1}, where there will be few reflections to carry the valence density information. For very large atoms the valence density information content of the X-ray data becomes hopelessly small.

In much of the sequel, we assume a specimen crystal of suitable quality and size, mounted on the diffractometer with the orientation matrix known at least approximately. "Suitable quality" means that the crystal must be single - untwinned, without adhering satellite crystals, or inclusions, bubbles, or other foreign matter. The crystal need not be beautiful. Indeed, very beautiful specimens often tend to a degree of perfection that gives rise to serious extinction effects. On the other hand, if the linear absorption coefficient is large, then beauty is a virtue, since an accurate absorption correction requires correctly indexed and accurately measured crystal faces. "Suitable size" means large enough to yield a sufficient number of significant reflections - a common rule of thumb is ten reflections per parameter - and small enough to be uniformly illuminated by the incident X-ray beam and not require excessively large absorption or extinction corrections.

Instrument Performance and Crystal Quality Checks

The months of experimental and computational effort that charge density analyses entail can be wasted if not preceded by suitable calibrations. A rational sequence of checks to perform before measuring a diffraction data set is given in Table 3. This sequence is intended to, first, establish that the quantum detector is well-behaved; then, use the detector to investigate the size, shape and uniformity of the X-ray source;

Table 3. Instrument and crystal assessment scans

---

1. Detector scans
    1.1. Area of uniform response
    1.2. Pulse energy distribution
    1.3. Count-rate linearity
2. Source scans
    2.1. Spatial flux distribution
        2.1.1. From the X-ray tube focal spot
        2.1.2. From a monochromator crystal
    2.2. Spectral intensity distribution
3. Crystal scans
    3.1. Mosaic width and anisotropy
    3.2. Effective source width (take-off angle)
    3.3. Spectral dispersion
4. Mathieson (1982a) scans
    Experimental deconvolution of the
    source (σ), spectral (λ), and crystal mosaic (μ)
    components of the two-dimensional I(ω,2θ)
    reflection profile

---

and, finally, use the detector and source to establish the mosaic quality of the specimen crystal. Crystal scans should be done for each and every specimen; the detector and source calibrations are usually quite stable for some months with present day instruments.

To perform the instrument and crystal check scans, first select an incident beam tunnel aperture large enough to ensure that the whole of the specimen crystal can "see" the whole of the source focal spot, but not so large that there will be excessive extraneous radiation, which can only raise the background and lower the signal-to-noise ratio due to scattering by an unnecessarily large volume of air around the crystal. An appropriate aperture can be selected using a simple geometrical optics ray construction, as shown in Figure 4.

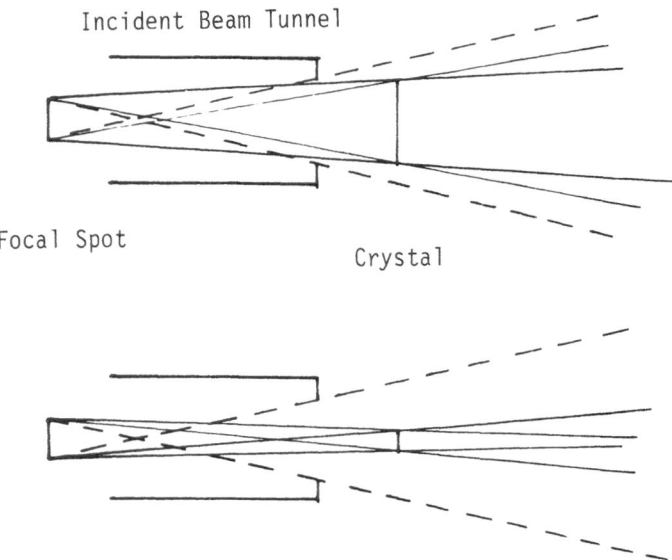

Figure 4. Divergence and crossfire in the beam from the X-ray tube target for maximum target and crystal, and minimum beam tunnel aperture diameters.

For the <u>detector area scan</u>, use a narrow source (minimal tube target take-off angle) and tune the crystal rocking angle ω to peak intensity for a low-angle reflection using a wide detector aperture, and perform a 2θ scan. Keep the crystal stationary (constant ω), and move only the detector (2θ). This sweeps the narrow reflected beam across the width of the detector aperture, and the response should be uniform over the maximum area that will be exposed during the data set measurements.

For the <u>pulse energy scan</u>, use a narrow source and wide detector aperture, and tune ω to peak intensity for a moderately strong low angle reflection to obtain a good signal, monochromated by Bragg reflection. Set a narrow electronic window on the pulse energy analyzer, and scan the threshold, lower level voltage to establish the correct lower level and window settings (Parrish and Kohler, 1956). Note that the pulse energy analyzer spectrum for a NaI(Tl) scintillation detector is broadened and energy resolution is degraded at high quantal flux rates (Mathieson, 1982b), so that beam attenuation for very strong reflections is needed to preserve not only the count rate linearity, but also the energy analyzer resolution.

For the <u>count-rate linearity scan</u>, tune ω to peak intensity for a moderately strong reflection, and measure the count rate with and without a beam attenuator as a function of tube current to obtain the dead time and attenuator factor (Chipman, 1969; Jennings, 1980). It is not possible to determine the attenuator factor to a precision of < 1% using peak-top stationary crystal count rates, because the attenuator absorption can vary by up to several percent across the spectral width of a reflection. After the dead time is determined, the attenuator factor should be redetermined from ratios of integrated reflection measurements with and without the attenuator. Dead time corrections are easy to apply during data processing if the step-scan count-rate profiles of the reflection measurements are recorded and analyzed. Errors in attenuator and dead time corrections can be mistaken for extinction (Mathieson, 1982c).

For the <u>source focal spot scans</u>, adjust the target take-off angle α to obtain an effectively square source,

$$\alpha = \arctan\,(1\ \text{mm}/10\ \text{mm})\ = 5.7° \quad \text{for a "normal" focus tube, and}$$

$$\alpha = \arctan\,(0.4\ \text{mm}/8\ \text{mm})\ = 2.9° \quad \text{for a "fine" focus tube.}$$

(Optimum take-off angles, to be determined as described below, might typically be about half these values.) To scan the focal spot, adjust to LOW power (MINIMUM kV and mA), ATTENUATE the direct beam, and position the detector at 2θ = 0. Then, scan the horizontal source width by moving the focal spot across 2θ = 0 using the tube shield horizontal (Y) translation adjustment with a narrow vertical slit (~100 μm) as detector aperture. Scan the vertical source height by moving the X-ray tube across the equatorial plane using the the tube shield height (Z) adjustment with a narrow horizontal slit as detector aperture.

Equally or more informative is a pinhole focal spot photograph, which is conveniently obtained using a 100 or 200 μm electron microscope aperture between the source and detector and a dental X-ray film at the detector aperture. The effective magnification is given by the ratio of the source and detector radii; it will be ~1:1 if the pinhole is mounted on a goniometer head at the crystal position, and ~10:1 if the pinhole can be placed at the source end of the incident beam tunnel. Typical focal spot images are sketched in Figure 5, which indicates that the separations of the cathode filament windings can sometimes be seen in the horizontal source profile, especially if the anode take-off angle is large; that the vertical flux profile is a doublet, because the projected current density of electrons from the sides of filament coil is greater than from the middle; and that the filament is often deformed so that the source is not square or rectangular. Filament deformation changes with age in use, and can be mistaken for changes in the mosaicity of a test specimen crystal or monochromator crystal.

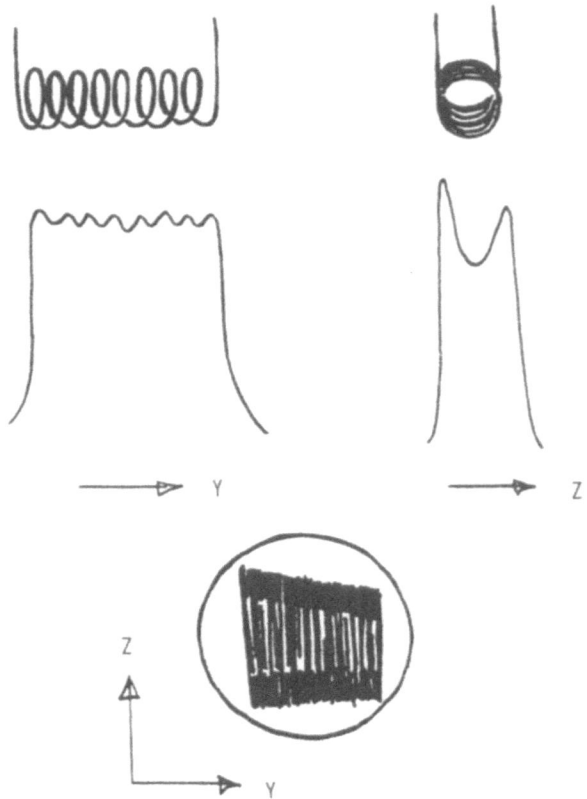

Figure 5. Schematic illustrations of various X-ray tube focal spot images: a horizontal (Y) intensity profile at large take-off angle, a vertical (Z) profile, and a pinhole photograph for a take-off angle $\alpha = \arctan (Z/Y)$ giving a nominally square source.

The effective focal spot from an incident beam monochromator crystal cannot be examined using the tube shield translations, but pinhole photographs are instructive, and the spatial flux distribution near the instrument center should be mapped using an electron microscope aperture mounted on a goniometer head. This is conveniently done by using the goniometer head height adjustment to displace the aperture from the instrument center in steps of ~10 μm at $2\theta = \omega = \varphi = 0$, and record the count rates at each displacement step as χ is rotated through 360° in steps of ~30°. Depending upon whether the monochromator geometry is perpendicular (as on Picker and Enraf-Nonius instruments) or parallel (as on Siemens-Nicolet née Syntex instruments), the flux distribution in the beam from the monochromator crystal will be narrowed in the vertical or horizontal direction, respectively.

This effect (Coppens, Ross, Blessing, Cooper, Larsen, Leipoldt, Rees, and Leonard, 1974; and Katrusiak and Ryan, 1988) is due to a mosaic focusing by the monochromator crystal, or ray selection from the divergent beam from the tube target, as indicated in Figure 6.

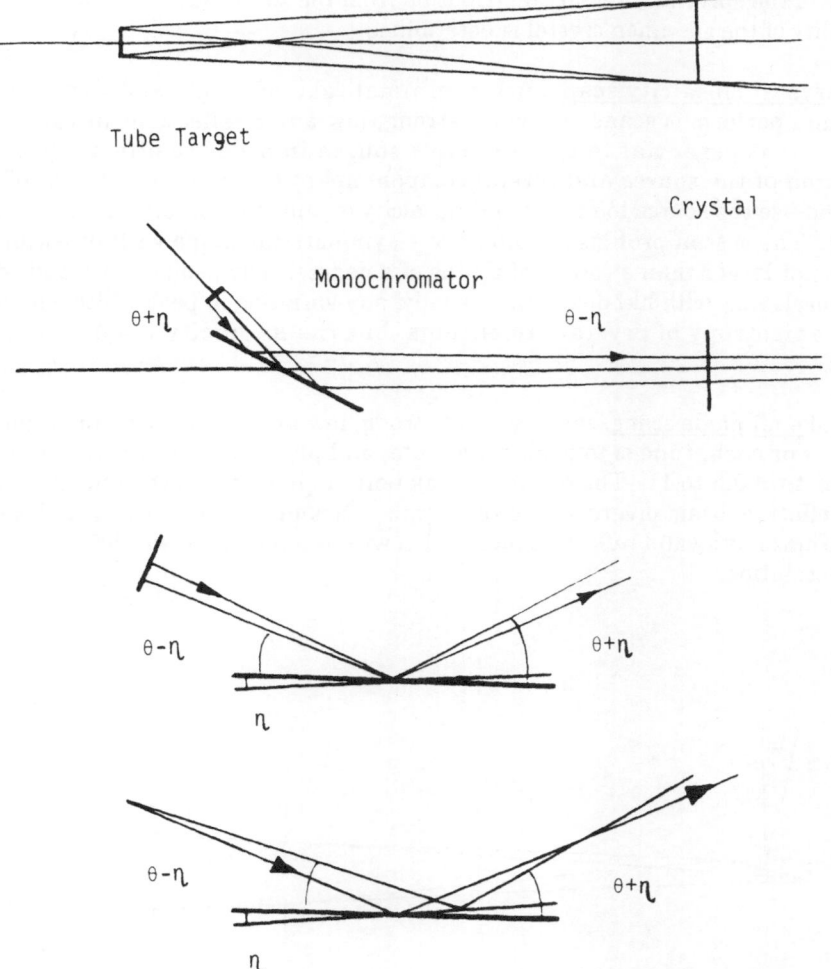

Figure 6. Geometrical optics of reflection of rays from a tube target by a flat mosaic monochromator crystal with Bragg angle θ and mosaic spread η.

Rays from the target incident on the monochromator at angle θ + η are reflected at θ - η, and rays incident at θ - η are reflected at θ + η, so the monochromated beam is convergent by 2η, and its diameter is limited to the minimum projected diameter of the spot illuminated on the monochromator crystal. The η values for commonly used pyrolytic graphite monochromator crystals are of the order of tenths of a degree, so the monochromated beam is practically parallel. This means that divergence and crossfire constructions like those shown in Figure 4 do not apply, and the area of uniform illumination at the instrument center may be only ~200 or 300 μm in minimum diameter. If any specimen crystal diameters exceed the minimum beam diameter, a non-homogeneous beam correction is required, based on a flux distribution calibration and a Gaussian quadrature grid calculated in the absorption correction program (Harkema, Dam, van Hummel, and Reuvers, 1980). This of course also requires correct indexing and accurate measurement of the specimen crystal faces.

Scans of the spectral intensity distribution from the source are best made after the mosaic quality of the specimen crystal is determined.

For crystal mosaicity scans, use a minimal take-off angle and wide detector aperture, and perform ω scans of several strong, low-angle reflections in various hkl directions. It is essential to use a narrow source in order to effect a practical deconvolution of the source and crystal components of the instrumental resolution function, and use the source to resolve the mosaicity or sub-microscopic fault structure of the crystal. The ω scan profiles should have a symmetrical shape with full-widths at half-height not larger than a couple of tenths of a degree. The peak shapes and widths will, in general, vary with hkl direction. Usually, any variation in peak width will be due mainly to anisotropy of crystal dimensions, but the mosaicity itself can also be anisotropic.

For take-off angle scans, select several strong, low-angle reflections in various hkl directions. For each, tune ω to peak count rate, and plot peak count rate versus α at intervals of Δα = 0.5 to 1°. The optimized take-off angle matches the effective source width and effective beam divergence to the crystal size and mosaic width, as indicated in Figure 7. This allows valid reflection integration with minimal scan widths and maximal reflection resolution.

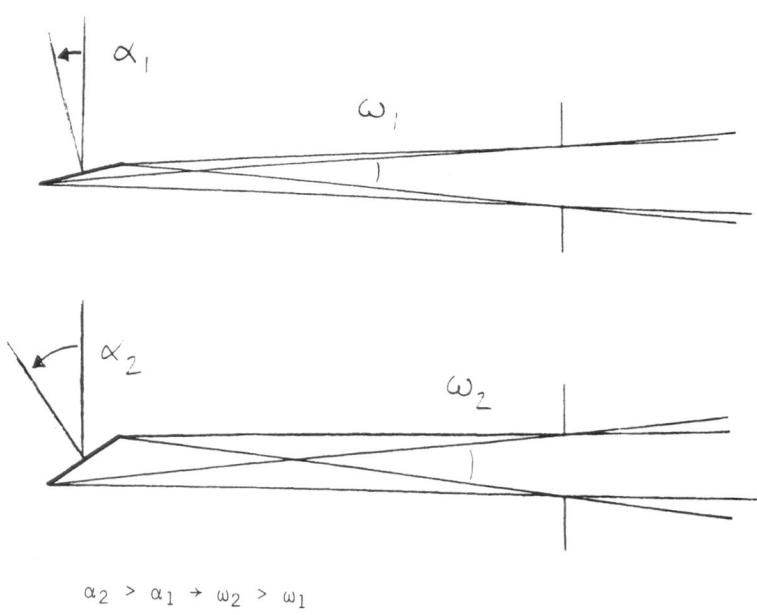

$$\alpha_2 > \alpha_1 \rightarrow \omega_2 > \omega_1$$

For a given aperture or specimen diameter
effective beam divergence ω increases with
take-off angle α

Figure 7. Schematic illustration of variation of effective beam divergence ω with tube target take-off angle α.

An example of a set of take-off angle scans is shown in Figure 8. In this example, the optimum take-off angles had nominal values of α = 4.0, 2.5, and 4.0° for the hkl = 200, 040, and 004 reflections, respectively, all with 2θ of ~12°. The plots showed up an error of 0.5° for the nominal zero of the α scale, and a value of α 3.25° (nominal) or 2.75° (actual) was chosen for measuring the data set. This represented a compromise between minimizing the necessary reflection scan widths and maximizing the efficiency of illumination of the specimen crystal. In the example shown, the differences among optimum α values corresponded to the differences among the peak widths found in the crystal mosaicity ω scans, which in turn corresponded to the differences among the [100], [010] and [001] crystal dimensions. Note that valid measurements of relative integrated intensities can be made with take-off angles either smaller or larger than optimum. In the former case, the integrated intensities will be uniformly scaled down, and narrower scans will give valid reflection integrations; in the latter case, the intensities will be uniformly scaled up, but wider scans will be needed for valid integrations.

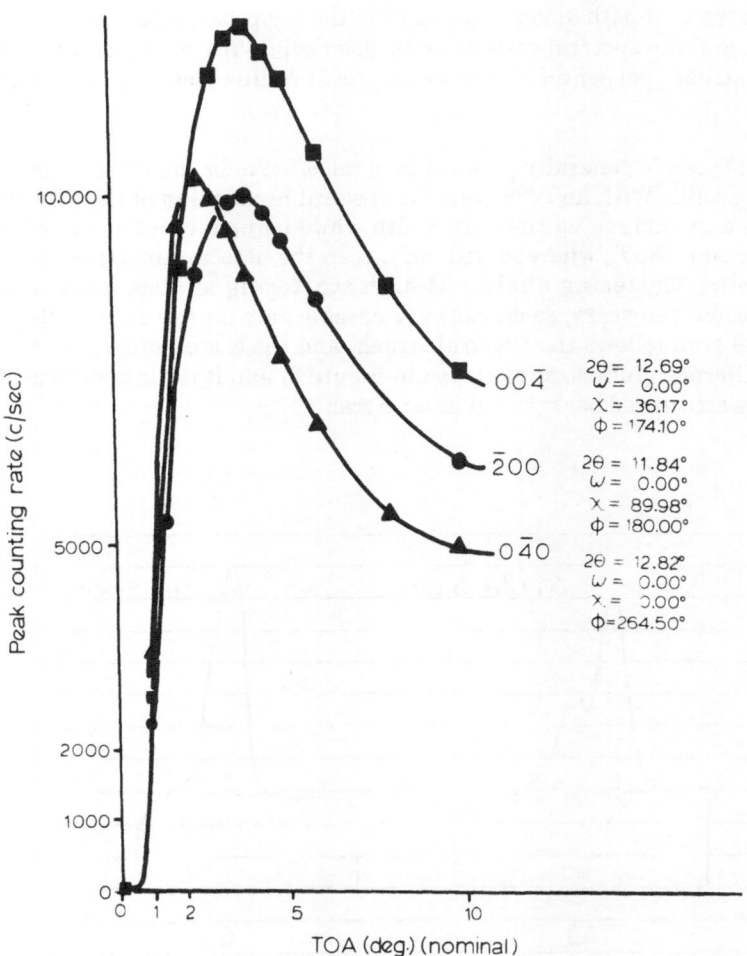

Figure 8. Take-off angle scans. Peak count rate plotted vertically <u>vs.</u> take-off angle plotted horizontally.

Unfortunately, neither Enraf-Nonius nor Siemens-Nicolet (née Syntex) diffractometers have a mechanism to adjust a displaced target focal spot to the α-axis. Present day tube manufacturing standards assure that the focal spot displacement will be quite small, but not always negligible. Therefore, it might be necessary to re-tune ω to recover peak count rate after each α-adjustment in the take-off angle scan, and finally to adjust the horizontal tube translation to re-align the source after setting optimum α, using the specimen crystal as an alignment crystal. If a monochromator with parallel geometry, as on the Siemens-Nicolet instrument, is being used, the take-off angle is fixed according to the monochromator Bragg angle. In this case, the beam is narrowed in the horizontal direction, as described above, and the experimenter cannot vary the horizontal beam width.

For <u>spectral dispersion scans</u>, set to the optimized take-off angle, select a wide detector aperture, and record ω/2θ scans of increasing orders <u>n</u>h <u>n</u>k <u>n</u>l, <u>n</u> = 1,2,3,... in several hkl directions, using generous scan widths. The recorded spectra should be interpretable as compared with a standard spectrum such as given by Roberts and Parrish (1962, Fig. 2.3.1, p. 73). An ω/2θ scan is necessary for this experiment, because the ω/2θ scan traces a path along a radius from the reciprocal lattice origin in the Ewald construction, and the spectral dispersion is also radial. An ω scan traces a path across the spectral streak, perpendicular to a reciprocal radius (see, e.g., Stout and Jensen, 1968, p. 182).

The ω/2θ scan is generally preferable to the ω scan for the data set measurements for several reasons. With an ω/2θ scan the spectral broadening of the reflection peak is accommodated by increasing the scan width while using a constant detector aperture width (Mathieson, 1983), whereas with an ω scan the detector aperture must be opened with increasing scattering angle. At high scattering angles, unreasonably wide apertures become necessary, so ω scans are useable only for low angle reflections. Also, since the ω/2θ scan follows the spectral streak, and the ω scan crosses it, the two scans have quite different profiles, as sketched in Figure 9, and it is quite difficult to properly correct for the structured background in an ω scan.

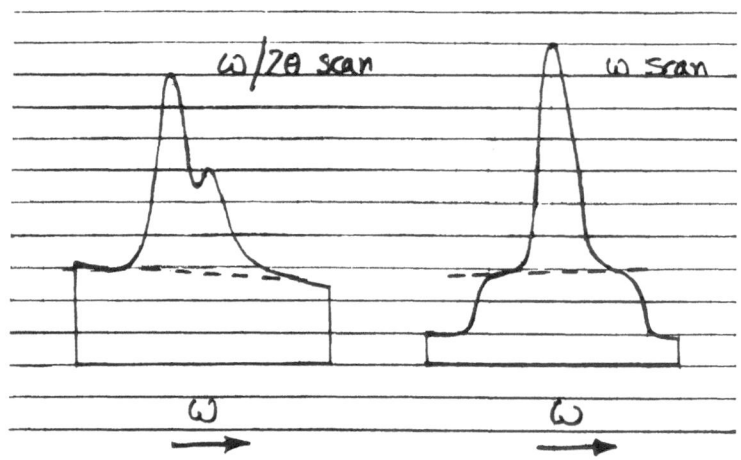

Figure 9. Schematic sketches of ω/2θ and ω scan profiles.

The spectral dispersion check scans should be repeated with several different detector apertures to select one that permits the detector to "see" all of the reflected beam, without "seeing" excessive extraneous background scatter. The spectral scans also provide the basis for empirically selecting suitable scan width parameters. For measurements with an unfiltered, β-filtered, or balanced filtered beam, or a beam from an incident beam monochromator with perpendicular geometry (as on Enraf-Nonius instruments), the scan widths should be

$$\Delta\omega = A + B \tan \theta \, ,$$

where the tan θ term allows for the spectral broadening. For a parallel geometry monochromator (as on Siemens-Nicolet instruments), the scan widths should be

$$\Delta\omega = [A^2 - B_1 \tan \theta + B_2 (\tan \theta)^2]^{1/2} \, ,$$

because the spectral dispersion by the monochromator crystal and by the specimen crystal both occur in the equatorial plane (see refs. cited by Blessing, 1986b). The negative sign for $B_1 > 0$ applies for the usual $\theta > 0$, parallel arrangement, and minimum $\Delta\omega$ occurs at $\tan \theta = B_1/(2B_2)$, which is in the range $1 < \tan \theta/\tan \theta_m < 2$ near $\theta = \theta_m$. (For a $\theta < 0$ antiparallel arrangement, $B_1 < 0$.)

If the diffractometer software does not allow for quadratic dependence of the squared scan width on tan θ for a parallel monochromator, very serious scan truncation errors for the high-angle measurements will occur. Even without a parallel monochromator, serious scan truncation errors are possible if the chosen value for B allows only for α_1-α_2 line separation and not for α_1 and α_2 line broadening, as shown in Table 4.

Table 4. Comparison of α_1-α_2 doublet separation with α_1 and α_2 line widths (Compton and Allison, 1935). All values in Å.

	Wavelengths λ		Doublet Splitting $\lambda\alpha_2 - \lambda\alpha_1$	Line Broadening $\Delta\lambda$	
	α_1	α_2		α_1	α_2
Cu Kα	1.54056	1.54439	0.00383	0.00058	0.00077
Mo Kα	0.70930	0.71359	0.00429	0.00029	0.00032
Ag Kα	0.55941	0.56380	0.00439	0.00028	0.00029

The line broadening, given as full-width at half-height (FWHH), is some 6 to 20% of the doublet splitting. A reasonable estimate for the base-width of the approximately Lorentzian-shaped spectral lines is ~5 × FWHH. Thus a value

$$B = \Delta\lambda/\lambda = (\lambda\alpha_2 - \lambda\alpha_1)/[(\lambda\alpha_1 + \lambda\alpha_2)/2] \, ,$$

which allows for the α_1-α_2 line separation but not the α_1 and α_2 line broadening, is too small by some 30 to 100%.

A particularly clear impression of the components of the convolution structure of Bragg reflection intensity distributions is provided by the technique described in detail in a noteworthy series of papers by Mathieson (1982a, 1989) and Mathieson and Stevenson (1986). For Mathieson scans, set to the optimized take-off angle, and install a narrow vertical slit at the detector aperture. This provides high resolution in ω and 2θ, with

angular aperture = $(180°/\pi) \times$ slit width / detector radius ,

0.1 mm = 100 μm $\Rightarrow$ ~0.02° for R_D = 250 mm.

At a series of ~100 steps in ω, with $\Delta\omega = 0.01°$ per step, perform a series of step scans of ~100 steps in 2θ, with $\Delta 2\theta = 0.02°$ per step. Contour plots mapping $I(\omega, 2\theta)$ over the ~100 $\times$ 100 points of the $\Delta\omega$, $\Delta 2\theta$ grid can be prepared using affine transformations of the plotting axes to reveal the two-dimensional intensity distributions corresponding to one-dimensional ω, ω/θ, and $\omega/2\theta$ scan profiles. Appropriate sections through these plots reveal the component contributions of the source breadth σ, the crystal mosaic breadth μ, and the wavelength spread λ to the composite convolution product $I(\omega, 2\theta)$. The plots provide an improved basis for the selection of scan variables for the bulk data set of one-dimensional $\omega/2\theta$ scan profile measurements (Mathieson, 1989).

Instrument Stability Checks

Before the diffractometer software is "released" for the several weeks to several months needed to measure a charge density data set - Of course, daily inspection, and sometime intervention, by the experimenter is required - a series of experimental stability checks is in order, as outlined in Table 5.

Table 5. Tests of instrument stability

---

1. Detector dark counts
2. Independent, stable X-ray source counts

   Fe-55 ($t_{1/2}$ = 2.5 y) sealed sample

   $$^{55}_{26}\text{Fe} \xrightarrow{\text{K-electron capture}} {}^{55}_{25}\text{Mn} + \gamma$$

   γ = Mn Kα, λ = 2.1031 A

3. LOW power, ATTENUATED direct beam counts
4. Stationary crystal, Bragg peak-top counts
5. Repeated scans of
   A triple of axial reflections, h00, 0k0, 00l, or
   A form of reflections {hkl}, or
   A set of scans of a single reflection hkl at $\Delta\psi$ = 30 or 60°.
6. Dietrich (1976) analysis of repeatedly measured profiles

This sequence is intended to establish: (1) and (2) the stability of the detector electronics; (3) the stability of the X-ray tube output; (4) the stability of the cryo-gas stream temperature and flow velocity, and of the crystal mount for various orientations in the gas stream, which typically has a considerable flow velocity; (5) the reproducibility of the crystal positioning; and (6) the combined effects of mechanical and electronic reproducibility of reflection count rate profiles. Non-statistical variations might be due to: room temperature variations, which can affect counter electronics performance; atmospheric or air conditioning temperature, pressure, and humidity variations, which affect air absorption and possibly electronics performance; cooling water temperature or flow variations, which affect X-ray tube output; photochemical radiation sensitivity of the specimen crystal; χ-dependent absorption by the crystal mount; non-uniform crystal illumination; or mechanical malfunction of the goniostat. The diagnostic power of the statistical analysis programmed by Dietrich (1976) can uncover problems that might require manufacturer's service for the diffractometer.

Data Collection Strategies

For a given amount of beam time per unique hkl, one gets more normally distributed - and probably smaller - errors in I(hkl) from an average of $\underline{n}$ symmetry equivalent or ψ-rotation equivalent measurements, scanned $\underline{n}$ times faster than from a single measurement scanned slowly. It is, of course, prudent to measure a unique data set first, and then proceed to measure set after set of equivalent data. To improve data averaging, and to check inter-set scaling, one should re-measure the axial and zonal reflections along with each equivalent general reflection set.

If possible, measure a full sphere of reflections, but do not measure the sphere $\pm$h, $\pm$k, $\pm$l, $\theta \leq \theta_{max}$ in a single run. Rather, in several runs, measure the several unique sectors of data to θ_{max}, repeating the boundary zones of data in each run. For example, for a monoclinic crystal, measure in turn unique sets

$\pm$h, k $\leq$ 0, l $\leq$ 0
$\pm$h, k $\leq$ 0, l $\geq$ 0
$\pm$h, k $\geq$ 0, l $\leq$ 0
$\pm$h, k $\geq$ 0, l $\geq$ 0

to obtain four measurements of not only each general reflection, hkl, with h, k, and l $\neq$ 0, but also each zonal reflection with h, k, or l = 0. Alternatively, for a triclinic crystal, measure at, say, ψ = 0, 60, 120, and 180°, but measure separate, successive data sets, each at a constant ψ. For crystals of orthorhombic or higher symmetry, one can measure only a hemisphere or less, but one should aim for no less than three measurements per unique datum, especially for the low-angle data, which carry most of the valence electron density information.

To reduce decay corrections for crystals susceptible to radiation damage, one should measure in shells of 2θ, from high- to low-angle, in order to measure the more sensitive high-angle data earlier, rather than using the common zigzag index-increment loops. To reduce absorption, extinction, and multiple reflection errors, one can measure overlapped data sets from two crystals - a small crystal for the low-angle data, and a larger one for the high-angle. This, of course, introduces the problem of inter-crystal scaling, which requires a good overlap of the two data sets.

To manage the very large number of high-angle data, one can measure only the high-angle reflections with E(hkl) > E(threshold) as calculated from the crystal structure determined from the low-angle, unique data, e.g., a Cu sphere. Although charge density analyses do generally require that the data set be a complete to a resolution corresponding to (sin θ)/λ of ~0.8 Å^{-1}, using only a partial data set at higher resolution introduces little bias.

DATA REDUCTION AND ERROR ANALYSIS

The computational processing of the diffraction measurements prior to the charge density analysis by least-squares modelling and Fourier mapping - colloquially termed "Bragg filtration" by Professor Robert Farrell Stewart - has been described in several recent publications (Blessing, 1986a, 1986b, 1986c, 1989, 1990; Blessing and Langs, 1987, 1988) that report results of an extensive data processing programming effort. The FORTAN source codes are available on request. They are commented in detail, and come accompanied by detailed user's notes, which supplement the published program descriptions.

ACKNOWLEDGEMENTS

RHB thanks his colleagues, especially George DeTitta, David Langs, and Grant Moss, for numerous helpful discussions of the subjects of this chapter, and he gratefully acknowledges support of research related to these subjects by USDHHS PHS NIH grants no. GM34073 and DK19856. CL thanks his collaborators and students at the Université de Nancy, and is grateful for support from the Centre National de la Recherche Scientifique and the Université de Nancy I. A special note of thanks goes to Dale Swenson and Mohamed Souhassou for their good work as teaching assistants for the tutorials on this chapter at the study institute in Sant Feliu de Guíxols.

REFERENCES

Becker, P. J. (1980a). In Electron and Magnetization Densities in Molecules and Crystals, NATO Advanced Studies Institute, Series B, edited by P. J. Becker, pp. 213-236. New York: Plenum.

Becker, P. J. (1980b). In Computing in Crystallography, edited by R. Diamond, S. Ramaseshan, and K. Venkatesan, pp. 3.01-3.24. Bangalore: Indian Academy of Sciences.

Becker, P. J., and Coppens, P. (1974). Acta Cryst. A30, 129-147.

Blessing, R. H. (1984). Acta Cryst. A40, C-156.

Blessing, R. H. (1986a). Am. Cryst. Assoc. Meeting, Hamilton, Ontario, Canada, June 1986, Abstr. No. V4.

Blessing, R. H. (1986b). Crystallogr. Rev. 1, 3-58.

Blessing, R. H. (1986c). J. Appl. Cryst. 19, 412.

Blessing, R. H. (1989). J. Appl. Cryst. 22, 396-397.

Blessing, R. H. (1990). IUCr Congress, Bordeaux, France, July 1990, Abstr. No. PS-02.02.27.

Blessing, R. H. and Langs, D. A. (1987). J. Appl. Cryst. 20, 427-428.

Blessing, R. H. and Langs, D. A. (1988). Acta Cryst. A34, 517-525.

Bond, W. L. (1967). In International Tables for X-Ray Crystallography, Vol. II, edited by J. S. Kasper and K. Lonsdale, pp. 299-301. Birmingham, England: Kynoch Press.

Chidambaram, R. (1980). In Computing in Crystallography, edited by R. Diamond, S. Ramaseshan, and K. Venkatesan, pp. 2.01-2.20. Bangalore: Indian Academy of Sciences.

Chipman, D. R. (1969). Acta Cryst. A25, 209-213.

Clementi, E. (1965). "Tables of Atomic Functions," IBM J. Res. and Dev., Vol. 9, Supplement.

Coppens, P. (1968). Acta Cryst. A24, 253-257.

Coppens, P., Ross, F. K., Blessing, R. H., Cooper, W. F., Larsen, F. K., Leipoldt, J. G., Rees, B., and Leonard, R. (1974). J. Appl. Cryst. 7, 315-320.

Compton, A. H., and Allison, S. K. (1935). X-rays in Theory and Experiment, 2nd ed., pp. 744-745. New York: Van Nostrand.

Cromer, D. T., and Waber, J. T. (1974). In International Tables for X-Ray Crystallography, Vol. IV, edited by J. A. Ibers and W. C. Hamilton, pp. 71-151. Birmingham, England: Kynoch Press.

DeTitta, G. T. (1985). J. Appl. Cryst. 18, 75-79.

Dietrich, H. (1976). J. Appl. Cryst. 9, 205-210.

Dunitz, J. D., Schomaker, V., and Trueblood, K.N. (1988). J. Phys. Chem. 92, 856-867.

French, S., and Wilson, K. (1978). Acta Cryst. A34, 517-525.

Harkema, S., Dam, J., van Hummel, G. J., and Reuvers, A. J. (1980). Acta Cryst. A36, 433-435.

Helmholdt, R. B., Braam, A. W. M., and Vos, A. (1983). Acta Cryst. A39, 90-94.

Hirshfeld, F. L. (1976). Acta Cryst. A32, 239-244.

Hubbell, J. H., McMaster, W. H., Kerr Del Grande, N., and Mallett, J. N. (1974). In International Tables for X-Ray Crystallography, Vol. IV, edited by J. A. Ibers and W. C. Hamilton, p. 50. Birmingham, England: Kynoch Press.

James, R. W. (1982). The Optical Principles of the Diffraction of X-Rays, pp. 49-50, 268-299.

Jennings, L. D. (1980). Symposium on Accuracy in Powder Diffraction, U. S. National Bureau of Standards Special Publication 567, pp. 73-81.

Johnson, C. K. (1970). In Crystallographic Computing, edited by F. R. Ahmed, S. R. Hall, and C. P. Huber, pp. 227-230. Copenhagen: Munksgaard Publishers.

Katrusiak, A., and Ryan, T. W. (1988). Acta Cryst. A44, 623-627.

Kato, N. (1980). In Electron and Magnetization Densities in Molecules and Crystals, NATO Advanced Studies Institute, Series B, edited by P. J. Becker, pp. 237-253. New York: Plenum.

Lehmann, M. S., and Larsen, F. K. (1974). Acta Cryst. A30, 580-584.

Le Page, Y., and Gabe, E. J. (1979). Acta Cryst. A35, 73-78.

Mathieson, A. McL. (1982a). Acta Cryst. A38, 378-387.

Mathieson, A. McL. (1982b). J. Appl. Cryst. 15, 98-99.

Mathieson, A. McL. (1982c). J. Appl. Cryst. 15, 99-100.

Mathieson, A. McL. (1983). J. Appl. Cryst. 16, 572-573.

Mathieson, A. McL. (1989). Acta Cryst. A45, 613-620, and earlier papers cited therein.

Mathieson, A. McL, and Stevenson, A. W. (1986). Acta Cryst. A42, 435-441, and earlier papers cited therein.

Parrish, W. and Kohler, T. R. (1956). Rev. Sci. Instrum. 27, 795-808.

Post, B. (1979). Acta Cryst. A35, 17-21.

Roberts, B. W., and Parrish, W. (1962). In International Tables for X-Ray Crystallography, Vol. III, edited by C. H. MacGillavy, G. R. Rieck, and K. Lonsdale. Birmingham, England: Kynoch Press.

Sakata, M., Stevenson, A. W., and Harada, J. (1983). J. Appl. Cryst. 16, 156-156, and references cited therein.

Spackman, M. A. (1987). J. Appl Cryst. 20, 256-258.

Stevens, E. D. (1974). Acta Cryst. A30, 184-189.

Stout, G. H. and Jensen, L. H. (1968). X-Ray Structure Determination. New York: Macmillan.

Tanaka, K., and Saito, Y. (1975). Acta Cryst. A31, 841-844.

Willis, B. M. T., and Willis, A. W. (1975). Thermal Vibrations in Crystallography. Cambridge, England: Cambridge University Press.

Zacharaisen, W. H. (1967). Acta Cryst. 23, 558-564.

BIBLIOGRAPHY

The foregoing list of specific references provides entry points into the large literature on our subject. We have also found the following general references to be especially helpful and instructive.

Reviews on techniques for accurate diffraction data:

T. C. Furnas, Jr., Single Crystal Orienter Manual, General Electric Co., 1957.

Feil, D. (1977). In "Electron Density Mapping in Molecules and Crystals," edited by F. L. Hirshfeld. Israel J. Chem. 16(2, 3), 149-153.

Rees, B. (1977). Ibid., pp. 154-158 and 180-186.

Coppens, P. (1978). In Neutron Diffraction, edited by H. Dachs, pp.71-111. Berlin: Springer-Verlag.

Seiler, P. (1985). In Lecture Notes, International School of Crystallography, Static and Dynamic Implications of Precise Structural Information, edited by A. Domenicano, I. Hargittai, and P. Murray-Rust, pp. 79-94. Rome, Italy: CNR.

Seiler, P. (1987). Chimia, 41, 104-116.

ACA and IUCr symposium proceedings:

"Accuracy in X-Ray Intensity Measurement," Trans. Am. Cryst. Assoc., Vol. 1, edited by S. C. Abrahams, 1965. Dayton, Ohio: Polycrystal Book Service.

"Proceedings of the International Meeting on Accurate Determination of X-ray Intensities and Structure Factors," edited by U. W. Arndt and A. McL. Mathieson, Acta Cryst. (1969). A25(1), 1-276.

"International Symposium on Accuracy in Structure Factor Measurement," Australian J. Phys., (1988). 41(3), 337-523.

"Internationa Symposium on X-Ray Powder Diffractometry," Australian J. Phys., (1988). 41(2), 101-336.

Earlier NATO-ASI proceedings:

"Lecture Notes on Experimental Aspects of X-Ray and Neutron Sinlge Crystal Diffraction Methods," NATO Advanced Studies Institute, Aarhus University, Denmark, 31 July to 11 August 1972.

Lehmann, M. S. (1980). In Electron and Magnetization Densities in Molecules and Crystals, NATO Advanced Studies Institute, Series B, edited by P. J. Becker, pp. 287-322 and 355-372. New York: Plenum.

Notebooks from tutorial workshops organized by the Continuing Education Committee of the ACA:

Lecture Notes, Diffractometry Tutorial, edited by H. Berman, Am. Cryst. Assoc. Meeting, Norman, Oklahoma, March 1978. Dayton, Ohio: Polycrystal Book Service.

Lecture Notes, Tutorial on Low Temperature X-Ray Diffraction Apparatus and Techniques, edited by R. Rudman, Am. Cryst. Assoc. Meeting, Asilomar, California, February 1977. Dayton, Ohio: Polycrystal Book Service.

Lecture Notes, Tutorial on Accurate Single-Crystal Diffractometry, edited by R. H. Blessing, Am. Cryst. Assoc. Meeting, New Orleans, Louisiana, April 1990. Dayton, Ohio: Polycrystal Book Service.

NECESSITY AND PITFALLS OF LOW-TEMPERATURE MEASUREMENTS

Finn Krebs Larsen

Chemistry Department
Aarhus University
DK-8000 Aarhus C
Denmark

INTRODUCTION

The electron density distribution, $\rho_T(\mathbf{r})$, of a crystalline solid may be calculated as the Fourier summation of X-ray structure factors, $F_T(\mathbf{H})$

$$\rho_T(\mathbf{r}) = \frac{1}{V} \sum_{\mathbf{H}} F_T(\mathbf{H}) \exp(-2\pi i \mathbf{H} \cdot \mathbf{r})$$

Structure factors are derived from intensities, $I_T(\mathbf{H})$ obtained from a diffraction experiment conducted at temperature T, and to a first approximation $F_T(\mathbf{H}) \propto \mathrm{sqrt}(I_T(\mathbf{H}))$. The $\rho_T(\mathbf{r})$ formula immediately shows, that the degree of detail which can be achieved in the experimentally determined charge density depends on how extensive and complete the set of structure factors is. The total electron density is the sum of atomic contributions,

$$\rho_T(\mathbf{r}) = \sum_{j=1}^{N} \rho_{T,j}(\mathbf{r})$$

For a structure of N atoms at positions $\mathbf{r}_j$ in the asymmetric unit the theoretical structure factor can be written as a sum of atomic scattering factors, $f_j(\mathbf{H})$

$$F(\mathbf{H}) = \sum_{j=1}^{N} f_j(\mathbf{H}) \exp(2\pi i \mathbf{H} \cdot \mathbf{r}_j)$$

Observed X-ray structure factors fall off with increasing scattering angle as the combined effect of the functional shapes of individual scattering factors and temperature factors, $T_{T,j}(\mathbf{H})$

$$F_T(\mathbf{H}) = \sum_{j=1}^{N} f_j(\mathbf{H}) \cdot T_{T,j}(\mathbf{H}) \exp(2\pi i \mathbf{H} \cdot \mathbf{r}_j)$$

$T_{T,j}(\mathbf{H})$ has an exponential fall off with scattering angle and in the isotropic approximation

The Application of Charge Density Research to Chemistry and Drug Design
Edited by G. A. Jeffrey and J. F. Piniella, Plenum Press, New York, 1991

$$T_{T,j}(\mathbf{H}) = \exp(-8\pi^2 <u_{T,j}^2> \sin^2\theta/\lambda^2)$$

where $<u_{T,j}^2>$ is the mean-square displacement of atom j. The effects on the electron density distribution of chemical bonding and thermal motion are thus highly correlated.

Coppens (1967) suggested that atomic positional and displacement parameters could be determined in a parallel neutron diffraction study. A deformation density calculated as the difference between the total density and the density corresponding to superimposed spherical atoms with parameters from the neutron study

$$\Delta\rho_{X-N}(\mathbf{r}) = \frac{1}{V} \sum_{\mathbf{H}} (F_{T,X}(\mathbf{H}) - F_{T,N}(\mathbf{H})) \exp(-2\pi i\mathbf{H}\cdot\mathbf{r})$$

will display redistributions of valence electron density caused by chemical bonding. The X-N method is very illustrative and has especially given better insight in hydrogen and lone-pair regions, than studies based solely on X-ray diffraction data can. However, for practical reasons the number of X-N studies are rather limited. Even for a neutron diffraction study at a high-flux reactor the necessary crystal size is of the order of mm^3, which often is a forbidding factor, and, furthermore, measuring time at neutron diffractometers may be difficult to get.

NECESSITY OF LOW-TEMPERATURE MEASUREMENTS

In many cases the atomic thermal motion can be adequately deconvoluted from analysis of X-ray data alone. The catch is that practically only the valence electron distribution is affected by chemical bonding. In other words, the charge density near the atomic nuclei, the so called core density can be taken to represent the atom, and so atomic positional and displacement parameters can be obtained from refinement of high-order data using a spherical-atom model. Such parameters can be used to calculate an X-X$_{high}$ deformation density in the same way as the X-N density. It appears that atomic parameters which are largely free of bias can be obtained by excluding low-order reflections from the least-squares refinement but there is no universally agreed upon value for $\sin\theta/\lambda$ cut-off value above which bonding effects are believed to be insignificant. It is seldom possible, especially for room-temperature data sets, to play safe and go to very high cut-off values because the remaining data will be so few in numbers and of so small intensity that standard deviations for the spherical-atom model parameters increase intolerably.

The answer to this dilemma has been to introduce a model for the structure factor calculation which allows for the electron redistribution caused by bonding formation, the aspherical-atom model. All measured intensities can then be retained in a simultaneous refinement of the conventional atomic posi-

tion and displacement parameters and the charge density param-
ters like populations of valence shell, atomic dipoles,
quadrupoles, octupoles, or even higher poles. A number of models using different types of functions in the multipole expansion has been developed and the most widely applied model of
this kind (Coppens, 1989) consists of an expansion in atom-centered spherical harmonic functions. The functional forms ensure
that in the least-squares refinement position and displacement
parameters primarily are determined by the high-angle data and
charge density parameters by the low-angle data. The deconvolution of vibration and bonding effects makes it possible to sum
the multipolar charge density functions and obtain the *static*
density distributions referring to an ensemble of non-vibrating
atoms. Such distributions are sometimes compared with theoretical results, where available.

A prerequisite for multipole analyses is an extensive data
set because usually many more parameters are introduced than
are used in spherical atom refinements. In the latter case generally 3 position and 6 displacement parameters are used per
atom while the aspherical refinement easily adds of the order
of 20 extra population parameters per atom (depending on
symmetry and level of poles necessary to describe the
redistribution of the electron density). In most cases, a room
temperature diffraction study simply gives too few significant
intensities for a satisfactory aspherical analysis.

Atoms in organic molecular crystals very often at room
temperature have an isotropic temperature factor parameter, $B = 8\pi^2 <u_T{}^2> \simeq 4$ Å^2 and give few significant intensities for $\sin\theta/\lambda$
bigger than 0.7 Å^{-1}. The number of obtainable reflections
allows refinement of a spherical atom model with anisotropic
description of the atomic displacement parameters. There will
be approximately 5 observations per parameter for the least-
squares refinement of such a model. The structure determination
thus may be an excellent conformation study but the size of the
data set does not allow a proper electron density study since a
multipole model will easily have three times as many
parameters as the spherical atom model. Actually, for a
detailed electron density study one should strive to have a
quality data set so extensive that there are more than 5,
possibly 10, observations per parameter for the least-squares
refinement of the aspherical model.

By lowering the temperature of the sample crystal the
atomic temperature factors, $T_{T,j}(\mathbf{H})$ increase and more reflections gain significant intensity, making it meaningful to collect data to $\sin\theta/\lambda$ values bigger than 0.7 Å^{-1}. Temperature
factor parameters for atoms in molecular crystals around room
temperature are to a good approximation proportional with temperature. Fig.1 shows equivalent isotropic temperature factor
parameters, $B_{eq} = 8\pi^2 (u1\ u2\ u3)^{2/3}$ for benzoylacetone, $C_{10}H_{10}O_2$
depicted as function of temperature; *u1*, *u2* and *u3* are the
root-mean-square amplitudes in axial directions. The observed
B_{eq} values are typical for atoms in molecular crystals.
Correspondingly, overall isotropic B values typically are, at
300K $B \simeq 4$ Å^2, at 100K $B \simeq 1.5$ Å^2, and at 20K $B \simeq 0.75$ Å^2.

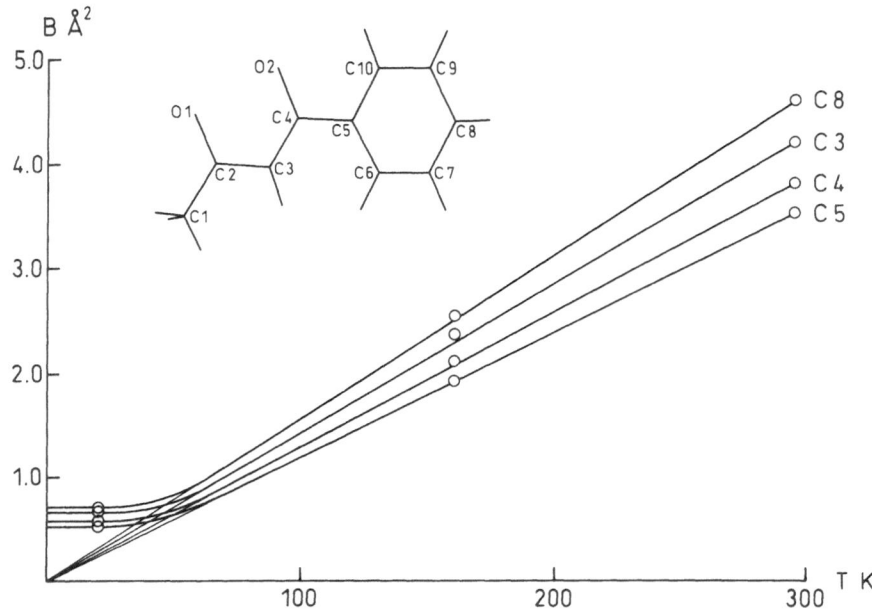

Fig.1. Temperature dependence of B_{eq}, the isotropic temperature
   factor parameters for a selection of atoms in
   benzoylacetone from X-ray studies at 295, 160 and 20K.

   The overall isotropic B value can be used to estimate the
reduction in intensity of the Bragg reflections. Examples of
such average Debye-Waller factors, $DW(B) = \exp(-2B\sin^2\theta/\lambda^2)$ are
given in Table 1 for different combinations of B and $\sin\theta/\lambda$
values. Also shown in Table 1 are examples of the enhancement
in intensity resulting from lowering the B factor from 4.0 to
1.5 Å^2 and from 1.5 to 0.75 Å^2 for reflections at different
$\sin\theta/\lambda$ values.

Table 1. Average Debye-Waller Factor, $DW(B) = \exp(-2B\sin^2\theta/\lambda^2)$
         and ratios.

$\sin\theta/\lambda(\text{Å}^{-1})$	$DW(B)$			$\dfrac{DW\ (1.5)}{DW\ (4.0)}$	$\dfrac{DW\ (0.75)}{DW\ (1.5)}$
	$B(\text{Å}^2)$ 0.75	1.5	4.0		
0.7	0.4795	0.2299	0.0198	11.6	2.1
1.0	0.2231	0.0498	0.0003	149.4	4.5
1.3	0.0793	0.0063	0.0000	4675	12.6

The enhancement is quite dramatic at high $\sin\theta/\lambda$ values, e.g. at $\sin\theta/\lambda = 1.0$ Å^{-1} the increase in intensity by lowering the sample temperature from 300 to 100K, and hereby lowering the B factor from 4.0 to 1.5 Å^{-1} gives a 150 fold increase in intensity. By further lowering the temperature to 20K a further increase of a factor between 4 and 5 is achieved. By lowering the temperature, high-order data become available and makes precise electron density studies feasible for many crystal structures which are too agitated at room temperature and simply give too few significant intensities for other than mere conformation studies. By lowering the temperature to 100K usually so many significant reflections can be measured out to $\sin\theta/\lambda \simeq 1.0$ Å^{-1} that there will be approximately 5 observations per parameter for the least-squares refinement of a multipole model, and it has been amply proved that convincing electron density studies can be performed with 100K data. The precision of the studies may be improved by lowering the temperature to about 10K which will for a complete set of data measured to a $\sin\theta/\lambda$ value 1.3 Å^{-1} give approximately 10 observations per parameter. In the next paragraph the most successful cooling methods are presented. First the gas-stream cooling technique which can give data collection temperatures down to about 100K and then helium cryo refrigerator cooling technique which can bring the temperature even below 10K.

COOLING TECHNIQUES AND APPARATUS

Diverse cooling techniques have been used for studying the low-temperature properties of materials and numerous types of cooling equipment have been built. An extensive survey is given by Reuben Rudman (1976) in his book *Low-Temperature X-ray Diffraction. Apparatus and Techniques*. Much useful information can also be found in the American Crystallographic Association publication of lecture notes from a *Low-Temperature X-Ray Diffraction Tutorial*, held at the Asilomar, California, meeting (Rudman et al., 1977).

An obvious way to conduct a literally cold experiment is to place the entire diffractometer in a cold room. This possibility has been explored successfully a few times in rooms at moderately cold temperatures as described by Rudman (1976). The more radical construction to place a Eulerian cradle inside a helium cryostat has been realized by Elf et al. (1984) with the aim of conducting neutron diffraction investigations. Similar apparatus constructed for X-ray diffraction purpose has not been reported and certainly such a solution is technologically difficult and expensive and is not likely to play an important role in low-temperature crystallography. For accurate crystallographic studies at low temperatures in particular two methods for cooling the sample have been developed to great sophistication.

1. The gas-stream method, in which a stream of cold nitrogen gas is directed at the sample, and
2. The conduction method, in which the crystal is cooled through contact with a good thermal conductor, which is part of a cooling unit like a coolant reservoir or a Joule-Thomson expansion device.

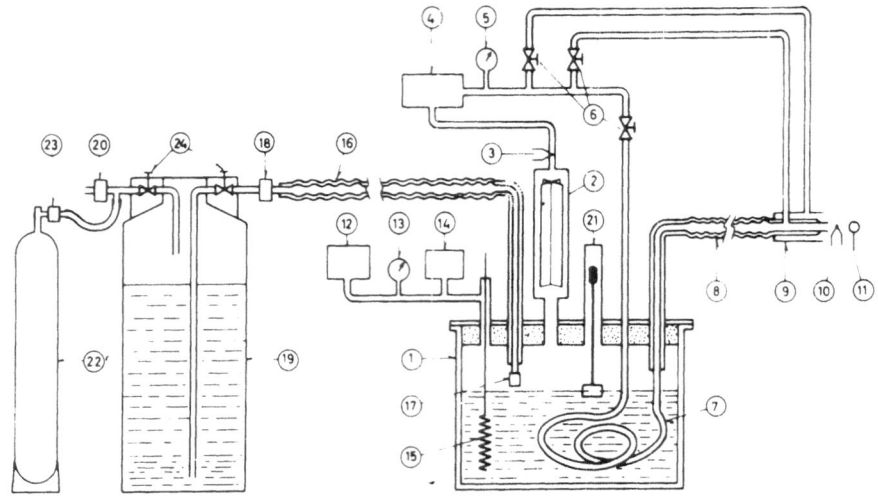

Fig. 2. Schematic diagram of the Risoe gas-stream low-temperature system.

Both types of cooling equipment are commercially available
and are in common use. Both have been developed to keep the
sample crystal at a chosen temperature within narrow limits for
long periods of time and so are suitable for electron density
determinations. General principles in the construction of the
two types of devices will be described and exemplified by a few
presently successful designs.

Gas-Stream Cooling

The cooling effect is achieved by directing a stream of
cold gas at the sample. Usually for economic reasons nitrogen
gas is used and liquid nitrogen is used as the coolant medium.
The concept is thus quite simple and scores of modifications of
gas stream cooling equipment have been designed. Most common
features of the technique are exemplified in Fig. 2 (Merisalo,
M., Nielsen, M.H., and Henriksen, K., 1973). This design is
based on a system conceived by Silver and Rudman (1971). The
cold gas generator (1) is a dewar vessel with liquid nitrogen.
Gaseous nitrogen is boiled off (15), heated to room temperature
(2) and is passed through a pressure regulator (4), which
maintains a constant pressure at the outlet and in this way
realizes a constant gas flow rate. The main stream of the gas
is redirected into the dewar vessel and is cooled in a copper
coil heat exchanger (7) and escapes via a super-insulated
transfer line (8) and a gas delivery nozzle (9) to flow past
the sample crystal (11). The two other streams can via manuel
valves (6) be used to obtain intermediate temperatures by
mixing warm and cold gas and to surround the cold gas stream,
exiting from the nozzle,with a concentric stream of dry room-
temperature gas in order to prevent condensation of atmospheric
gases on the crystal. An automatic refilling system ensures con-

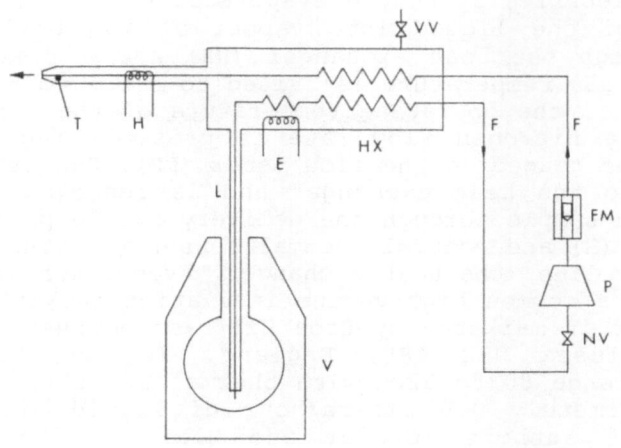

Fig. 3. Schematic diagram of the *Cryostream Cooler* system.

tinuous operation over any desired period of time. A liquid-level controller (21) opens and closes a solenoid valve (18). In the open position liquid nitrogen from a big supply dewar (19) will flow into the cold gas generator through a vacuum jacketed transfer line (16). The pressure is generated by a nitrogen gas cylinder (22) connected via a pressure regulator (23) and valves (24) to the dewar. The pressure can be released via the solenoid valve (20). While filling, surpressure triggers the pressure switch (14) and is vented through the relief valve (12). The phase separator (17) decreases the turbulence incurred during the filling process. The pressure regulator (4) ensures that the cold gas flow rate is increased minimally while the cold gas generator is being filled. Pressures in the system are checked on manometers (5) and (13). The crystal temperature as monitored by a thermocouple (10) in the gas stream in this design can be kept constant to about ± 1 K for days on end, provided the supply dewar is replenished.

Though there usually is a very short term decrease of temperature of a couple of degrees at the start of every refill, depending on how well the pressure regulator (4) works. The device operates at temperatures down to about 100K. A system (*Nitrogen Cryostat FR 558 NH*) of pretty much this design is marketed by Enraf-Nonius, Delft, The Netherlands, for use with their CAD 4 single-crystal diffractometer.

The problem of keeping the temperature constant through periods of filling up the cold gas generator has been solved in a different way in the *Cryostream Cooler* system (Cosier and Glazer, 1986). The principle of this device is to use a pump instead of a pressure regulator to separate effectively the liquid-nitrogen in the cold gas generator from the gas outflow, so that no instabilities can be generated in the gas stream, the design is represented in Fig. 3.

Liquid nitrogen is drawn up by the suction of a small dia-phragm pump (P) from an unpressurised dewar vessel (V) through a vacuum insulated supply line (L). The liquid nitrogen passes through an electrically heated evaporator coil (E), which converts most of the liquid into vapour at the boiling point. Flowing through the heat exchanger (HX) and a flexible nylon tube (F) the gas temperature is raised to close to room temperature, which is the operating temperature of the constant-flow pump (P). The nitrogen flow rate is preset using the needle valve (NV) and gauged at the flow meter (FM). Gas from the pump flows back to the heat exchanger and is recooled. It is directed at the sample through the delivery nozzle passing over a heating coil (H) and control thermal sensor (T). The liquid-nitrogen supply line, the heat exchanger, evaporator and delivery nozzle share a common high-vacuum insulation jacket. The *Cryostream Cooler* is marketed by Stoe Diffraction Systems, Norwood Green, Middlesex UB2 4HF, England. It operates in the temperature range 80 to 320K with thermal stability of ± 0.1 K using approximately 0.6 liters/hour of liquid nitrogen. The claim by the authors (Cosier and Glazer, 1986) that the *Cryostream Cooler* represents a major improvement in gas-flow cooling devices seems warranted. The performance with respect to temperature stability and low consumption is at the top and the feature that heat-exchanger case and delivery-nozzle is one assembly insensitive to orientation gives interesting possibilities for attaching this apparatus to diffraction equipment.

A very convenient feature of the gas-cooling technique is that the sample crystal can be seen even when cooled. Centering the crystal on the diffractometer is straightforward and also the interesting possibility of growing single crystals from liquids contained in a capillary mounted at the sample position, is feasible. However, the nozzle construction and its positioning as close to the sample crystal as possible is tricky and essential for how low a temperature can be reached and for preventing frosting. The principle of making an envelope of warm gas in laminar flow with the cold gas usually takes pretty well care of the frosting problem, but even so it is often deemed necessary to surround the entire diffractometer with a dry box. This upsets some of the advantages of the open construction of the gas-cooling equipment.

Another essential difficulty of this cooling method is to determine the absolute value of the sample temperature. The temperature is usually monitored by a thermocouple, which is positioned in the cold-gas stream near the mouth of the nozzle. This opening can only be brought within a few millimetres from the sample crystal in order not to interfere with the incoming and diffracted X-ray beams. Therefore there is bound to be a temperature gradient between the tip of the thermocouple and the crystal. Experience from the literature shows that sample temperature easily can be misjudged by more than 10°, and this can be a highly disturbing effect when results are compared or worse if one wants to refine simultaneously diffraction data from independent experiments as in X-N studies (Coppens and Vos, 1971).

Conduction Cooling: Flow Cryostats

Conduction cooling has been realized in the form of flow

cryostats in which the sample crystal is mounted on a cooling block through which the coolant passes. In order to prevent condensation the crystal and the cold parts of the cryostat have to be inclosed in a vacuum, which for X-ray diffraction usually means one outer beryllium shroud and inside this two further beryllium shrouds to reduce heating of the crystal by radiation. The wall of the inner confinement is kept in thermal contact with the cooling block and ensures that the crystal is part of the temperature equilibrium. The crystal temperature is measured very reliably by a sensor embedded in the cooling block. The coolant can be either liquid nitrogen or liquid helium. Consumption of cryogen is comparable to that of open gas-flow systems but recuperation of spent cryogen gas is relatively straightforward for a flow cryostat and therefore makes it economically more feasible to use liquid helium, opening up the possibility of cooling to temperatures even below 10K.

Flow cryostats can be made rather compact and therefore can be fitted onto four-circle diffractometers (Coppens et al., 1974, and Zeyen et al., 1984), and even Chi-geometry diffractometers (Albertson et al., 1979). Flow-cryostats have generally proven difficult to work with and in spite of some notable successes seem largely given up in accurate X-ray diffraction work. The major problem has been to position the stiff vacuum jacketed lines which are used to bring the cryogen from the supply dewar to the flow-cryostat. The cryogen lines often impair the diffractometer movements. The flow-cryostat designs used for single-crystal diffraction all seem to incorporate rotating seals separating the coolant at cryogenic temperatures from the atmosphere at ambient temperature. It has turned out to be an exceedingly difficult problem to make these seals tight, yet without putting excessive forces on the mechanical parts of the diffractometer. The slightest leak in seals of this type will ice it up and impair the rotational movement of the crystal.

A very sophisticated new invention, a *magnetically coupled crystal holder* recently reported by Argoud and Muller (1989) circumvent these difficulties by allowing the cryostat to remain stationary. The crystal is mounted on a small, delicate, gimballed crystal holder, which allows crystal positioning to any orientation by easy rotation about three axes. A magnet is also fixed to the crystal holder which sits inside a cavity, cooled by a helium-flow cryostat. The crystal can be centered on the diffractometer. The whole assembly is rigidly fixed relative to the base of the diffractometer. The crystal holder is not in mechanical contact with the axes of the diffractometer. Actually, the coupling between crystal and diffractometer axes is achieved by mounting a strong magnet in place of the classical goniometer head, i.e. outside the vacuum and thermal screens of the cryostat. The magnetic coupling to the magnet on the crystal holder makes the crystal follow the orienting movements of the diffractometer axes.

Helium-flow cryostats have a consumption of at least 0.5 liter per hour of liquid helium when maintaining the sample temperature at 10K. Liquid helium is quite expensive and not easily accessible most places. Therefore, even if some of the major obstacles in the use of flow cryostats can be removed, the inconvenience of handling and the considerable expense of buying liquid helium will still exist and make helium-flow cryostats unattractive. Fortunately, there is an alternative cooling device that allows crystallography at very low temperatures. Joule-Thomson coolers in the form of two-stage closed-cycle helium refrigerators, which use no cryogen, have been developed into compact, reliable devices capable of maintaining the temperature of a sample crystal at 10K.

Conduction Cooling. Closed-Cycle Helium Refrigerators

In the mid-seventies a new development in the pursuit of crystallographic studies at temperatures below 77K began when a closed-cycle helium refrigerator was first adapted to a diffractometer. The closed-cycle helium refrigerator comprises a compressor unit and an expander module connected with flexible gas lines. The compressor is of air conditioner type, which pressurises the helium gas for the expander where refrigeration is accomplished by adiabatic expansion of the helium. The Fig. 4 shows a simplified diagram of a *Displex* (manufactured by APD, Cryogenics Inc., Allentown, PA 18103, U.S.A.) two-stage Gifford-McMahon type expander module. The valve motor turns the valve disc which determines an operating cycle of gas intake and exhaust. High-pressure helium admitted by the rotating valve disc during the intake period flows through the slack cap and into the regenerators. The regenerators, cooled during the previous exhaust stroke, cool the incoming gas as it flows through. Gas flowing through raises the slack cap and lifts the displacer, creating expansion space at the heat stations for gas that has passed through the regenerator. Also as the displacer lifts, gas above the slack cap is partially compressed and pushed through the orifice into the surge volume. Before the displacer reaches the valve stem, the valve closes. Compression of gas above the slack cap decelerates and stops the displacer before it can collide with the valve stem. When the valve opens to exhaust, high-pressure gas at the heat stations is free to expand and hereby refrigerate them. The exhausting gas also cools the regenerators. As the pressure drops, partially compressed gas bleeds from the surge volume, pushes the slack cap and displacer towards the heat stations, forces exhaust, and positions the displacer for the next cycle. In about one hours operating time a two-stage expander usually will have decreased the temperature to about 70K at the first-stage heat station and to less than 10K at the second station.

Development in the Cryocooler Technique

Three-stage closed-cycle helium cryocoolers capable of reaching 3.6K are commercially available but, unfortunately, they are still rather bulky, heavy and expensive, and not reported in use for crystallographic studies. However, it seems that also the two-stage Gifford-McMahon cryocooler has potential for reaching working temperatures in the vicinity of the boiling point of helium. The conventional two-stage cryocooler usually has densely packed finemeshed copper nets in the first stage regenerator. In the second stage regenerator lead in the form of lead spheres (0.2 to 0.3 mm in diameter) is preferred

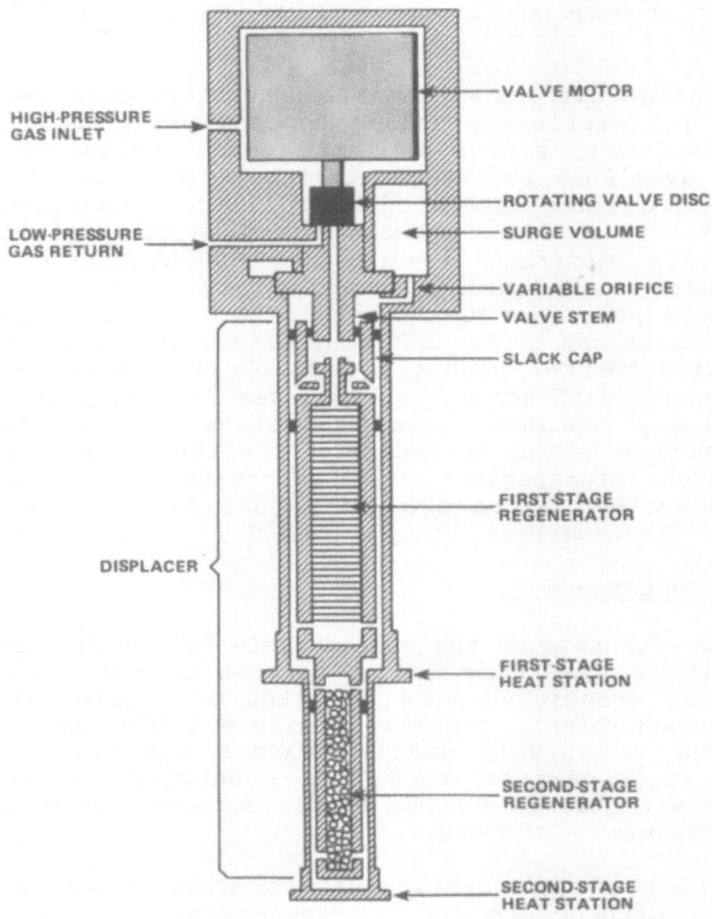

HIGH-PRESSURE GAS INLET

VALVE MOTOR

ROTATING VALVE DISC

LOW-PRESSURE GAS RETURN

SURGE VOLUME

VARIABLE ORIFICE

VALVE STEM

SLACK CAP

FIRST-STAGE REGENERATOR

DISPLACER

FIRST-STAGE HEAT STATION

SECOND-STAGE REGENERATOR

SECOND-STAGE HEAT STATION

Fig. 4. Schematic diagram of two-stage *Displex* expander module. Actual height is 30 or 40 cm.

because of its high heat capacity below about 80K. The specific heat for lead, however, decreases rapidly with decreasing temperature, making the refrigeration power and efficiency so small around 10K that the lowest practically attainable temperature is limited to ~ 8K. Recently, materials have been found and explored, notably Er_3Ni, which have much higher heat capacities per unit volume than lead has below 10K (Sahashi et al., 1989). In a non-magnetic substance, such as lead, only the lattice vibrations and the conduction electrons make a contribution to the specific heat. For magnetic materials the spin interactions may give considerable contribution to the specific heat near the magnetic phase transition. Er_3Ni has complicated spin configurations around 7K, resulting in a considerable volumetric specific heat right down to 5K, which is higher than that of Pb below 15K and comparable with it at higher temperatures. Kuriyama et al. (1989) report reaching a no-load lowest temperature of 4.5K using Er_3Ni as second stage regenerator material in a two-stage Gifford-McMahon refrigerator. The same refrigerator equipped with a lead filled second stage regenerator and under similar running conditions reached a no-load lowest temperature of 6.28K. The refrigeration capacity below 10K at the second stage is approximately 2 W better for Er_3Ni than for Pb.

Two-stage Gifford-McMahon refrigerators have mainly been developed for stationary use in application where weight and size of equipment is of minor importance. Typically, the commercially available two-stage refrigerators weigh around 5 kg and have an overall length of about 40 cm. Consequently, the first applications of such cooling devices for low-temperature single crystal diffraction work have been at physics research centers where traditionally big and sturdy four-circle diffractometers are part of the equipment. More and more university laboratories choose to buy big four-circle diffractometers (the *HUBER* diffractometer with the big Eulerian cradle, type No. 512, is such a diffractometer) in order to facilitate mounting of auxilliary equipment like two-stage refrigerators. The finding of more efficient regenerator materials will hopefully result in the miniaturization of two-stage refrigerators to a degree that they may be used with just few modifications of existing diffractometers.

Mounting Cryocoolers

A big advantage of the closed-cycle helium Gifford-McMahon cryocoolers is that they use no cryogen thereby avoiding the encumbrances associated with the flow of liquid nitrogen or helium through pipes, jacketed flexible tubing and rotatable joints. The relatively small expander unit with the cold stations is separated from but connected to the heavy compressor with flexible tubes which transport the presurrised, room-temperature helium gas.

Cooling of the crystal must be achieved by conductance when using cryocoolers for low-temperature crystallography. Furthermore, the crystal and the cold stations must be surrounded by a vacuum in order to prevent frosting. Therefore mounting a cryocooler say on a four-circle diffractometer presents somewhat complicated problems which have resulted in intricate designs. One solution (Samson et al., 1980) involved allowing the crystal to rotate relatively to the cold station of the cryocooler and let the thermal conduction take place through a standard, flexible, very-high-purity copper cable. The construction has the nice feature that the incident and diffracted X-ray beam cuts through narrow kapton-foil covered slits in the spherical vacuum shroud, i.e. absorption by the X-ray windows is small and identical for all reflections. However, mechanically it is a very difficult and expensive solution, which includes different types of dynamic vacuum seals between moving parts separating the vacuum from the exterior. The seals have to be adjusted well as a compromise between excessive friction for the moving parts and leakage into the vacuum. In other solutions (Henriksen, Larsen, and Rasmussen, 1986, and Archer and Lehmann, 1986) the crystal is mounted rigidly on the 10K cold station, which means that the cryocooler must follow all the crystal movements and rotating seals may be avoided completely. Conceptionally, the mount is just a giant goniometer head without arcs and the cryocooler with vacuum shroud is an integral part hereof. The Fig. 5 shows a two-stage *Displex* cryocooler mounted on a *HUBER* diffractometer in a

Fig. 5. Two-stage *Displex CS201* cryocooler mounted
in Henriksen et al. (1986) centering device
on the φ bearing of a *Huber 512* four-circle
diffractometer. Helium gas lines and electrical connections are supported from the
ceiling by a swinging link suspension.

Henriksen et al. (1986) centering device. The lower part of the
vacuum shroud is a beryllium cup which serves as X-ray window.
Absorption in the cylinder wall is easily calculated but relies
on uniform wall thickness. There will also be powder scattering
from the beryllium cylinder walls, but by having a cylinder of
sizeable radius it is possible to collimate the X-ray beam so
that powder scattering from the off-center beryllium walls will
not reach the counter.

Crystal Centering

        In the gas stream cooling case crystal centering may be
checked optically at all temperatures. For flow-cryostats and
Gifford-McMahon cryorefrigerators after initial optical crystal
centering at room temperature radiation shields and a vacuum

shroud are mounted obstructing the view to the crystal, so centering has to be done by equalizing positive and negative 2θ for pairs of reflections at different Chi and Phi values. A prerequisite is that the zeroes of the goniostat are accurately known. For cryorefrigerators mounted along the Phi-axis, usually the x- and y-translations (perpendicular to the Phi-axis of the four-circle instrument) change little by cooling. However, sizeable height adjustment z-translation is necessary upon cooling because of the thermal contraction of the displacer housing and the crystal support. For instance for the small *Displex CS201* (Henriksen et al., 1986) a height adjustment of 0.66 mm is necessary after cooling from 295K to 10K.

Temperature Measurement, Regulation and Stability

As an example the instrumentation described in Henriksen et al. (1986) is given. The sample crystal is mounted on the end of a thin, 2 mm long glass fibre which is glued to a 15 mm long pointed aluminium pedestal. Crystals which can evaporate in the vacuum instead are put into a capillary. Two thermal sensors, a platinum resistor and a calibrated germanium diode are located immediately beneath the base of the pedestal in a 40 mm long aluminium extension which is mounted in good thermal contact with the 10K cold station. The *Displex* cryorefrigerator always works at its highest capacity, so intermediate temperatures between 10K and room temperature are attained by passing a current through a 100 ohm resistance wire wound around the aluminium extension at the base of the crystal pedestal.

A temperature controller keeps the temperature stable within ± 0.1°. Temperature differences between sample temperature and sensor temperature have been investigated by determining the temperature reading for well known phase transitions. The observed difference is judged to be within one degree. Having the crystal in a vacuum provides for accurate temperature measurement in contrast to the problems of establishing the accurate temperature when using gas-cooling technique.

PITFALLS IN LOW TEMPERATURE CRYSTALLOGRAPHY

Centering Problems Caused by Defect Vacuum

A poor vacuum will impair the minimum obtainable temperature in a conductance cryocooler because of increased heat conductance through the vacuum. Usually, a system with no dynamic seals is easily evacuated to better than 10^{-3} Torr and when the system is cooled below the melting points of nitrogen and oxygen cryopumping sets in and reduces the pressure to less than 10^{-4} Torr. If leakage of the system is sufficiently small continuous pumping is not necessary at the lowest temperatures. In a data collection on S_4N_3Cl at 60K with the Aarhus University *Displex CS201* setup a deteriorating vacuum had a *surprising* effect.

The vacuum problem was realized but the leakage was so small that temperature could be controlled easily for one day at a time before the vacuum had deteriorated so much that heat conductance overcame the cryorefrigerators cooling power at the

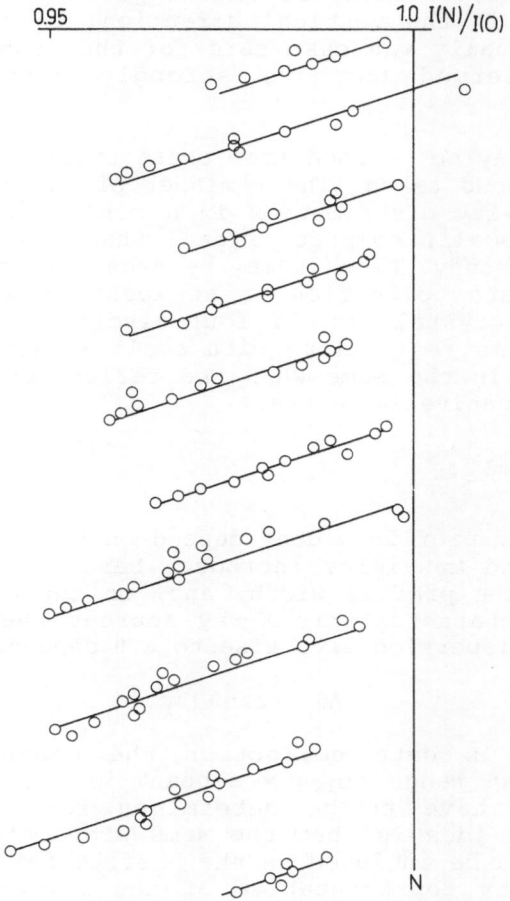

Fig. 6. Relative intensity of low χ value standard
         reflection shown as function of time with
         the time axis running down the page.

set temperature of 60K. Before that happened, the system was
reevacuated and a full data collection with the sample crystal
at 60K was eventually obtained. Data reduction analysis
revealed that one standard reflection exhibited the curious
trend shown in Fig. 6. After each reevacuation the standard
intensity had a value close to its starting value. However,
during data collection the intensity decreased progressively
until the next reevacuation. Therefore, although the sample
crystal undoubtedly remained at 60K, obviously pressure changes
systematically influenced the intensities. The following
explanation is suggested.

    Although the aluminium pedestal with the sample crystal
remained at 60K the displacer housing experiences systematic
temperature differences when the vacuum deteriorates, which
will result in systematic changes in the contraction of the
displacer housing, and therefore the sample crystal will undergo systematic height changes. A pyrolytic-graphite incident
beam monochromator, in the perpendicular arrangement of
monochromator and specimen diffraction planes, was used in producing MoKα radiation for the experiment. Unfortunately, the

beam homogeneity in the vertical direction has a much less flat distribution than in the horizontal direction (for illustration of this problem see Fig. 5 in Coppens et al., 1974). The tolerance for displacement of the sample crystal therefore is especially low in the vertical direction. In the S_4N_3Cl study the χ-value actually was near zero for the standard reflection which was observed to have strongly varying values of intensity.

A lesson may be learned from these observations and a couple of precautions taken. The reminder of the inherent problem with the intensity distribution in a monochromated beam urges to strive for small, compact crystal shapes when attempting a charge-density study. Also it may be necessary to pump continuously during data collection or at least regularly check the height of the crystal. For a four-circle diffractometer one should check some reflections with small χ values to make sure that these remain the same when the reflection is centered at positive and negative 2θ values.

Scan Range Pitfall

Diffraction profile widths depend on the scattering angle. Crystal size and mosaicity introduce basic, angle-independent components of the profile width, and, on top of that, for the commonly used characteristic X-ray sources the spectral $\alpha_1-\alpha_2$ splitting and dispersion give rise to a θ dependent component,

$$\Delta\theta = \tan\theta \cdot \frac{\Delta\lambda}{\lambda}.$$

Very commonly in data collection the conventional linear formula for scan range, $\omega_{LIN} = a+b\tan\theta$ is used. The factors a and b ideally have to be determined for each particular experiment. The idea is that the same proportion of the Bragg intensity should be included in the profile for the chosen scan width and points representative of the background should be measured towards each end of the scan. Far too often, too narrow scan widths are attempted, resulting in errors in the integrated intensities. Destro and Marsh (1987) made an analysis of scan truncation errors which made evident that the profile width has other angle dependent components than the spectral dispersion. They introduced an angle-dependent "aberration" function, which presumably includes instrumental effects such as beam divergence. Mathieson (1989) suggests for scan range an expression

$$\omega_{RMS} = [(p')^2 + (q'\tan\theta)^2]^{1/2} + (p'' + q''\tan\theta)$$

based on an analysis of the intensity distribution functions. If Gaussian shaped contributions are involved in the convoluted profile width the root-mean-square part of the ω-expression is appropriate and drastic implications for the scan range at high values of $\sin\theta/\lambda$ follow. The ω_{RMS} expression predicts a relatively faster increase in scan range with θ than does the ω_{LIN} expression, i.e. truncation errors are predicted to increase with θ if scan range and data reduction are based on an ω_{LIN} expression.

However, if all reflection profiles are recorded as step scans and the scan ranges are sufficiently wide over the whole

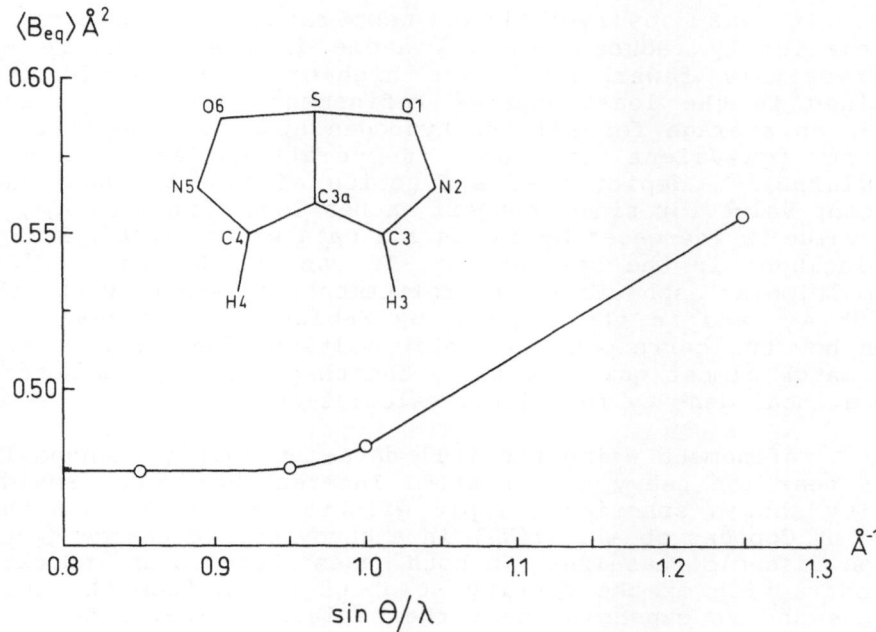

Fig. 7. The average equivalent isotropic temperature
factor, $<B_{eq}>$ for different $\sin\theta/\lambda$ cut-off
values in the 11K study of 2,5-diaza-1,6-
dioxa-6a-thiapentalene by Fabius et al. (1989).

θ range no matter which scan width expression is used for data
collection, intensities can be calculated at a separate data
reduction. Full widths of the profiles may be calculated from
step-scanned data in a systematic manner by the $(\sigma I/I)$
algorithm (Lehmann and Larsen, 1974), e.g. as implemented and
described by Blessing (1987). Such a data reduction procedure
is particularly useful when the crystal exhibits scan width
anisotropies. Peak profiles are learned during data reduction
or rather reflection integration limits are obtained as
function of setting angles from a least-squares analysis of
peak profile widths of the highly significant reflections and
these reflection integration limits are used when calculating
intensities of weak reflections. A prerequisite for a
successful outcome is that the scan ranges are sufficiently
wide at all angles so that measurements were made so far into
the flanks of the profile that intensities of the same bias may
be calculated at all scattering angles.

At low temperatures it becomes feasible to measure very
many high-order reflections although they still may be rather
weak. Therefore in order to improve counting statistics for the
peak intensity it is tempting to cut the scan range as narrow
as possible to spend relatively more time counting peak
intensity. Such considerations unfortunately resulted in too
narrow scan widths at the very highest $\sin\theta/\lambda$ values in a
charge-density study of 2,5-diaza-1,6-dioxa-6a-thiapentalene,
$SC_3N_2O_2H_2$ (Fabius et al., 1989). For the weak high-order
reflections it is difficult to judge if profile scans are
sufficiently wide, so how does a case of narrow high-order scan
widths manifest itself. In particular thermal parameters will
be influenced. Fig. 7 shows the misery for the above mentioned

case. It was observed that temperature parameters are systematically reduced for all atoms in the structure when progressively fewer and fewer high-order reflections are retained in the least-squares refinement. The Fig. 7 shows $\langle B_{eq} \rangle$, an average for all non-hydrogen atoms in the structure of the equivalent isotropic temperature factors, $B_{eq} = 8\pi^2 (u1u2u3)^{2/3}$, depicted as a function of $\sin\theta/\lambda$. $\langle B_{eq} \rangle$ has a constant value for $\sin\theta/\lambda$ cut-off values less than 0.95 Å^{-1}, but this value is increased by 15% if all data with $\sin\theta/\lambda \leq 1.25$ Å^{-1} are included in the refinements. It was decided to retain in the multipolar aspherical atom refinements only data with $\sin\theta/\lambda \leq 0.95$ Å^{-1} and in the article by Fabius et al.(1989) it is shown how the corresponding static multipole-deformation model maps match almost quantitatively the theoretical maps obtained from a local density functional calculation.

A refinement using the full data set mainly changed features near the heavy sulfur atom. Interestingly, the residual density showed spherical ripple effects reminiscent of those found by Coppens et al. (1977) in a study of the charge distribution in cyclooctasulfur. In both cases there is an indication of contraction of the density at about 0.6 Å from the sulfur nucleus and an expansion near the nucleus. Unfortunately, one cannot have confidence in the analogy, since the 2,5-diaza-1,6-dioxa-6a-thiapentalene high-order data obviously have systematically too small intensities.

CONCLUSION

Electron density distribution studies are usually undertaken with the aim of analysing very fine details when comparing structures of related compounds or similar molecules in different surroundings. Often the study comprises a comparison with a theoretically calculated electron density distribution. In the beginning of the article it was argued that low-temperature data acquisition is necessary for such studies in order to get sufficiently many observations for the refinement of the many parameters of a flexible model which is capable of deconvoluting vibrational and bonding effects.

Coping with Hydrogen Atoms

However, even a low-temperature X-ray study will yield distorted position and displacement parameters for hydrogen atoms. The hydrogen atom has no core electrons and the single valence electron is part of a strong covalent bond, which shifts the electron density maximum from the core position into the bonding region. The well known result is that X-ray studies typically show C-H, N-H and O-H distances which are about 0.1 Å shorter than the actual internuclear distances. A patchy solution to this problem is to start out concluding that the direction between the hydrogen atom and the atom to which it is bonded is correctly determined. A literature value for the X-H distance is then used to deduce the position of the hydrogen nucleus. It is more involved to deduce proper hydrogen vibrational parameters since hydrogen exerts pronounced internal vibrations because of its light weight and terminal position. Besides, it also takes part in the intermolecular vibrations. The displacement caused by the latter movement may be estimated from a rigid body analysis of the X-ray diffraction data while the contribution from the internal vibrations might be inferred from spectroscopic evidence. The total displacement tensor for each hydrogen atom is the sum of the two contributions. For interpretation of atomic displacement parameters from diffraction studies of crystals, see Dunitz et al. (1988). Hydrogen atoms sit on the perifery of molecules and so are extremely important for the molecular electrostatic properties and the chemical reactivity. Accurate knowledge of hydrogen positional and displacement parameters therefore is extremely important for electron density studies of molecules which contain hydrogen atoms. The proper remedy to above mentioned problems really is to conduct a parallel neutron diffraction study, whenever feasible and not rely on the more indirect methods.

It is a generally agreed upon recommendation (Blessing, 1987) to base an electron density study on a data set which comprises more asymmetric units. It is also advisable to inspect every intensity profile for abnormalities (spikes, peaks off-center, etc.) and carefully make sure that integration and further data reduction is correctly performed. A set of representative scaling reflections sometimes reveal peculiarities like the case with the vacuum pitfall and separate refinements of high-order data may reveal systematic errors in the data like in the example with the scan-range pitfall. The data set should be checked so carefully because the experimental electron density distribution is difficult to verify. Rees (1977) has described how standard deviations can be assigned to any point in the experimental density. Theoretically calculated difference density for a smaller molecule of light atoms often fits very well the experimental density. They are rarely quantitatively identical, though, with respect to all features as judged from estimated standard deviations in the experimental density, and since the small differences could be indication of necessary changes of the theory behind the theoretical calculation an independent verification of the experimental results would be very desirable.

The *rigid bond* test

For molecular crystals bond lengths and the general geometry must look reasonable compared with literature values unless peculiar packing may give a possible explanation for deviations. Close inspection of the anisotropic displacement parameters has also proven useful and most revealing. They are known to soak up all sorts of errors, systematic errors in the data, like it was found in the scan range pitfall, but also errors and insufficiencies in the model. Hirshfeld (1976) has reasoned that the relative vibrational motion of a pair of bonded atoms has an effectively vanishing component in the direction of the bond. If $z^2_{A,B}$ denotes the mean square displacement amplitude of atom A in the direction of atom B, then for every covalently bonded pair of atoms A and B

$$\Delta_{A,B} = z^2_{A,B} - z^2_{B,A} = 0$$

Conversely, if in parts of the molecule this postulate is not fulfilled, one may deduce that the structural model is insufficient. Hirshfeld (1976) terms this test the *rigid bond* postulate

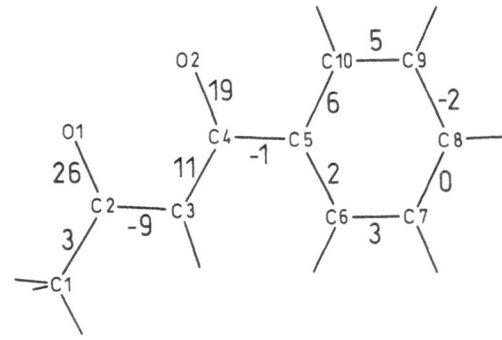

Fig. 8. *Rigid bond* test on results of 20K study
of benzoylacetone. $\Delta_{A,B}$ values are given
in pm^2. the esd of $\Delta_{A,B}$ is about 4 pm^2.

and has estimated that for atoms at least as heavy as carbon
$\Delta_{A,B}$ normally should be smaller than 0.001 $Å^2$ = 10 pm^2. Fig. 8
shows results from an ongoing 20 K X-ray study of benzoylacetone. This molecule at room-temperature has a disordered *cis*-
enol system in the solid state. The low-temperature study was
undertaken with the aim of establishing if an ordering of the
hydrogen atoms takes place at reduced temperature. It turns out
that the two C-O bond distances remain identical at 20K, C2O1 =
1.293(2) Å and C4O2 = 1.299(2) Å and, what is more interesting
in the present context, the $\Delta_{A,B}$ values as shown in Fig. 8 significantly exceed 10 pm^2 for those bonds indicating that the
model with single atoms at the oxygen positions does not
describe adequately the enol group of benzoylacetone even at
20K.

Inspection of $\Delta_{A,B}$ values are particularly revealing for
low-temperature studies. The effect on $\Delta_{A,B}$ of model insuffi-
ciency is usually not temperature dependent but the anisotropic
displacement parameters become smaller and are determined with
better precision at low temperature. Therefore the $\Delta_{A,B}$ peculiarities become more significant. The big advantage of render-
ing probable the model as well as the values of the anisotropic
displacement parameters is that it is made likely that the mutual bias is small and therefore that more trust can be put
into the experimentally determined electron density distribution.

How Cold?

The general conclusion is that detailed electron density
determinations are only possible, or at least are greatly improved, as low-temperature studies. But which data collection
temperature should one aim for ? Coppens and Vos (1971) made an
analysis of the temperature dependence for mean-square displacements for a number of frequencies for a vibrator with the
mass of the cyanuric acid molecule. They concluded that all

modes above about 100 cm⁻¹ are practically in their ground state at liquid nitrogen temperature. For such modes, the advantage of decreasing the temperature below 78K is minimal. However, for typical molecular crystals the frequencies of many lattice modes are below 100 cm⁻¹ and it is advantageous to reduce the temperature to 10K in order to deexcite as completely as possible most of the low frequency lattice modes. An added advantage is that the displacement parameters become less temperature dependent at the very lowest temperature as can be seen in Fig. 1. The constancy of the experimental temperature is less critical at 10 than at 100K.

Setting up a successful low-temperature system takes effort and most likely involves disappointments before the reward in the form of beautifully detailed electron density maps are at hand. Fortunately, eminent cooling equipment, well suited for diffraction purposes, is commercially available. In particular the gas-cooling equipment is easy to adapt to most existing diffractometers while it may be more of a problem to mount a two-state cryocooler. Initial costs for setting up a gas-cooling equipment may be less than for setting up the two-stage cryocooler but the running expense for liquid-nitrogen will soon upset this advantage; so long term economy under most circumstances will count in favour of the two-stage cryocooler. There is no doubt that low temperature work should be recommended for electron density studies, but the advantages are so great that low-temperature work should be made the standard mode of data acquisition in which case the acquired technical proficiency undoubtedly would make the electron density quality study more the rule than the exception pushing crystallography forward.

REFERENCES

Albertson, J., Oskarsson, Å. and Ståhl, K.(1979).
    J. Appl. Cryst., **12**, 537-544.
Archer, J.M. and Lehmann, M.S. (1986).
    J. Appl. Cryst., **19**, 456-458.
Argoud, R. and Muller, J. (1989).
    J. Appl. Cryst., **22**, 584-591.
Blessing, R.H. (1987).
    Cryst. Rev., **1**, 3-58.
Coppens, P. (1967).
    Science, **158**, 1577-1579.
Coppens, P. and Vos, A. (1971).
    Acta Cryst., **B27**, 146-158.
Coppens, P., Ross, F.K., Blessing, R.H., Cooper, W.F., Larsen,
    F.K., Leipoldt, J.G. and Rees, B. (1974).
    J. Appl. Cryst., **7**, 315-319.
Coppens, P., Yang, Y.W., Blessing, R.H., Cooper, W.F. and
    Larsen, F.K. (1977).
    J. Am. Chem. Soc., **99**, 760-766.
Coppens, P. (1989).
    J. Phys. Chem., **93**, 7979-7984.
Cosier, J. and Glazer, A.M. (1986).
    J. Appl. Cryst., **19**, 105-107.
Destro, R. and Marsh, R.E. (1987).
    Acta Cryst., **A43**, 711-718.

Dunitz, J.D., Schomaker, V. and Trueblood, K.N. (1988).
J. Phys. Chem., **92**, 856-867.

Elf, F., Will, G., Chatzipetros, J. and Dujka, B. (1984).
Revue Phys. Appl., **19**, 793-794.

Fabius, B., Cohen-Addad, C., Larsen, F.K., Lehmann, M.S. and Becker, P. (1989).
J. Am. Chem. Soc., **111**, 5728-5732.

Henriksen, K., Larsen, F.K. and Rasmussen, S.E. (1986).
J. Appl. Cryst., **19**, 390-394.

Hirshfeld, F.L. (1976).
Acta Cryst., **A32**, 239-244.

Kuriyama, T., Hakamada, R., Nakagome, H., Tokai, Y., Sahashi, M., Li, R., Yoshida, O., Matsumoto, K. and Hashimoto, T. (1989).
Proc. Intl. Conf. on Cryogenics and Refrigeration
International Academic Publishers, Beijing, China, 91-99.

Lehmann, M.S. and Larsen, F.K. (1974).
Acta Cryst., **A30**, 580-584.

Mathieson, A.McL. (1989).
Acta Cryst., **A45**, 613-620.

Merisalo, M., Nielsen, M.H. and Henriksen, K. (1973).
Risoe Report **No. 279**, Risoe National Laboratory, DK-4000 Roskilde, Denmark.

Rees, B. (1977).
Israel Journal of Chemistry, **16**, 180-186.

Rudman, R. (1976).
Low-Temperature X-ray Diffraction. Apparatus and Techniques. Plenum Press, New York and London.

Rudman, R., Hope, H., Stevens, E.D. and Petsko, G.A. (1977).
LTXRD Tutorial, Asilomar, California. American Crystallographic Association publication.

Sahashi, M., Tokai, Y., Kuriyama, T., Nakagome, H., Li, R., Ogawa, M. and Hashimoto, T. (1989).
Proc. Intl. Conf. on Cryogenics and Refrigeration.
International Academic Publishers, Beijing, China, 131-139.

Samson, S., Goldish, E. and Dick, C.J. (1980).
J. Appl. Cryst., **13**, 425-432.

Silver, L. and Rudman, R. (1971).
Rev. Sci. Instrum., **42**, 671-673.

Zeyen, C.M.E., Chagnon, R., Disdier, F. and Morin, H. (1984).
Revue Phys. Appl., **19**, 789-791.

THE USE OF SYNCHROTRON RADIATION

AND ITS PROMISE IN CHARGE DENSITY RESEARCH

Philip Coppens

Chemistry Department
State University of New York at Buffalo
Buffalo, New York, 14214, USA

INTRODUCTION

Many chapters in this Volume attest to the fact that experimental charge densities and one-electron properties derived directly from the diffraction data provide important information, much of which is not otherwise available experimentally. Nevertheless, the accuracy currently achieved is insufficient for a large number of issues on which experimental information would be very valuable. There is, for example, undisputed evidence for intermolecular effects on the electron density, but the majority of such effects are just below our present limit of accuracy. The difference between molecules in different solid phases, the effects of substituents on the charge density, the changes in ligand density on complex formation, the detailed analysis of theoretical results, are all accuracy limited at the present state of the field, in which, away from the nuclear positions, *standard deviations* in the deformation density of about $0.05 e\text{Å}^{-3}$ are achieved with careful diffractometer techniques. This means that in order for differences to be significant they must not be smaller than say $0.1 - 0.2 e\text{Å}^{-3}$. What can be done to reduce this limitation so that charge density analysis may become a more generally useful analytical technique?

The successes of the past decades were made possible by technical developments, including the availability of abundant computing facilities.[1] One of the recent significant developments in diffraction is the advent of very intense, narrowly collimated, and tunable synchrotron sources. Their suitability for charge density research is the subject of this Chapter.

The Application of Charge Density Research to Chemistry and Drug Design
Edited by G. A. Jeffrey and J. F. Piniella, Plenum Press, New York, 1991

209

Our present results are limited by a number of factors, such as the inaccuracy of the weaker reflections due to counting statistics, and errors due to effects such as absorption and extinction, which can be reduced by selecting a shorter wavelength or using a smaller crystal. In particular in the vicinity of the nuclear positions the deformation density is sensitive to inaccuracies in the thermal and positional parameters. Their accuracy can be enhanced by the addition of very high order data, made possible by the use of shorter wavelengths at reduced sample temperatures.

THE USE OF SYNCHROTRON RADIATION

Intensity considerations

With second generation sources such as the National Synchrotron Light Source (NSLS) and the Stanford Synchrotron Research Laboratory (SSRL) monochromatic-beam intensities at the sample table of 10^{11} photons/mm^2/sec can be obtained. In the third generation of synchrotron sources of 6-7GeV electron beam energy, now under construction, this intensity will be increased even further.

An often more appropriate measure of the beam characteristics is its brightness, defined as the intensity per mrad per mm^2 for a given energy bandwidth. A high brightness corresponds to an intense bean with low divergence, which is of particular importance when samples with low mosaic spread are being studied, or in the measurement of weak reflections.

The integrated intensity of an X-ray beam diffracted by an ideally imperfect sample of volume V is a function of the angular velocity of the sample rotation ω. For an incident beam of intensity I_0 it is given by:

$$E\omega/I_0 = QV_{crystal}/A \qquad [1.1]$$

$$\text{with } Q = \frac{N_c^2 \lambda^3}{\sin 2\theta} |F^2| \left(\frac{e^2}{mc^2}\right)^2$$

where N_c is the number of unit cells per unit volume, and e^2/mc^2 is the Classical Thomson scattering amplitude of an electron (=0.281 x 10^{-12} cm),† and A is an absorption factor.

The low divergence of the synchrotron beam in the diffraction plane leads to extremely narrow peaks, and the concentration of the intensity in a very small ω range. To illustrate the diffracted beam intensity that is obtainable let us take as example a crystal of 100μ linear dimension, 1000Å^3 unit cell and a structure factor F equal 400 electrons. With photons of a wavelength of 1.5 Å , a value of about 10^{-7} is obtained for $E\omega/I_0$. This leads to a diffracted beam of about 10^4 c/sec with an incident beam of 10^{11} photons/mm^2/sec. However, such a calculation ignores the high brightness of synchrotron radiation which concentrates all scattering in a very small solid angle. The brightness at a bending magnet beam at NSLS is about 10^{14} photons s^{-1} mm^{-2} mrad^{-2} for a 0.1% energy-bandwidth (compared with about 10^7 for a 1kW X-ray tube). Such a beam, if fully reflected by the mosaic blocks in the crystal leads with the parameters discussed above to a peak intensity of the order 10^7 c/s, well outside the linear range of conventional scintillation counters. The ability to study much smaller amounts of scattering material is the basis for *microcrystallography*, the study of very small micron-sized crystals, and for the rapid development of *surface crystallography*.

Alternatively, very weak reflections of normal sized crystals may be measured with good accuracy. Examples of such studies are the measurements of reflections of silicon and germanium, which are forbidden in the spherical atom model,[2] and the 15K study of the structure of TTF-TCNQ (tetrathiofulvalene - tetracyanoquinodimethanide) below its metal-to-insulator transition.[3] The modulated structure, which is a result of the Peierls transition in the one-dimensional conductor, gives rise to satellite reflections with intensity only 10^{-4} times those of the main reflections.

In particular for more complex structures many reflections are weak, and just above background when conventional sources are used. Under these conditions the weak reflections contribute considerably to the experimental uncertainty in the charge density, as discussed further in the section on standard deviations in the experimental densities.

The measurement of very narrow diffraction peaks

As a result of the low divergence of the synchrotron beam from a perfect crystal monochromator (typically fractions of milliradians) and its low energy spread, the rocking curve is often limited only by the mosaic spread of the sample crystal. For a ruby sphere measured at the CHESS A2 beamline, for example, the background-to-background width was only 0.02°.

The narrow width of the diffraction profile makes it difficult to use standard search techniques for locating reflections of a crystal of unknown orientation, as the step-width has to be taken very small to avoid passing over the reflection. Film techniques for locating reflections can be used to bracket the angular range, which may then be searched with a very narrow step interval.

The mosaic distribution often shows a narrow peak with rather broad tails. A step scan with equidistant steps for collecting the integrated intensity would cause most of the measurement time to be spent on the tails of the distribution, where the intensity is low. Therefore, a non-equal step scan mode, or *odd-scan* was introduced by Nielsen et al (1986). In this scan mode the angular position P is related to the step number n, the number of steps N, the width of the scan W, and the scan center P_0 by the expression:

$$P = P_0 + W\,X' \qquad\qquad [1.2]$$

with $X' = (1-O)\,X + 4\,O\,X^3$

X is the fractional step number defined as $X = n/N - 1/2$ for the nth step, and the starting point of the scan is defined as n=0. The relation between step number and counter positions is illustrated in Fig. 1.

By varying the oddness parameter O a continous transition from 'normal' to 'very odd' is achieved. A saving of data collection time by a factor of 2-3 can often be obtained by implementation of such a scanning procedure.

Counter deadtime and its relation to the time structure of the synchrotron beam

Synchrotron radiation is not only very bright and tunable, but also has a pulsed time structure, because the emitting electron beam consists of distinct bunches travelling around the circular path. X-ray flashes are observed when a bunch passes the beam port. The time structure is illustrated in figure 2, in which a repeat rate typical for normal operation at the X-ray ring of the National Synchrotron Light Source at Brookhaven is shown.

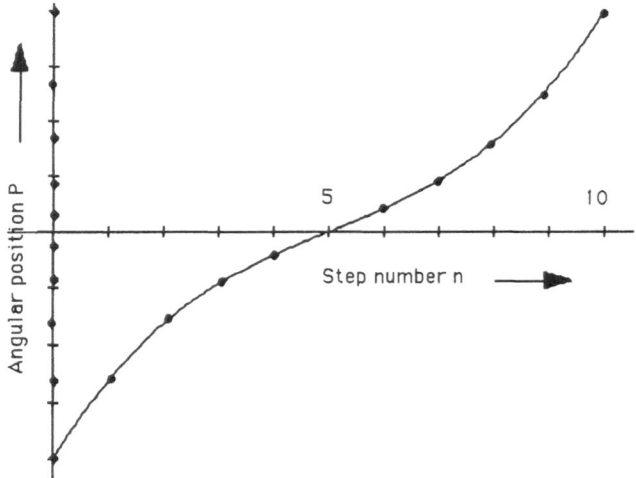

Fig 1.   Illustration of a 10 step odd-scan with $O = 0.5$. The spacing on the horizontal axis is equi-distant. The actual angular position is found on the vertical axis.

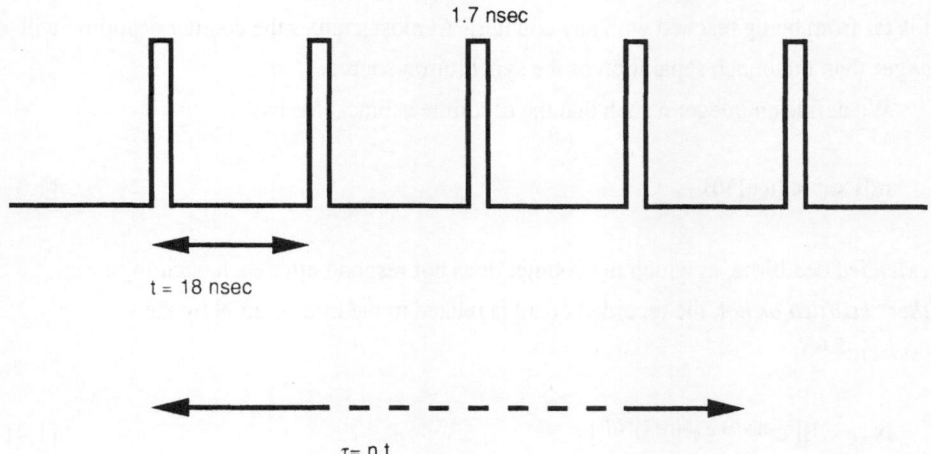

Fig. 2. Illustration of the time structure of the synchrotron beam. The values given are for NSLS with 30 bunch operation. The counter deadtime τ is expressed as a multiple of the bunch period.

The counters used in most accurate crystallographic measurements are based on photon induced scintillations in Tl-doped NaI. They typically have deadtimes τ of the order of 1 μs. At best their deadtime is limited by the light emission of NaI, which decays with a lifetime of 0.23μs. This is not sufficient to accomodate the count rate of the medium and strong reflections of a crystal that is of adequate size for the recording of the weaker reflections with good counting statistics. Selection of two crystals of different size for the weaker and stronger reflections respectively, is, in our experience a source of additional error in scaling, and to be avoided. But much faster counters are now becoming available, based on plastic scintillators with light decay times of the order of nanoseconds.[4]

It is clear that no advantage can be gained by using counters with deadtimes faster than the separation between the bunches of the storage ring, unless the deadtime could be reduced below the length of a single bunch, which is typically less than 1 nsec. But this limit is far from being reached with any counter! At most sources the counter deadtime will be *longer* than the bunch separation of the synchrotron source.

We define an integer n such that the deadtime is bracketed by:

$$n/B < \tau < (n+1)/B \qquad\qquad\qquad [1.3]$$

For *extended* deadtime, in which the counter does not respond after each occuring event, *whether recorded or not,* the recorded count is related to the true count N by the expression:[5,6,7]

$$N_{rec} = B[\, e^{-nN/B} - e^{-(n+1)N/B}] \qquad\qquad [1.4]$$

where B is the number of bunches per second.
Differentiation gives a maximum at:

$$[N_{rec}]_{max} = (n/n+1)^n - (n/n+1)^{n+1} \qquad\qquad [1.5]$$

The function [1.4] is plotted for n= 123, 50, 2 in Fig. 3. The occurence of a maximum for the slower counting chains is evident, the counter 'chokes' when the beam becomes very intense.

The top curve in the figure corresponds to the fast plastic scintillation counters now commercially available, the bottom curve to much of the equipment presently in use. The implications for accurate data collection are evident.

STANDARD DEVIATIONS IN EXPERIMENTAL DEFORMATION DENSITIES

The *deformation electron density* is defined as the difference between the total density and the density of a reference state, which may be a superposition of spherical atoms

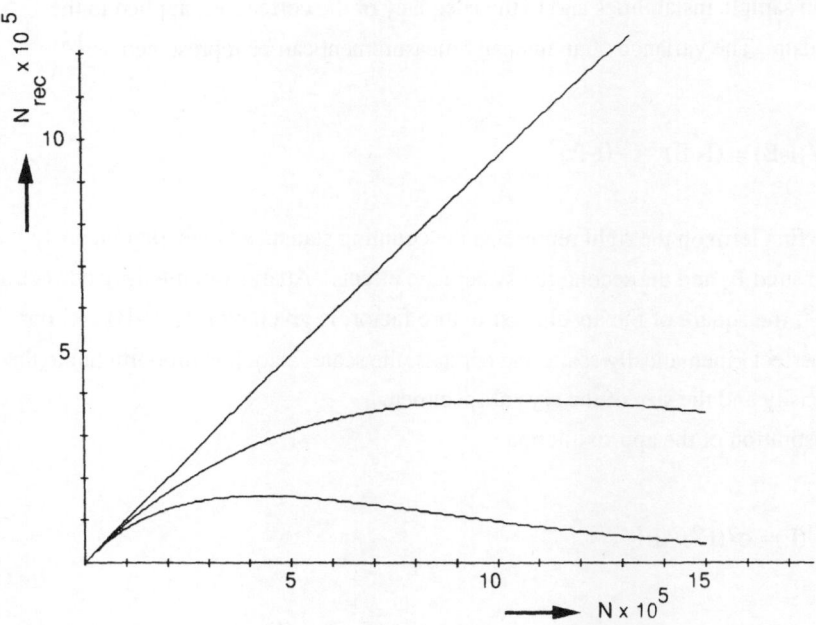

Fig. 3. Recorded counts versus true number of events, for n=123 (lower curve), n=50 middle curve, and n=2 (top curve), integral counting mode and a number of bunches per second of $5.29 \times 10^7 \ sec^{-1}$.

located at the atomic positions,[8] a superposition of oriented atoms[9] or molecular fragments,[10] or some other state against which the comparison is to be made. In general:

$$\Delta\rho = \rho_{obs}/k - \rho_{ref} = \rho'_{obs} - \rho_{ref} \qquad [1.6]$$

where the scale factor k is defined by $F_{obs} = k\,F_{calc}$.

The accuracy of $\Delta\rho$ is affected by both the errors in the experimental observations and the errors in the reference density, which depend on the positional and thermal parameters used in its construction.[11,12] The experimental observations are subject to source and sample instabilities and to the adequacy of the corrections applied to the intensity data. The variance of an intensity measurement can be represented as:[13]

$$\sigma^2(I-B) = (I+B) + C^2(I-B)^2 \qquad [1.7]$$

where the first term on the right represents the counting statistics in the total intensity I and the background B, and the second the systematic effects. Apart from intensity correction factors, F^2, the square of the absolute structure factor, is given by $F^2 = (I-B)/k^2$. For small imperfect kinematically scattering crystals, the scale factor k is proportional to the beam intensity and the size of the crystal specimen.

Substitution of the approximation

$$\sigma^2(F) = \sigma^2(F^2)/4F^2 \qquad [1.8]$$

gives:

$$\sigma^2(F) = 1/(4k^2) + B/(2k^4 F^2) + C^2 F^2/4$$

In general the first two terms are dominant for the weak reflections, and reflect the importance of high-intensity sources, as k^2 is directly proportional to beam intensity.

The error at a general position in the scaled density ρ'_{obs} $(= \rho_{obs}/k)$ is given by:[14] $\sigma^2(\rho'_{obs}) = 4/V^2 \sum \sigma^2(F)$, where the sum is over a hemisphere of reciprocal space. Substitution of [1.8] gives for the error in the observed density:

$$\sigma^2(\rho'_{obs}) = 1/V^2 \{N/(4nk^2) + 1/(2k^4)\sum(B/nF^2) + C^2\sum F^2/(4n)\} \qquad [1.9]$$

where N is the total number of observable reflections, and n is the number of symmetry-equivalent reflections that has been averaged to obtain the value of F used in the density calculation. Expression [1.9] illustrates the importance of a low background and high intensity in accurate electron density work.

The proportionality constant C in the second term of [1.9] represents errors in the measurements proportional to the measured intensity, which can be minimized by the use of smaller crystals or shorter wavelength radiation. Since such a course of action also reduces the scale factor k, a compromise between the two requirements is necessary.

Additional contributions to the errors in the deformation density are due to the error in the scale factor k, and to errors in the structural parameters used in the calculation of the reference density ρ_{calc}, assuming the reference density functions tro be error-free. The combined error in $\Delta\rho$ is given by:

$$\sigma^2 (\Delta\rho) = \sigma^2(\rho'_{obs}) + \sum_m (\frac{\partial\rho_{calc}}{\partial u_m})^2 \sigma^2(u_m) +$$

$$2 \sum_m (\frac{\partial\rho_{calc}}{\partial u_m}) \sigma(u_m) \rho'_{obs} \frac{\sigma(k)}{k} \gamma (u_m, k) + (\rho'_{obs})^2 [\sigma^2(k)/k^2] \qquad [1.10]$$

where u_m represent the set of structural parameters, and the last term accounts for the contribution from the error in the scale factor.

The use of short wavelength radiation allows the collection of large numbers of high-order intensity data. But even at low temperatures the decrease in the atomic form factor with $\sin\theta/\lambda$ limits the range in which the aditional effort is warranted. As the standard deviations are derived directly from the elements of the inverse least-squares matrix, their value is inversely proportional to the elements of the least squares matrix, which are a sum over the products of structure factor derivatives: $A_{ij} = \sum_k w_k \frac{\partial F}{\partial u_i} \frac{\partial F}{\partial u_j}$, where u_i are the parameters and the sum is over all reflections. Apart from the weights w, the contribution D_{ij} of each reflection to an element of this matrix is given by: $D_{ij} = S_i S_j F^2$ and $D_{ij} = S_i S_j S_k S_l F^2$, for the positional and thermal parameters respectively. where S_i is the projection of the scattering vector on the ith reciprocal axis. To analyze the dependence of D_{ij} on S, we may replace F^2 by its statistical average over a shell in reciprocal space: f^2, and $S_i S_j$ and $S_i S_j S_k S_l$ by S^2 and S^4 respectively.

D_{ij} goes through a maximum as a function of $\sin\theta/\lambda$, because the form factor f decreases, while S increases. Values for the C and Sn atoms with a thermal parameter B =

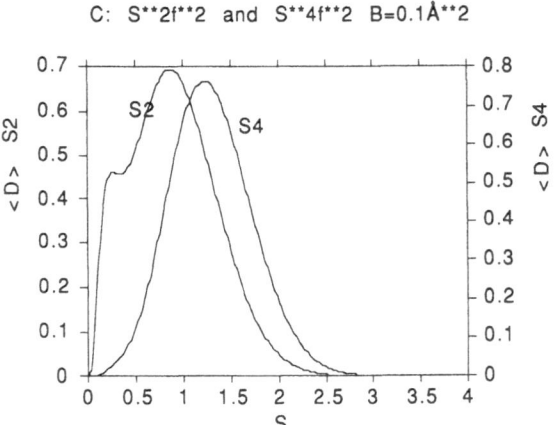

Fig. 4a. The functions S^2f^2 and S^4f^2 for a spherical carbon atom with $B=0.1\text{Å}^2$

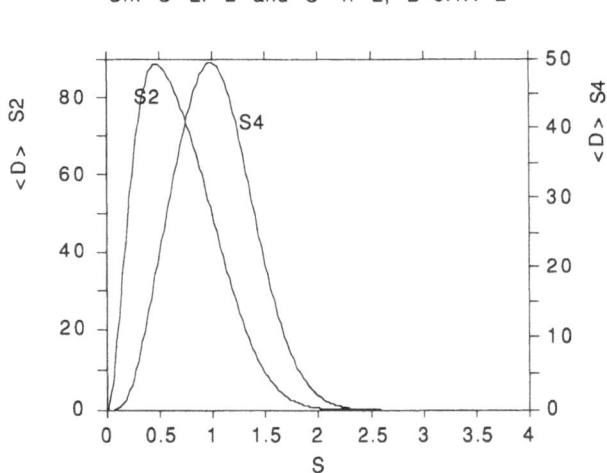

Fig. 4b. The functions S^2f^2 and S^4f^2 for a spherical tin atom with $B=0.1\text{Å}^2$

0.1Å^2, typical for very low values of the temperature parameters are plotted in Fig. 4. It is clear that little is to be gained with regard to the standard deviations by extending data collection beyond 2Å^{-1}. However, other effects on the charge density, such as core deformations may become accessible beyond this limit.

STUDY OF HEXAMINECHROMIUM(III) HEXACYANOCHROMATE(III), $Cr(NH_3)_6Cr(CN)_6$ [15]

Notwithstanding all the obvious advantages of synchrotron radiation, problems of positional and intensity stability are evident to even the occasional user of synchrotron beam time. It is therefore not a foregone conclusion that such advantages can be exploited in accurate studies.

A first feasibility study was performed on crystals of $Cr(NH_3)_6Cr(CN)_6$ at the A2 wiggler beamline at CHESS. The synchrotron at Cornell has a relatively large emittance, which implies a larger electron beam size, and less sensitivity to positional fluctuations of the beam.

Chromium hexamine hexacyano chromium is an ionic compound. Thermal motion in the crystals is moderate. The compound crystallizes in the rhombohedral space group $R\bar{3}c$, with a = b = c = 7.4635 Å, and $\alpha = \beta = \gamma = 97.83(1)°$. A wavelength of 0.302 Å, somewhat below the critical wavelength of 0.55 Å, was selected as a compromise between intensity and extinction and absorption considerations. 7225 reflections were collected at room temperature in a period of 9 days, using a crystal of about 100μ size. A scan width of 0.5° was used with a non-isometric ω step scan as described above, with steps varying between 0.005° at the center and 0.02° in the tails of the reflection. A piezocrystal adjusting the angular position of the monochromator was controlled through a feedback mechanism which continuously optimizes the incident beam intensity.

The symmetry-equivalent reflections were averaged after corrections for second order contamination, polarization and deadtime effects, and analysis of the experimental profiles. Upon elimination of 236 reflections collected during beam instabilities, an internal agreement factor between the intensities of symmetry related reflections of 0.012 was obtained. This value compares well with the best data sets collected with conventional instrumentation, and provided, for the first time, evidence that accurate data can be collected with synchrotron radiation.

Deformation density maps in the $Cr(CN)_4$ plane, before and after averaging over the symmetry of the molecular ion, shown in Fig. 5, clearly show the bonding density in the $C\equiv N$ triple bond and the lone pairs at both the carbon and nitrogen atoms. The asphericity of the density around the chromium atom is visible in the negative region around the chromium atom. The maps have a satisfactory low background in regions away from the atoms.

In order to fully exploit the potential of short wavelength radiation, data collection should be performed at reduced temperatures, rather than at room temperature. Furthermore, the intensity/background ratio at the CHESS A2 beamline was not sufficient to collect many of the weak reflections to which the chromium atoms do not contribute. The room-temperature study described in the next section is on harder materials, and includes careful collection of the weak reflections.

CHARGE DENSITY ANALYSIS OF Al_2O_3 AND Cu_2O [16]

Charge density studies of relatively simple small-unit cell inorganic materials are often more difficult than those of more complicated materials. The small size unit cell results in the valence electron scattering being concentrated in a small number of low-order reflections, which must be measured with great accuracy and without systematic errors. In many cases the high symmetry leads to groups of weak reflections to which only part of the scattering density contributes.

Both solids studied by Kirfel and Eichhorn, cuprite (Cu_2O) and α-corundum (Al_2O_3), fall in this class. Much attention was paid to the measurement of the weak reflections. A wavelength of 0.56 Å was used from the storage ring DORIS II at the Hamburg synchrotron facility HASYLAB. The crystals used were of conventional size (spheres of about 200μ diameter), as a result extinction effects were not negligible. The narrowness of the synchrotron reflection profile and the advantage gained in the use of synchrotron radiation is illustrated in Fig. 6. Comparison with the earlier analyses of conventional data[17,18] on the same solids shows very clearly the increased accuracy obtained in the weak reflections, which show much better agreement between observed and calculated values (Fig. 7). The corresponding improvement in the electron density maps is pronounced. In Cu_2O the omission of the spherical atom-harmonic vibration model forbidden, even-even-odd reflections, and the weak (oxygen spherical atom only) odd-odd-even reflections leads to apparent translational symmetry in the density maps. This artifact is effectively removed by the inclusion of the synchrotron data. For Al_2O_3 there is a distinct difference between the conventional and combined data set charge densities for the plane containing three oxygen atoms (Fig. 8). The accumulation of charge within the triangle evident from the conventional result is now completely absent, while the peak accumulations in the "back" of the oxygen atoms appear in both maps.

It would appear, however, that the authors conclusion that "nothing is gained in the measurement of the strong and medium intensity reflections with synchrotron radiation, since the signal-to-noise ratio does not profit from the small beam divergence" is overly pessimistic. The accuracy of the strong and medium reflections tends to be limited by the use of too large crystals and too soft radiation. We have argued that synchrotron radiation has much to contribute in this respect.

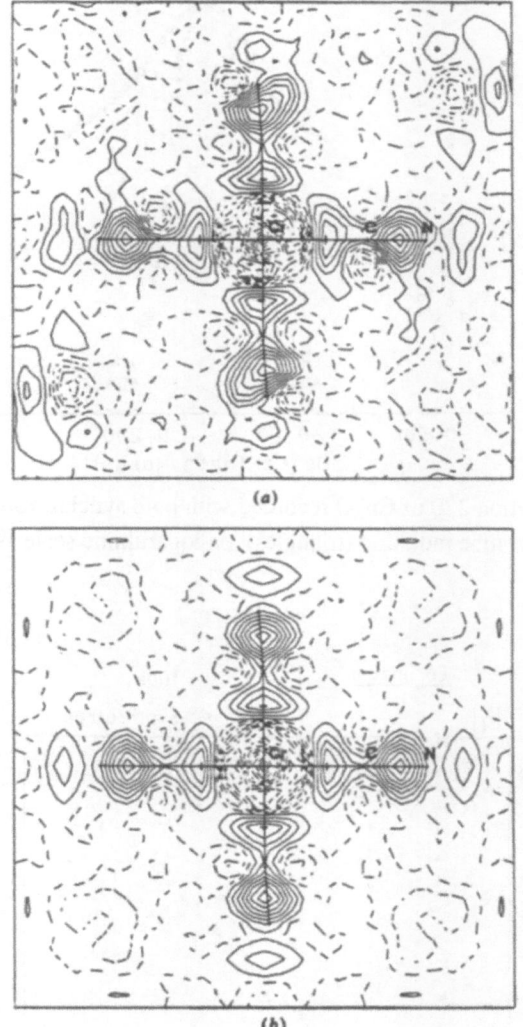

Fig. 5. X-X deformation maps in the $Cr(CN)_4$ plane based on parameters from the high-order refinement ($\sin \theta/\lambda > 0.7$ Å^{-1}), including all reflections with $\sin \theta/\lambda > 0.8$ Å^{-1}. Contours at 0.05 eÅ^{-3}, negative contours broken. (a) Before averaging, (b) after averaging over chemically equivalent regions.[15]

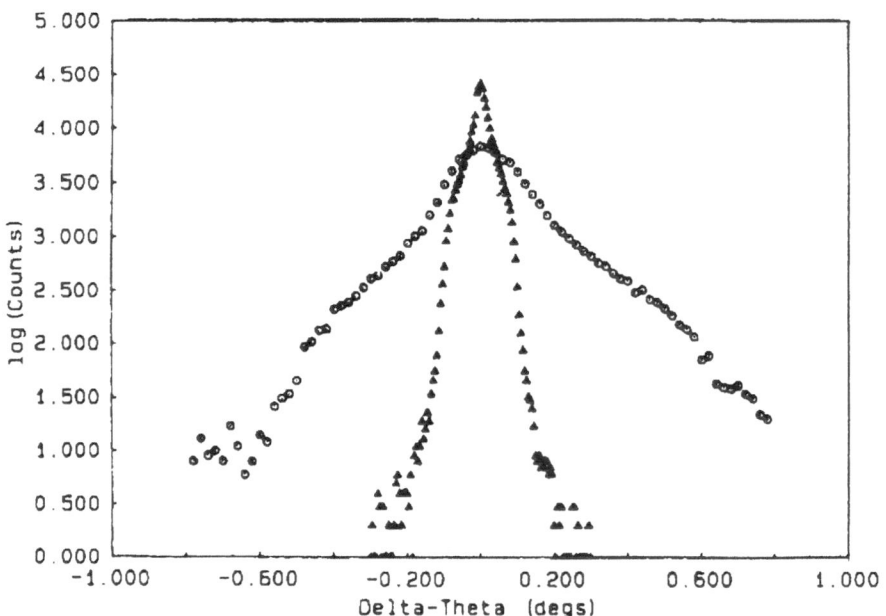

Fig. 6. Reflection 220 of Cu$_2$O recorded with both synchrotron radiation (circles) and AgKα tube radiation (triangles), on logarithmic scale.[16]

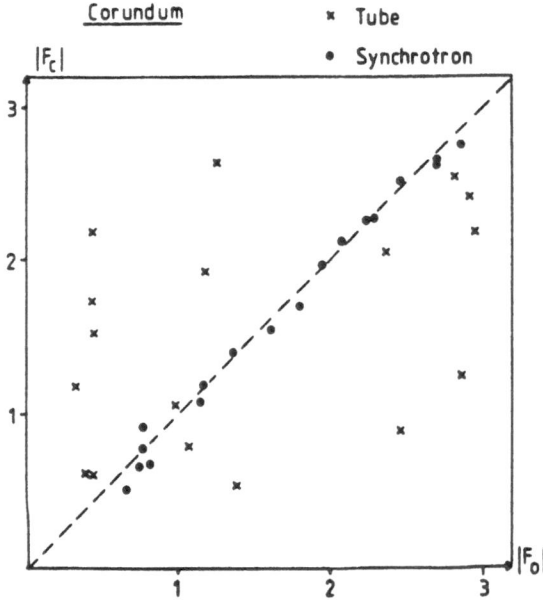

Fig. 7. Plot of F$_{obs}$ vs F$_{calc}$ for weak Al$_2$O$_3$ reflections (|F|<3). Crosses: X-ray tube; circles: synchrotron data.[16]

(a)   (b)

 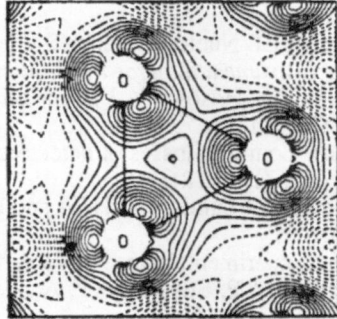

Fig. 8. Static model deformation density maps for Al_2O_3. (a), (b): plane (001) through oxygen layer at z=1/4; (a) Synchrotron data; (b) X-ray tube data. Monopoles included; contours at 0.025 e$Å^{-3}$, negative broken, positive full lines.[16]

CONCLUSION

Further experimental improvements, such as control of beam stability, reduction of background and use of faster counting equipment, are necessary before the many advantages of synchrotron radiation will be fully realized. However the technology to achieve this goal is well known. We may expect that with the proper effort charge density analysis may become faster, more routine, and sufficiently accurate for the study of effects such as intermolecular interactions and electron correlation. An electron-density microscope is not around the corner, but is is certainly within our reach in the next decades.

ACKNOWLEDGEMENTS

Support of our work by the National Science Foundation (CHE8711736) is gratefully acknowledged. The SUNY X3 beamline is operated through support of the Division of Basic Energy Sciences of the Department of Energy (DEFG0286ER45231).

REFERENCES

1. P. Coppens and R. Feil, The past and future of experimental charge density analysis, this volume
† In this expression polarization perpendicular to the diffraction plane has been assumed, so that the polarization factor equals 1. The synchrotron X-ray beam is in fact strongly polarized in the horizontal plane.
2. D. Mills and B.W. Batterman, Synchrotron-radiation measurements of forbidden reflections in silicon and germanium, <u>Phys. Rev.</u> B22:2887-2897 (1980).

3. P. Coppens, V. Petricek, D. Levendis, F.K. Larsen, A. Paturle, Y. Gao, and A.D. LeGrand, Synchrotron-radiation study of the five-dimensional modulated phase of tetrathiafulvalene-tetracyanoquinodimethanide at 15K, Phys. Rev. Lett. 59:1695-1697 (1987).

4. J.L. Radtke, IEEE Nuclear Science Symposium, Jan. 1990. Takahashi, T. and Kikuta, S. in *X-ray Instrumentation for the Photon Factory: Dynamic Analysis of Microstructure in Matter,* Eds, S. Hosoya, Y. Litak and H. Hashizumi; KTK Scientific Publishers, Tokyo, 1986, p.137.

5. U.W. Arndt, Counting losses of detectors for x-rays from storage rings, J. Phys. E. Sci. Instr. 11:671 (1978).

6. B.W. Batterman, Counting statistics and loss corrections for CHESS, (1978). CHESS Tech. Mem. #6

7. P. Coppens, Experimental aspects of a diffraction beamline, in: "Synchrotron Radiation Crystallography," P. Coppens, ed., Academic Press, Manchester, UK (to be published).

8. P. Coppens, Concepts of charge density analysis: the experimental approach, in: "Electron Distributions and the Chemical Bond," P. Coppens and M.B. Hall, eds., Plenum Press, New York (1982).

9. W.H.E. Schwarz, K. Ruedenberg, and L. Mensching, Chemical deformation densities. 1. Principles and formulation of quantitative determination, J. Am. Chem. Soc. 111:6926 (1989). W.H.E. Schwarz, P. Valtazanos, K. Ruedenberg, Electron difference densities and chemical bonding, Theor. Chim. Acta. 68:471 (1985).

10. M.B. Hall, Computation and interpretation of electron distributions in inorganic molecules in: "Electron Distributions and the Chemical Bond," P. Coppens and M.B. Hall, eds., Plenum Press, New York (1982).

11. B. Rees, Variance and covariance in experimental electron density studies and the use of chemical equivalence, Acta Cryst. A32:483-488 (1976).

12. E.D. Stevens and P. Coppens, A priori estimates of the errors in experimental electron densities, Acta Cryst. A32:915 (1976).

13. L.E. Mc Gandlish, G.H. Stout, and L.C. Andrews, Statistics of derived intensities, Acta Cryst. A31:245-249 (1975).

14. D.W.J. Cruickshank, The accuracy of electron density maps in X-ray analysis with special reference to dibenzyl, Acta Cryst. 2:65 (1949).

15. F.S. Nielsen, P. Lee, and P. Coppens, Crystallography at 0.3Å: Single crystal study of $Cr(NH_3)_6Cr(CN)_6$ at the Cornell High Energy Synchrotron Source, Acta Cryst. B42:359 (1986).

16. A. Kirfel and K. Eichhorn, Accurate structure analysis with synchrotron radiation. The electron density in Al_2O_3 and Cu_2O, Acta Cryst. A46:271-284 (1990).

17. J. Lewis, D. Schwarzenbach, and H.D. Flack, Electric field gradients and charge density in corundum, α-Al_2O_3, Acta Cryst. A38:733-739 (1982).

18. R. Restori and D. Schwarzenbach, Charge density in cuprite, Cu_2O, Acta Cryst. B42:201-208 (1986).

THE USE OF NEUTRON DIFFRACTION IN CHARGE DENSITY ANALYSIS

Hartmut Fuess

Strukturforschung, Fachbereich Materialwissenschaft
Technische Hochschule Darmstadt
Petersenstr. 20, 6100 Darmstadt

1. Introduction

Charge density in crystalline solids reflects the distribution of electrons in this solid. Neutrons, however, are essentially scattered by the nucleus of an atom. At a first glance it seems therefore slightly puzzling to include neutron scattering into a course on charge densities. In fact there are two areas where neutrons contributed considerably to the experimental evidence now available on charge densities (i) precise determination of charges by the X-N method introduced by Phil Coppens about twenty years ago (ii) direct observation of the distribution of magnetic moments in the outer shells of transition elements by polarized neutron techniques.

Progress in experimental X-ray techniques has the contribution of neutrons to charge distribution studies diminished, most studies published nowadays are obtained by X-X-methods as already pointed out by previous speakers.

The experimental determination of magnetization densities has contributed considerably to our understanding of the localization of magnetic moments in a solid and to the interaction between central atoms and ligands. Overlap of these outer electrons of a central atom introduces covalent contributions to bonding in otherwise ionic systems. Most studies of magnetization densities were performed on salt – like compounds of 3d or 4f elements, some studies were performed on free radicals which exhibit a magnetic moment due to an unpaired localized electron. Very few studies comparing charge and magnetization densities for the same compounds have been performed so far.

2. Production and basic properties of the neutron

2.1. Fission, moderation

In research reactor neutrons are produced by fission. Uranium–235 undergoes fission with thermal neutrons with the production of about 2.5 fast neutrons.

$$n_{therm} + U^{235} \rightarrow 2 \text{ fission fragments} + 2.5 \text{ n} + 200 \text{ MeV}$$

The neutrons produced are high energy, fast neutrons. They are slowed down, moderated, by collision with light elements, e.g. D_2O or graphite. Research reactors are optimized to produce large quantities of neutrons, characterized by the flux. The high flux reactor at the Institute Laue-Langevin has a flux $\emptyset_{th} = 2 * 10^{15}$ n cm^{-2} sec^{-1} in the core. The development of neutron sources is shown in Fig. 1.

The Application of Charge Density Research to Chemistry and Drug Design
Edited by G. A. Jeffrey and J. F. Piniella, Plenum Press, New York, 1991

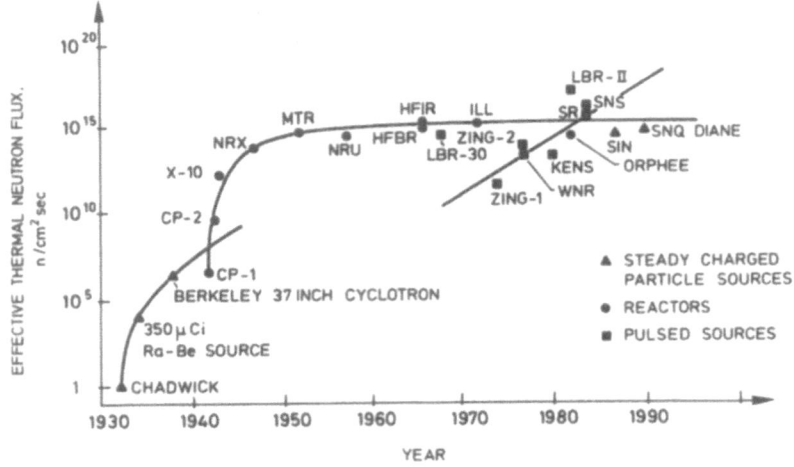

Fig. 1 Development of neutron sources

Most of those neutrons never reach the sample, the flux at the sample position may in some applications reach 10^8 n cm^{-2} s^{-1}. Moderation at room temperature produces so-called thermal neutrons (wavelength between 0.5Å$< \lambda <$ 2.0Å). This range is extended by cold (long wavelength) and hot (short wavelength) sources.

The distribution of tubes which conduct neutrons out of the reactor is presented in Fig. 2 for the Brookhaven high flux reactor.

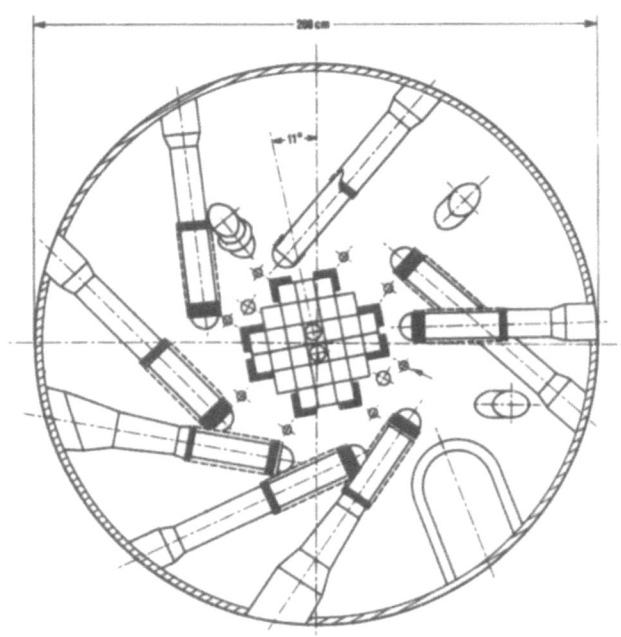

Fig. 2 Collimators in a neutron beam reactor

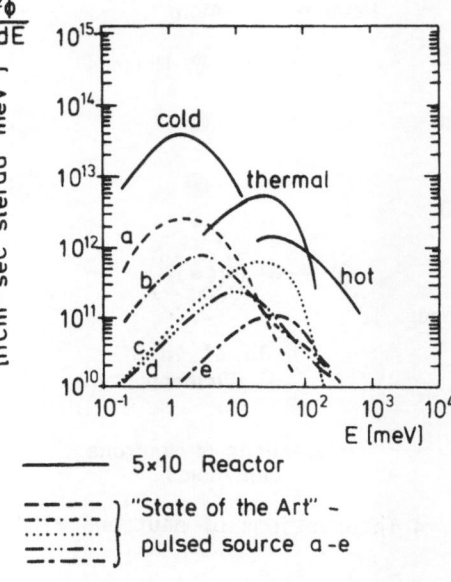

$\frac{d^2\phi}{d\Omega dE}$

$[n\,cm^{-2}\,sec^{-1}\,sterad^{-1}\,meV^{-1}]$

5×10 Reactor

"State of the Art" –
pulsed source a –e

Fig. 3 Energy distribution in a reactor and a spallation source

A typical distribution of neutrons after moderation is given in Fig. 3. The distribution may be described by Maxwellian curves.

2.2. Properties of the neutron

The neutron has no charge, but a magnetic moment which is due to its internal structure. The neutron is characterized by one of the quantities: wavelength (λ), momentum (p), energy (E), velocity (v) or wavenumber (k).

$$\lambda = 2\pi/k = h/mv = h/p = h/\sqrt{2m_0 E} \ [\text{Å}]$$

h is Planck's constant. The magnetic moment is $\mu = -1.913_n$ ($\mu_n = eh/2mc$).

$_n$ is a Bohr magneton. The neutron energy is in the range of vibrational energies in solids. Therefore energy changes by interaction processes between neutrons and samples are easily detectable.

2.3. Interaction of neutrons with matter

The interaction of neutrons with matter is an interaction of the neutron with single particles. The main reaction is between a neutron and a nucleus where absorption or scattering may occur. Due to the magnetic moment of the neutron an additional interaction between neutrons and magnetic moments of 3d-, 4f-or other transition elements takes place and produces magnetic scattering. I should be mentioned that a scattering process between the magnetic moment of the nucleus and the neutron occurs

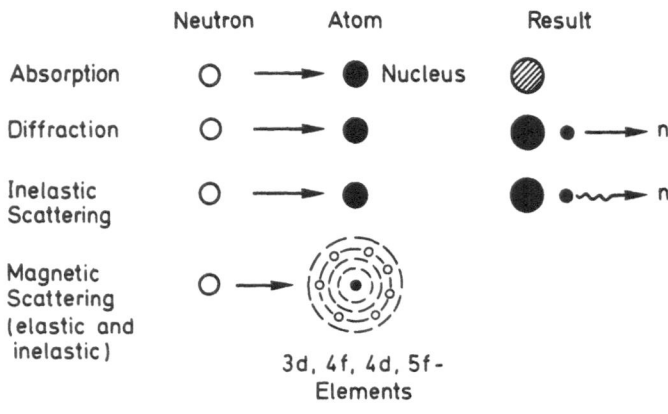

Interactions of Neutrons
with Atoms

Fig. 4 Interactions of neutrons with atoms

which, however, is fairly small in intensity. The basic interaction processes are summarized in Fig. 4.

The strength of the interaction is described by a quantity called cross-section (σ). The total cross-section includes both absorption and scattering:

$$\sigma_t = \frac{\text{outgoing current of scattered neutrons + absorbed neutrons}}{\text{incident neutron current per cm}^2}$$

$$\sigma_{total} = \sigma_{scattering} + \sigma_{absorption}$$

Absorption produces a different nucleus and may give all kinds of results for nuclear physics experiments.
The scattering power of an individual nucleus is given by the scattering length b which is connected with the cross-section by

$$\sigma_s = 4\,b^2$$

The scattering length is a quantity connected with each individual nucleus and is therefore different for isotopes of the same element.

The variation of b (given in units of 1 Fermi = 10^{-15}m) is represented in Fig. 5. A comparison of these scattering powers of neutrons as compared with X-rays is schematically given in Fig. 6.

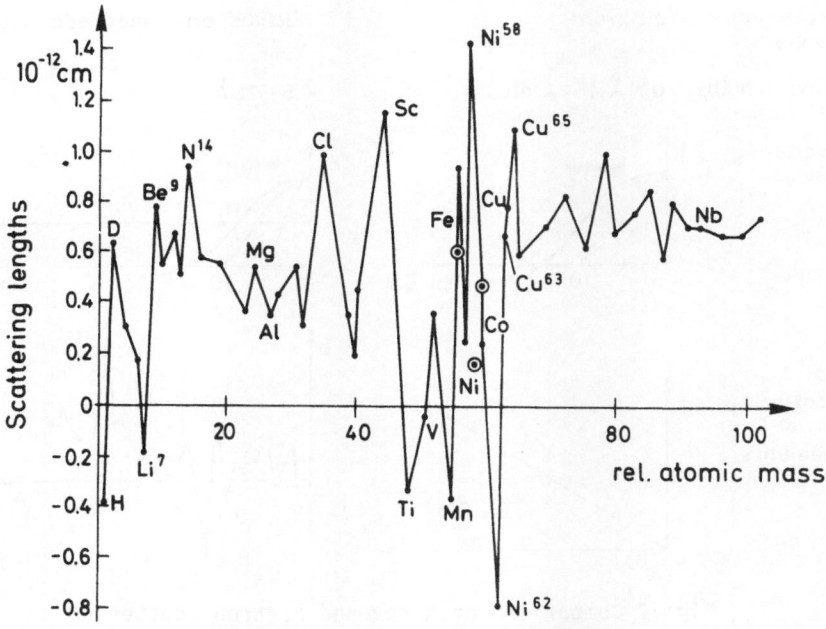

Fig. 5 Variation of scattering lengths b for different nuclei

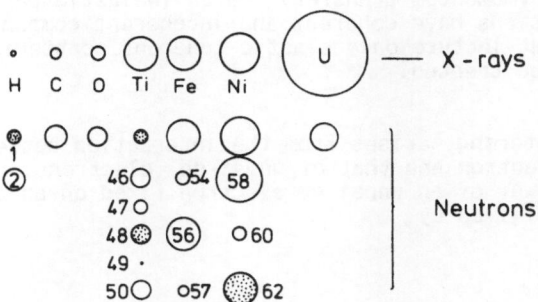

Radii are proportional to the scattering amplitude b

⊛ ⊛ negative values for b

Fig. 6 The relative scattering power of elements for X-rays and neutrons

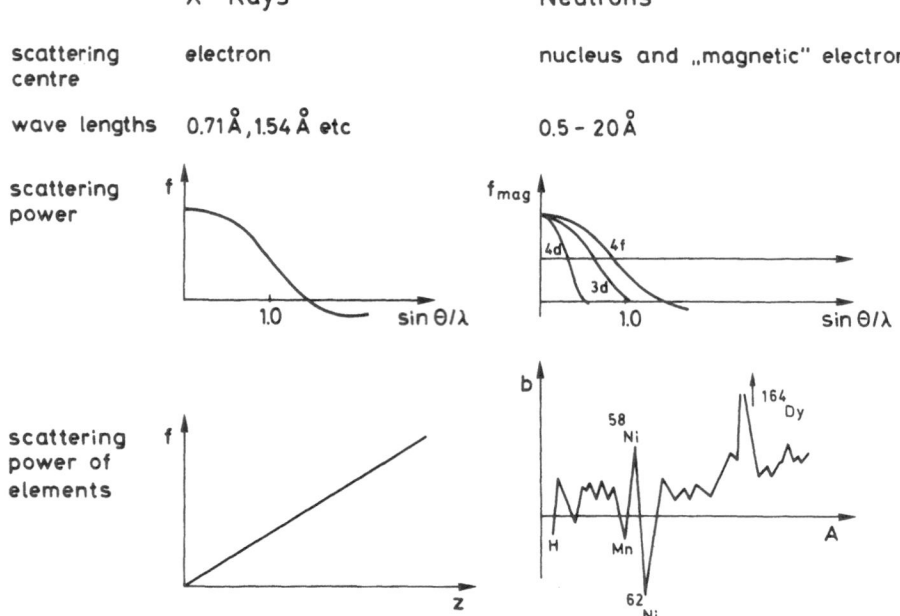

Fig. 7 Comparison of X-ray and neutron scattering

The scattering length of neutrons is independent of the scattering angle whereas the X-ray scattering is governed by a form factor which decreases with increasing scattering angle.

Fig. 7 compares the scattering processes in both cases.

2.4. Scattering processes

The complete process in a neutron scattering experiment is composed of an elastic part (momentum transfer) and an inelastic part (energy transfer). Both interactions have coherent and incoherent components. In the context of the present lecture only elastic coherent scattering (nuclear and magnetic) will be treated.

Magnetic scattering arises from the interaction between the magnetic moment of the neutron and that of unpaired electrons. Let us consider the scattering power of an unpaired electron fixed on an atom. Its scattering length may be given by

$$a = \frac{e^2}{2m_0 c^2} \gamma \quad = 2.7 * 10^{-15} \, m$$

with e and m_0 charge and mass of an electron: c velocity of light, γ magnetic moment of the neutron. This scattering power is in the same order of magnitude as the nuclear scattering. Magnetic scattering is governed by a form factor which rapidly falls off with increasing scattering angle (see Fig. 7).

3. Charge Density Studies by the X-N-Method

The combination of X-ray and neutron diffraction data is based on the assumption that a non-bonded, free atom is spherical and that any deviation from sphericity is due to chemical bonding. The method was successfully applied to a great number of compounds. Two different diffraction experiments have to be performed. The total electron distribution is determined by an X-ray analysis, the precise atomic position (=position of the nucleus) is obtained by neutron diffraction. The next step is the calculation of a promolecule, that is a calculation of an electron density using scattering factors of a spherical non-bonded atom. Thus all deformation in electron density due to chemical bonding is obtained in a difference calculation which presents the total observed density (X-ray experiment) minus the promolecule density with spherical atoms (calculated at the position of the neutron experiment). This explains the term X-N-method.

Fig. 8 presents the example of the deformation density in water molecules in $MgS_2O_3*6H_2O$. Clearly electron density in the bond between oxygen and hydrogen is seen; furthermore residual density in the region of the lone pairs at the oxygen atom is displayed on the right hand of Fig. 8. This electron density differs for the three crystallographically different water molecules. Whereas the upper one presents two distinct maxima in the lone pair region, a broad, rather diffuse density is observed in the middle one and a single peak for the bottom. It is concluded that this density in the lone pair region is influenced by the neighbouring Mg^{2+}-cation. In the top picture not only a Mg^{2+}-cation but also an H-atom of the next water molecule are at distances near enough to influence the density. In the two other cases only one Mg-cation is present.

A direct comparison between an observed and a calculated density is displayed in Fig. 9 for the thiosulfate group. The agreement for the bond between sulfur and oxygen is fairly good but the observed density for the lone pair region is much less than the calculated one.

A more recent example which is close to the main stream of discussion of this study institute is shown in Fig. 10. It displays electron density in the main plane of a 18 crown structure (Luger et al, 1991). The complex encloses two cyanamid groups, hydrogen bonded to the crown. Differences between oxygen atoms are more clearly seen in the multipole development (Fig. 10c) than in the deformation X-N-map (Fig. 10b).

Most charge density studies in recent years were based on X-X-data. We shall try to discuss some of the corrections necessary to produce an X-N-difference map. Whereas the agreement in the positional parameters between refinements based on X-ray and neutron diffraction are fairly small, differences in thermal parameters may be considerable and corrections have to be applied. Differences for six compounds are presented in Table 1.

Table 1. Differences in thermal parameters. (X-ray-neutrons)

	ΔU_{11}	ΔU_{22}	ΔU_{33}
NaHCOO	0.0012(1)	0.0006(1)	0.0003(1)
KClO$_4$	-0.0003(2)	-0.0009(2)	0.0007(1)
KClO$_3$	0.0013(1)	0.0020(1)	0.0016(1)
MgS$_2$O$_3$*6H$_2$O	0.0017(1)	0.0005(1)	0.0018(1)
MgSO$_3$ *6H$_2$O	0.00061(6)	0.00061(6)	0.00061(6)
Fe$_2$SiO$_4$	-0.00002(10)	0.00017(10)	0.00018(10)

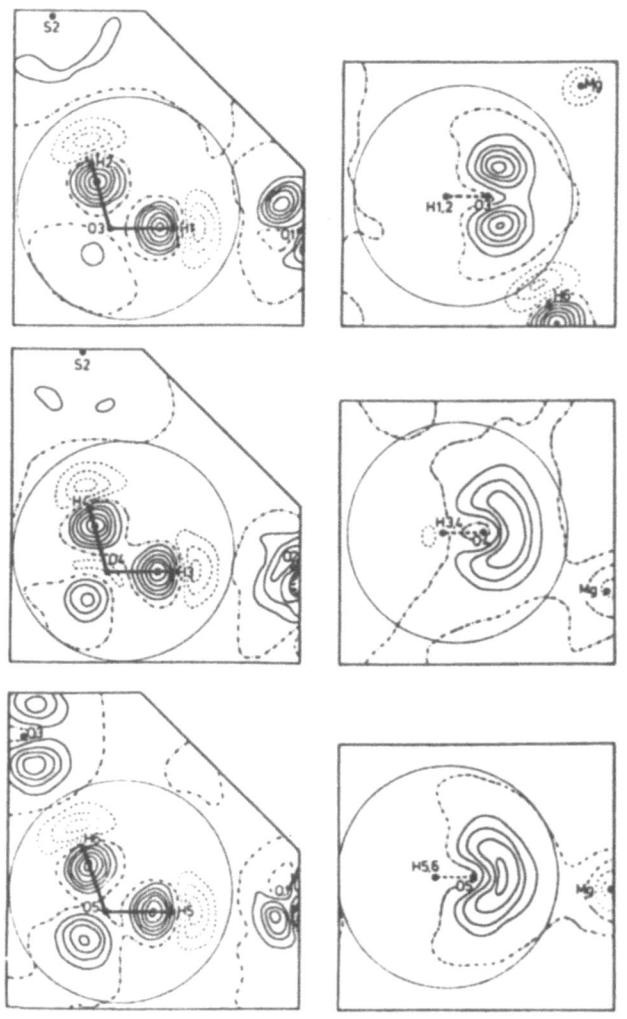

Fig. 8 Deformation electron density (X-N) for water molecules in
MgS₂O₃*6H₂O

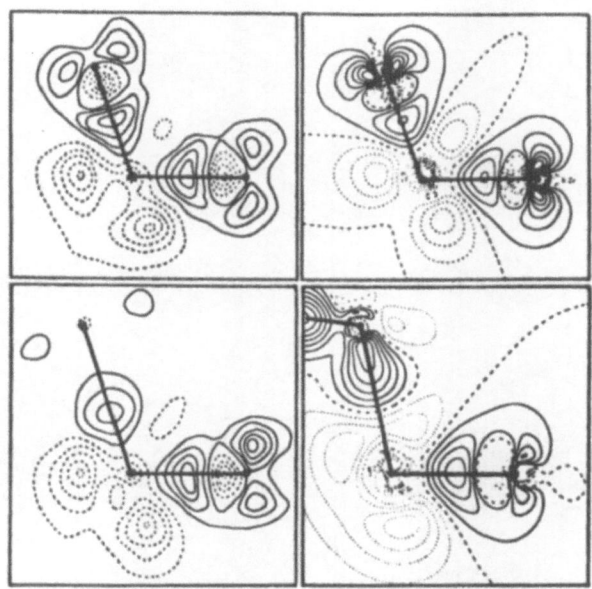

Fig. 9 Left: experimental static deformation density in $Na_2S_2O_3$, resolution $1.0Å^{-1}$. Right: theoretical static deformation density in H_3NSO_3. Contour interval $0.1e/Å^{-3}$. The zero contour (dashed) is omitted in the experimental map but included in the theoretical map. Top: Electron deformation density in the O-S-O plane. Below: electron deformation density in the X-S-O plane (X=S,N). Solid lines and dotted (or dashed) lines represent positive and negative densities, respectively.

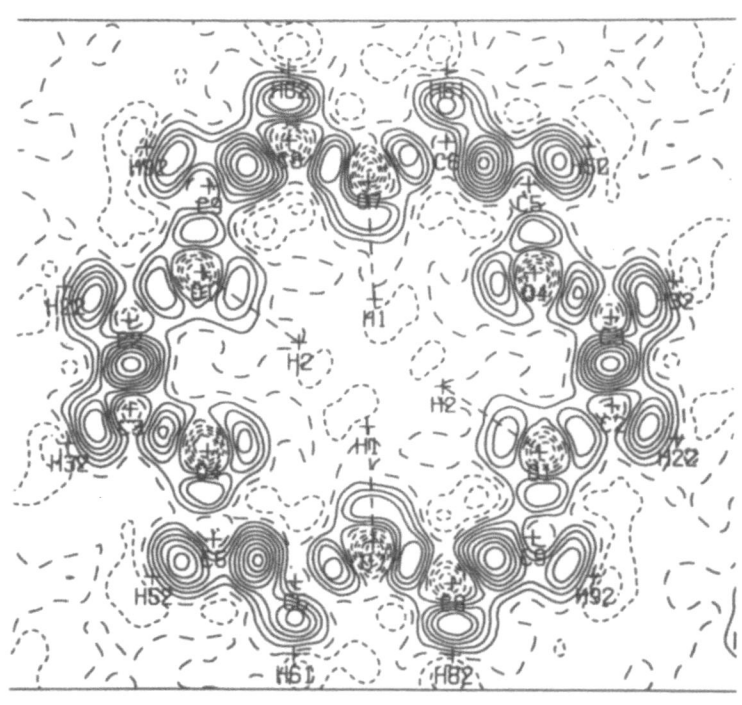

Fig. 10a Experimental deformation density in the main plane of the
complex 18 crown-6 with 2 cyanamide (after Luger et al., 1991)

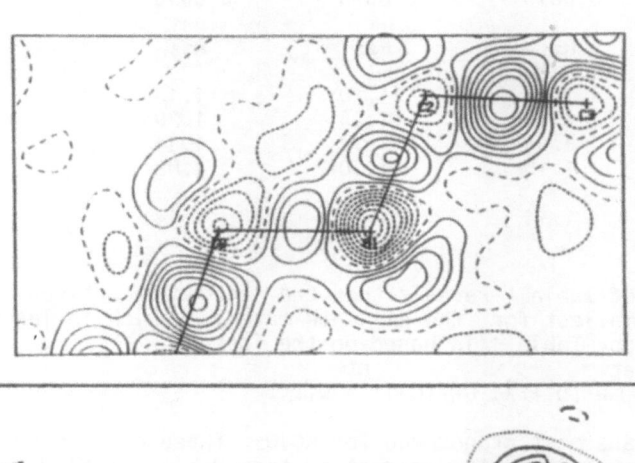

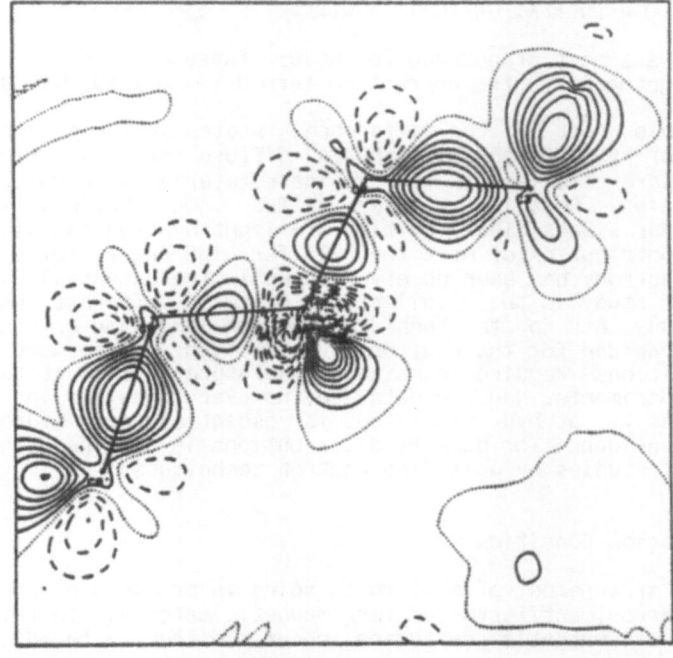

Fig. 10b Experimental deformation density for planes of the three non-equivalent oxygens (above) and multipole refinement (below) (Luger et al., 1991).

Table 2. Difference in thermal parameters KClO$_3$

	ΔU_{11}	ΔU_{22}	ΔU_{33}	$\langle \Delta U_{ii} \rangle$atom
K	0.0017(3)	0.0019(3)	0.0013(3)	0.0016
U	0.0012(1)	0.0021(1)	0.0016(1)	0.0017
O(1)	0.0004(3)	0.0031(4)	0.0019(3)	0.0018
O(2)	0.0019(2)	0.0017(2)	0.0015(2)	0.0017
$(\Delta U_{ii})_{dir}$	0.0013	0.0020	0.0016	

	R_{11}	R_{22}	R_{33}	$\langle R \rangle$atom
K	1.22	1.20	1.13	1.18
Cl	1.15	1.25	1.20	1.20
O(1)	1.05	1.14	1.14	1.11
O(2)	1.15	1.20	1.08	1.14
$\langle R \rangle_{dir}$	1.14	1.20	1.14	

Inspection of Table 1 reveals that the agreement between thermal parameters is smallest for the silicate material which is least affected by thermal motion. Table 1 is based on the formula

$$\Delta U_{ii} = \sum w \; [U_{ii}^{at}(X) - U_{ii}^{at}(N)] / \sum w_i.$$

Differences are most pronounced for KClO$_3$. These differences are shown in Table 2, together with the correction term $R_{ij} = U_{ij}(X)/U_{ij}(N)$.

A considerable part of the differences stated in Tables 1 and 2 may be accounted for by differences in thermal diffuse scattering (TDS). In fact differences are smallest in the silicate material with relatively small thermal motion. Detailed studies of TDS contributions in Ca(HCOO)$_2$ (Fuess, Burger & Bats, 1981) by time-of-flight neutron techniques presented a TDS contribution of more than 20% for high-angle reflections. Pronounced anisotropy has been observed as well. The essential difficulty in this type of study is the collection of two different data sets. No complete study has so far been reported on the same crystal, a larger specimen is needed for the neutron investigation. Furthermore the experimental conditions require identical temperature, difficult to realize in separate experiments. Neutron data are, however, required in cases where electron density at hydrogen atoms is essential (e.g. hydrogen bonds). More direct evidence for density distributions is obtained in magnetization density studies by polarized neutron techniques.

4. Magnetization Densities

An ordered arrangement of electronic spins is provided by a crystal of a saturated ferro-, antiferro- or ferrimagnetic material. In this case both nuclear and magnetic scattering occur and the scattered coherent intensity is

$$I = F_N{}^2 + 2F_N F_M \; \vec{q} \, \vec{\lambda} + p^2 \, F_M{}^2$$

F_N is the structure factor for nuclear Bragg scattering. F_M is the structure factor for magnetic scattering. The vector $\vec{\lambda}$ presents the polarization of incoming neutrons, $\vec{q}$ is a vector which presents the difference in direction between the electronic spins and the diffraction vector. In the usual experiment with unpolarized neutrons $\lambda = \pm 1/2$ and the cross-section reduces to

$$I = F_N{}^2 + q^2 \, F_M{}^2$$

a superposition of intensities.

For polarized neutrons $\lambda = \pm 1$ and in an experiment with polarized neutrons the geometry is generally chosen so that q = 1. The scattered intensity is thus different for the two spin states of the neutron

$$I^+ = (F_N + F_M)^2$$

$$I^- = (F_N + F_M)^2$$

These formulae indicate an interference of nuclear and magnetic scattering and not a superposition of intensities. The quantity actually measured is the flipping ratio

$$R = \frac{I^+}{I^-} = \frac{(F_N+F_M)^2}{(F_N-F_M)^2} = \frac{(1+\gamma)^2}{(1-\gamma)^2} \text{ with } \gamma = \frac{F_M}{F_N}$$

This is the ratio between a measurement with neutrons with incoming spin up ($\lambda = +1$) and down ($\lambda = -1$). These intensities are both composed by nuclear and magnetic components. The intensities are registered in a stationary count, therefore magnetic and nuclear Bragg reflections have to be superposed in reciprocal space, a prerequisite naturally fulfilled in ferro- and ferrimagnetic materials, in many cases, however, not for antiferromagnetic compounds. Most studies of magnetization densities were therefore carried out on ferromagnetic materials. We present the investigations on rare earth iron garnet (YIG, ErIG and TbIG). Yttrium iron garnet $Y_3Fe_2Fe_3O_{12}$ contains Fe-atoms on two non-equivalent crystallographic sites. Three Fe^{3+} per formula unit are tetrahydrally coordinated, two are in an octahedral environment. The magnetic spins of each sublattice are ferromagnetically coupled and the two sublattices are aligned antiparallel to give a ferrimagnetic structure (T_c = 559K). The polarized neutron study (Bonnet et al., 1979) produced a magnetization density on the two iron ions and in addition density of unpaired electrons on the position of oxygen. Fig. 11 presents the total magnetization density (left) together with a difference density (right). This difference was calculated by the total density minus a density calculated only from magnetic moments on oxygen. Complete studies on ferrimagnetic erbium (ErIG) and terbium iron garnet (TbIG) confirmed the magnetic moments on the oxygen atoms in additon to magnetic density on the rare earth site. This study is therefore an illustration of the theoretical model of superexchange proposed by Neel to explain antiferromagnetic interactions.

A study of magnetization densities by polarized neutrons is a fairly time consuming procedure. In the majority of cases a complete set of reflections cannot be collected. In addition to the time argument measurements are excluded when the nuclear intensities are too small to give a reasonable flipping ratio. In such cases a direct Fourier synthesis would contain artefacts introduced by the incompleteness of the data. A possible model for such cases is that of multipoles. Techniques based on multipole expansion have now been used to determine the orbital populations and the distribution of the moment for a number of compounds, mainly transition metal complexes.

A successful application of the multipole method was the interpretation of the distribution of unpaired spin in the nitroxide group of some free radicals. The aim of the polarized neutron study by Brown et al., 1979 and Boucherle et al., 1987 was the localization of unpaired spin. Most theoretical calculations showed a preference for the oxygen site. The first compound selected for study was the biradical molecule tanolsuberate $C_{26}H_{46}N_2O_8$ which becomes ferromagnetic at very low temperatures. Fig. 12a presents the spin density from the measured 70 reflections, whereas Fig. 12b shows a reconstruction from the results of the multipole fit to all the measured data. The conclusion to be drawn is an equal distribution of spin density between oxygen and nitrogen.

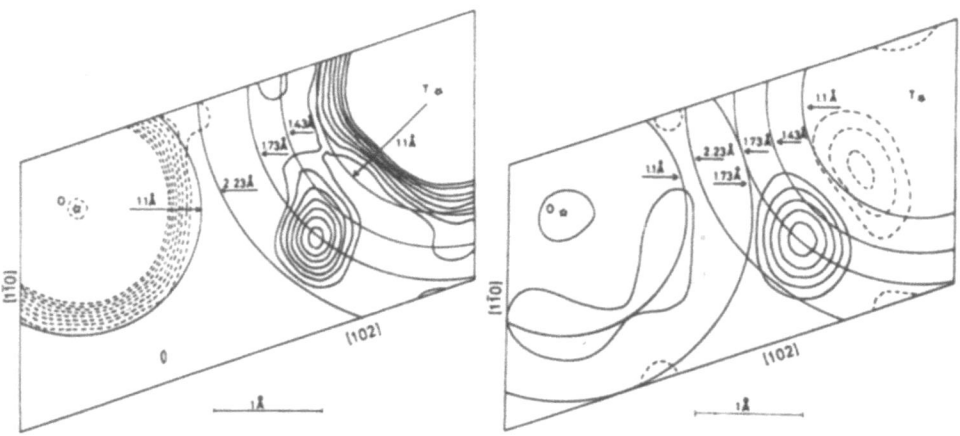

Fig. 11   Total magnetization density (superexchange) in the plane of
          Fe(octahedral)-oxygen-Fe(tetrahedral) at left and a difference
          density (right).

A second example of these series of compounds is the free radical α,α-
diphenyl-ß-picryl-hydrazyle   (DPPH): The   measurement of the magnetic
structure factors was performed at 1.5K in a field of 4.6 Tesla. A large
amount of the spin density (61%) remains localized on the central hydrazyle group (Fig. 13).

The   remaining part of the spin density is delocalized on the three
aromatic rings of DPPH,   the amount of spin transferred onto a ring
depends on the twist angle of the ring with respect   to the hydrazyle
backbone. Only very few studies which compare results of charge and
magnetization density on the same compound were published so far. In a
polarized neutron study on Mn_2SiO_4 Fuess et al. (1988) determined spin
density for one oxygen atom which is on the superexchange path between

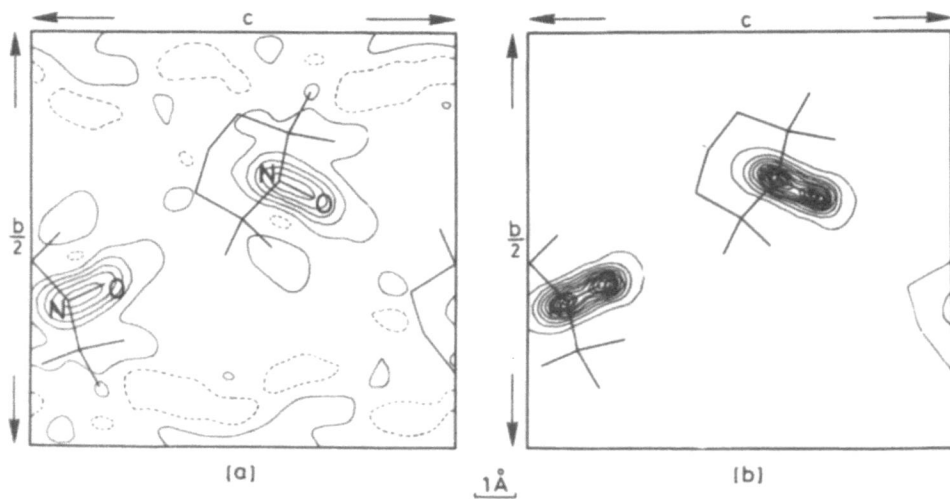

Fig. 12 Projection of the spin density in di(2,2,6,6-tetramethyl-4-
piperidinyl-1-oxyl)-suberate down [100].
a) A Fourier synthesis containing 70 measured structure factors.
b) Multipole expansion up to hexadecapoles on nitrogen and oxygen. The
contour interval in both maps is $0.06_{\mu B}Å^{-2}$ (after Brown et al., 1979)

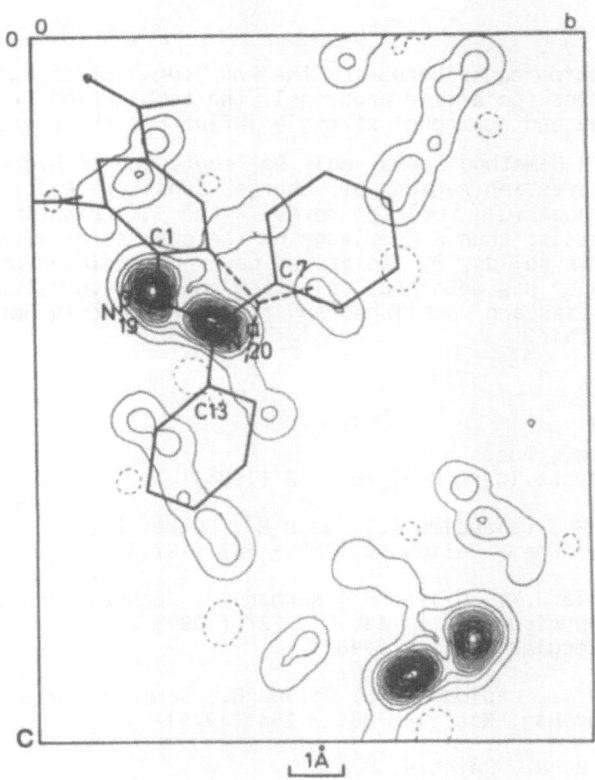

Fig. 13 Spin density in the hydrazyl group of DPPH. Contour intervals
        0.03 μB/Å²

two Mn^{2+}-ions. Charge density studies on the olivine structures Fe_2SiO_4, Mn_2SiO_4 and Co_2SiO_4 by various authors revealed an excess of charge density for the same oxygen atom.

A more quantitative treatment has been proposed by Figgis and coworkers (e.g. Figgis and Reynolds, 1986). They studied simple Cr(III), Ni(II) and Co(II) complexes by X-ray diffraction and polarized neutrons and compared the results with theoretical calculations. They determined occupation numbers for different orbitals by multipole expansions from the experimental data.

For $Ni(NH_3)_4(NO_2)_2$ the results reveal that 27% of the spin is transferred to the ligand atoms by covalent interactions. By a simple ligand field model qualitative but not quantitative agreement with the experimental data was obtained.

5. Summary

Neutron diffraction contributes to the knowledge of charge (X-N-method) and spin densities (polarized neutrons). The combination of two different data sets (X-ray and neutrons) strongly influences the precision of data.

Therefore the X-N-method seems only be suggested if hydrogen atoms are involved in interesting regions of charge density. Polarized neutron studies are a powerful tool to reveal the distribution of magnetic moments in crystals, thus a knowledge of the spatial arrangement of 3d-, 4f-electrons in solids. By polarized neutron studies experimental evidence of covalency was obtained. Only few comparative studies of charge and spin densities are published so far. Agreement in between and with theory is only fair.

References

Bats J.W., Fuess H.:
Acta Crystallogr. B42, 26 - 32 (1986)

Bonnet M., Delapalme A., Fuess H., Becker P.:
J. Phys. Chem. Solids 40, 863 - 867 (1979)

Boucherle J.X., Gillon B., Marhani J., Schweizer J.:
J. de Physique, Coll, 43, C7, 227 (1982)
and Molecular Physics (1987)

Brown P.J., Capiomont A., Gillon B., Schweizer J.:
J. Magn. Mag. Mat. 14, 289 - 294 (1979)

Figgis B. N., Reynolds P.A.:
Intern. Rev. Phys. Chem. 5, 265 - 272 (1986)

Fuess H., Ballet O., Lottermoser W.:
in "Structural and magnetic phase transitions in minerals",
Springer-Verlag, 185 - 207 (1988)

Fuess H., Bats J.W., Cruickshank D.J.W., Eisenstein M.:
Angew. Chem. 97, 511 - 512 (1985)

Fuess, H., Bats J.W., Dannöhl H., Meyer H., Schweig A.:
Acta Crystallogr. B38, 736 - 743 (1982)

Fuess H., Burger H., Bats J.W.:
Z. Kristallogr. 156, 255 - 263 (1981)

Hock R.: Thesis, University Frankfurt, 1990

Luger P., Buschmann J., Koritsanszki T., Knöchel A., Patz M.,
J. Amer. Chem. Soc., 1991, in press

STRUCTURAL CHEMISTRY AND DRUG DESIGN

Gastone Gilli and *Pier Andrea Borea*[+]

*Dipartimento di Chimica and Centro di Strutturistica Diffrattometrica,
Università di Ferrara, Via Borsari 46, I-44100 FERRARA (Italy)
[+]Istituto di Farmacologia, Università di Ferrara, Via Fossato di Mortara
64/B, I-44100 FERRARA (Italy)

INTRODUCTION: DRUGS, RECEPTORS AND STRUCTURAL CHEMISTRY

A <u>drug</u> is any xenobiotic able to modify the equilibrium conditions of living beings. Most drugs act by interacting with specific macromolecular targets which are called <u>receptors</u> and are located inside or on the surface of cells. The drug-receptor interaction has been shown to be extremely specific, in a stereochemical sense, i.e. only drugs having a definite shape, distribution of functional groups and absolute configuration may fruitfully interact with a given receptor.

It is this stereospecifity of action which makes receptorial drugs ideal subjects for <u>structural chemistry</u>. It seems reasonable to assume that a detailed knowledge of the drug-receptor interactions would allow us not only to know how and why the drug acts but also to develop new drugs more potent, more selective and having far less unwanted side effects (<u>drug design</u>). The practical realization of this concept is unfortunately hindered by our lack of knowledge of the real structure of receptors. Although the information obtained by X-ray crystallography on the structure of biological macromolecules, both proteins and nucleic acids, has grown at a fantastic rate in the last decades, receptors have so far resisted any attempt to clarify their structure.

Thus, out of the two partners of the interaction, drug and receptor, only one is actually known and the only strategy accessible to molecular pharmacological studies remains that of trying to guess the structure of the

The Application of Charge Density Research to Chemistry and Drug Design
Edited by G. A. Jeffrey and J. F. Piniella, Plenum Press, New York, 1991

241

receptor or, better, of the small part of the receptor responsible for the interaction (the receptor binding site) from all available information on structure and properties of drugs specifically recognizing that particular receptor. Using a somewhat worn out analogy we have to guess the shape of a lock from the shape of keys opening it more or less easily in such a way to be able to plan the perfect working key.

PART 1: BIOLOGICAL AND PHARMACOLOGICAL ASPECTS

Receptors and Ligands: Agonists, Antagonists, Partial Agonists

Our present views on the role of receptors in living organisms imply that: (i) the drug binds to the receptor in a reversible way, (ii) the bonded drug produces conformational changes in the receptor itself triggering the chain of events leading to the final biological response. A drug behaving in such a way, that is binding and producing the expected effect, is called an agonist because it is able to mimic, at least in some respects, some endogenous substance which is the natural ligand (hormone, neurotransmitter or autacoid) of that specific receptor binding site. There are other chemicals which are able to bind to the receptor site without producing any per se biological effect. These compounds are called antagonists since they antagonize the action of agonists and can produce a biological effect only because they are able to inhibit the binding and the consequent action of endogenous ligands and their agonists.

Not all agonists can reproduce completely the behaviour of the endogenous substance. These are called partial agonists and have the property that they cannot give the maximum response of a full agonist independently of their concentration at the site of action. This observation leads to the definition of intrinsic activity of an agonistic drug as the ratio between the maximum response achievable from the drug and that produced by the naturally occurring agonist or by the most active compound known. Of course all antagonists have intrinsic activity of zero, full agonists of one and partial agonists in between zero and one.

Another important concept is that of affinity, which is the ability of the drug (either agonist or antagonist) to bind to the receptor. The affinity can be measured by the value of the constant, K_A, of the association reaction

$$\text{RECEPTOR + DRUG} \rightleftharpoons \text{DRUG-RECEPTOR} \qquad (1)$$

242

or by the value of its dissociation constant K_D, being $K_D=1/K_A$. In pratice the affinity of a drug is measured as its ability to inhibit the binding of a radiolabelled specific and high affinity ligand, that is as an inhibitory affinity constant, K_i; this is obtained as (Cheng and Prusoff, 1973)

$$K_i = \frac{IC_{50}}{1+[L^*]/K_D^*} \qquad (2)$$

where IC_{50} is the inhibitor concentration which displaces 50% of the label-led ligand, $[L^*]$ the concentration of the labelled ligand and K_D^* its dissociation constant. It may be shown that K_i practically coincides with the value of K_D of the inhibitor under examination.

The affinity of most endogenous ligands and of their best agonists or antagonists is very high and, accordingly, their K_i or K_D are very small, of the order of magnitude 10^{-8}-10^{-10} M.

The biological role of receptors

Most receptors are localized on cell membranes (membrane receptors) though some of them are known to reside inside the cells (intracellular receptors). They are physiological regulatory molecules which have the function of recognizing their endogenous ligands and transmitting their regulatory signals to the target cell. The transmission of information through the cell membrane is achieved in two main ways: (i) conformational changes of membrane receptors which modify their permeability towards positive or negative ions (ion channels), (ii) promotion of the synthesis of other intracellular transmitter molecules (second messengers).

Fig.1 shows a summary of all receptor types so far known. The scheme is adapted from Berridge (1987) and Ross and Gilman (1985). Only the receptors marked R1 are intracellular; they are deputed to induce the synthesis of specific proteins by enhancing the transcription of their genes and are characteristically activated by steroid and thyroid hormones.

R2 and R4 receptors are ion channels and the effect of their ligands is that of opening the channel for a specific ion. Some classes of channels are operated by receptors (ROC=receptor operated channels) but other classes are simply operated by membrane potential variations (VOC= voltage operated channels). It may be remarked that there are other very important classes of VO channels, the most famous being the Na^+ and K^+ channels which, with their opening and closing, allow the propagation of the action potential (nerve impulse) along the axons of neurons.

R3 receptors are more complex. Their final effect is that of producing an increase ($R3_s$) or a decrease ($R3_i$) of the intracellular concentration of cyclic adenosine-3',5'-monophosfate (c-AMP) which, in turn, activates c-AMP dependent protein kinases (kinase A). In general protein kinases are enzymes which catalyze the transfer of phosphate groups from adenosine triphosphate (ATP) to serine or threonine residues of other enzymes which are so switched to their active forms (or vice versa) with final effects strongly dependent on the nature of the cellular system. Both $R3_s$ and $R3_i$ act by modulating ,in positive or negative sense, the synthesis of c-AMP operated by the enzyme adenylate cyclase (AC) by means of two guanosine triphosphate (GTP) binding regulatory proteins (G_s and G_i) belonging to the general class of proteins G as operated by the hydrolysis of GTP to guanosine diphosphate (GDP).

Activation of R5 by its ligand (through another G protein) induces phospholipase C (PLC) to hydrolize PIP_2 to inositol-1,4,5-triphosphate (IP_3) and diacylglycerol (DAG). The presence of IP_3 is linked to the release of Ca^{2+} from its intracellular stores (mainly mitochondria and endoplasmic or sarcoplasmic reticulum) while DAG in presence of Ca^{2+} activates a distinct protein kinase C. Other related events are: (i) the hydrolysis of arachidonic acid ($C_{20:4}$), coming from the membrane phospholipids, by the Ca^{2+}-activated phospholipase (PL) with consequent formation of prostaglandins and related compounds and (ii) the activation of the guanylate cyclase (GC) with production of cyclic guanosine-3',5'-monophosphate (c-GMP) from GTP and consequent further activation of another class of c-GMP dependent protein kinases (kinase G).

R6 is stimulated by insulin and several polypeptide growth factors to phosphorilate proteins in correspondence with tyrosyl residues, i.e one of its internal constituent is essentially a tyrosine kinase.

The general scheme depicted in Fig. 1 can be used as a first criterion of classification for most receptorial drugs known. Table 1 reports some important classes of drugs and shows the types of receptorial systems they are most probably connected to. Since most of the following discussion is dealing with neurotransmitters and their receptors, while receptors of R1 and R6 type are mainly related to growth and metabolic processes, the latter have been excluded from the table and will not be taken into any further consideration throughout this chapter.

An Example of Receptorial System: Neurotransmitter Receptors in the Brain

One of the classes of receptors which has been more extensively studied in the last fifteen years is that deputed to the control of the interneuronal communications. The central nervous system (CNS) of mammalians contains a very large number of neurones, for instance some ten billion in the human CNS. A neuron (Fig. 2) is a specialized cell where it is possible to

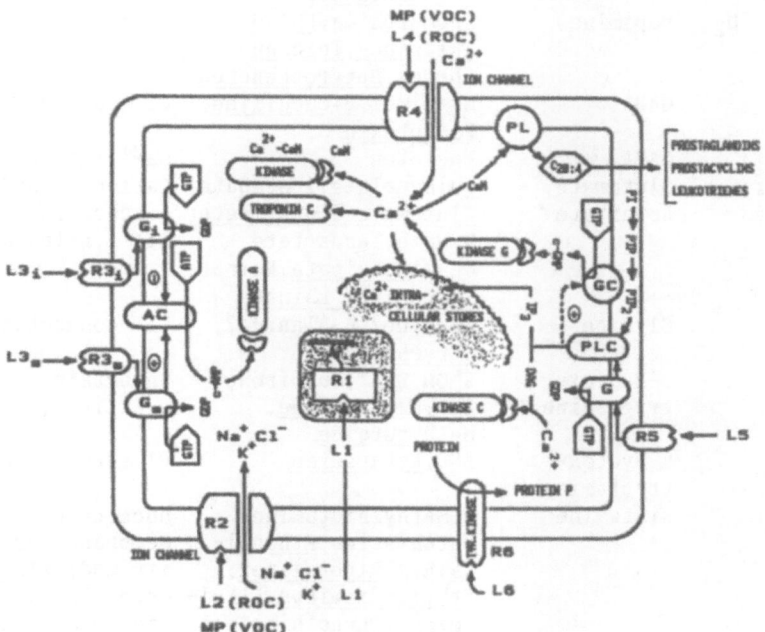

Fig. 1. Graphical synopsis of the main receptor systems and their
regulatory mechanisms. Abbreviations: R1-R6=Receptors, L1-L6=Recep-
tor Ligands, MP=Membrane Potential; AC=Adenylate Cyclase, ATP=Aden-
osine-5'-triphosphate, CaM=Calmodulin, c-AMP=Cyclic Adenosine-3',5'-
monophosphate, c-GMP=Cyclic Guanosine-3',5'-monophosphate, $C_{20:4}$=
Arachidonic Acid, DAG=Diacylglycerol, G=Protein G, GC=Guanylate Cyclase, GDP=Guanosine-5'-diphosphate, GTP=Guanosine-5'-triphosphate,
IP_3=Inositol-1,4,5-triphosphate, PI=Phosphatidylinositol, PIP=Phos-
phatidylinositol-4-phosphate, PIP_2=Phosphatidylinositol-4,5-biphos-
phate, PL=Phopholipase, PLC=Phospholipase C.

Table 1. An overview of the biochemical pharmacology of some physiologica-
ly relevant receptor systems and their tentative classification according to
the classes given in Fig.1 . Abbreviations used are: mod=modulation, act=ac-
tivation, dec=decrease, inc=increase, inh=inhibition, stm=stimulation; AC=
Adenylate Cyclase, PI=Phosphatidylinositol, PIP_2=Phosphatidylinositol-4,5-
biphosphate, PAF=Platelet Activating Factor, PL=Phospholipase.

Receptor	Endogenous Ligand	Typical Agonists and <u>Antagonists</u>	Main Agonistic Effects Quoted	Type
Muscarinic M_1	Acetylcholine	Muscarine,McN-A-343, <u>Pirenzepine,Atropine</u>	K^+ conductance mod	R2
Muscarinic M_2	Acetylcholine	Muscarine,<u>Atropine</u>	K^+ conductance mod AC inh	R2 R3
Nicotinic	Acetylcholine	Nicotine,<u>Mecamyl-amine</u>	Cationic conduc-tance inc	R2
Dopamine D_1	Dopamine	<u>Phenothiazines,</u>	AC act	R3

(continued)

245

Dopamine D_2	Dopamine	Thioxanthenes Apomorphine, Phenothiazines, Thioxanthenes, Butyrophenones	AC inh ?	--
GABA A	GABA	Muscimol, Bicuculline, Picrotoxin	Cl^- conductance inc	R2
GABA B	GABA	Baclofen	Ca^{2+} influx dec	R4
Excitatory aminoacids	Glutamate, Aspartate	Quisqualate, Ibotenate, Glutamate diethylester,	Cationic conductance inc	R2
		N-Methylaspartate, α-Aminoadipate, Kainate, Lactonized kainate	Ca^{2+} influx dec	R4
Glycine	Glycine	Taurine?, β-Alanine?, Strychnine	Cl^- conductance inc	R2
$5\text{-}HT_1$	5-Hydroxytryptamine	8-OH DPAT, Buspirone, LSD, Metergoline, Methysergide	Uncertain: AC inh	--
$5\text{-}HT_2$	5-Hydroxytryptamine	LSD, Ketanserin	PI turnover mod	R5
H_1	Histamine	2-Methylhistamine, Betahistine, Ethanolamines, Piperazines, Ethylendiamines, Alkylamines, Phenothiazines	Uncertain: Membrane permeability mod; PIP_2 breakdown; Intracellular Ca^{2+} mobilization	(R5)
H_2	Histamine	4-Methylhistamine, Cimetidine, Ranitidine	AC act	R3
α_1	Norepinephrine (Epinephrine)	Phenylephrine, Prazosin	Ca^{2+} influx inc Ca^{2+} mobilization by PIP_2 breakdown	R4 R5
α_2	Norepinephrine (Epinephrine)	Clonidine, Rauwolscine, Yohimbine	AC inh K^+ conductance inc	R3 R2
β_1	Norepinephrine (Epinephrine)	Dobutamine, Metaprolol, Practolol	AC act	R3
β_2	Norepinephrine Epinephrine	Terbutaline, Butoxamine	AC act	R3
Opiates	Enkephalins Endorphins	Morphine, Etorphine, Leu-enkephalin, D-Ala^2-enkephalin, Methadone, Naloxone, Naltrexone	AC inh K^+ conductance inc Ca^{2+} influx dec	R3 R2 R4
Adenosine A_1	Adenosine	Cyclohexyladenosine, NECA, R-PIA, 2-CADO, PACPX, Xanthines	AC inh	R3
Adenosine A_2	Adenosine	NECA, CGS 21680, CGS 15943, Xanthines	AC act	R3
Calcium channels	---	BAY K 8644, Dihydropyridines, Verapamil, Diltiazem	inh of Ca^{2+} entry into cells by antagonists	R4
PAF	PAF	Thienotriazolodiazepines (WEB 2086), Ginkolides (BN 52021)	PI turnover stm PL A2 act	(R5)
Benzodiazepines	---	Benzodiazepines, Zopiclone, Ro15-1788, PrCC, βCCE and DMCM (inverse agonists)	Mediated by GABA	R2

Moreover the following substances or class of substances have been shown to be coupled to PI turnover in CNS: Muscarinics, α_1 Adrenergics, H_1 Histaminergics, $5\text{-}HT_2$ Serotoninergics, $ACTH_{1-24}$, β-Endorphins, Substance P, Substance K, V_1-Vasopressin, TRH, CCK, Neurotensin, Bradykinin, Nerve Groth Factor, Glutamate and therefore may be directly or indirectly connected to R5 receptors (Snider et al., 1987).

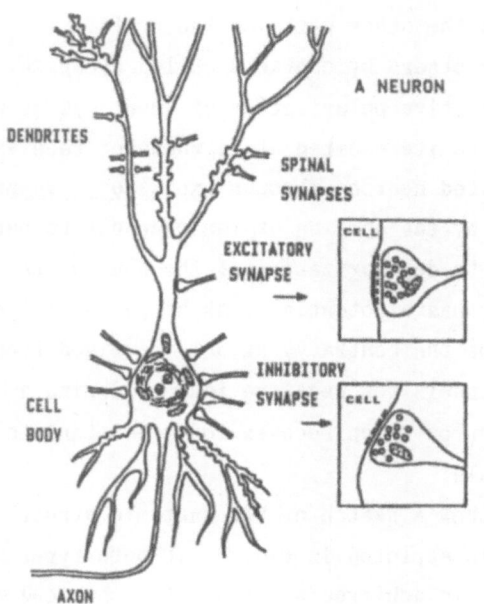

Fig. 2. Schematic representation of a neuron and, inside the inserts, of the neuronal synapses.

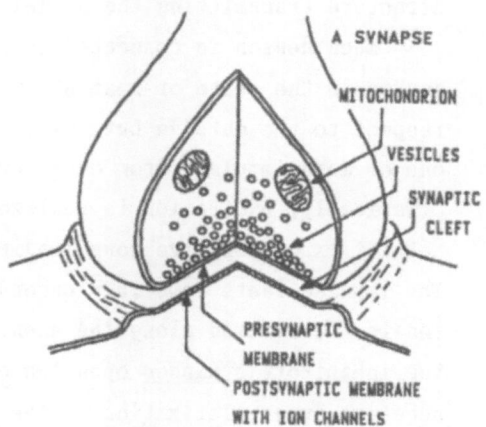

Fig. 3. Section of a synaptical contact showing the distribution of vesicles in proximity of the intersynaptic cleft.

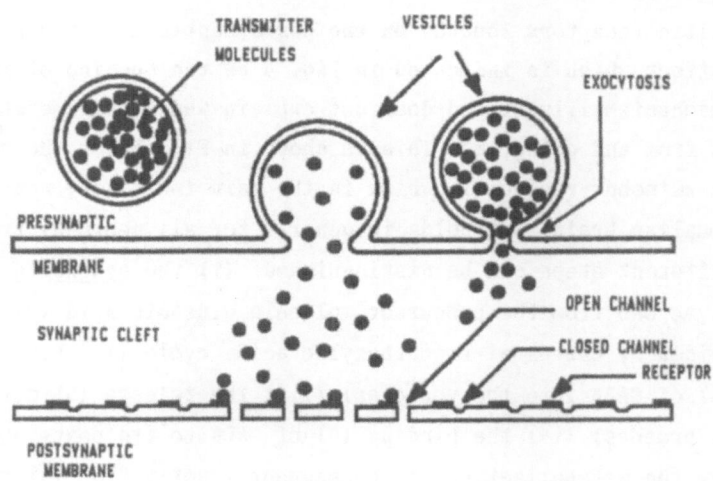

Fig. 4. An enlarged schematic view of presynaptic and postsynaptic endings encompassing the synaptic cleft where the neurotransmitter is poured into by the exocytosis process.

discern three main regions. The <u>cell body</u> contains the nucleus and is the
site for metabolic activities; the <u>dendrites</u> are thin extensions which provide for collecting signals coming from other neurons and the <u>axon</u> is the
structure transmitting the signal to the other parts of the brain.

Each neuron is connected to the others by contacts called synapses. The
neuron in the state of rest has a negative polarization of about -70 mV with
respect to the outside but switches to its excited state whenever receiving
one or more impulses from other excited neurons through <u>excitatory synapses</u>;
classically, excitation is realized by the opening of ion channels to permit
net influx of positive ions leading to depolarization of the neuron itself.
The impulse lasts some 10^{-3} seconds, has a potential peak height of 100 mV
and is transmitted along the axon. On the contrary, signals received from
the <u>inhibitory synapses</u> open ion channels for negative ions producing a
negative hyperpolarization of the neuron which becomes more resistant to be
fired by subsequent excitatory signals.

The smaller inserts of Fig. 2 show a sketch of the anatomic structure
of synapses while an enlarged view is depicted in Fig. 3. In both types of
synapses the transfer of information is achieved by means of a chemical substance, called <u>neurotransmitter</u>, released by the presynaptic neuron ending.
The neurotransmitter is generally stored in <u>vesicles</u> inside the presynaptic ending. When the presynaptic neuron is excited the vesicles move towards the membrane and, by a mechanism called <u>exocytosis</u> (Fig. 4) and probably controlled by the entrance of Ca^{2+} ions through potential operated channels stimulated by the nerve impulse, pour their content into the <u>intersyn-</u>
<u>aptic cleft</u>. Finally the released neurotransmitter molecules interact with
their specific receptors located on the postsynaptic membrane producing the
expected effect which is indicated in Fig. 4 as the opening of ion channels.

The mechanism illustrated does not explain where the neurotransmitter
is "coming from and going to". This is shown in Fig. 5 for the special case
of GABA (γ-aminobutyric acid), which is the main inhibitory neurotransmitter
in the mammalian brain, but holds in general for all neuronal transmitters.
Several different steps can be distinguished: (i) the <u>synthesis</u> (2) operated
by the enzyme GAD from the precursor molecule glutamic acid (ga), the energy
being provided by the usual tricarboxylic acids cycle (1:T.A.C.); (ii) the
<u>storage</u> (4) of GABA into the vesicles; (iii) the <u>release</u> (5) governed by the
exocytosis process; (iv) the <u>binding</u> (6) of GABA to its postsynaptic receptor; (v) the <u>metabolization</u> and consequent <u>inactivation</u> (3) by means of
the specific enzyme GABA-T; (vi) the <u>uptake</u> (7) which is a process of absorption of the transmitter by the neuron that released it (7A) or by other
neurons (7C) or by the glia, the cells of support of the CNS (7B).

248

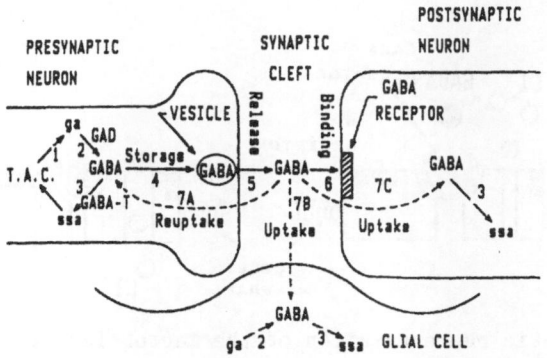

Fig. 5. Schematic representation of the biochemical processes occurring at a synaptic contact of the GABA inhibitory system in CNS.
Abbreviations: T.A.C.=Tricarboxylic Acids Cycle, GAD=Glutamic Acid
Decarboxylase, GABA=γ-Aminobutyric Acid, GABA-T=GABA Transaminase,
ga=glutamic acid, ssa=succinic semialdehyde.

Another aspect of receptors which is relevant to the present discussion
is that receptors can be linked to others in receptorial systems which are
substantially complexes of macromolecules. One of the best studied cases is
the macromolecular complex sketched in Fig. 6 and constituted by the {GABA
receptor-BDZ receptor-Barbiturates receptor-Cl^- ion channel} association
(BDZ=benzodiazepines). In such a case complex interrelations among ligands
of the different receptors can occur. GABA is the neurotransmitter deputed
to the opening of Cl^- channels but both barbiturates and BDZs can modulate
its action with a substantial modification of the original definition of
agonistic and antagonistic action. For BDZs in particular, agonists turn out
to be drugs which help GABA to open the Cl^- channel (or better, increase the
frequency of opening of the channel) and antagonists those which bind to the
BDZ receptor without affecting the GABA action. Of course, another class of
drugs becomes theoretically possible, that of inverse agonists or drugs
binding to the BDZ receptor hindering the GABA action, and it is interesting
to notice that these inverse agonists at BDZ receptors have actually been
found (Braestrup and Nielsen, 1983).

It is clear that such a receptorial association is equivalent to an
allosteric modulation of the GABA receptor by the binding site of BDZs as
suggested for the first time by Braestrup et al. (1983). Their allosteric
model of BDZ receptor assumes that the receptor occurs in two forms which
are in equilibrium under normal conditions. One form is active in the sense

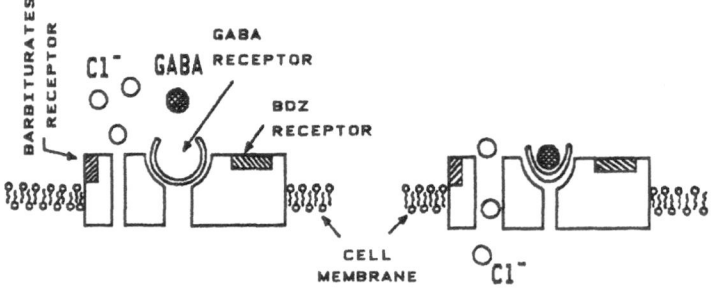

Fig. 6. Schematic representation of the macromolecular complex {GABA receptor-BDZ receptor-Barbiturates receptor-Cl^- ion channel} which is supposed to operate the opening of the Cl^- channels at the postsynaptic membrane of the inhibitory GABA-ergic synapses.

that the binding of GABA can cause the opening of the channel and the other one is inactive and unaffected by GABA binding. Agonists at the BDZ receptor are therefore drugs which stabilize the active form of the GABA receptors, inverse agonists stabilize the inactive form and antagonists bind to the receptor without affecting the equilibrium.

Fundamentals of Molecular Pharmacology and Drug Design

Different drugs producing the required effect on a specific receptorial system often have definite chemical and stereochemical similarities among them. The molecular fragment that all these molecules have in common and is responsible for their pharmacodynamic properties (and putatively for their recognition by the receptorial system) is called pharmacophore or chemical vector of that specific class of effects.

In general, drug design can be defined as all the methods (chemical synthesis, biological testing, structure-activity relationships, structure determination methods, molecular graphics,...) used for producing more potent and selective drugs by means of chemical changes in the pharmacophore.

From a practical point of view, however, the situation can be much more complicated than that described by the rather misleading pharmacophore concept. First of all the right receptor must be identified with certainty, a goal which can be sometimes not so easily accomplished. Even if it is known, there remains the problem that the receptorial system can be influenced in different stages of its operation mechanism.

Reconsidering the GABA-ergic synapse of Fig. 5 (a system which can be considered to be representative of most neurotransmitter systems and thus of most neurodrugs) it is apparent that we can obtain a GABA-like agonistic effect at least by:

 -increasing the GABA synthesis,

 -decreasing the GABA metabolization,

 -inhibiting the GABA uptake by the glial cells and postsynaptic neurons,

 -increasing the GABA reuptake by the presynaptic neuron,

 -stimulating the postsynaptic receptor with drugs having high affinity
  and/or high intrinsic activity.

A good example of the interplay of these different factors comes again from the GABA system which has been extensively studied by Krogsgaard-Larsen and collaborators (Krogsgaard-Larsen et al., 1979; Krogsgaard-Larsen, 1980). GABA (1) has three conformational degrees of freedom around its C-C sigle bonds and, in principle, could act at different points of the system in a different conformation. The project of these authors included, therefore, the synthesis of many conformationally restrained compounds which were tested for different properties, in particular, inhibition of GABA receptor binding, GABA uptake and inhibition of GABA-T (GABA transaminase). It was found that the active GABA conformation at the postsynaptic receptor was probably that shown in (1) and this in analogy with other very potent re-stricted ligands such as muscimol (2), THIP (3) and isoguvacine (4). However it was observed that other restricted GABA-analogues where rather weak re-ceptor ligands but far better inhibitors of GABA uptake (5,6) or of GABA-T (7,8).

1 GABA   2 Muscimol   3 THIP

4 Isoguvacine   5 Guvacine   6 Nipecotic acid

7 Ethanolamine-   8 γ-Acetilenic
   -o-sulphate   GABA

It seems reasonable to conclude that a correct definition of what is biological activity is the crucial point in molecular pharmacology. Keeping to the neurodrug example, such a drug could be a better agonist for one (or more than one) of the following reasons:
- it fits the receptor better (better binding);
- after binding it is more suited for producing the desired conformational changes in the receptor (better efficacy);
- it is metabolized less quickly to a less active compound;
- it is metabolized more quickly to a more active compound;
- its vehiculation in the body is improved;
- its ability to cross interposed biological membranes is increased;
- its uptake is decreased;
- its reuptake is increased;
- it binds to a different receptor which is functionally related to that we are studying.

The only way for overcoming the above difficulties is that of having at our disposal accurate biological testing methods able to operate a kind of biochemical dissection of the receptorial system. This problem appears to be one quite difficult to be resolved; however some methods for obtaining at least a part of the required biological and biochemical information will be underlined in the next section.

Fundamentals of Drug Transport

To produce its effect a drug must, after administration, arrive at its site of action in appropriate concentration. Essentially its arrival depends on three main factors: the ability of being vehiculated by the blood stream, the capacity of crossing the interposed membranes and the property of not being easily metabolized before reaching the final target. In particular the brain is protected against the invasion of xenobiotics, that is of foreign chemical substances, by a very efficent anatomic structure of the capillary vessels, called blood brain barrier and representing an important boundary in the form of a permeability barrier preventing the passive diffusion of chemicals from the blood stream into the various regions of the CNS (central nervous system).

For the scope of the present discussion it is important to single out the physico-chemical factors determining vehiculation and barrier crossing. In both cases the most important parameter of the molecule is its lipo- philicity, that is the property of being soluble in lipidic phases rather than in water. Different tabulations of drug lipophilicities are available in the form of values of oil/water or octanol/water partition coefficients.

The quantity most used at present has been originally proposed by Hansch (Hansch, 1971; Fujita et al., 1964) and is the "logP"=$\log_{10}P$, P being the octanol-water partition coefficient of the substance. Since the free energy of the partition equilibrium is actually $\Delta G^\circ = -2.303 \cdot RT \cdot \log P$, the logP quantity is a free energy related property. Several attempts have been made to prove that logP is an addictive property of the chemical groups (each having a logP called π) constituting the molecule, in such a way as to easily calculate the total logP of related series of drugs from tabulated data. The topic will be discussed further in LFER/QSAR section.

Vehiculation in the blood is caused by complexation of the drug with plasma proteins; albumins, globulins and lipoproteins are known to be more or less specific carriers of drugs whose lipophilicities vary within a relatively large spectrum of logP values.

On the contrary, the drug barrier crossing phenomenon appears to occur in a much narrower range of drug lipophilicities in consequence of the intrinsically lipophilic character of cell membranes which are mainly built up of phospholipids. Though membranes do have specific mechanisms of active transport for some physiological compounds, drugs, in general, cannot take advantage of them and their transport through the membrane can be only passive, that is, driven by the concentration gradient. Drugs which are not electrolites can diffuse across the biological barrier at a rate proportional to their partition coefficents; weak electrolites, a class including the majority of drugs, diffuse more quickly the larger their partition coefficent is and the smaller their ionization constant.

In general it is observed that the rate of barrier crossing increases with the value of logP up to the point that molecules become so large that their diffusion coefficent starts to decrease again because of the increased drug-lipid interactions. Accordingly several authors (for a review see Leo et al., 1971) assume a parabolic dependence of biological properties on logP of the form

$$\log(\text{biological response}) = a + b \cdot \log P + c \cdot (\log P)^2 \qquad (3)$$

where $\underline{b}$ is a positive and $\underline{c}$ a negative coefficient.

Most drugs are transformed both before and after arriving at the site of action. In general, biotransformation before action is a negative factor causing a decrease of drug concentration near the final target, although for a small number of drugs it is well documented that the metabolites are as active as or more active then the parent drug. Drug action is eventually interrupted by processes of biotransformation and subsequent excretion mainly through the kidneys and liver.

Some Concepts on Biological Activity Measurements

The development of drugs starts with the recognition that some particular pharmacological effect may be useful therapeutically (Barlow, 1980). Once some potentially useful compound has been found it is necessary to know not only what such a substance does but how well it does it, so that comparisons can be made and better drugs discovered. This is a difficult task as the effects the pharmacologist is called upon to study are quite different. Drug developement depends critically upon adequate testing and this demands an appreciation of the problems of quantitative pharmacology.

Many drug effects involve physical or chemical changes which are susceptible to quantitative measure; these can be classified into: (i) changes in mechanical properties, e.g. in length, volume, pressure, flow, rates of contraction; (ii) changes in electrical properties, e.g. in polarization or conductance of cell membranes, in rate of firing of neurons; (iii) chemical changes, e.g. in the concentration of substances involved in cell functions or in metabolism.

In measuring an effect the first point to make clear is how the size of the response depends on the dose of drug administered. If a complex effect is to be measured (e.g. effect of drug on behaviour) it is only possibile to devise some scoring system allowing effects to be ordered and compared. In toxicity tests the response can be all or none: a nerve cell either conducts or it does not; an animal lives or dies. These responses are called quantal. In different situations the response is graded (e.g effects involving muscle contraction or chemical changes) and in these cases is called graded.

The second point concerns where to measure the effect. Many tests are carried out in vivo on animals, mainly rats or mice of selected strains but even, with decreasing frequency, guinea-pigs, rabbits, hamsters, dogs, cats, monkeys and pigs; the testing is intended to simulate the drug effect on man though animal genetic specificities may cause trouble in some instances. In vitro testing carried out by working with single cells or even pieces of tissue makes it possible to semplify the nature of the response by reducing the number of variables on which it depends and the complicating effects of time which are more evident in living animals.

Quantitative measurements are aimed to quantify the concentration [C] which must be administered to provoke a given effect in a fixed percentage of the sample (animals, tissues, cells...). The most common expressions of [C] are: (i) LD_{50}, median lethal dose, the drug concentration able to kill 50% of the animals or cells treated; (ii) ED_{50}, median effective dose, the concentration producing the effect in 50% of the animals or anatomic preparations treated; (iii) IC_{50}, median effective inhibitory concentration, the concentration (e.g. in receptor binding experiments) of inhibitor which displaces 50% of a labelled ligand.

The problem of measuring the biological activity of a drug is quite a complex one which can only be tackled by combining several different in vivo and in vitro tests. Many of them concern the possibility of a therapeutical application of the drug, dealing with its toxicity and other unwanted side effects. Of more interest for the scope of the present discussion are the tests which are intended to define the drug pharmacological profile and the in vitro experiments able to clarify the biological and biochemical aspects of the drug action, in particular its ability to bind to a specific receptor site.

The in vitro technique which can be more useful for the characterization of both receptors and their ligands is that of receptor binding, already outlined in the short discussion on equations (1) and (2). This technique allows the direct study of the specific binding ability by directly labelling a given receptor site and is essentially the only method presently available for establishing a biunivocal correspondence between a drug and a specific receptor.

This method will be further considered in the experimental binding thermodynamics section.

PART 2: PHYSICAL CHEMISTRY OF THE DRUG-RECEPTOR INTERACTION

An Overview of Molecular Interactions

Physiological processes require receptors to be able to operate in a reasonably short time and this is particulary true for interneuronal communications which are known to occur in a few milliseconds. Therefore ligands must bind to the receptor, produce their effect and come out quickly so that the receptor is returned to its original waiting state and all the process must be as reversible as possible. Common chemical bonds cannot guarantee such speed and reversibility, so that nature has chosen to use only weaker intermolecular forces for securing drug-receptor binding while xenobiotics which bind covalently to receptors or their related enzymes produce poisoning of organisms and a long list of natural poisons of such sort is known.

Since at least an approximate knowledge of the different intermolecular forces and a semiquantitative appraisal of the interaction energies involved seem to be indispensable to any serious approach to molecular biology and pharmacology, we have thought it worthwhile to summarize here the main concepts of this quite complex subject. Intermolecular forces can be approximately classified as follows.

References. Intermolecular forces: Kaplan, 1986; Kitaigorodsky, 1973; Pertsin and Kitaigorodsky, 1987. Hydrogen Bonding: Hamilton and Ibers, 1968; Pimentel and McClellan, 1969; Pimentel and McClellan, 1971; Schuster et al., 1976; Kollman and Allen, 1972; Emsley, 1980; Gilli et al., 1989. Charge transfer: Gutmann, 1978; Kitaigorodsky, 1984; Huheey, 1983.

Repulsion or exchange forces. Almost all stable molecules have closed electron shells (that is all their molecular orbitals are doubly occupied) which cannot accept other electrons from another molecule without violating the Pauli principle. For this reason they oppose an almost vertical barrier of potential to the approach of other molecules at a distance shorter than the sum of the van der Waals radii (r_{vdW}) of the outer atoms. Table 2 reports the van der Waals radii of the most common atoms.

Electrostatic forces. Molecules may possess several electric multipoles: (i) monopoles in simple ions (Na^+, Cl^-) or ionic molecules (protonated amines, anions of carboxylic acids); (ii) small dipoles located on heteronuclear bonds whose magnitude increases with the electronegativity difference between the two bonded atoms and whose vectorial sum constitutes the total dipole moment of the molecule; (iii) higher order multipoles such as quadrupoles; (iv) small transient multipoles induced by the movements of the surrounding molecules which are originated by instantaneous displacements of electron clouds with respect to nuclei and whose magnitude tends to increase with the square of the molecular polarizability, α.

Most molecules are neutral or zwitterionic, do not possess net charges and often have no or rather small dipole moments so that their basic attraction forces are essentially due to interactions among transient induced dipoles; they are called dispersion interactions or dispersion forces or sometimes London forces from the name of the physicist who gave them the first theorical interpretation (London, 1930) and have the property that the dispersion interaction energy is inversely proportional to the sixth power of the distance, r, between molecules, i.e. $E=kr^{-6}$.

Interactions implying quadrupoles or higher moments are believed to be small and are commonly neglected while dipole-dipole interactions may be

Table 2. Atomic van der Waals radii (A).

						H	1.20
C	1.70	N	1.55	O	1.52	F	1.47
Si	2.10	P	1.80	S	1.80	Cl	1.75
		As	1.85	Se	1.90	Br	1.85
						I	1.98

H in aromatic rings 1.00 A; C perpendicularly to the benzene ring 1.77 A. Data according to Bondi (1964).

important in some cases. For lowering the energy they require a correct orientation of the interacting dipoles and in this sense are often called orientation forces. In liquids the average kinetic energy 3RT/2 of some 3.7 kJ mol^{-1} at room temperature perturbates in such a way the correct mutual orientations that the total resulting attraction is very small. When drugs fix themselves to the receptor site, however, the attraction caused by dipolar interactions may be considerable provided the relative orientations of the dipoles on drug and receptor are correct or, in other words, if the electrostatic potential spanned by the drug is complementary to the potential generated by that specific part of the receptor. In this respect it has to be remarked that α-helix coils in proteins may have relevant global dipole moments in consequence of the parallel alignment of many isooriented peptidic dipoles.

Charge transfer. Intermolecular charge transfer or donor-acceptor interactions occur between electron donors (Lewis' bases) and acceptors (Lewis' acids). They establish at least a partial covalent bond between highly polarizable groups which is often described, within the HOMO-LUMO formalism, as the formation of a molecular orbital (MO) by donation from the higher occupied MO of the donor to the lower unoccupied MO of the acceptor. Classical examples are the molecular complexes $NH_3 + BF_3 = H_3N{-}{>}BF_3$ or $I_2 + I^- = [I{-}I{-}I]^-$. In biological systems interactions can only occur among barely polarizable hard acids or bases so that it can be reasonably assumed that they do not include any significant charge transfer component with the only exception of molecules containing iodine (i.e. thyroid hormons).

Hydrogen Bonding. H-bonding occurs when a hydrogen atom is bonded to two (or sometimes more) other atoms. This situation may be depicted schematically as D-H--A, where D is the H-bonding donor and A the acceptor. In principle all atoms more electronegative then hydrogen (C,N,O,F,S,Cl,Se,Br, I) can play the role of A and D though stronger hydrogen bonds are associated with the most electronegative ones (N,O,F,Cl).

Several quantum mechanical calculations have been devoted to clarifying the nature of the bond in H-bonding complexes and in particular the relative contributions of the different interaction forces to the total interaction energy. Probably the most popular and quoted partitioning scheme is that developed by Morokuma for the treatment of $(H_2O)_2$ and $(HF)_2$ dimers (Umeyama and Morokuma,1977) which makes use of the energy decomposition analysis developed within the ab initio SCF-MO theory (Morokuma, 1971; Kitaura and Morokuma, 1973, 1975). The total hydrogen bonding energy is partitioned according to

$$E_{HB} = E_{ex} + E_{es} + E_{pl} + E_{ct} \qquad (4)$$

where E_{ex} is the repulsion or exchange energy while E_{es}, E_{pl} and E_{ct} are, respectively, the electrostatic, polarization and charge transfer interaction energies. The authors were able to conclude that E_{es} is the main attractive term while the contribution of E_{ct} is small, so that the hydrogen bonded complexes can be qualitatively classified as "electrostatic > charge transfer" or simply "electrostatic" complexes.

H-bonds can be topologically classified as intra and intermolecular; from the point of view of their energy they are often divided into neutral medium strength and ionic strong hydrogen bonds. Such a classification is not very efficient because it cannot take into account that the strength of the H-bond is governed by two different and widely independent factors at the same time, that is its geometry (a H-bond can be strong only if the correct geometry can be achieved) and the nature of the interacting atoms. A different and more articulate classification has been adopted here.

Weak H-bonding can be observed for any couple of donor and acceptor atoms whenever the two groups cannot achieve the correct geometry of approach for some steric reason. The main factor is usually associated with the D-H--A angle which, for maximizing the electrostatic interaction between the D-H dipole and the negatively charged acceptor, must be in the range of some 160-180°. A good example comes from the intramolecular H-bonds closing five, six or seven membered rings; for the specific case of the O-H--O bond it can be shown (Gilli and Bertolasi, 1990) that the O-H--O angle increases with the ring annellation, being respectively some 115, 150 and 170° with parallel shortening of the O---O distance and strengthening of the H-bond. The H-bond closing a five membered ring is therefore to be classified as a weak H-bond and in fact it is so weak that the hydrogen of the O-H group forms, whenever possible, a second hydrogen bond (bifurcated H-bonding) with another acceptor.

A second reason why a H-bonding should be classified as weak comes from the small intrinsic electronegativities of the H-bonded partners, the most classical case being that of the C-H--A interactions (Taylor and Kennard, 1982). And not by chance the A atom which is acceptor of the weak H-bonding from the C-H group is usually found to accept another stronger H-bond of more traditional type.

Medium H-bonding is typical of water, alcohols, amines, amides and carboxylic acids and determines the properties of many systems of biological interest, in particular the structure of water and proteins. Its geometry is rather well defined: (i) the O-H--O group tends to be linear; (ii) the D-H distance is not significantly lengthened with respect to that observed in the absence of H-bond; (iii) the D---A contact distance is practically identical to the sum of the van der Waals radii of A and D, that is the van der Waals radius of the interleaving H atom is almost zero. A wealth of thermodynamical data are available which show that its enthalpy of formation is in the range $-\Delta H$= 2-8 kcal mol^{-1}. More in detail the $-\Delta H$ of H-bonding formation is some 4.8 kcal mol^{-1} in water and the average $-\Delta H$ values for carbonyls, amides, ethers, amines acceptors with the phenol as donor are 4.7, 5.4, 5.8, 8.6 kcal mol^{-1}, respectively (Pimentel and McClellan, 1971).

Medium-strong H-bonding can be thought to derive from medium H-bonding strengthened by the presence of positive and/or negative charges. A possible range of energies is 8-20 kcal mol^{-1}, though there is a substantial lack of reliable data. It seems reasonable to distinguish two different cases: (i) charge assisted and (ii) resonance assisted H-bonding. Charge assisted H-bonding (CAHB) arises from net ionic charges on the donor and/or the accep-tor groups and is relatively common (e.g.) in proteins owing to the occur-rence of charged side-chain donors ($-NH_3^+$, $-NH-C(NH_2)_2^+$, Histidine-H$^+$) and acceptors ($-COO^-$) or of the two terminal $-NH_3^+$ and $-COO^-$ groups. The effect of charge in strengthening the H-bond can be documented by the data collect-ed (Görbitz, 1989) on the H-bonds formed between the intrachain or terminal groups of oligopeptides and aminoacids: the average N---O contact distance is 2.840, 2.908, 2.912 and 2.929 A respectively for H-bonds $-COO^-$---H_3N^+-, >C=O---H_3N^+-, $-COO^-$---H-N< and >C=O---H-N<. Resonance assisted H-bonding (RAHB) is similar to CAHB but with the difference that the strengthening charges arise from polar resonance forms (Gilli et al., 1989; Gilli and Ber-tolasi, 1990); the only system extensively studied is that of β-diketone enols where the resonance O=C-C=C-OH <---> $^-$O-C=C-C=OH$^+$ strengthens the intramolecular O-H---O bond in such a way that the O---O distance can become 2.417 A and the calculated H-bond energy 12.7-19.7 kcal mol^{-1} (values that have to be compared with those for the water dimer: d_{O--O}=2.74 A, $-\Delta H$=4.8 kcal mol^{-1}). In these compounds an appreciable lengthening of the O-H bond and a shift of the proton towards the middle point between the two oxygens are also observed. It has been suggested (Gilli et al., 1989) that RAHB can be a relevant factor in the H-bonds of biological importance which cause the pairing of bases in DNA and the secondary structure of proteins.

Very strong H-bonding appears to be inevitably associated with ions. It is characterized by relevant lengthening of the D-H distance, nearly central position of the proton, D---A distances definitely shorter than the sum of van der Waals radii and bond energies intermediate between true H-bonding and covalent bonding. Complexes like [F--H--F]$^-$ or [H$_2$O--H--H$_2$O]$^+$ have $-\Delta H$ of 150-250 and 130-150 kcal mol^{-1} and must be considered true chemical spe-cies where the charge transfer (covalent bonding) is relevant and we may speak of H-bonding only in a formal sense. In general there is some confu-sion in the field of very strong H-bonds and a clarification could be achieved only by fixing a reference level for the energies involved. The minimum level could be reasonably fixed at some 30 kcal mol^{-1}, which is a little less than that of the weakest chemical bonds (see the compilation in Huheey, 1983). Very strong H-bonds are of little or no importance in determining the drug-receptor interactions.

Force fields. Molecular pharmacology makes large use of semiquantitative calculations of intermolecular interaction energies as well as of the energies of the preferred conformations of the interacting molecules. Both types of calculation require what is called a force field, that is a set of constants describing the interatomic interaction. Semiempirical force fields are obtained from theoretical considerations or quantum mechanical calculations on sample systems while the more used empirical force fields make use of coefficients which are empirically determined in such a way that some physically observable quantity can be more or less accurately predicted, e.g. the lattice energy for intermolecular interactions and the correct molecular geometries for intramolecular ones. No field universally applicable to both intra and intermolecular calculations is presently known; here only intermolecular fields will be briefly discussed (see Burkert and Allinger, 1982, for more details on intramolecular force fields to be used in molecular mechanics and conformational energy calculations).

All fields known make use of the atom-atom approximation, that is the total interaction energy, E_{int}, is considered to be the sum of interaction energies of all couples of atoms i and j located on different molecules

$$E_{int} = 1/2 \ \Sigma_{i,j} \ E_{ij} \qquad (5)$$

The interaction energy E_{ij} between atoms i and j separated by the distance r is usually expressed as

$$E_{ij} = A_{ij} \ r^{-12} - C_{ij} \ r^{-6} + q_i q_j r^{-1} \qquad (6)$$

where the first term is the short range repulsion energy between atoms i and j and is sometimes written in the alternative forms $A'_{ij}exp(-B_{ij}r)$ or $A''_{ij}r^{-9}$, while the second term rapresents the dispersion attraction energy between the two atoms. The sum of the first two terms is usually called non bonded or van der Waals interaction energy. The last term accounts for the electrostatic interaction energy between permanent multipoles calculated from the partial charges, q, carried by the single atoms; partial charges can be obtained from quantum mechanical ab initio or semiempirical calculations or derived from experimental measurements of deformation densities. The equation (6) requires extensive tabulations of the coefficients A,B and C for all couples of atoms i,j involved. A simpler tabulation can be achieved using the expression

$$E_{ij} = \epsilon^*_{ij} [(r/r^*_{ij})^{-12} - 2(r/r^*_{ij})^{-6}] + q_i q_j \, r^{-1} \qquad (7)$$

where the two parameters of the interaction, ϵ^*_{ij} and r^*_{ij}, are calculated as $\epsilon^*_{ij} = (\epsilon^*_i \cdot \epsilon^*_j)^{1/2}$ and $r^*_{ij} = r^*_i + r^*_j$ starting from the single atom constants ϵ^*_i and ϵ^*_j, which are the energy parameters for the two atoms $\underline{i}$ and $\underline{j}$ having van der Waals radii r^*_i and r^*_j.

The most controversial point concerns the field which has to be used for H-bonding D-H--A interactions. A common approximation consists in computing the non bonded energy between D-H and A by the usual atom-atom potentials but having care of giving a value of zero to the van der Waals radius of the hydrogen atom; the H-bonding energy is then optimized by putting suitable partial charges on the three atoms in such a way that the total energy calculated by equation (7) fits the experimental H-bonding energy. More sophisticated treatments have made use of partial charges obtained from quantum mechanical SCF calculations or from experimental deformation densities (Berkovitch-Yellin and Leiserowitz, 1982).

For a complete and recent review on intermolecular force fields the reader is referred to Pertsin and Kitaigorodsky (1987).

Enthalpic Contribution to Drug-Receptor Binding

It seems useful to have at least the order of magnitude of the energies involved in the binding process. In general, the drug-receptor binding process can be considered as the transfer of a drug molecule from its cage of water molecules inside the solvent to a second cage, the binding site, which is, most probably, already filled up with other water molecules. At the end of the transfer the drug will occupy the binding site cavity while water molecules previously organized around the drug or within the binding site are released according to the equilibrium

Drug-Water + Binding Site-Water $\rightleftharpoons$ Drug-Binding Site + Water $\qquad$ (8)

The process corresponds to a double substitution reaction where old and new bonds are simply interchanged. In such interchange covalent or charge transfer bonds are to be excluded for the reasons given in the previous section and what we are left with is essentially represented by dispersion and dipolar interactions together with hydrogen bonding.

The order of magnitude of the dispersion interaction energy of a molecule with its surrounding can be evaluated from the lattice energy of crystals of non polar molecules. It increases with molecular dimensions and, for instance, is some 9.4, 16.1 and 23.1 kcal mol^{-1} for benzene, naphtalene and anthracene, respectively. Since dispersion interactions are sums of a large

number of small and isotropic atom-atom potentials, it seems unlikely that
they are so selective as to distinguish between drug-water and drug-receptor
interactions and therefore it can be assumed with little error that the differences of dispersion energy in the binding equilibrium average to zero.

The contribution of dipole-dipole interactions may be not neglegible if
a drug with dipole moment greater than 2-3 D happens to interact with a
large and properly oriented dipole on the receptor binding site. The reason
is that in this case, at variance with dispersion energies, there is no com-
pensating effect on the side of the solvent because dipolar water-drug interactions are practically averaged to zero by the thermal motion (in spite
of the water dipole moment of 1.85 D). A semiquantitative appraisal of the
energy involved can be done in the following way. Let us assume a drug molecule with μ of 3 D and a length of 10 A. Its dipolar interaction energy with
another identical molecule approaching it at 2 A (the shortest H--H contact
distance) in the most favourable orientation is only 0.5 kcal mol^{-1}. However
the same favourable approach to an α-helix coil of thirty residues gives
much more interesting results: (i) the coil has a dipole moment of 70 D if
the moment of an isolated peptide unit of 3.5 D is assumed; (ii) it becomes
some 100 D if the dipole moment of 5.0 D per unit is taken, a greater value
which accounts for the mutual polarization of the residues within the helix
and is in agreement with experiments and theoretical calculations (Hol and
Wierenga, 1984, and references therein); (iii) the dipole-dipole interaction
energy with our 10 A molecule (Table 3) is calculated to range from -4.1 to
-5.9 kcal mol^{-1} at 2 A and from -3.1 to -4.5 kcal mol^{-1} at 2.5 A contact
distances, values which are comparable with H-bond enthalpies.

It may be concluded that dipole-dipole interactions can be determining
factors in the drug-receptor interaction only if the polar drug matches a
macro-dipole of a receptor helix. Nothing is presently known on the actual
occurrence of such macro-dipoles in proximity of receptor binding sites
though their presence can be reasonably extrapolated by the fact that enzymes of known molecular structure often display α-helices of 10-20 residues

Table 3. Dipole-dipole interaction energy of an α-helix of 20 residues with
a drug molecule having a length of 10 A and μ of 1.0 D and approaching the
helix with the most favorable orientation up to the contact distance, d,
indicated. The α-helix is 30 A long and may have μ values of 70 or 100 D;
enegies are in kcal mol^{-1}.

μ(α-helix) (Debye)	Contact distance in A						
	2.0	2.5	3.0	3.5	4.0	4.5	5.0
70	-1.37	-1.05	-0.84	-0.67	-0.57	-0.49	-0.43
100	-1.96	-1.50	-1.20	-0.98	-0.82	-0.70	-0.60

pointing to the active sites with the probable function of helping the binding of phosphate or diphosphate anions (Hol and Wierenga, 1984). Previous
considerations show that dipolar interactions, if present, should contribute
for not more than 5-6 kcal mol^{-1} to the total binding enthalpy.

As far as H-bonding is concerned it has to be taken into account that
all H-bond donors and acceptors are already saturated by water molecules
before the binding process. With this in mind only the following two cases
can be distinguished: (i) all H-bond donors and acceptors on the drug match
complementary acceptor and donor groups on the binding site; (ii) some or
all do not match and their H-bonds are going to be lost. Let us now assume
for the sake of simplicity that: all H-bonds with water have an enthalpy of
-5 kcal mol^{-1}, those assisted by a single charge of -8 kcal mol^{-1} and those
assisted by two opposite charges of -10 kcal mol^{-1}, which is probably a reasonable estimate for the H-bonds of biochemical interest. It is now possible
to have an estimate of the $\Delta H°$ of the binding equilibrium. In case (i) there
is a contribution ranging from -3 to -5 kcal mol^{-1} for any H-bond which is
substituted by a more efficient one at the binding site, while the $\Delta H°$ will
be nearly zero if no drug-receptor H-bond is charge assisted. In case (ii)
there is simply a contribution of +5 kcal mol^{-1} for any drug-water H-bond
which is not paralleled by a drug-receptor one.

The general conclusion of this paragraph should be an estimate of the
$\Delta H°$ of the binding equilibrium. It has been shown by the use of simple arguments that such a $\Delta H°$ value depends on two easily determined molecular
parameters, the total dipole moment and the number of H-bonding donor and
acceptor groups present on the drug. Taking 3 D for the first and four for
the second, it is clear that the $\Delta H°$ can range from +20 to some figure in
between -12 and -20 kcal mol^{-1} as far as H-bonding is concerned and that a
contribution between zero and -6 kcal mol^{-1} has to be added to account for
dipole-dipole interactions.

Experimental Binding Thermodynamics: Methods and Results

In the previous section we have underlined the enthalpic factors which
could affect the drug-receptor binding but the entropic ones have been
neglected and will be considered here. Calling R the receptor and D the drug
a generic binding equilibrium and its affinity constant are written

$$D + R \rightleftharpoons DR \tag{9}$$

$$K_A = \frac{[DR]}{[R][D]} = \frac{[DR]}{[B_{MAX} - DR][D_{TOT} - DR]} = \frac{1}{K_D} \tag{10}$$

where [D$_{TOT}$] is the total drug concentration added, [B$_{MAX}$] the total concentration of binding sites in the sample and K_D the dissociation constant. As $[DR]/[D_{TOT} - DR] = [Bound]/[Free] = [B_{MAX}] K_A - K_A[Bound]$, the K_A and B_{MAX} values can be obtained from the slope and the intercept of the Scatchard plot [Bound]/[Free] vs. [Bound]. The standard free energy is calculated from $\Delta G° = -RT\ln K_A$, usually at 298.15 K, the standard enthalpy $\Delta H°$ from the van't Hoff plot $\ln K_A$ vs. $1/T$, whose slope is equal to $-\Delta H°/R$, and the standard entropy as $\Delta S° = (\Delta H° - \Delta G°)/T$. This simple theory shows that the thermodynamic parameters of the binding can be obtained by measurements of drug affinity constants if the labelled drug is available or of inhibition constants of a reference labelled ligand (which we have already shown to be essentially the same) if the drug under examination is not available as a labelled compound. The measurements are carried out at four or five temperatures in the range of 4-35 °C and final parameters obtained through a van't Hoff plot. Such a method is known to be affected by several flaws but is the only one of practical application because the very low receptor concentration in membrane preparations (of the order od 1-100 pM/g of tissue for most neurotransmitter receptors) hinders direct microcalorimetric measurements.

Table 4 reports a tabulation of thermodynamical parameters obtained by different authors on some receptorial systems (for a review see Raffa and Porreca, 1989). All $\Delta G°$ values are negative in relation to the fact that only drugs of high affinity are studied; they range from -13.7 to -6.2 with an average value of -10.3 kcal mol^{-1} corresponding at room temperature to a mean dissociation constant of drug-receptor complex (K_D of equation 10) of only 28.8 nM. Values of $\Delta H°$ are in the range from -21.2 to 13.5 kcal mol^{-1} which corresponds reasonably well to the range estimated in the previous paragraph, and have mean value of -5.0 kcal mol^{-1}.

The most interesting aspect of Table 4 concerns the observed values of $\Delta S°$ which range from -36 to 84.4 with an average of 17.7 cal mol^{-1} K^{-1} . The standard entropy values are not independent of those of enthalpy and this is shown in Fig. 7 reporting the scatter plot of $\Delta H°$ vs. $T\Delta S°$ for all the 48 compounds of the table together with their regression line. Data are related by the equation $-T\Delta S° = -10.6 - 1.075\Delta H°$ with a correlation coefficient of 0.977. It is presently difficult to say whether this correlation reflects a general law of binding thermodynamics, although it must be pointed out that the values of Table 4 represent almost one half of all data available in the scientific literature for membrane neuronal receptor systems and that the sample has been chosen at random. The reason for such a behaviour arises from the fact that all dissociation constants of practical interest fall in the range 10 pM $\leq K_D \leq$ 100 µM and therefore all equilibrium data must be encompassed by the two parallel lines having equations $\Delta H° - T\Delta S° = -15.0$ kcal mol^{-1} for $K_D = 10$ pM or $\Delta H° - T\Delta S° = -5.5$ kcal mol^{-1} for $K_D = 100$ µM.

Table 4. Thermodynamical parameters of drug-receptor binding equilibrium for a selection of drugs and receptors. Agonists non-underlined and antagonists underlined; energies in kcal mol^{-1}, entropies in cal mol^{-1}K^{-1}; logk$_W$= log$_{10}$ of the capacity factor determined by inverse phase HPLC.

Drug	Receptor	logk$_W$	$\Delta G°$	$\Delta H°$	$\Delta S°$
6,7-ADTN	Dopaminergic	-0.4	-9.6	-16.4	-24.0
Dopamine	Dopaminergic	-0.8	-8.2	-8.7	-1.8
Pergolide	Dopaminergic	3.5	-10.3	0.7	37.4
Apomorphine	Dopaminergic	3.0	-9.2	0.8	34.4
Sulpiride	Dopaminergic	2.2	-10.6	-21.2	-36.0
Piquidone	Dopaminergic	1.4	-11.3	-17.4	-20.7
Tiapride	Dopaminergic	-0.5	-6.2	-14.2	-27.0
Metoclopramide	Dopaminergic	1.1	-10.1	-13.1	-10.3
Clebopride	Dopaminergic	2.9	-12.4	-12.8	-1.4
Alizapride	Dopaminergic	2.5	-10.2	-12.2	-6.6
(-)-Sultopride	Dopaminergic	0.8	-11.4	-12.2	-3.0
Zetidoline	Dopaminergic	2.1	-11.0	-11.5	-1.3
YM 09151-2	Dopaminergic	3.5	-13.4	2.9	54.8
Raclopride	Dopaminergic	2.4	-10.6	3.6	48.3
Haloperidol	Dopaminergic	3.1	-12.1	-3.0	30.9
Clozapine	Dopaminergic	3.0	-9.2	0.4	32.5
cis-Flupenthixol	Dopaminergic	4.4	-10.7	3.6	48.9
[^{3}H]Spiperone	Dopaminergic	3.3	-13.3	5.2	62.6
(+)-Butaclamol	Dopaminergic	5.0	-11.2	13.5	84.4
Imipramine	Antidepressant	-	-9.4	-11.5	-6.8
Desipramine	Antidepressant	-	-8.6	-7.0	5.5
DAGO	Opiates	-	-12.9	4.3	55.1
DADLE	Opiates	-	-12.0	2.4	48.2
EKC	Opiates	-	-12.0	3.3	51.3
Ethorphine	Opiates	-	-13.7	2.3	51.8
CHA	Adenosine	-	-12.5	13.2	82.0
L-PIA	Adenosine	-	-13.0	10.5	79.0
DPX	Adenosine	-	-11.5	-10.5	3.4
IBMX	Adenosine	-	-7.6	-4.5	10.5
Alprazolam	Benzodiazepines	-	-11.2	-7.7	12.9
Chlordiazepoxide	Benzodiazepines	-	-8.6	-5.3	11.8
Clonazepam	Benzodiazepines	-	-11.9	-7.5	16.3
Diazepam	Benzodiazepines	-	-10.7	-8.4	8.4
Flunitrazepam	Benzodiazepines	-	-11.3	-8.6	9.9
DesMe-medazepam	Benzodiazepines	-	-6.9	5.6	45.8
Ro 15-1788	Benzodiazepines	-	-11.5	-9.9	5.1
Isoproterenol	β Adrenergic	-	-9.6	-12.0	-9.1
Epinephrine	β Adrenergic	-	-7.8	-11.8	-12.9
Norepinephrine	β Adrenergic	-	-7.1	-13.8	-21.8
Dobutamine	β Adrenergic	-	-7.7	-3.2	14.6
Terbutaline	β Adrenergic	-	-6.2	-4.1	6.6
Ephedrine	β Adrenergic	-	-9.3	-10.0	-2.4
Propranolol	β Adrenergic	-	-12.7	-6.5	20.0
Sotalol	β Adrenergic	-	-8.5	-6.1	7.7
Butoxamine	β Adrenergic	-	-7.8	-1.3	20.0
Sotalol	β Adrenergic	-	-7.8	-8.0	-0.7
Pindolol	β Adrenergic	-	-11.6	-5.5	19.6
Alprenolol	β Adrenergic	-	-11.4	-6.7	15.0

Data from the following references: Testa et al. (1987); Kilpatrick et al. (1986); Reith et al. (1984); Borea et al. (1988); Hitzemann et al. (1985); Murphy and Snyder (1982); Kockman and Hirsch (1982); Moehler and Richards (1981); Weiland et al. (1980); Contreras et al. (1986); Bree et al. (1986).

In some cases a relationship can be established between standard entropy and enthalpy values and another physical parameter of the drug, the
chromatographic $logk_W$, which is a measure of the drug lipophilicity (defined
in the next paragraph). Out of the 48 compounds of Table 4 the $logk_W$ values
are available only for the 19 dopaminergic drugs (Kilpatrick et al., 1986).
Fig. 8 reports a tridimensional scatter plot showing that the three quantities $logk_W$, enthalpy and entropy are mutually related. In particular the
correlation coefficients are respectively 0.989, 0.834 and 0.784 for the
regressions entropy-enthalpy, $logk_W$-entropy and $logk_W$-enthalpy.

Enthalpy and Entropy: A Simple Model of Drug-Receptor Interaction

Any attempt made to interpret the observed correlations among thermodynamical parameters requires a previous definition of what lipophilicity
may be representative of. From a pure physicochemical point of view the
lipophilicity of a molecule is measured by the energy needed to create a
cavity within the solvent which can allocate the molecule itself (Ben Naim,
1980); in this sense a logP (log_{10} of octanol/water partition coefficient)
of 3.0 for a specific drug has the meaning that the drug is one thousand
times more soluble in octanol than in water or that the difference between
the energies needed for creating the cavity in water and in octanol is equal
to $2.303 \cdot RT \cdot log_{10}(1000) = 4.1$ kcal mol^{-1} at room temperature and in favour of
octanol. Molecules having many H-bonding donors and/or acceptors perturb
less the water structure and are non-hydrophobic or hydrophilic, while molecules not having them (e.g. hydrocarbons) are repelled by water and are
strongly hydrophobic. Water is the biological solvent of election and logP
is used as a measure of molecular lipophilicity in water with reference to
octanol, but any other reproducible reference could be used; more recently,
hydrophobic chromatographic stationary phases have been used as a reference
instead of octanol and these give origin to two other lipophilicity scales
called R_M and $logk_W$ (or simply logk: the log of the capacity factor).

The aspect of lipophilicity which is important here is that, when the
drug molecule binds to its receptor binding site, the energy of cavity crea-
tion is released lowering the $\Delta G°$ of the equilibrium. For a reason which is
not really understood this free energy is not released as $\Delta H°$ but mainly as
$-T\Delta S°$, that is, it produces not a lowering of enthalpy but an increase of
entropy. A number of different theories have tried to explain this fact, but
for the scope of the present treatment it is sufficient to use the simple
picture that follows. Let us imagine a large hydrophobic molecule embedded
by the surrounding water molecules; these, unable to form H-bonds with their
guest, will strengthen the H-bonds among themselves closing the foreign mol-

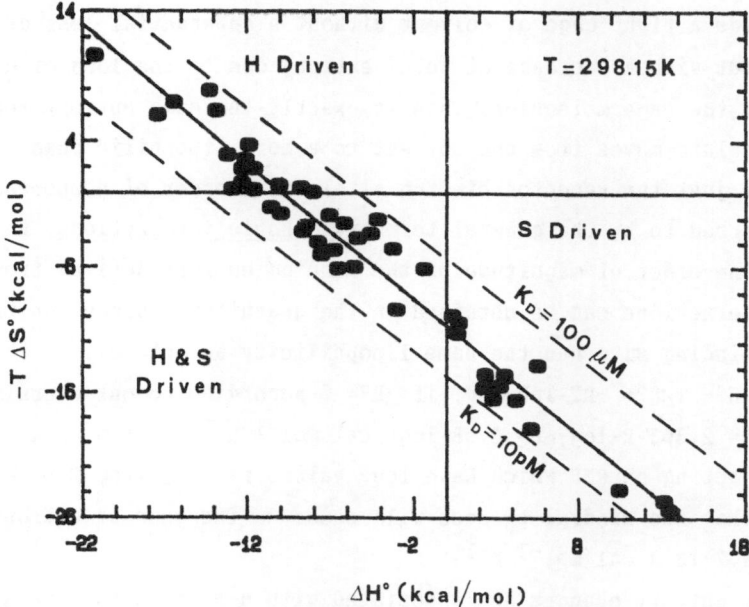

Fig. 7. Scatter plot of the standard entropy vs. standard enthalpy values for the binding equilibrium of a selection of agonists and antagonists to some receptorial systems. The regression line shown corresponds to a correlation coefficient of 0.977. Entropies are expressed as $-T\Delta S°$ values for T=298.15 K).

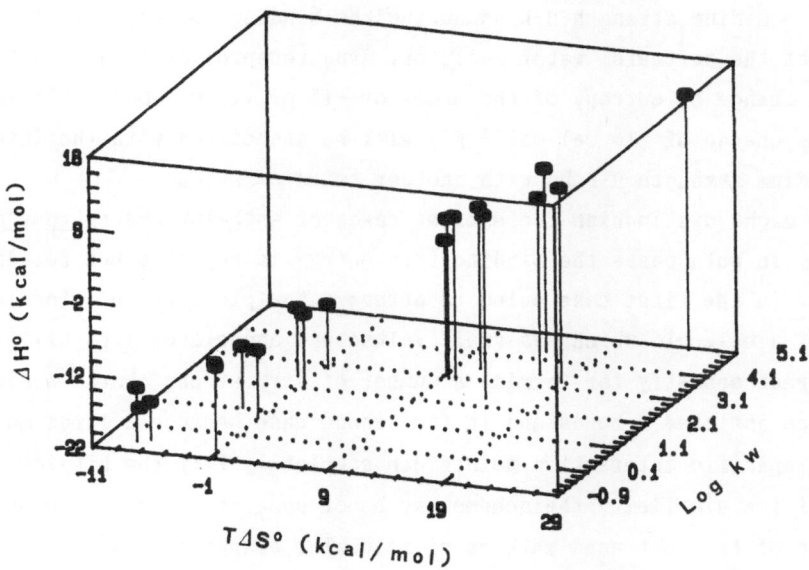

Fig. 8. Tridimensional scatter plot of the standard enthalpy vs. standard entropy and logarithm of the HPLC capacity factors for a selection of ligands of the dopaminergic receptor (data derived from Kilpatrick et al., 1986).

ecule inside a tight cage of solvent without a substantial loss of H-bonding enthalpy but with a decrease of total entropy due to the loss of degrees of freedom of the cage molecules. This is exactly the same entropy recovered when the solute moves from the solvent to a more lipophilic phase (the usual octanol or just the receptor binding site). This order of phenomena is sometimes referred to by the general term <u>hydrophobic interactions</u>. An evaluation of the order of magnitude of the binding entropy derived from hydrophobic interactions can be obtained in the gratuitous assumption that the receptor binding site has the same lipophilicity as octanol. In this hypothesis $\Delta G° = \Delta H° - T\Delta S° = -RT \cdot \ln P$ and, if $\Delta H° = 0$ according to our oversimplified model, $\Delta S° = 2.303 \cdot R \cdot \log_{10} P = 4.58 \cdot \log P$ cal mol^{-1} K^{-1}. Taking as an example the drugs acting on CNS which have logP values ranging from 2 to 4 we can estimate that the binding process will cause an increase of entropy of the order of 9.2-18.3 cal mol^{-1} K^{-1}.

Other entropy changes are associated with H-bonding in general. According to Pimentel and McClellan (1971) the formation (in CCl$_4$) of a H-bond having $\Delta H° = -5$ kcal mol^{-1} between the donor phenol and several acceptors (amides, amines, ketones and esters, aldehydes, ethers) calls for an average $-\Delta S°$ of 8.8-13.2 cal mol^{-1} K^{-1}; moreover every change of enthalpy of -1 kcal mol^{-1} is reported to cause an average change of entropy of -2.4 cal mol^{-1} K^{-1}. With the warning that data collected in CCl$_4$ are not properly comparable with those in water, these findings seem to imply that the gain or the loss of a medium strength H-bond during the binding (irrespectively of the nature of the partners: water-receptor, drug-receptor or drug-water) will cause a change of entropy of the order of -10 or +10 cal mol^{-1} K^{-1} and that the same change of -10 cal mol^{-1} K^{-1} will be associated with the interchange of a medium strength H-bond with another twice as strong.

We might distinguish two extreme cases of <u>enthalpy</u> and <u>entropy driven</u> binding. In both cases the binding free energy is negative but for different reasons: in the first case owing to strong enthalpic drug-receptor interactions of a molecule which has small volume and associated lipophilicity but the correct geometry for forming a number of very strong charge assisted or resonance assisted H-bonds and in the second case because a large molecule without specific interaction points can completely fill the binding site. In spite of its simplicity the scheme may be of some relevance in connection with one of the most used methods of molecular graphics, that of <u>molecular superimposition</u>. The method looks for the best superimposition of all drugs known to bind to a specific receptor in an attempt to discover the shape of its binding site. Quite sophisticated versions of such a procedure are known, including conformational energies and electrostatic potentials calculations. Our simplified considerations show, however, that <u>what is to be</u>

superimposed depends on binding thermodynamics, being probably the positions of H-bonding donors and acceptors if the control is enthalpic but the general shapes of molecules when the binding is controlled by entropy.

PART 3: QUANTITATIVE STRUCTURE ACTIVITY RELATIONSHIPS (QSAR)

The linear model proposed by Free and Wilson (1964) and the Hansch approach (Fujita et al., 1964) are the most widely employed methods in the field of quantitative structure activity relationships (QSAR). They differ somewhat in their logical foundations; the first is a _de novo_ method which does not assume any preconceived model of drug action and in this sense is just a statistical dissection of the global property (biological activity) among the different chemical fragments constituting the drug while the second requires a previous hypothesis on the physicochemical factors which may affect the biological response.

De Novo Analysis: The Free-Wilson Method

The Free-Wilson model is represented by the equation

$$A_n = \mu + \Sigma_i \Sigma_j \; p_{ij} \; \alpha_{ij} \; X_{n,ij} \tag{11}$$

where i is the chemical position index, j the substituent index, n the compound index, μ the regression constant, $X_{n,ij} = 1$ or 0 according to whether the substituent j is present or absent in the position i, α_{ij} the regression coefficients, p_{ij} the appearance frequency of substituent j in position i.

The model is based on the assumption that in a homologous series of drugs the biological activity A_n of each compound is given by a constant contribution μ (practically due to the molecular frame or pharmacophore) plus as many individual contributions α_{ij} as many are the substituents j in the different positions i. The aim of the analysis is to obtain the best estimate of the coefficients α_{ij}, which are supposed to measure the practical utility of the substitution of the group j in the position i for the goals of an efficient drug design.

The method has been of relatively scarce application in comparison with the Hansch analysis; this is to be mainly imputed to the larger number of different compounds that need to be synthesized and biologically tested for the analysis be statistically significant and not to particular flaws in the method itself which, as a matter of fact, may be considered less biased than others.

Table 5. Individual group contributions a_{ij} obtained from the Free-Wilson analysis of binding inhibitory constants of some benzodiazepines (n=39, r=0.968, F=10.76, P<0.01; regression const.=-1.797). Data from Borea (1983).

Position...R7 Group j	α_{7j}	Position.....R1 Group j	α_{1j}	Position............R3 Group j	α_{3j}
Cl	0.07	CONHMe	0.39	H	0.21
NO$_2$	-0.07	H	0.22	Me (S-isomer)	0.18
CF$_3$	-0.18	Me	0.13	OH	-0.05
CN	-1.11	CH$_2$CH$_2$X	-0.24	OEt	-0.15
		CH$_2$CF$_3$	-0.76	OCO(CH$_2$)$_2$COOH	-0.21
		CH$_2$cPr	-0.76	OCOC$_6$H$_2$(OMe)$_3$	-0.71
		L.C.	-0.97	OCOCH$_2$CH$_2$CH$_2$Me	-1.87
				Me (R-isomer)	-2.16

R_1 — N — R_2
R_7 — (benzodiazepine ring) — R_4, R_3
$R_{2'}$
$R_{4'}$

Position..R4 Group j	α_{4j}	Position...R2' Group j	$\alpha_{2'j}$	Position..R4' Group j	$\alpha_{4'j}$
N$_4$	0.22	Cl	0.56	H	0.10
N$_4$-R	-1.95	F	0.43	Cl	-3.66
		CF$_3$	0.13		
		H	-0.32		

Position...R2 Group j	α_{2j}
CO	0.13
CH$_2$	-2.46

Abbreviations: Me=methyl, Et=ethyl, cPr=cyclopropyl, L.C.=long chain, X=Cl or OH.

An example of application of the linear model to a practical case is shown in Table 5 which reports the individual group contributions $\underline{a}_{ij}$ obtained from the Free-Wilson analysis of the receptor binding affinity data, expressed as log(1/K$_i$), where K$_i$ values are the inhibitory constants for ^{3}H-diazepam binding to rat brain synaptosomal membranes for a series of 39 benzodiazepines (Borea, 1983) including most of those of current clinical use. The square of the correlation coefficient amounts to 0.937, which means the model explains 94% of the sample variance. The comparison of the individual group contributions obtained shows the primary importance of some substitutions for the biological activity: (i) presence of the carbonyl group at position 2 of the diazepine ring; (ii) presence of an unsubstituted nitrogen atom at position 4 of the same ring; (iii) presence of an electron withdrawing substituent of small steric hindrance at position 2' of the phenyl group. The results of this analysis have been used in the formulation of a model of benzodiazepine receptor binding site (Borea and Gilli, 1984) which will be discussed in more detail in the next part of this chapter.

Linear Free Energy Relationships (LFER) and the Hansch Method

Of more diffuse application is the Hansch method which in its more classical and familiar form can be expressed by means of the expression

$$\log 1/[C] = k_1 \log P + k_2 (\log P)^2 + \Sigma_i k_{3,i} \cdot \sigma_i + \Sigma_i k_{4,i} \cdot E_{S,i} + k_5 \quad (12)$$

where the $\underline{k}$'s are coefficients to be determined and [C] is the concentration which must be administered to provoke a given effect in a fixed percentage of the sample (animals, tissues, cells...). We have already seen that the most common expressions of [C] are LD_{50}, the median lethal dose, ED_{50}, the median effective dose and IC_{50}, the median effective inhibitory concentration. As far as the other parameters are concerned, P is the octanol/water partition coefficient discussed in a previous section, σ_i and $E_{S,i}$ (Jaffè, 1953; Exner , 1972; Taft, 1956) are respectively the Hammett constant and the steric Taft constant of the substituent in the i-th substitution position on the molecular frame of the pharmacophore.

The Hansch model tries to explain the drug biological response as a combination of effects due to its physicochemical properties, described either by overall molecular quantities (e.g. the partition coefficent, P, or the dipole moment, μ) or by individual parameters of the substituents (such as σ and E_S values or π values when the overall logP is disassembled in the lipophilic contributions of the single substituents, π).

The units of concentration and lipophilicity variables expressed in the form log(1/[C]) and logP are obviously free energies when multiplied by RT, R being the gas constant and T the absolute temperature. At constant temperature the two quantities log(1/[C]) and logP become simply proportional to free energies. This formulation is generally employed for all physicochemical parameters used in molecular pharmacology which are systematically expressed in such a way as to be energy related; accordingly, they are collectively called extrathermodynamical parameters or linearly free energy related parameters. In a similar way correlations of these quantities with molecular properties are called LFER or linear free energy relationships.

Table 6 summarizes the extrathermodynamic parameters which have found most frequent application in LFER studies. Extended tabulations of such parameters are available from different sources (see for instance Hansch and Leo, 1979; Pomona College Medicinal Chemistry Project. CLOGP3 Program,1986). For an interesting review on LFERs applied to CNS drugs see Gupta (1989).

The Meaning of LFER

To be scientifically founded the Hansch method must implicitly assume that some kind of correspondence between extrathermodynamical parameters and some crucial aspects of drug transport or binding or final efficacy can be established. Pharmaceutical chemists have applied the Hansch method to several series of related drugs, often with quite good results. In many cases

Table 6. A selection of physicochemical parameters most frequently used in linear free energy related studies (SUB= parameter of the single substituent group or single atom, MOL= parameter of the drug molecule as a whole).

A. Hydrophobic parameters

π	Hansch hydrophobic constant (SUB)
logP	log_{10} of the octanol/water partition coefficient (MOL)
R_M	Chromatographic parameter (MOL)
logk	Chromatographic capacity factor in HPLC (MOL)

B. Steric parameters

E_S	Taft steric parameter (SUB)
$E_S{}^{o,m}$, $E_S{}^p$	Modified Taft steric parameters for ortho and meta or para substituents (SUB)

C. Polarizability parameters

V_m	Molar volume (MOL and SUB)
R_m	Molar refractivity (MOL and SUB)
α	Electric polarizability (MOL)

D. Electronic parameters

σ	Hammett constant (SUB)
σ_m, σ_p	Hammett constants for meta and para substituents (SUB)
σ_I	Taft induction constant (SUB)
σ_R	Taft resonance constant (SUB)
F	Inductive field constant (SUB)
R	Resonance constant (SUB)
$\sigma^\star$	Taft polar constant (SUB)
pK_a	-log of acid dissociation constant (MOL and SUB)
μ	Dipole moment (MOL)

E. Quantum mechanical indices

$\varrho(x,y,z)$	Electron densities (MOL and SUB)
$V(x,y,z)$	Electrostatic potential (MOL and SUB)
ε	Atomic partial charges (SUB)
q	Atomic net charges (SUB)
f	Frontier electron densities (SUB)
E_{LUMO}	Energy of the lowest unoccupied molecular orbital (MOL)
E_{HOMO}	Energy of the highest occupied molecular orbital (MOL)

the k coefficients of the regression can be determined with remarkable statistical significance but, owing to the share statistical procedure used, they do not carry with them any obvious physical meaning. This is the main drawback of all correlative methods, which cannot be truly overcome by any scientific theory or model. However, the long experience cumulated over the years by many research groups allows us to formulate a tentative table of correspondences (Table 7) between the most common quantities used in LFER and their putative physicochemical and biochemical meaning. The main difference with previous tabulations of this sort concerns the role given to lipophilicity, which has been considered in the past mainly that of allowing or not the crossing of biological membranes while it is reported here also as a determinant factor of entropic nature in the drug-receptor binding. Such a hypothesis is the natural consequence of the ideas exposed in the section on binding thermodynamics and concerning the interrelations between drug lipophilicity and standard entropy of the binding equilibrium.

Table 7. Selected extrathermodynamical parameters used in LFER studies and their most plausible physicochemical effects in drug pharmacokinetics and pharmacodynamics. Abbreviation used: MOL=overall molecular parameter, SUB=substituent parameter.

Property	Parameter	MOL/ SUB	Symbol	Physicochemical Effect on Drug Action Mechanisms
Lipo- philicity	Hydrophobic constant	SUB	π	Modification of the ability of being transported by the blood proteins.
	Partition coefficient	MOL	logP	Modification of the ability of crossing the biological mem-
	Chromatogra- phic para- meters	MOL MOL	R_M logk	branes interposed from the point of administration to the site of action.
				Change of $\Delta S°$ of drug-receptor binding equilibrium.
Electro- philicity	Hammett con- stant	SUB	σ	Change of $\Delta H°$ of drug-receptor binding or drug-carrier binding
	HOMO energy	MOL	E(HOMO)	equilibria in consequence of
	LUMO energy	MOL	E(LUMO)	changes in charge transfer or
	Electron densities	SUB	$\varrho(x,y,z)$	dipolar interactions.
	Electrostatic potential	MOL	$V(x,y,z)$	Change in metabolization rates.
	Dipole moment	MOL	μ	
Polariza- bility	Molar Re- fractivity	MOL	R_m	Changes of $\Delta H°$ of drug-receptor binding or drug-carrier binding
	Molar Volume	MOL	V_m	equilibria in consequence of changes of van der Waals inter- actions.
Steric hindrance	Taft steric parameter	SUB	E_S	Modifications of stereochemical factors determining the drug- receptor interaction.

PART 4: AN EXAMPLE OF RECEPTOR MAPPING INVESTIGATION

It seems to be worthwhile to conclude the present chapter by illustrat-ing the different investigations carried out, over the years, on a specific receptorial system of acknowledged pharmacological and pharmaceutical impor-tance. We have chosen the receptor of benzodiazepines for two different rea-sons. The first is that we have been professionally involved in its study for a long time and the other that almost all different SAR (Structure-Activity Relationships) techniques have found their application in the study of such a receptor. Another fortunate aspect of this system is that the BDZ receptor and efficient binding techniques for the study of its ligands were

discovered in the middle of the last thirty years during which the study of
BDZs has gone on, so that it becomes possible to make a direct appraisal of
the improvements binding assays have produced in the field of drug design.

The Benzodiazepines (BDZ)

BDZs are psychotropic drugs of very large use. They were discovered
(for reviews on the subject see: Sternbach et al., 1968; Martin, 1987) more
or less by chance in the second half of the fifties and their use rapidly
spread owing to their wide spectrum of biological activities, including
anxiolytic, sedative-hypnotic, taming, anticonvulsant and muscle relaxant
properties associated with an extremely low toxicity. From a chemical point
of view all BDZs display the same fundamental 1,4-benzodiazepine frame of
Fig. 9 and can be further divided in "classical" 1,4-benzodiazepin-2-ones,
1,4-benzodiazepines and 1,2-anellated imidazo- or triazolo-benzodiazepines.

Early Structure-Activity Relationships on Benzodiazepines

Several in vivo tests on animals were developed and systematically employed (Sternbach et al., 1968) for a quantitative appreciation of the different pharmacological activities outlined above. The ED_{50} values obtained
from different tests were found to be widely intercorrelated and became the
object of several qualitative SAR investigations, whose results can be summarized, with reference to the scheme of Fig. 9, in the following way. Activities were found to be greatly enhanced by substitution of strong electron withdrawing groups (e.g. NO_2, Cl, Br, CF_3) in position 7 (ring A), moderately improved by methylation in position 1 (ring B), while they appear to
be strongly increased by the presence of electron withdrawing groups of
small volume in positions 2' of ring C (e.g. Cl, F). At the contrary, biological activity was found to decrease dramatically in consequence of any
substitution in positions 3' and 4' of ring C.

Some QSAR studies have been reported as well. All of them apply to relatively limited series of compounds (Yoshimoto et al., 1977; Blair and Webb,
1977; Borea et al., 1979; Biagi et al., 1980; Gilli et al., 1982) and substantially confirm the conclusions drawn from qualitative SAR in terms of
several extrathermodynamical parameters, such as lipophilicity (R_M, $\Sigma\pi$),
Hammett σ of the 7-substituent, field constant F of the group in 2', overall
dipole moment, partial charges on the carbonyl oxygen in 2 and HOMO/LUMO
energies. Of course many so called 'independent parameters' of the regression may happen to be strongly intercorrelated, so that apparently different
correlations by different authors can be often given the same meaning. For

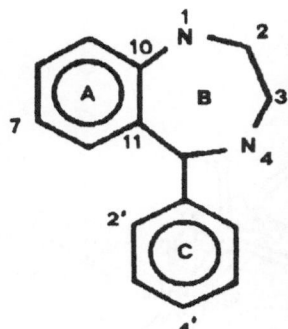

Fig. 9. The molecular frame common to all 1,4-benzodiazepines (BDZs). Several BDZs in clinical use have a carbonyl oxygen in position 2.

instance, it is clear that Hammett σ of the 7-substituents, dipole moment, HOMO/LUMO energies and partial charges on the carbonyl oxygen are all widely interdependent, a fact which does not seem to be universally acknowledged.

It is worth reporting that in the early seventies Camerman and Camerman (1971) proposed that the anticonvulsant behaviour of different classes of drugs, such as BDZs, barbiturates, hydantoins and succinimides, could be related to a common stereochemical pattern consisting of two bulky hydrophobic groups and two electron donating atoms in a fixed spatial arrangement. In the case of BDZs the four groups are represented by the two aromatic rings A and C, the carbonyl oxygen in 2 and the iminic nitrogen in position 4. It is interesting to notice that, while the model correctly fits most of the qualitative SAR findings for BDZs, it does not hold in its generality because the anticonvulsants of the other classes were successively found to be very weak binders of the BDZ receptor (Braestrup and Nielsen, 1983).

Discovery of the BDZ Receptor and Structure-Affinity Relationships

After the discovery of BDZ receptors in the mammal central nervous system (Squires and Braestrup, 1977; Moehler and Okada, 1977), binding affinities for a large number of BDZs soon became available and many researchers took an increasing interest in structure-affinity studies (Crippen, 1982; Borea, 1983; Borea and Bonora, 1983; Fryer, 1983; Braestrup and Nielsen, 1983; Borea and Gilli, 1984; Hamor and Martin, 1984; Loew, 1984). These studies are to be considered an improvement with respect to classical structure-activity relationships, which are always perturbed by uncertainties on the receptor the drug is actually binding to and by the mixing up of other factors, in particular drug transport and metabolization.

The results can be so summarized: (i) affinities of 1-alkylated or de-alkylated BDZs are similar, showing that the N_1-H is not primarily involved in the drug-receptor interaction; (ii) both N_4-substituted BDZs and BDZs

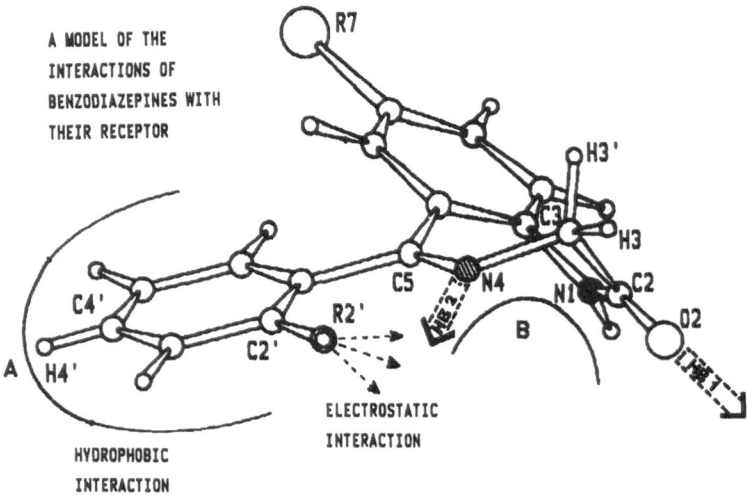

A MODEL OF THE
INTERACTIONS OF
BENZODIAZEPINES WITH
THEIR RECEPTOR

Fig. 10. A model of the main BDZ-receptor interactions
as determined from structure-affinity studies. Receptor
binding seems to be mainly governed by the two hydrogen
bonds accepted by the drug molecule. Adapted from Borea
and Gilli (1984). Reproduced with permission.

lacking the carbonyl oxygen at C_2 are weak or very weak binders, suggesting
that these two points are those mainly responsible for optimal binding;
(iii) R and S stereoisomers of 3-monosubstituted BDZs display dramatic affinity differences (Moehler and Okada, 1978), showing that the binding is
essentially stereospecific; (iv) previous SAR results are confirmed as regards the positive effects of overall lipophilicity and of small electrophiles in position 2' and the strong negative effect of substituents in 4'.

Structure-affinity studies have given two important new contributions.
The first concerns one of the most strongly held tenets from the early SAR
studies, i.e. that 7-substituents should be electron withdrawing; based on
receptor affinity data, it is now known that also compounds with electron
releasing groups (e.g. OCH_3 or OH) can display affinities in the nanomolar
range (Haefely et al., 1985). The second contribution is rather indirect but
perhaps of more general relevance: the BDZ receptor is one of the first systems for which it has been necessary to admit that drug-receptor binding affinities depend on the overall lipophilicity of the bonded molecule, a fact
which calls in question the old ideas that the role of drug lipophilicity
is limited to the control of drug transport and which has already been discussed in the section on binding thermodynamics.

Fig. 10 reports, with some minor modifications, a model of BDZ binding

(Borea and Gilli, 1984) which is a summary of what has been discussed above. It assumes that 1,4-benzodiazepines interact with positively charged receptor sites by means of two H-bonds accepted by the $C_2=O_2$ carbonyl and the N_4 ring atom; a third point of interaction would be the phenyl ring, which fits in a hydrophobic pocket so tightly that 4'-substituents can inhibit almost completely the binding. This model is in substantial agreement with those proposed by other authors (Crippen, 1982; Fryer, 1983; Loew, 1984). In particular, it accepts the idea suggested by Blount et al. (1983) that the active conformation of the 7-membered diazepine ring is that displayed in the figure and having the apical C_3 atom pointing upwards. In such a model monosubstitution at C_3 produces two stereoisomers for which it has been shown (Blount et al., 1983) that the isomer substituted at the equatorial hydrogen position H_3 (Fig. 10) binds more efficiently by two or three orders of magnitude. All this suggests that the binding site must be a planar cleft in the part allocating the benzodiazepine system, while the C_5-phenyl ring may or must be out of this plane.

Other BDZ Receptor Ligands and New Binding Assays

More recently many new substances, chemically unrelated to BDZs, have been discovered, which are able to bind with high or very high affinity to their receptor but display pharmacological properties sometimes different from those of BDZs. As previously discussed (pags. 9 and 10 and Fig. 6) they display a spectrum of biological activities which range continuously from agonist (or anticonvulsant or anxiolytic or BDZ-like) to inverse agonist (or convulsant or anxiogenic) properties through the true antagonist behaviour of drugs binding without producing any per se effect. A specific allosteric model of the BDZ receptor has been proposed (Braestrup et al., 1983) in an attempt at explaining this large spectrum of properties (see pag. 10). From a chemical point of view these new ligands, shown in Fig. 11, can belong to the classes of cyclopyrrolones (e.g. zopiclone, suriclone), triazolopyridazines (e.g. CL 218-872), phenylquinolines (e.g. PK 9084), pyrazoloquinolines (e.g. CGS 9896 and CGS 8216), imidazobenzodiazepines (e.g. Ro 15-1788) and β-carbolines (e.g. DMCM, βCCM, PrCC and ZK 93426). Table 8 reports some biochemical and pharmacological data for a selection of these ligands which will be used in the subsequent discussion.

The biochemical tests reported in the table deserve some attention. The classical binding techniques (pags. 2, 15, 24) allow appreciation of drug-receptor affinities but not of the nature of the effects produced (intrinsic activity). The discovery of ligands having non-agonist properties stimulated attempts at developing assays able to detect both affinity and intrinsic activity properties using _in vitro_ experiments only. Only two of them will be discussed here and for a short review see Karobath et al. (1983).

GABA ratio. It has been found (Moehler and Richards, 1981; Braestrup et al., 1982) that introduction of GABA in BDZ receptor binding assays enhances the affinity of BDZ receptor agonists, does not markedly alter that of antagonists and decreases that of inverse agonists. Calculated GABA ratios (i.e. IC_{50} without GABA/ IC_{50} with GABA) of several ligands have been found to correlate quite well with their _in vivo_ pharmacological profiles and thus appear to be reliable indicators of their intrinsic activities.

PAL ratio. It has also been observed (Hirsch, 1982) that UV irradiation of brain homogenates in the presence of flunitrazepam induces changes in the BDZ receptors, probably caused by covalent fixation of a ligand fragment (photo-affinity labelling= PAL). After this treatment the receptor binding

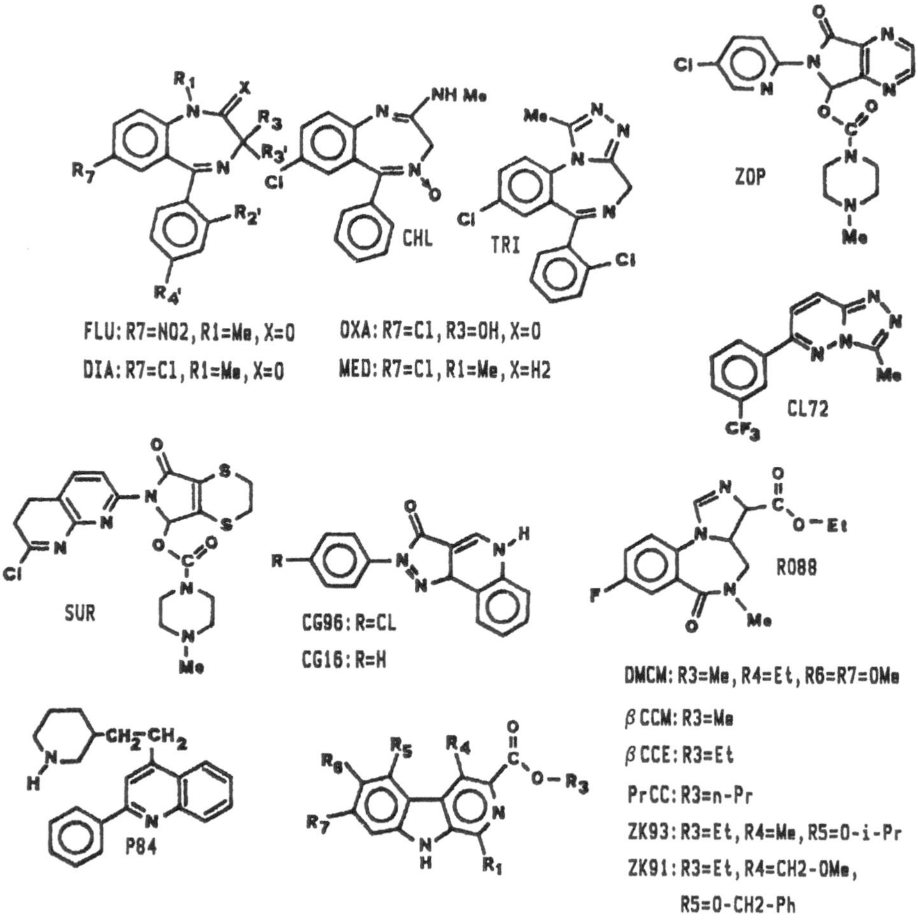

FLU: R7=NO2, R1=Me, X=O   OXA: R7=Cl, R3=OH, X=O
DIA: R7=Cl, R1=Me, X=O   MED: R7=Cl, R1=Me, X=H2

CHL TRI ZOP CL72

SUR CG96: R=CL CG16: R=H R088

DMCM: R3=Me, R4=Et, R6=R7=OMe
β CCM: R3=Me
β CCE: R3=Et
PrCC: R3=n-Pr
ZK93: R3=Et, R4=Me, R5=O-i-Pr
ZK91: R3=Et, R4=CH2-OMe,
      R5=O-CH2-Ph

P84

Fig. 11. Chemical formulas of BDZ receptor ligands belonging to the different chemical classes discussed in the text (for the codes used see Table 8).

Table 8. Synopsis of pharmacological and biochemical data for a selection of BDZ receptor ligands (adapted from Borea et al., 1987). Abbreviations: tBDZ and iBDZ=triazolo- and imidazo-BDZ, CYP=cyclopyrrolones, TPA=triazolopyridazines, PHQ=phenylquinolines, PQ=pyrazoloquinolines, βC=β-carbolines.

Compound	Chemical Class	Code	IC_{50} (nM)	GABA Ratio	PAL Ratio	Pharmacological Profile
Flunitrazepam	BDZ	FLU	5.	2.45	104.	agonist
Diazepam	BDZ	DIA	16.	2.30	166.	agonist
Oxazepam	BDZ	OXA	38.	2.35	97.	agonist
Triazolam	tBDZ	TRI	1.9	---	---	agonist
Zopiclone	CYP	ZOP	36.	1.53	5.5	agonist
Suriclone	CYP	SUR	2.2	---	---	agonist
CL 218-872	TPA	CL72	140.	1.98	6.1	agonist
Chlordiazepoxide	BDZ	CHL	1120.	2.23	---	agonist
Medazepam	BDZ	MED	6800.	---	---	agonist
PK 9084	PHQ	P84	459.	1.7	4.2	partial agonist
CGS 9896	PQ	CG96	0.6	1.20	1.2	partial agonist
ZK 91296	βC	ZK91	1.1	1.23	1.7	partial agonist
Ro 15-1788	iBDZ	RO88	3.3	1.22	1.1	antagonist
ZK 93426	βC	ZK93	0.4	1.39	1.5	antagonist
PrCC	βC	PrCC	12.	1.11	1.4	antagonist
CGS 8216	PQ	CG16	0.3	0.90	1.2	antagonist
βCCE	βC	βCCE	7.	0.86	0.9	part. inv. agon.
βCCM	βC	βCCM	8.	0.61	1.1	inverse agonist
DMCM	βC	DMCM	10.9	0.46	---	inverse agonist

IC_{50} for inhibition of specific BDZ binding from rat brain membranes; GABA Ratio=IC_{50} without GABA/IC_{50} with GABA; PAL Ratio=IC_{50} in photoaffinity-labelled membranes/IC_{50} in non-labelled membranes. Data from: Sqires and Braestrup,1977; Moehler and Okada,1977; Tallman et al.,1978; Karobath and Supavilai,1982; Braestrup et al.,1982; Braestrup et al.,1983; Braestrup and Nielsen,1983; Borea and Bonora,1983; Petersen et al.,1984; Jensen et al., 1984. For X-ray crystal structures see: Camerman and Camerman,1972; Gilli et al.,1978a and 1978b; Bertolasi et al.,1982, 1984, 1990; Butcher et al.,1983; Ferretti et al.,1985; Codding and Muir,1985.

affinities of β-carbolines, which are antagonists or inverse agonists, and of Ro 15-1788, a typical antagonist, remain practically unchanged while those of BDZs are drastically reduced; also other agonists, chemically different from BDZs, undergo affinity losses, though smaller than those for the BDZs. The PAL ratio is calculated as the quotient 'IC_{50} in photo-affinity labelled membranes/ IC_{50} in non-labelled membranes' and, by its use, a distinction of BDZ receptor agonists from antagonists and inverse agonists appears to be possible, at least in some cases (Karobath and Supavilai, 1982; Brown and Martin, 1983). In the present case, however, the most interesting fact seems to be that the PAL method is almost certainly capable of labelling the very binding site of BDZs, because for these compounds the binding affinities obtained after labelling are reduced by two orders of magnitude.

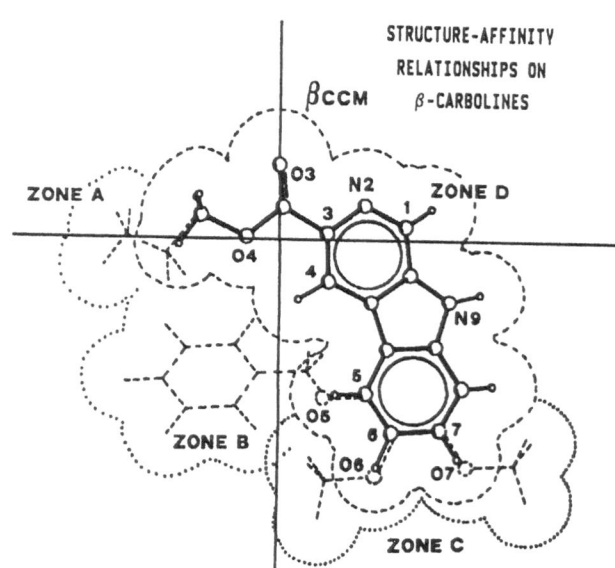

STRUCTURE-AFFINITY
RELATIONSHIPS ON
β-CARBOLINES

Fig. 12. A sketch of the molecular structure of βCCM with
its van der Waals envelope (dashed line). Substitution by
other chemical groups modifies the molecular volume as
shown by the dotted lines. Reproduced with permission from
Borea et al. (1987).

Some structure-activity studies on β-carbolines have been reported
(Loew et al., 1985; Codding and Muir, 1985; Borea et al., 1987; Allen et
al., 1988). Though different models have relevant similarities, they may
differ in several minor details and that reported here (Fig. 12) concerns
our views on the subject (Borea et al., 1987). Any reasonable model must
account for both binding affinities and the nature of intrinsic activities
which, in this case, may be agonistic, antagonistic or inverse agonistic.

As regards binding affinities, β-carbolines are high affinity ligands
only if the following conditions are fulfilled <u>at the same time</u>: (i) presence of an esteric, amidic (or other electronegative group such as -C≡N) in
position 3; (ii) full aromaticity of both 6-membered rings, producing the
planarity of all the molecule; (iii) absence of substituents in position 1.
All this suggests that the recognition site must be a planar cleft where the
main drug-receptor interactions are mediated by H-bonds accepted by the
group in position 3 (possibly strengthened by other H-bonds formed by the
two ring nitrogens). Conversely, any affinity disappears in consequence of
substitution in the zone D.

As far as intrinsic activities are concerned, they can be rationalized
according to the scheme depicted in Fig. 12. The molecule sketched in heavy

lines is that of βCCM, a typical inverse agonist. The surrounding space has
been divided into the four A, B, C and D subdomains. Substitution by groups
of increasing volume in zones A and B shifts the pharmacological properties
from full inverse agonists (βCCM) to partial inverse agonists (βCCE), to
antagonists (PrCC, ZK93), till agonists (ZK91). At the contrary substitution
of βCCM in zone C seems to increase inverse agonistic properties (DMCM).

Towards a General Model of BDZ Receptor Binding Interactions

The possibility of mapping a receptor site able to bind both BDZs and
β-carbolines (and of course the ligands of the other chemical classes, in
particular the classical antagonist RO88) has been investigated several
times (Codding and Muir, 1985; Fryer et al., 1986; Tebib et al., 1987) but
it has been generally frustrated as regards a fundamental aspect of the
problem, that of accounting for the agonist/inverse agonist properties of
the BDZs/β-carbolines couple. Starting from an analysis of the PAL ratio
measurements (see pag. 39), showing that β-carbolines cannot bind in the
same exact region of space as agonistic BDZs (even if the twos are able to
displace each other), we have suggested (Borea et al., 1987) a new model
assuming only a partial superposition of BDZs and β-carbolines at the binding site, the model which is illustrated in Fig. 13 for the two representative compounds βCCM and OXA and has the following main features:
1. An unique recognition site for all ligands of Table 8 is assumed.
2. The overall shape of the binding site is essentially a planar cleft which
can lodge either the condensed benzodiazepine system or the whole β-carbo-
line molecule. This planar cavity has a kind of out-of-plane bump that can
allocate the phenyl ring of BDZs or, in case of cyclopyrrolones (ZOP, SUR),
their N-carboxypiperazine group.
3. The specific drug-receptor binding points are B1 and B2 for BDZs and pyr-
roloquinolines (CG96, CG16), B1 and B3 for β-carbolines, RO88 and cyclopyrrolones (ZOP, SUR). X-H and X'-H are indicated in the figure as the putative
H-bonding donor groups to sites B1 and B2. S1 and S2 are regions of close
hydrophobic drug-receptor contact; substitution of any chemical group in
such positions destroys the binding ability of the derivative.
4. Different ligands bind at different positions of the same site, still
maintaining their ability to displace one another. Different locations within the site give origin to different kinds of intrinsic activity. With reference to Fig. 13, there is a continuous property shift in the order inverse
agonists, antagonists, agonists for a site location going from right to
left. The relative zones whose occupation confer agonistic or inverse agonistic properties to the ligand are indicated in the figure by the symbols

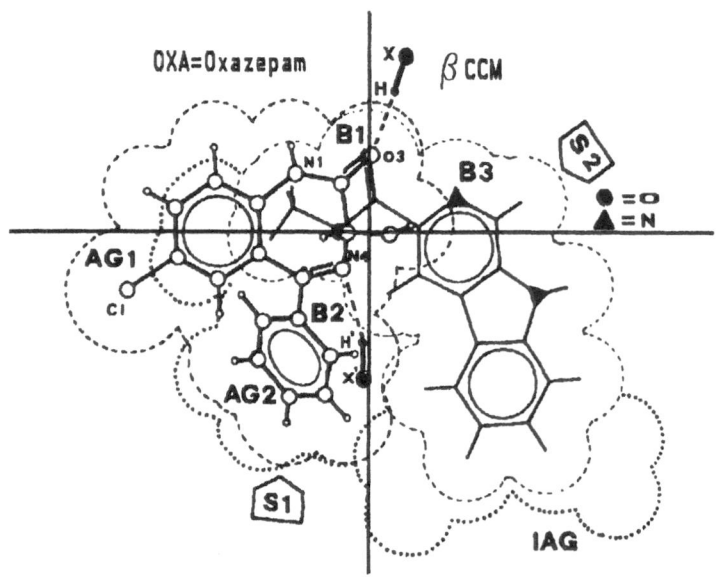

Fig. 13. Superposition of the structures of oxazepam (OXA) and
βCCM binding in different positions at the same receptor site.
B1 and B2 are the main drug binding points towards the hypothetical H-bonding donors X-H and X'-H and B3 is an additional
binding point for β-carbolines. S1 and S2 are regions of steric
hindrance for the binding of BDZs and β-carbolines, respectively. AG1 and AG2 are the zones whose occupation causes agonistic
behaviour; IAG could be a zone of increasing inverse agonistic
properties. Reproduced with permission from Borea et al. (1987).

AG1 and AG2 or IAG. The intermediate localization of a well known antagonist, the imidazobenzodiazepine RO88 (Ro 15-1788), is shown in Fig. 14 with
reference to the same system of coordinates. Analogue drawings for binders
of other classes are reported in the original paper (Borea et al., 1987).
5. In a more functional sense, a generic ligand is to be considered as an
effector able to modify the receptor conformation in different ways that, in
turn, determine the different sorts of intrinsic activity. In this case a
ligand, whose binding is strictly delimited by the few interaction points
B1-B3 and S1, will push against the receptor by means of its borderline substituents. Fig. 13 indicates that pushing leftwards and downwards on the
left causes agonistic behaviour and downwards on the right inverse agonistic
behaviour. In this sense a pure antagonist becomes a ligand for which the
two driving forces have attained a perfect equilibrium.

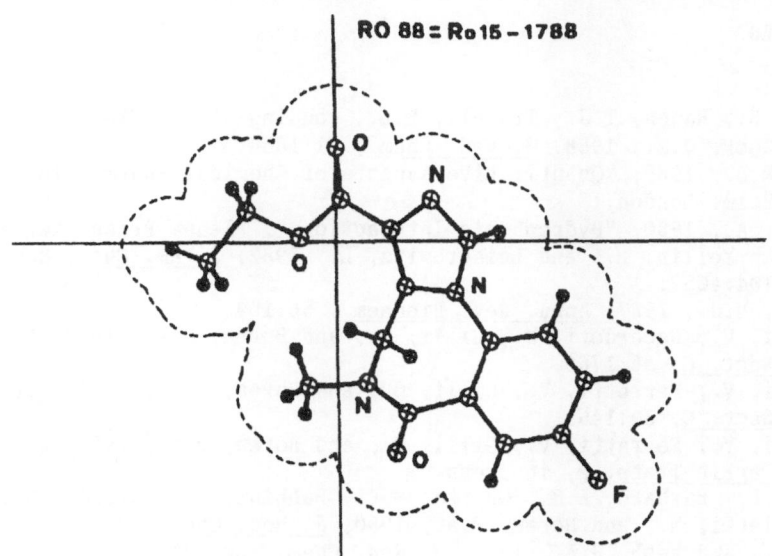

Fig. 14. The hypothesized way of binding of the antagonist RO88
(Ro 15-1788) in the same reference frame as Figs. 12 and 13. The
figure concerns the N(imidazole),O(carbonyl)-_cis_-conformation.
The N,O-_trans_-conformer allows another similar superimposition,
where the seven-membered ring points upwards. Reproduced with
permission from Borea et al. (1987).

6. The proposed model is based on the evidence coming from a variety of
biochemical, pharmacological and stereochemical data and we do not know of
any BDZ receptor ligand which does not fit in it. Of course nothing prevents
such a compound being found tomorrow, though we feel reasonably optimistic
on the matter in view of the fact that new ligands discovered later (and
which were not, but could be foreseen on the grounds of the present model)
have been so far easily incorporated in the model itself. In particular
β-carboline-BDZ hybrids (Dodd et al., 1987) and pyridodiindoles (Trudell et
al., 1987) are large molecules almost covering the total area occupied in
Fig. 13 by OXA and β-CCM _together_; it seems encouraging that some of them
were actually found to bind to the BDZ receptor in the nanomolar range,
providing direct evidence of the fact (previously simply imagined) that the
size of the BDZ receptor binding site is much larger than BDZs themselves.

Acknowledgments. The authors thank Prof. Lorenzo Beani for useful discussion on several pharmacological aspects of this paper, Mr. Stefano Gilli
for the computer drawings and the C.N.R. (Rome: Progetto Finalizzato Chimica
Fine II) for financial support.

REFERENCES

Allen, M.S., Hagen, T.J., Trudell, M.L., Codding, P.W., Skolnick, P., and Cook, J.M., 1988, J. Med. Chem., 31:1854.

Barlow, R.B., 1980, "Quantitative Aspects of Chemical Pharmacology", Croom Helm, London.

Ben-Naim, A., 1980, "Hydrophobic Interactions", Plenum Press, New York.

Berkovitch-Yellin, L., and Leiserowitz, L., 1982, J. Am. Chem. Soc., 104:4052.

Berridge, M.J., 1987, Annu. Rev. Biochem., 56:159.

Bertolasi, V., Sacerdoti, M., Gilli, G., and Borea, P.A., 1982, Acta Cryst. Sect. C, 38:1768.

Bertolasi, V., Ferretti, V., Gilli, G., and Borea, P.A., 1984, Acta Cryst. Sect. C, 39:1469.

Bertolasi, V., Ferretti, V., Gilli, G., and Borea, P.A., 1990, J. Chem. Soc. Perkin Trans. 2, in press.

Biagi, G.L., Barbaro, A.M., Guerra, M.C., Babbini, M., Gaiardi, M., Bartoletti, M., and Borea, P.A., 1980, J. Med. Chem., 23:193.

Blair, T., and Webb, G.A., 1977, J. Med. Chem., 20:1206.

Blount, J.F., Fryer, R.I., Gilman, N.W., and Todaro, L.J., 1983, Mol. Pharmacol., 24:425.

Bondi, A., 1964, J. Phys. Chem., 68:441.

Borea, P.A., Gilli, G., and Bertolasi, V., 1979, Il Farmaco, 34:1073.

Borea, P.A., 1983, Arzneim. Forsch., 33:1086.

Borea, P.A., and Bonora, A., 1983, Biochem. Pharmacol., 32:603.

Borea, P.A., and Gilli, G., 1984, Arzneim. Forsch., 34:649.

Borea, P.A., Bertolasi, V., Ferretti, V., and Gilli, G., 1987, Mol. Pharmacol., 31:334.

Borea, P.A., Bertelli, G.M., and Gilli, G., 1988, Eur. J. Pharmacol., 146:247.

Braestrup, C., Schmiechen, R., Nielsen, M., and Petersen, E.N., 1982, Science, 216:1241.

Braestrup, C., and Nielsen, M., 1983, in: "Handbook of Psychopharmacology", L.L. Iversen, S.D. Iversen and S.H. Snyder, eds., Plenum Press, New York.

Braestrup, C., Nielsen, M., Honorè, T., Jensen, L. H., and Petersen, E.M., 1983, Neuropharmacology, 22:1451.

Bree, F., El Tayar, N., Van de Waterbeemd, H., Testa, B., and Tillement, J. P., 1986, J. Receptor Res., 6:381.

Brown, C.L., and Martin, I.L., 1983, Neurosci. Lett., 35:37.

Burkert, V., and Allinger, N.L., 1982, "Molecular Mechanics", American Chemical Society, Washington.

Butcher, H., Hamor, T.A., and Martin, I.L., 1983, Acta Cryst. Sect. C, 39:1469.

Camerman, N. and Camerman, A., 1971 ,in: "Molecular and Quantum Pharmacology", E. Bergmann and B. Pullman, eds., Reidel, Dordrecht.

Camerman, A., and Camerman, N., 1972, J. Am. Chem. Soc., 94:268.

Codding, P.W., and Muir, A.K.S., 1985, Mol. Pharmacol., 28:178.

Cheng, Y., and Prusoff, W.H., 1973, Biochem. Pharmacol., 22:3099.

Contreras, M.L., Wolfe, B.B., and Molinoff, P.B., 1986, J. Pharmacol. Exp. Ther., 237:154.

Crippen, G., 1982, Mol. Pharmacol., 22:11.

Dodd, R.H., Ouannes, C., Poitier, N.C., Prado de Carvahlo, L., Rossier, J., and Potier, P., 1987, J. Med. Chem., 30:1284.

Emsley, J., 1980, Chem. Soc. Rev., 9:91.

Exner, O., 1972, in: "Advances in Linear Free Energy Relationships", N.B. Chapman and J. Shorter, eds., Plenum Press, New York.

Ferretti, V., Bertolasi, V., Gilli, G., and Borea, P.A., 1985, Acta Cryst. Sect. C, 41:107.

Free, S.M., and Wilson, J.W., 1964, J. Med.Chem., 7:395.

Fryer, R.I., 1983, in: "Benzodiazepines from Molecular Biology to Clinical Practice", E. Costa, ed., Raven Press, New York.

Fryer, R.I., Cook, C., Gilman, N.W., and Walser, A., 1986, Life Sci, 39:1947.

Fujita, T., Iwasa, J., and Hansch, C., 1964, J. Amer. Chem. Soc., 86:5175.

Gilli, G., Bertolasi, V., Sacerdoti, M., and Borea, P.A., 1978a, Acta Cryst. Sect. B, 34:2826.

Gilli, G., Bertolasi, V., Sacerdoti, M., and Borea, P.A., 1978b, Acta Cryst. Sect. B, 34:3793.

Gilli, G., Borea, P.A., Bertolasi, V., and Sacerdoti, M., 1982, in: "Molecular Structure and Biological Activity", J.F. Griffin and W.C. Duax, eds., Elsevier, Amsterdam.

Gilli, G., Bellucci, F., Ferretti, V., and Bertolasi, V., 1989, J. Am. Chem. Soc., 111:1023.

Gilli, G., and Bertolasi, V., 1990, in: "The Chemistry of Functional Groups", S. Patai ed., John Wiley & Sons, New York, in press.

Goerbitz, C.H., 1989, Acta Cryst. Sect. B", 45:390.

Gupta, S.P., 1989, Chem. Rev., 89:1765.

Gutmann, V., 1978, "The Donor-Acceptor Approach to Molecular Interactions", Plenum Press, New York.

Haefely, W., Kyburz, E., Gerecke, M., and Moehler, H., 1985, Adv. Drug. Res., 14:165.

Hamilton, W.C., and Ibers, J.A., 1968, "Hydrogen Bonding in Solids", Benjamin, New York.

Hamor, T.A., and Martin, I.L., 1984, in: "X-Ray Crystallography and Drug Action", A.S. Horn and C.J. De Ranter, eds., Clarendon Press, Oxford.

Hansch, C., 1971, in: "Drug Design", Vol. 1, E.J. Ariens, ed., Academic Press, New York.

Hansch, C., and Leo, A., 1979, "Substituent Constants for Correlation Analysis in Chemistry and Biology", John Wiley & Sons, New York.

Hirsch, J.D., 1982, Pharmacol. Biochem. Behav., 16:245.

Hitzemann, R., Murphy, M., and Curell, J., 1985, Eur. J. Pharmacol., 108:171.

Hol, W.G.J., and Wierenga, R.K., 1984, in: "X-Ray Crystallography and Drug Action", A.S. Horn and C.J. De Ranter eds., Clarendon Press, Oxford.

Huheey, J.E., 1983, "Inorganic Chemistry", 3rd Ed., Harper & Row, New York.

Jaffè, H.H., 1953, Chem. Rev., 53:191.

Jensen, L.H., Petersen, E.N., Braestrup, C., Honoré, T., Kehr, W., Stephens, D.N., Schneider, H., Seidelmann, D., and Schmiechen, R., 1984, Psychopharmacology, 83:249.

Kaplan, I.G., 1986, "Theory of Molecular Interations", Elsevier, Amsterdam.

Karobath, M., and Supavilai, P., 1982, Neurosci. Lett., 31:65.

Karobath, M., Supavilai, P., and Borea, P.A., 1983, in: "Benzodiazepine Recognition Site Ligands: Biochemistry and Pharmacology", G. Biggio and E. Costa, eds., Raven Press, New York.

Kilpatrick, G.J., El Tayar, N., Van de Waterbeemd, H., Jenner, P., Testa, B., and Marsden, C. D., 1986, Mol. Pharmacol., 30:226.

Kitaigorodsky, A.I., 1973, "Molecular Crystals and Molecules", Academic Press, New York.

Kitaigorodsky, A.I., 1984, "Mixed Crystals", Springer-Verlag, Berlin.

Kitaura, K., and Morokuma, K., 1973, J. Am. Chem. Soc., 95:7563.

Kitaura, K., and Morokuma, K., 1975, J. Am. Chem. Soc., 97:4786.

Kockman, R.L., and Hirsch, J.J., 1982, Mol. Pharmacol., 22:335.

Kollman, P.A., and Allen, L.C., 1972, Chem. Rev., 72:283.

Krogsgaard-Larsen, P., Honoré, T., and Thyssen, K., 1979, in : "GABA-Neurotransmitters. Pharmacochemical, Biochemical and Pharmacological Aspects", Alfred Benzon Symposium 12, P. Krogsgaard-Larsen, J. Scheel-Krueger and H. Kofod, eds., Munksgaard, Copenhagen.

Krogsgaard-Larsen, P., 1980, "Specific GABA Receptor Agonists and Uptake

Inhibitors: Design, Development and Structure-Activity Studies",
    FADL'S Forlag, Copenhagen.

Leo, A., Hansch, C., and Elkins, D., 1971, Chem. Rev., 71:525.

Loew, G.H., Nienow, J.R. and Poulsen, M., 1984, Mol. Pharmacol., 26:19.

Loew, G.H., Nienow, J., Lawson, J.A., Toll L., and Uyeno, E.T., 1985, Mol.
    Pharmacol., 28:17.

London, F., 1930, Z. Phys. Chem. Abt., B11:222; Z. Phys., 63:245.

Martin, I.L., 1987, Neuropharmacology, 26:957.

Moehler, H., and Okada, T., 1977, Science, 198:849.

Moehler, H., and Okada, T., 1978, Life Sci., 22:985.

Moehler, H., and Richards, J.G., 1981, Nature, 294:763.

Morokuma, K., 1971, J. Chem. Phys., 55:1236.

Murphy, K.M., and Snyder, S.M., 1982, Mol.Pharmacol., 22:250.

Pertsin, A.J., and Kitaigorodsky, A.I., 1987, "The Atom-Atom Potential
    Method", Springer-Verlag, Berlin.

Petersen, E.N., Jensen, L.H., Honorè, T., Braestrup, C., Kehr, W., Stephens,
    D.N., Wachtel, H., Seidelmann, D., and Schmiechen, R., 1984, Psycho-
    pharmacology, 83:240.

Pimentel, G.C., and McClellan, A.L., 1969, "The Hydrogen Bond", Freeman,
    San Francisco.

Pimentel, G.C., and McClellan, A.L., 1971, Annu. Rev. Phys. Chem., 22:347.

Raffa, R.B., and Porreca, F., 1989, Life Sci., 44:245.

Reith, M.E.A., Jersen, H., and Lajtha, A., 1984, Biochem. Pharmacol.,
    33:4101.

Ross, E.M., and Gilman, A.G., 1985, in: "Goodman and Gilman's The
    Pharmacological Basis of Therapeutics", A. Goodman Gilman, L.S.
    Goodman, T.W. Rall and F. Murad, eds., Mac Millan Publishing Co.,
    New York.

Schuster, P., Zundell, G., and Sandorfy, C., eds., 1976, "The Hydrogen
    Bond", Vols. 1-3, North-Holland Publishing Co., Amsterdam.

Snider, R.M., Fisher, S.K., and Agranoff, B.W., 1987, "Psychopharmacology:
    the third Generation of Progress", H.Y. Meltzer, ed., Raven Press,
    New York.

Squires, R.F. and Braestrup, C., 1977, Nature, 266:732.

Sternbach, L.H., Randall, L.O., Banziger, R. and Lehr, H., 1968, in: "Drugs
    Affecting the Central Nervous System", vol. 2, A. Burger, ed.,
    Marcel Dekker, New York.

Taft, R.W., 1956, in: "Steric Effects in Organic Chemistry", M.S. Newman,
    ed., John Wiley and Sons, New York.

Tallman, J.F., Thomas, J.W., and Gallager, D.W., 1978, Nature, 274:383.

Taylor, R., and Kennard, O., 1982, J. Am. Chem. Soc., 104:5063.

Tebib, S., Bourguignon, J.J., and Wermuth, C.G., 1987, J.Comp. Aided
    Design., 1:153.

Testa, B., Jenner, P., Kilpatrick, G.J., El Tayar, N. Van de Waterbeemd, H.,
    and Marsden, C.D., 1987, Biochem. Pharmacol., 36:4041.

Trudell, M.L., Basile, A.S., Shannon, H.E., Skolnick, P., and Cook, J.M.,
    1987, J. Med. Chem., 30:456.

Umeyama, H., and Morokuma, K., 1977, J. Am. Chem. Soc., 99:1316.

Yoshimoto, M., Kamioka, T., Miyadera, T., Kabaiashi, S., Takagi, H., and
    Tachikawa, R., 1977, Chem. Pharm. Bull., 25:1378.

Weiland, G.A., Minneman, K.P., and Molinoff, P.B., 1980, Mol. Pharmacol.,
    18:341.

Correlation of Crystal Data and Charge Density with the Reactivity and Activity of Molecules: Towards a Description of Elementary Steps in Enzyme Reactions

Gerhard Klebe,

Hauptlaboratorium of BASF AG, Carl-Bosch-Strasse,

D-6700 Ludwigshafen/Rh., F.R.G.

Abstract

The understanding of chemical reactivity and biological activity requires knowledge of the three dimensional structure of molecules and information about their conformational flexibility and their mutual interaction. Chemical reactions in biological systems are mostly performed in enzyme catalysts. Along the reaction pathway elementary steps like a nucleophilic addition or substitution, an electron transfer, or a change of coordination states occur. The structure determination of enzyme/substrate or inhibitor complexes give some ideas on the possible reaction mechanisms. Due to the limited resolution of large molecule structure determinations any detailed conclusions on the geometrical transition along the reaction coordinate remain speculative. The correlation of high resolution data from small molecule structure determinations and results from charge density and computational studies allow for a more detailed description of the elementary steps performed along the pathway. The enzyme catalyst introduces small but distinct structural and electronic changes of the ground state structure of the substrate molecule which tend towards the transition state. These perturbations may be large enough to account for a significant increase in reactivity. An effective way to intervene with biological systems is to influence the turn-over rate of an enzyme reaction. This control can be achieved by the inhibition of the enzyme, one way how many drugs act. The reasons for an effective inhibition can be many-fold. They are determined by the structural and electronic properties of the reacting species, by the transportation phenomena and by the thermodynamic aspects which contribute to the free energy of the entire process.

1. Introduction

A knowledge of the structure of molecules and their conformational flexibility is a prerequisite for understanding their chemical reactivity and biological activity [1].

The Application of Charge Density Research to Chemistry and Drug Design
Edited by G. A. Jeffrey and J. F. Piniella, Plenum Press, New York, 1991

During a chemical reaction a reactant molecule is transformed into a product molecule. This transformation involves the change of structural and energetic parameters along the reaction pathway. In order to discuss these transformations in terms of a reaction mechanism, or to predict the product formation, we have to obtain information on the atomic arrangement of the reacting molecules, not only at the incipient and final step but also along the entire pathway.

The energetic requirements for this transformation have to be determined and one has to consider whether or not the molecules need to be transferred into a special conformation favorable for the reaction.

Small molecules act in clefts or grooves at the surface of large receptor molecules, e.g. proteins. In the incipient step the drug molecule is recognized by a protein through long-range interactions (e.g. Coulombic interactions). As it becomes bound to the protein through hydrogen bonds, electrostatic and hydrophobic contacts (more directional and short-range interactions) it loses conformational degrees of freedom and it has to adopt a certain conformation which fits the geometrical requirements of the binding site of the target protein. The energetic conditions for this conformational transition as well as the energy balance of the whole substrate/receptor interaction (enthalpic and entropic contributions) finally determine the binding constants or in vitro biological activity of the drug.

Depending on the biological function of the protein various processes could be influenced by the binding of a small molecule to its protein receptor. If we consider the special case of enzyme reactions the role of the protein is to catalyze a certain biochemical reaction. Normally the small molecule (substrate) gets chemically modified inside the protein and is released from the receptor in its altered form. As elementary steps along the enzyme reaction different pathways could be proceeded, e.g. nucleophilic addition or substitution, bond formation or cleavage, oxidation (electron transfer) or valence shell expansion.

If the rate of an enzyme reaction should be influenced, e.g. to reduce the product formation of an undesired biochemical reaction (overproduction resulting in a disease) a compound is required which binds more tight to the receptor and which does not perform the enzyme reaction. In this case the compound acts as an inhibitor for the enzyme. In most cases these compounds ("drugs") share similarities with the natural substrate or resemble the molecular geometry of the transition state which occurs along the enzyme reaction pathway.

The "rational approach" to drug design makes use of this information to propose new and potent compounds which are able to intervene with biological pathways. The discussion of both, chemical reaction and drug/receptor interaction, requires information on the three dimensional structure of the interacting molecules, preferably for all accessible and energetically favorable conformations. In addition their electronic structures and the potential energy surface which underlies the conformational transitions and the whole transformation process is demanded.

Utilizing three selected examples it will be shown how crystallographic data can be used to give insight into chemical reactions as performed in enzyme catalysts. The first example concerns about the creatine amidinohydrolase which cleaves the creatine molecule with water by a nucleophilic addition into sarcosine and urea. The second case describes the hydrolysis of single-stranded RNA in the enzyme ribonuclease A. The reaction proceeds through a nucleophilic substitution at phosphorous via a pentavalent trigonal bipyramidal transition state. The last example involves the oxidation of organic substrates at an iron heme center in the class of P450-enzymes which involves an intermediate change of coordination and electronic state at iron. In all cases possibilities to inhibit the particular enzyme reaction are discussed.

2. Methods to Obtain Information on Structure/Reactivity and Activity Relationships

2.1 Contributions from X-Ray Structure Analysis

2.1.1 Determination of Static Ground State Structures

The most powerful method to elucidate the three dimensional structure of molecules is X-ray crystallography: unfortunately it is limited to the crystalline state where most molecules are at rest, whereas nearly all chemical transformations and biologically interesting substrate/receptor interactions are performed in solution or in the gas phase. Only a limited number of solid state reactions are known where the investigation by diffraction methods led to a detailed description of the pathway [2-6]. However, the precise crystal structures of starting material and product may give an indication on how the structural changes during a chemical reaction might proceed and at least help to reduce the number of possible mechanisms to be discussed [7]. For many reactions a number of intermediates are involved along the pathway which can sometimes be trapped, crystallized and investigated by X-ray structure analysis. Often they are of help in the reduction of the number of mechanistic possibilities [8-10]. Isolable intermediates are often kinetically inactive, stable, and relatively long-lived, and special care is needed to confirm that product formation really does involve these intermediates.

In several cases the knowledge of the crystal structure of a receptor protein allowed for a rational design of small molecule inhibitors [11]. However, even if the receptor structure is unknown the X-ray structures of small molecules give substantial insight in the planning and optimizing of new drugs [12]. Compared with crystals of small molecules protein crystals contain a large amount of water (up to 70 %). Thus, in many cases the conformation in a protein crystal is similar to that in solution, as the close vicinity of the biopolymer is equivalent in both phases. Furthermore, many examples are known where drug molecules diffuse into protein crystals and bind to their receptors [13]. In other cases enzyme reactions are performed in the crystal [14]. Diffraction studies of these processes (with a fast data collection using synchrotron radiation) allow insight into subsequent steps along an enzyme reaction [15]. Co-crystallizing an enzyme with a possible transition state analog of a particular substrate intermediate may give evidence for a reaction mechanism [16]. However, as the precision of protein structure determination (see below) is limited and these reactions mostly require information on proton positions (which are never detected by X-rays in these structures) for interpretation the experiment often leaves some ambiguities [17].

The prediction of drug/receptor interaction requires information on the preferred contact geometry between a drug molecule and the amino acid residues oriented towards the active site of the protein receptor. As hydrogen bonds often play a dominate role in drug binding, aside from hydrophobic interactions, it has to be considered, that the modification of the polar groups either in the protein (mutant) or in the drug (bioisoster) can substantially alter the binding geometry due to a deviating H-bond directionality between the exchanged atoms. Crystallographic data from structure determinations of small molecules appear to be a reliable source of determining H-bond geometries [18] in particular since most of the approximative computational approaches show some discrepancies with the experimental results [19]. Nonpolar contact geometries can just as well be extracted from data base information [20].

Routine structure determinations reveal molecular dimensions for small molecules with a precision of < 0.01 Å for distances, $< 0.5^0$ for bond angles and $< 5^0$ for torsion angles. As X-ray data collection requires hours for a full data set the measurements provide time-averaged information. Furthermore, the determined geometries are based on coordinates which correspond to the mean positions of electron density peaks averaged over the whole crystal. As a consequence any dynamic transition during data collection or any dynamic or static disorder in the crystal will end up in a smearing of the electron density and thus in less precisely defined dimensions between the centers of these peaks. Furthermore, in some cases the space group used for structure refinement is ambiguous and could lead to errors (e.g. artificially superimposed disorder) in structure determination.

Diffraction data from crystals of large molecules often suffer from their low diffraction power and thus result in structure determinations of less resolution. To overcome an unsatisfying parameter to observation ratio during refinement some of the "rigid" molecular dimensions are constrained and/or the refinement is combined with an energy optimization [21]. The refinement of protein structures received a substantial speed-up by the introduction of simulated annealing [21] since the radius of convergence is much larger than that of conventional restrained refinement which tries to minimize the difference between observed and calculated structure factors. The new technique is searching the conformational space by molecular dynamics in those regions allowed by the diffraction data through the introduction of an effective potential energy including the differences in structure factors ("R-factor"). As a consequence the need for manual model rebuilding between subsequent refinement cycles is reduced. However, if the method is not conscientiously applied, the danger exists that details in the obtained protein structures possess a certain geometry because they are in perfect agreement with the force field used in the molecular dynamics. This fact has to be considered if protein data is used to parameterize computational techniques, since the outlined procedure anticipates some of the results concerning molecular dimensions. Thus, the conformation of a small molecule embedded into a protein has to be regarded as much less precisely defined as compared to the results from small molecule studies. A major deficiency of X-ray crystal structure determination is the fact that only little information is available on the energy content of the molecule under consideration.

2.1.2 Information on Dynamic Properties and Molecular Flexibility from Crystal

Data

From the analysis of thermal displacement parameters (as determined by X-ray crystallography) we can learn about the mobility of atoms in a molecule and about the potential in the neighborhood of the atoms [22], large amplitudes of motion can indicate a rather flat local potential [23-25]. In the final stage of protein refinement temperature factors are applied. Often these parameters give an indication of the flexibility in different areas of the protein, and could be used to understand some mechanistic properties of these molecules. For example, the residues which form a structurally flexible channel through which a substrate molecule can diffuse into the active site inside the protein show elevated thermal motion (e.g. residues in retinol binding protein [26] or camphor-P450 surrounding the entrance to the active site [27]).

As already mentioned, the results of X-ray diffraction are influenced by errors. The thermal displacement parameters are especially affected by inappropriate absorption correction or improper weighting in the least squares refinement. Any smearing of the electron density peaks tends to be absorbed by the thermal parameters. Nevertheless, the statistical evaluation of a large data set of these parameters reveal significant and convincing conclusions.

Conformational analysis has to be considered as an important aspect in the discussion of reactivity and biological activity. From the statistical analysis of crystal data we can deduce the most stable arrangements of a certain molecular fragment [28] and sometimes information on the interconversion pathway between the low energy conformations can be derived [29]. The conformation in which a particular fragment is found in the crystal is determined by its internal energy and the interactions with its environment. As X-ray analysis does not supply much information on the energy content the impact of the crystal field is unknown and its influence can only be estimated. Nevertheless, different conformations can only occur if their energy difference is roughly of the same order of magnitude as the influence of the crystal packing energy. For a large data set the distribution of crystal data in conformational space might reflect some information on the shape of the potential energy surface which corresponds to the properties of the isolated fragment. These findings have to be validated against computational results. However, the molecular conformation of a fragment in a crystal is determined by an anisotropic environment, where molecules are embedded with directed interactions. Energetically favored conformations achieved under these conditions appear to be more relevant to the question of chemical reactivity and drug/receptor interactions than geometries present in an isotropic environment (e.g. solution or gas phase).

2.1.3 Structure Correlation Method

So far we have only focused on aspects which concern distortions of ground state structures or their mutual interconversion. During a chemical reaction the reactants as well as the intermediates, which at best could be trapped, might undergo considerable structural changes on their way to the top of the potential barrier (transition state) separating them from the product states. Such changes may be analyzed by using the principle of structure correlation [22, 30, 31]. This method is an attempt to derive detailed structural information on the reaction pathway of a particular molecular fragment from a set of closely related crystal structures . These structures have to be arranged in the right sequence and can be regarded as snapshots of "frozen-in" states along the pathway. The result of these correlations visualized in terms of scatter plots depend on the selection of the data set and the structural parameters used for the comparison. The conclusions are extracted by a statistical evaluation of the data taken from various crystal structures and it is assumed that the derived geometrical transformation is representative for the behavior of a particular molecular system during a chemical reaction. Thus, these findings have to be interpreted alongside experimental observations gathered from other techniques recorded during the dynamic process.

Bürgi and Dunitz [22] mentioned the tempting idea that the sample points of such a structure correlation tend to congregate in the low-lying region of the potential energy surface which underlies the idealized pathway under consideration. The structure correlation method assumes that the crystal field deforms the fragment under consideration comparable with the forces which act on a molecule along its transformation to the transition state of a chemical reaction, or conformational interconversion. This basic assumption gives rise to the following questions. Does the data scatter in the correlation diagrams somehow mimic a Boltzmann distribution at a certain temperature and how much information on the geometry and energetic properties of the transition state structure can be extracted? As tempting as the comparison with a Boltzmann distribution might be there is no straight-forward way to determine for each crystal structure the absolute deformation energy which operates on a particular molecular fragment in its crystal environment [32]. In addition, such an analysis requires a well distributed and statistically unbiased data set (no overrepresentation of certain classes of compounds). However, the data distribution in the correlation diagrams provide qualitative information on the shape of the potential energy surface, at least in its low energy regions.

In the last years some data became available on series of compounds undergoing a similar reaction where accurate structure determinations in the solid state were performed in conjunction to kinetic and energetic measurements. In two recent studies crystallographic data of closely related compounds were related to experimentally determined activation barriers (ring inversion in metallocenes and acetal hydrolysis [33, 34]).

The structural changes along the reaction coordinate with the energetic changes are expressed by use of an empirical relationship of a particular functional form (e.g. polynomials). This relationship, parameterized for the reference molecule by experimental data (structural, vibrational, electronic, kinetic etc.), holds for the related molecules in the series after applying a perturbation on the energy surface of the reference molecule. Following this approach the changes of the ground and transition state structure with the activation energy can be estimated [34].

For many of the organic reactions the differences in the ground state structures may be quite small whereas the associated reaction rate differences may be relatively large. As an estimate the resulting rates of change may amount to as much as 10 KCal in free energy of activation per 0.05 Å of ground state structural change (shift of the minima of the potential energy surface in the series of related compounds, not to be mixed up with the displacement of atoms from their particular equilibrium positions) [34].

Regarding these estimated values the role of an enzyme catalyst which binds and immobilizes a substrate molecule in a reactive conformation is to introduce small but distinct structural and electronic changes in the ground state structure of the substrate [34]. These perturbations if they tend towards the transition state of the reaction may be large enough (s. range mentioned above) to account for a significant increase in reactivity. If a difference in free energy of activation of 10 KCal is achieved the corresponding reaction in the enzyme would be accelerated at room temperature by approximately seven orders of magnitude compared to the uncatalyzed situation [34].

Equally well this aspect explains why modifications in the substituent pattern of a substrate molecule which slightly alters its ground state structure may have a dramatic effect on the turn-over rate of the enzyme reaction. This aspect could be exploited for the design of potent inhibitors. In many cases an inhibitor is structurally related to the natural substrate. Once the mechanistic details of an enzyme reaction are well understood compounds could be designed which differ from the structure of the natural substrate only by small geometrical changes tending "away" from the transition state structure. As a consequence the turn-over rate of the enzyme reaction is slowed down (inhibition).

2.2 Contributions from Complementary Techniques

Besides X-ray crystallography electron diffraction and microwave spectroscopy allow precise structure determination in the gas phase. Many ideas about reaction mechanisms can be gleaned from various kinetic and spectroscopic techniques. NMR-spectroscopy allows, within a certain time window, the observation of molecular topology [35] and by means of two-dimensional data the approximate determination of interatomic distances [36]. For dynamic processes which are performed within the time scale of this method, line shape analysis of temperature or pressure dependent recorded spectra permits the determination of kinetic and activation parameters (such as rate constants, $\Delta G^{\#}$, $\Delta H^{\#}$, $\Delta S^{\#}$, $\Delta V^{\#}$). Other spectroscopic methods as IR- and UV/VIS-spectroscopy could be used to some extend to monitor the proceeding of a chemical reaction or a drug/receptor interaction. It has to be considered that each of these techniques observes the molecular system on a different time scale.

Thermodynamic measurements give insight into enthalpic and entropic properties of a chemical reaction. These data result in global values for an entire ensemble and no information on the microscopic make-up of the investigated compounds is simultaneously available. Various techniques have been applied to monitor kinetic properties of chemically and biologically reacting systems. In these cases the concentration of the reaction partners are recorded to determine rate constants, activation parameters or binding constants.

The reactivity of molecules cannot be discussed without taking the electronic structure into account (cf. the powerful tool supplied by the Woodward-Hoffmann rules). Aside from spectroscopic methods (UV, ESR, NMR, and PE), which give some hints on the electron distribution in molecules, the X-X and X-N techniques in the solid state offer insight into electron distribution. It allows to evaluate the electron distribution between atomic centers. Extreme care must be taken in these studies during data collection and processing to ensure the significance of each final difference density peak. Many investigations were published where these experimental results were compared to elaborate ab-initio calculations (s. below) [37]. Computational methods are well established and are used to predict, simulate and examine models. They can by-pass and overcome the gap or the shortcomings between the results of some experimental determinations, e.g. the three-dimensional geometries taken from X-ray analysis, their energetic content as derived from spectroscopic, kinetic, and thermodynamic data, and their electronic structure as indicated from spectroscopy or deformation density studies. The computational methods as quantumchemical calculations (semiempirical or ab-initio), molecular mechanics and molecular dynamics allow to combine structural data with energetic properties, conformational flexibility (entropic aspects), electronic properties and information on intermolecular interactions.

The precision of these computational techniques is very much dependent on the method, its inherent assumptions and the computer power on hand. The most precise methods are ab-initio calculations, which are "free" from any use of experimental parameters. In these methods the Schrödinger equation is solved numerically whereas in the approximative techniques (s. below) the most time consuming parts of the calculation (e.g. two electron integrals) are approximated by simplified expressions calibrated at experimental data. However, due to the limited computational resources, the quality of the achieved results depend on the basis set selected for the study. Even with the largest computers available they are limited to molecules up to ca. 50 atoms. A number of semiempirical methods (e.g. SINDO [38], MNDO [39] (AM1 [40], PM3 [41]), MINDO/3 [42], CNDO [43] etc.) are available, which require, reciprocal to an increased speed and decreasing precision, an increasing amount of experimental data for parametrization. Today these methods seem to be the state of the art for the simulation of properties of small molecules (with 20 - 100 atoms) [44] and are often applied in the calculation of reaction coordinates and transition states. Larger systems and the incorporation of environmental effects are normally treated by molecular mechanics [45], which are about 1 - 2 orders of magnitude faster. However, in this case the results depend even more on the quality of the parametrization of the force field. As a rule, one must check whether a particular force field or semiempirical method is known to be reliable for the system to be simulated [46]. Without a special parametrization molecular mechanics are not suited for transition state structures. Any of the above mentioned energy optimization methods relaxes a certain start structure into the next local minimum (cf. steepest decent following the direction of the energy gradient). To investigate a larger range of conformational space special care is needed to generate a (hopefully) complete set of starting structures for energy minimization. Studies concerned with the conformational flexibility of molecules should employ molecular dynamics [47], which might be based on an empirical force field or a quantum chemical approach. These methods make a certain range of conformational space available dependent on the simulation temperature and the length of the simulation run.

3. Nucleophilic Addition and Amide Bond Fission in Creatinase

The first example to be discussed describes a nucleophilic attack of an activated nucleophile (H_2O) towards a carbonyl or imino carbon of an amide or imino bond to be cleaved. In the enzyme creatinase the natural substrate creatine is metabolized to urea and sarcosine [48].

The crystal structure of creatine amidinohydrolase (P.monia) was solved by X-ray crystallography [48]. The enzyme has been co-crystallized with the competitive inhibitor carbamoyl sarcosine, closely similar to the natural substrate (replacement of the guanidino by a carbamoyl moiety, loosing by that its positive charge).

creatine   carbamoyl sarcosine

Crystallographic evidence proves a similar binding of the natural substrate and the inhibitor. In *Fig. 1* it is shown how the carbamoyl sarcosine is embedded into the enzyme.

The residual electron density map is satisfactorily accounted for by the assignment of carbamoyl sarcosine with a fourth ligand, presumably a water molecule added to or in the close neighborhood of the carbonyl carbon atom. The amino group is H-bonded to GLU B262 and GLU B358 and the water molecule on the opposite side is close to HIS B232. Recent studies enabled this unique assignment of the atoms to the electron density [49].

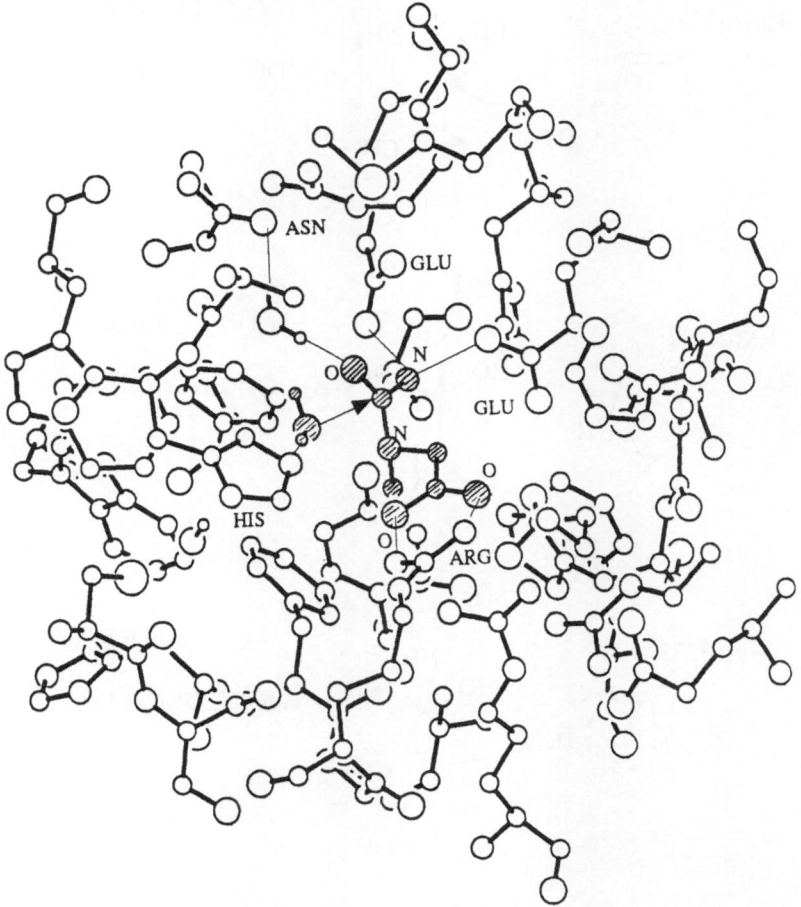

Fig. 1. Active site of Creatine Amidinohydrolase with the inhibitor sarcosine and a water molecule which acts as nucleophile. The structure was determined by X-ray structure analysis.

To facilitate the amide hydrolysis the carbonyl oxygen of sarcosine (or the imino carbon of creatine, resp.) is tightly bound through hydrogen bonds to the protein catalyst. This electron withdraw from the carbonyl carbon enables the nucleophilic attack and its pyramidal distortion. Furthermore the carbamoyl or guanidino moiety resp. is twisted away from a totally planar arrangement which corresponds to the energy minimum of the isolated molecule. This distortion, which should be fairly high in energy (since a peptide-like bond is twisted!) transfers the geometry around the tertiary nitrogen into an arrangement suited for the formation of a hydrogen bond to HIS-B232 (cf. step 3). Thus, the enzyme catalyst introduces specific perturbations of the ground state structure of the substrate which distort it towards the transition state of the nucleophilic substitution.

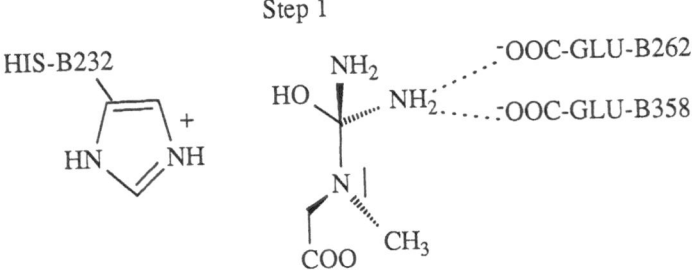

HIS-B232 H_2N + NH_2 $^-$OOC-GLU-B262

H O $^-$OOC-GLU-B358

HN N N CH_3

H COO$^-$

Step 1

HIS-B232 NH_2 $^-$OOC-GLU-B262

HO NH_2 $^-$OOC-GLU-B358

HN NH + N CH_3

COO CH_3

Step 2

HIS-B232 NH_2 $^-$OOC-GLU-B262

HO NH_2 $^-$OOC-GLU-B358

HN NH + N CH_3

COO$^-$

Step 3

HIS-B232 NH_2 $^-$OOC-GLU-B262

H O NH_2 $^-$OOC-GLU-B358

HN N HN + CH_3

COO$^-$

Step 4

HIS-B232 O NH_2 $^-$OOC-GLU-B262

H_2N $^-$OOC-GLU-B358

HN NH +

HN CH_3

COO$^-$

Step   5

The position of the water molecule is between the guanidino moiety and HIS B232. A close inspection of the electron density suggests a distance about 2.0 Å between the carbonyl carbon and the oxygen of water. In the most likely mechanism the opposing HIS is involved in the different steps of hydrogen transfer, assisting the formation of a tetrahedral intermediate (s. scheme), protonating the sarcosine (or guanidino) nitrogen during the bond rupture to form the reaction products. The two GLU residues in the active site provide a recognition site for the cationic guanidino group (natural substrate or carbamoyl group in the inhibitor) and stabilize the tetrahedral intermediate by hydrogen bonding. The whole mechanism involves an inversion of configuration at nitrogen.

It appears reasonable that the main driving force for the formation of the tetrahedral intermediate is due to an electrostatic destabilization (electron withdrawal) of the ground state structure of the guanidino group which facilitates a nucleophilic attack. Natural substrate (guanidino) and inhibitor (carbamoyl) are distinct by a positive charge on this part of the molecule. Thus, the addition of a water (as OH^-) to the carbamoyl moiety and the hydrolysis of the inhibitor would result in the development rather than in a dispersal of charge (as for the natural substrate!) in the partially hydrophobic environment of the active site. This fact explains why carbamoyl sarcosine acts as an inhibitor for creatinase.

However, the structure determination gives evidence for two important aspects with respect to a mechanistic interpretation: the electronic ground state structure of the reactant is perturbed by close contacts to the protein environment which facilitate the formation of a tetrahedral intermediate and the approach direction of the nucleophile to an activated double bond is indicated.

As the data from protein structure determinations do not allow for an interpretation of geometrical and electronic details the results from small molecule structure determinations and deformation density studies should be consulted.

Already a comparison of the C=O bond length in structures where this group is embedded in a network of H-bonds with those which lack these interactions reveals a substantial influence of hydrogen bonding on the electronic nature of this bond [50, 51].

C=O	Distance	e.s.d.	entries
no	1.215	0.015	94
H-bond	1.237	0.025	20

data from small molecule structure determinations
measured at T≤ 120 K taken from the Crystallographic Data
File [50], all distances in Å

A correlation of the C=O bond length with the O...N,O distance in crystal structures containing an intramolecular hydrogen bond gives clear evidence that with an increasing strength of the H-bond (expressed by a decreasing distances) the C=O bond is elongated (s. *Fig. 2*).

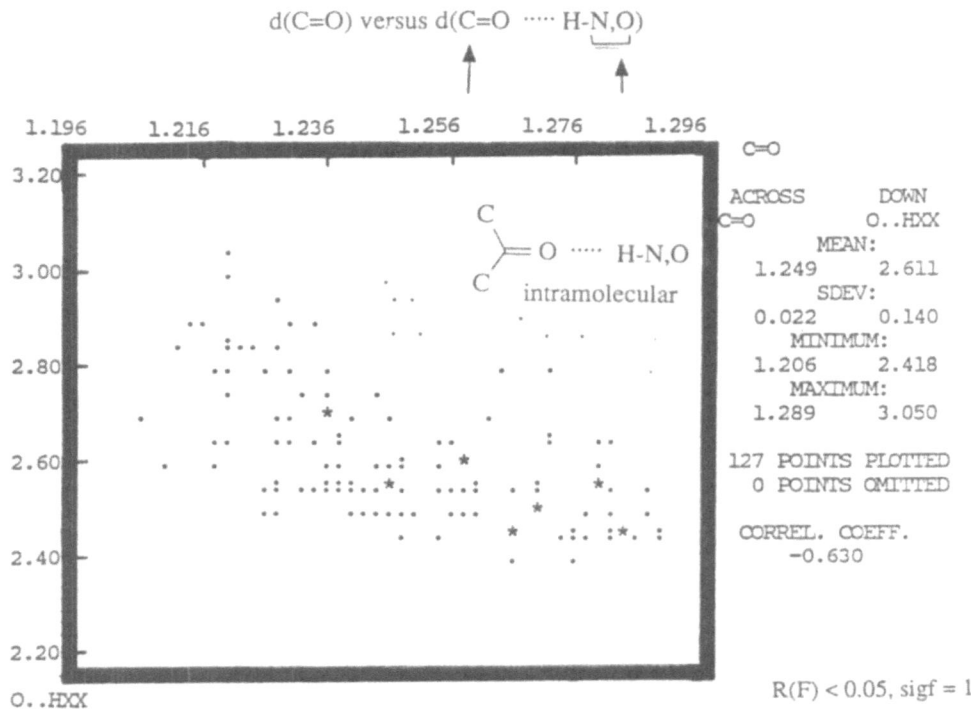

Fig. 2. Correlation of the C=O bond length versus the O...N,O distance in crystal structures containing intramolecular hydrogen bonds. The data of structures with an R(F) < 0.05 and a sigma flag of 1 [50] were considered.

The twisting of the guanidino moiety results in an additional redistribution of the electron density in the molecule. A search through the Cambridge File reveals most guanidino fragments to be close to planarity, e.g. in the crystal structure of creatine (CREATH03) the value amounts to 10.6^0). On the average in these structures the three C-N connections are fairly equivalent in length. However, in those structures where this portion of the molecule deviates from planarity (> 30^0), the central C-N bond length is significantly longer than the other two which display somewhat smaller values compared to the planar examples [51].

Detailed information on the electron distribution in molecules is available from ab-initio calculations or elaborate diffraction experiments. In the latter case the results are always superimposed by the perturbations which rise from the surrounding crystal field. It has been shown to achieve quantitative agreement between ab-initio calculations and diffraction experiments the effect of the next-nearest neighbors and the influence of the crystal field has to be incorporated into the calculations [52]. Thus, the comparison of these studies allows to estimate on the influence of an anisotropic environment (e.g. a crystal field or an active site of an enzyme) on the electron density in molecular fragments.

In detail the electron rearrangement for the water molecule in different environments has been evaluated. Essentially the following common features are detected for the redistribution in the water molecule. An enhancement of the molecular dipole moment is achieved by an electron depletion at the hydrogen sites and the "lone"-pair region close to the oxygen nucleus in conjunction with a slight electron excess in an extended region between the oxygen atom and the neighboring cation or hydrogen bond donor [52].

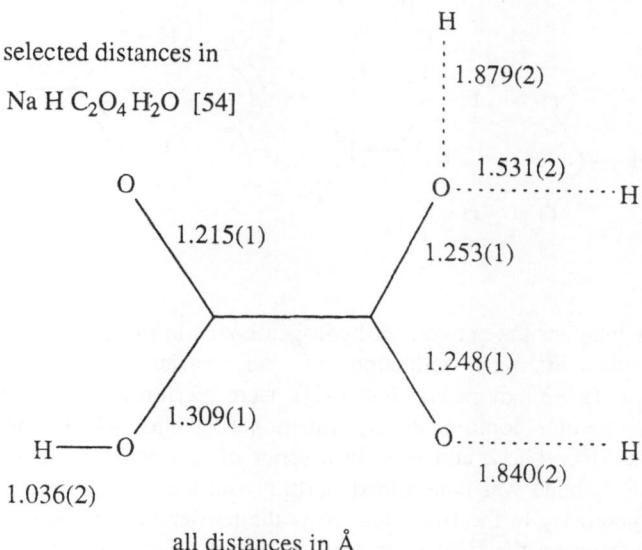

selected distances in

Na H C$_2$O$_4$ H$_2$O [54]

all distances in Å

These findings could indicate similar trends for the perturbation of a C=O fragment embedded into a network of hydrogen bonds. Several studies were performed on formiates, oxalates and oxalic acid [53 and ref. therein] which contain carbon oxygen bonds in various environments and allow to extract a clear correlation between bond strength and bond density peak height [53]. As a quantitative measure of the bond strength (order) the integral of the positive bond density in the CO bond was estimated numerically. As long as we keep in a series of closely related compounds a meaningful correlation can be expected.

A very recent study on Na HC$_2$O$_4$ H$_2$O [54] reveals the density distribution in four different CO bonds simultaneously. Already a comparison of the bond lengths(based on neutron diffraction) displays a spread from a "single " bond up to a "double" bond. Accordingly, the oxygen in the longest connection is directly bound to a hydrogen (besides C), the oxygen in the shortest CO bond shows no further short contacts (any long range contacts to Na are not considered) in the crystal. The two remaining CO groups are involved in an increasing number of hydrogen bonds, which parallels a growing elongation of their bond lengths. The integral of the bond density for the four CO bonds of increasing length corresponds to 0.11, 0.08, 0.05, and 0.04 e. Interestingly, theoretical calculations performed on formiates [55] reveal similar trends for a bond order/bond density correlation. However, the study on sodium hydrogen oxalate gives some indication that hydrogen bonding can substantially alter the electron density in a CO bond and thus can perturb the electronic structure of a substrate molecule to facilitate for a nucleophilic attack.

The twist of the guanidino moiety in creatine or the carbamoyl moiety in carbamoyl sarcosine resp. appears surprising since any computational method performed on the isolated molecules reveal these portions to be planar. The semiempirical techniques [40, 41] determine for creatine (as zwitter ion) a difference between planar and twisted (80^0) arrangement of more than 20 kcal/mol. An ab-initio calculation [56] (DZP-basis) results in a difference of 46 kcal/mol.

To get some ideas whether the network of hydrogen bonds in the active site could serve as an explanation for the dramatic reduction of the barrier to rotation the following MNDO-simulations (PM3-parametrization [41]) were performed (s. scheme). Creatine was surrounded in its zwitter ionic state by functional groups which should simulate the neighborhood of a ARG, GLU, and HIS. In a series of consecutive calculations the rotation about the central C-N bond was determined starting with the twisted arrangement and ending up with a planar geometry in the final step. Now the barrier to rotation only amounts to 0.5 kcal/mol (slightly favoring the final geometry). If the arrangement in the incipient (80^0) and the final step (0^0) is used and the "counter" groups are withdrawn to a distance of 10 Å the energy difference between the two geometries increases towards the difference observed for the isolated zwitter ion (difference for an energy calculation (no full optimization) reveals 10 kcal/mol). Regardless the question whether the absolute values of these energy calculations are very precise they indicate that hydrogen bonding could have a dramatic influence on the conformation of a substrate molecule at the receptor site.

Further insight into the geometrical conditions of the amide hydrolysis may be obtained from small molecule crystal structures. Following the structure correlations of Bürgi and Dunitz [22] the reaction proceeds via an approach of the nucleophile from above and behind the C=O(N) group in a local mirror plane of the R_2C=O(N) fragment. This approach direction is indicated by structural data of 31 N...R_2C=O fragments contained in crystal structures [57]. In Fig. 3 these fragments were matched so that the three atoms directly bound to the central carbon are superimposed. All experimentally observed positions of N are indicated (+). To compare these results with a computer simulation the model reaction of NH_3 with H_2C=O was investigated by MNDO using the PM3 parametrization [41]. To stabilize the charge transfer during the reaction an H_2O molecule was added in a position suitable for a H-bond to the carbonyl group (cf. activation of the polarized double bond by the hydrogen bond network in the protein catalyst). As reaction coordinate the N...C approach distance was varied from 3.2 to 1.45 Å. To record the potential energy dependence on the approach angle, the N...C=O angle was systematically altered between 80^0 and 150^0. Fig. 4 shows the computed energy surface together with the reaction path of minimal energy. In a projection along the energy axis the N...C and the approach angle found in the 31 crystallographic examples were included. The diagram indicates that most entries fall into the low-lying area of the potential energy surface of the model reaction [57].

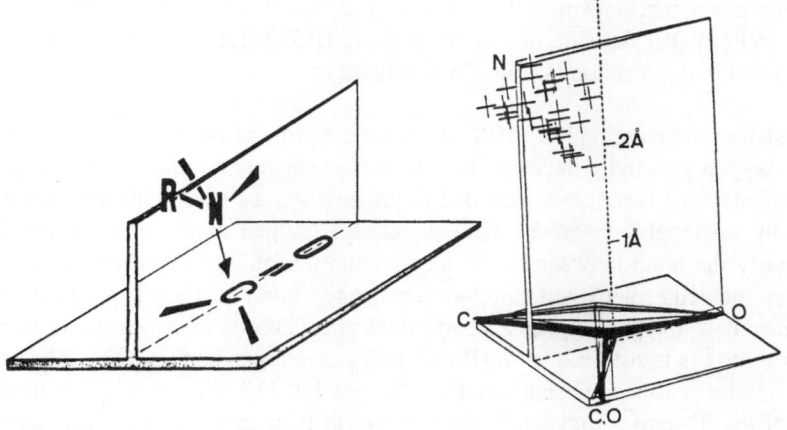

Fig. 3. Superposition of 31 N...$R_2C=O$ fragments, extracted from crystal structures, giving experimental evidence for the approach direction of the nucleophile.

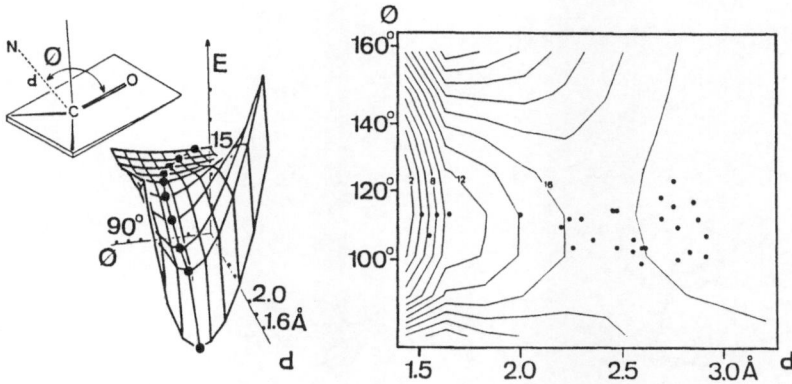

Fig. 4. Simulated model reaction *(left)* of NH_3 with $H_2C=O$ computed by MNDO (using PM3 parametrization). To stabilize the charge transfer during the reaction a water molecule was added in the neighborhood of the carbonyl group. The pathway of minimal energy is indicated (•). The potential energy surfaces was obtained by varying the N...C=O approach distance between 3.20 and 1.45 Å in dependence of the approach angle (systematically altered between 80^0 and 150^0). In the contour diagram *(right)*, contour interval is 2 KCal, the approach geometry as found in crystal structures (•) was included.

4. Nucleophilic Substitution at Phosphorous via a Pentavalent Intermediate

Single-stranded RNA is hydrolyzed by the enzyme ribonuclease A (RNase). The reaction cleaves a phospho-diester linkage and requires a pyrimidine base on the 3'-side. The protein catalyst was studied by X-ray crystallography [58-61] and the structures of the native enzyme as well as those of several protein/inhibitor complexes were determined. One of these inhibitors, a pentavalent vanadate may be regarded as a transition state analog to the formation of the cyclic 2',3'-phosphate via a pentavalent transition state [60].

Although the exact mechanism of the enzyme reaction is not fully resolved it is generally recognized [60-62] that the side chains of GLN11, HIS12, LYS14, THR45, and HIS119 are closely involved in the reaction (s. *Fig. 5* and scheme).

In the first step of the reaction [60], HIS 12 removes a proton from the 2'-oxygen. Through this step the 2'-oxygen gets more nucleophilic and is than capable to attack phosphorous to form a cyclic phosphate. The reaction is assumed to proceed via a pentavalent intermediate with the 2'- (attacking nucleophile) and 5'- (leaving group) oxygen atoms both in apical positions. Simultaneously the bond between the 5'-oxygen and the ribose is cleaved. During the second step a water molecule, activated through a hydrogen bond to HIS 119, attacks P from the opposite side of 2'-oxygen. Again a pentavalent phosphorous intermediate is formed. In the final step a proton is transferred from HIS 12 to the 2'-oxygen and the O2'- P bond is cleaved. According to the outlined mechanism HIS 12 and HIS119 should orient a nitrogen in the proximity of the 2'- and 5'-oxygen in order to fulfill their catalytic role. The additional polar residues LYS41, GLN11, and THR45 facilitate this reaction by keeping the reacting molecules in position through a network of hydrogen bonds. Furthermore these contacts introduce a polarization of the substrate required for the reaction sequence.

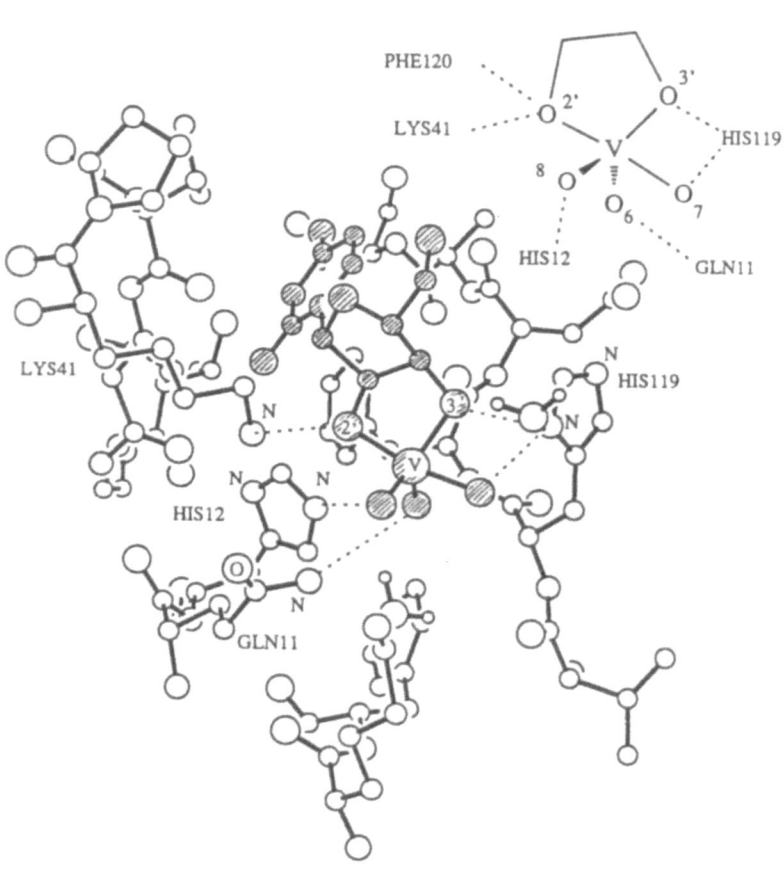

Fig. 5. Active site of Ribonuclease A with the transition state analog uridine vanadate as determined by structure analysis

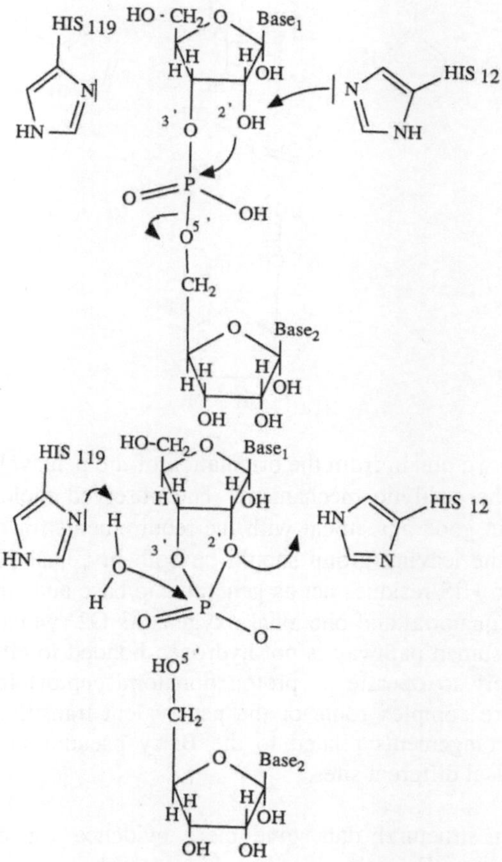

HIS 119   HO·CH₂ O Base₁   HIS 12

3' 2'

O

P

OH

O 5'

CH₂

O Base₂

OH OH

Step   1

HIS 119   HO-CH₂ Base₁   HIS 12

3' 2'

O

P

O⁻

HO 5'

CH₂

O Base₂

OH OH

Step   2

The active site of ribonuclease A is shown in *Fig. 5* together with the inhibitor uridine vanadate. The vanadium atom occupies the center of a distorted trigonal bipyramid with the oxygen atoms O2' and O7 in the apical positions ($\alpha=156^0$) while O3', O6 and O8 lie in the basal plane. Since the structure was studied by neutron diffraction detailed information on the hydrogen bond pattern is available. LYS41 is hydrogen bonded to O2' and NE2 of HIS12 forms a close contact to O8. ND1 of HIS119 is located in the vicinity of the other apical oxygen O7, furthermore this residue forms a H bond to the equatorial O3'. Additional hydrogen bonds occur between PHE120 and O8 and GLN11 and O6. A water molecule is located in the neighborhood of O7 and O3'and is presumably involved in the cleavage reaction of the natural substrate.

Step 3

How much insight can we obtain from the orientation of the pentavalent transition state analog uridine vanadate on the catalytic mechanism? The observed molecular arrangement of the trigonal bipyramid is in good agreement with the requirements for a S_N2-mechanism that the incoming ligand and the leaving group should be both in apical positions. It appears most reasonable that the two HIS-residues act as general acid/base and catalyze the proton transfer. They coordinate one equatorial and one axial oxygen. As O2' which receives a proton during the first step of the assumed pathway is not hydrogen bonded to either of the two HIS but to LYS (which is unlikely to operate as proton donator/acceptor) it may be that the proton transfer follows a more complex route or the pentavalent transition state structure performs some low-energy rearrangements related to the Berry pseudorotation which exchanges the ligands on the topological different sites.

However, the protein structural data give clear evidence for a S_N2 mechanism via a pentavalent transition state. From small molecule crystal data using the principle of structure correlation information on the geometrical transition from tetra- to pentavalency is available. Single crystal data of 46 independent molecular fragments containing penta-coordinate Si [63] were cast into the form of symmetry displacement coordinates [64]. These coordinates are appropriate for comparing various nuclear arrangements of a certain molecular fragment that can be regarded as a distorted version of a more symmetrical structure (here: D_3h-symmetrical TBP). The correlation of S_1 (reflecting the difference between the lengthening of one axial bond and the contraction of the other) versus S_2 (describing the "degree of pyramidalization" at silicon) is shown in *Fig. 6*. From the scatter data, and the fitted regression curve, the detailed changes along the transition from four- to five-coordination during the formation of the S_N2-transition state can be extracted.

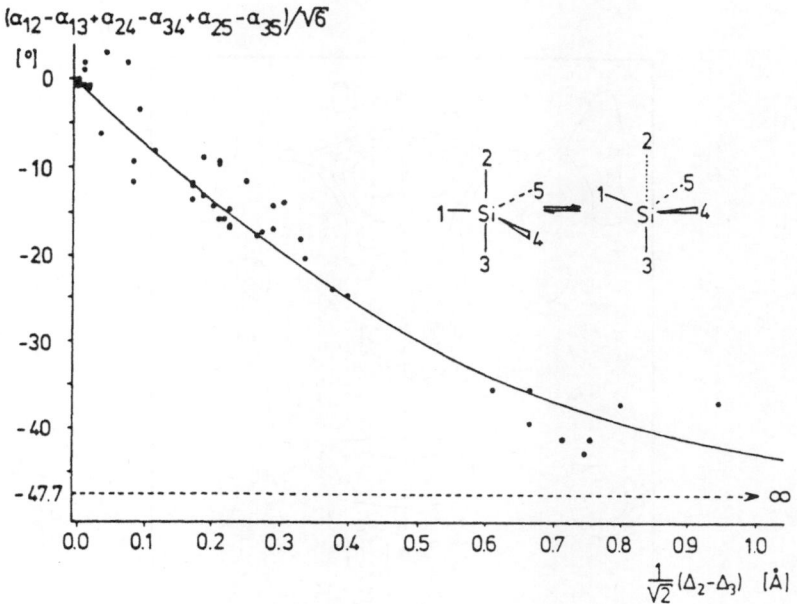

$$(\alpha_{12}-\alpha_{13}+\alpha_{24}-\alpha_{34}+\alpha_{25}-\alpha_{35})/\sqrt{6}$$

$\frac{1}{\sqrt{2}}(\Delta_2-\Delta_3)$ [Å]

Fig. 6. Correlation of the symmetry displacement coordinates S_1 and S_2 mapping the expansion of the coordination state at silicon from four to five.

1

To elucidate the electron distribution in the pentavalent S_N2-transition state the results of a deformation density study on *1* should be considered [65]. In this compound a pyridine-type nitrogen and a chlorine occupy the two apical positions of a trigonal bipyramid and could be assigned to the incoming ligand and the leaving group. As there are additional Cl and N atoms bound to silicon on the equatorial positions an internal bond length reference is present. Compared to the equatorial bonds the axial Si-N distance is 14.2 % expanded, the expansion of the axial Si-Cl contact amounts to 2.7 %. A section through the linear arrangement of the incoming pyridine-nitrogen, the central atom Si, and the leaving group chlorine is shown in *Fig. 7*. It gives an idea of the electron distribution between substrate, nucleophilic center, and the atom being displaced. The "attack" of the basic center might be described by the "lone"-pair of the pyridine-nitrogen, which is heavily deformed by the central atom.

As mentioned above a pentavalent intermediate may perform positional rearrangements of its ligands through a low energy barrier. This interconversion pathway which permutes ligands on topologically different sites is assumed to proceed via the Berry pseudorotation. Evidence for this pathway is available from the correlation of crystal and electron diffraction data of pentavalent phosphorous compounds. As reaction coordinates of the C_2v-symmetrical Berry pathway, the arithmetic means of the angles between the pivot atom and two ligand atoms in each of the perpendicular mirror planes respectively were used [66].

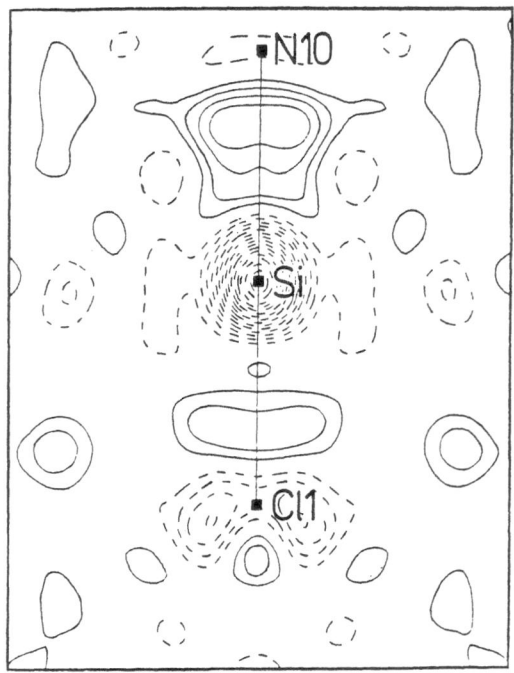

Fig. 7. Deformation density in *1*. Section through substrate
pyridine-type nitrogen), central atom Si, and leaving group (Cl) of
the TBP-arrangement; contour interval is 0.1 e/Å^3, zero contour
omitted, negative contours broken.

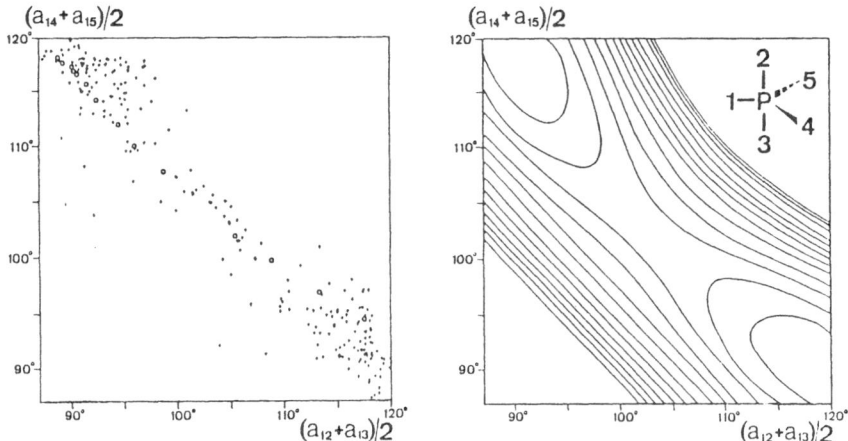

Fig. 8. Correlation of experimentally determined pentavalent phosphorous
structures (+) along the Berry pathway *(left)*. Potential energy surface for
PF$_5$, computed by MNDO, contour interval is 1 KCal *(right)*.

5. Expansion of Coordination and Oxygen Transfer in Camphor-P450

The cytochrome P450-enzymes represent a class of b-type-heme proteins which catalyze the hydroxylation of aliphatic and aromatic substrates (e.g. biochemical oxidation of lanosterol to ergosterol). The structure of camphor-P450 has been solved by X-ray crystallography [13,27,67-69]. Inside the protein the substrate camphor is hydroxylized at the 5-position by molecular oxygen. The reaction pathway is determined by the coordination of oxygen at the iron located in the center of the heme group.

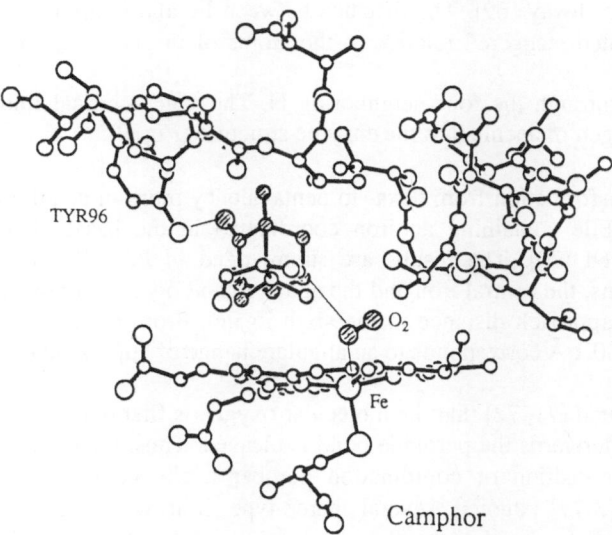

In the substrate free situation the iron is found in a low-spin state which indicates a hexa-coordination at Fe. The crystal structure gives evidence that several water molecules are present in the active site and one of them occupies the sixth coordination site at iron. If the substrate camphor diffuses into the enzyme a change in electronic state towards a high spin complex is spectroscopically recorded. A transition from octahedral to pentavalent square pyramidal coordination at iron is crystallographically confirmed. The camphor molecule is bound through a hydrogen bond between its carbonyl group and TYR 96 (*Fig. 9*).

Fig. 9. Active site of Camphor-P450 with the natural substrate camphor. An oxygen molecule was model-build at the sixth coordination site of iron.

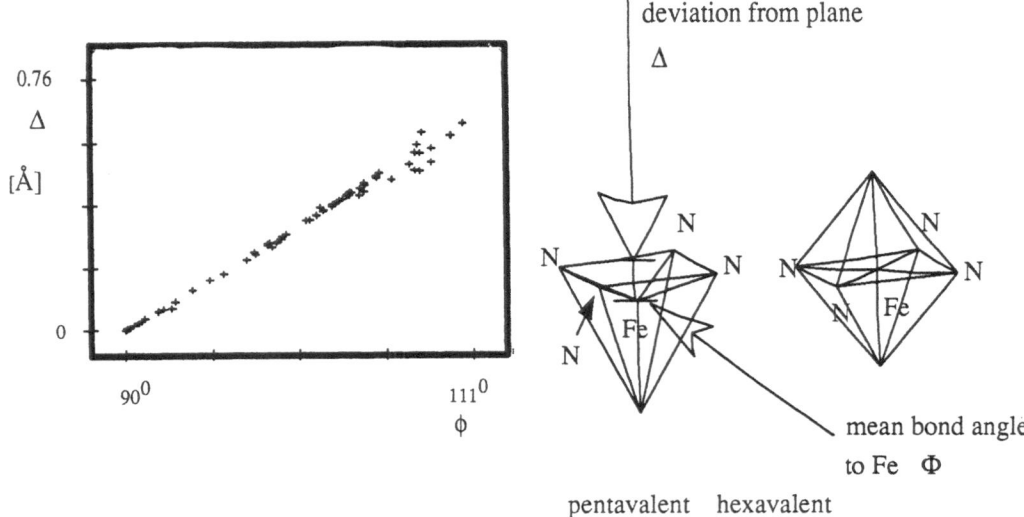

Fig. 10. Expansion of coordination state at Fe in heme-type complexes. The transition from penta- to hexa-coordination was mapped by using crystal data. The correlation coordinates Δ and φ are indicated (*left*).

Furthermore, hydrophobic interactions with the aliphatic side chains in the active site orientate the camphor molecule in a way that the carbon C5 is located adjacent to the binding site of the oxygen molecule at iron. After oxidation the iron returns to a hexavalent low-spin state.

How could we get further insight into the structural and electronic changes which are involved in this reaction? A small molecule crystal structure of a heme-type iron complex which shows molecular oxygen coordinated to iron may be regarded as a model for an intermediate step along the reaction pathway [69]. The distance between Fe and O amounts to 1.898(7) Å and the oxygen is oriented staggered relative to the atoms of the heme group. The iron deviates only little
from a best plane through the four neighboring N. These geometrical findings were used to model-build an oxygen molecule into the enzyme structure (*Fig. 9*).

To describe the transformation from hexa- to pentavalency more in detail all crystal structures in the Cambridge File containing an iron coordinated in the basal plane to four N were structurally correlated [70]. The results are summarized in *Fig. 10* where the mean angle between the nitrogens, the central iron and the apical ligand opposite to the sixth binding site is plotted against the approach distance to the sixth ligand. From the diagram we can estimate that the approach of 0.1 Å corresponds to an angular change of approximately 3^0.

It is generally accepted [71, 72] that the molecular oxygen is first reduced to the peroxide level in P450 and then afterwards the peroxide bond is cleaved. Thus, the enzyme reaction involves together with the transition of coordination number a change in oxidation state at iron. Coppens et al. [73-77] studied several heme-type iron complexes in the penta- and hexa-coordinated state to elucidate the electron density distribution around the central atom.

The investigated hexa-coordinated complexes possess Fe(II) in low- and high-spin state. Whereas the low spin compound shows density accumulation around Fe in the direction bisecting the iron-pyrrole N bonds, the high-spin complex exhibits a density pattern with peaks close to iron in direction towards the pyrrole-N ligands, indicating an effect of covalency superimposed on the cylindrical distribution of the d^6 ion [77].

The electron density of a pentavalent Fe(III) high spin complex [77] was experimentally determined. With respect to the Fe-N pyrroles-bonds a bending of the "lone pairs" at N out of the pyrrole plane towards the iron atom is detected. The iron atom is located above a plane through the atoms of the porphyrin ring (cf. *Fig. 10*). The fifth ligand, an oxygen atom, is coordinated to iron and it appears from the deformation maps that its hybridization state "is closer" to sp^2 than sp^3, thus suggesting a partial π-interaction with Fe.

A number of compounds are known to inhibit the oxidation of camphor in camphor-P450. For four of these (*2 - 5*) the crystal structure of the enzyme/inhibitor complex has been determined [13]. The relative dissociation constants are given [78, 79], that of camphor amounts to 2.13 10^{-3} mM.

2	*3*	*4*	*5*
2.23 10^{-6}mM	1.0 10^{-4}mM	4.0 10^{-2}mM	7.0 10^{-3}mM

If we consider the biological and spectroscopic properties of these complexes experiment gives evidence that in all cases iron is hexa-coordinated since all compounds occur in a low-spin state. Metyrapone (*2*) is the strongest inhibitor. The three stereoisomeric phenyl-imidazoles possess pronounced differences in biological activity (*3* > *5* > *4*) which parallel the experimentally determined dissociation constants of the complexes. Surprisingly 2-phenyl-imidazole (*5*) is appreciably active despite the fact that a direct coordination to iron should be prohibited due to steric repulsion. Compared to the spectroscopy data of the substrate free enzyme the properties of this inhibitor complex are very similar. An estimate of the binding constants for *5* and the natural substrate camphor leads to similar values. Furthermore the pronounced differences in activity between 1- and 4-phenyl-imidazole (factor 400!) appears unreasonable. A value of that magnitude normally separates a "potent" drug from one "not worth to be considered".

An interpretation of these data purely based on a comparison of the molecular structures of the inhibitors (e.g. "active analog approach") would fail. If we regard the structure determinations of the enzyme/inhibitor complexes the data can be explained consistently.

Metyrapone is coordinated to iron and a hydrogen bond to TYR 96 is found, analog to camphor (*Fig. 11*). Even if this compound reflects the sterically most bulky inhibitor in the series the least structural rearrangements of the enzyme structure are detected (compared to the substrate free case). Thus, these favorable interactions result in the high binding affinity of *2*.

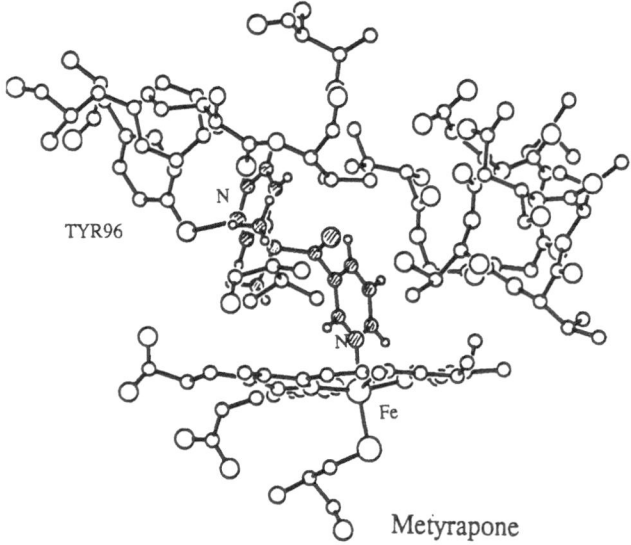

Fig. 11. Active site of Camphor-P450 with the inhibitor metyrapone **2** co-crystallized

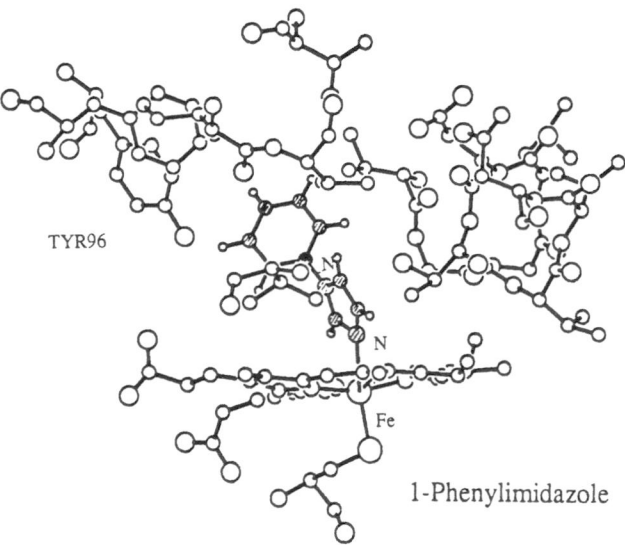

Fig. 12. Active site of Camphor-P450 with the inhibitor 1-phenyl imidazole **3** co-crystallized

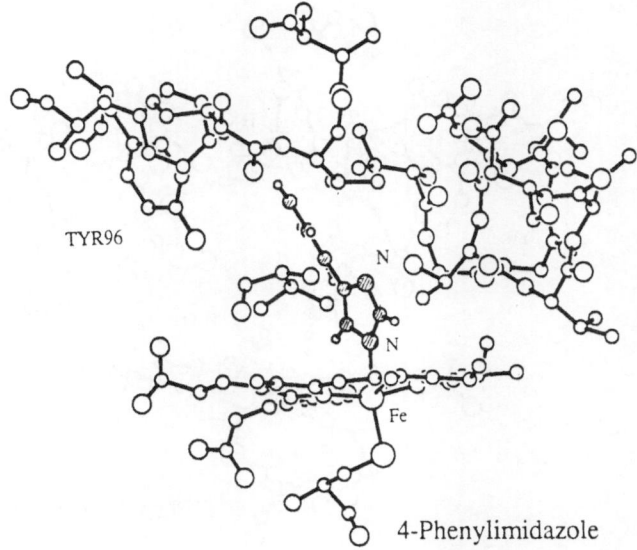

TYR96

N

N

Fe

4-Phenylimidazole

Fig. 13. Active site of Camphor-P450 with the
inhibitor 4-phenyl imidazole *4* co-crystallized

Both, 1- and 4-phenyl-imidazole coordinate to iron but a comparable H-bond to TYR 96
cannot be performed (*Fig. 12, 13*). Any additional contribution to binding is due to
hydrophobic contacts. Some rearrangements of the residues in the active site are registered as a
result of an induced fit. The pronounced differences of ca. 400 in the dissociation constants
gets obvious if we regard the number of interactions both compounds can perform in the
solvent (e.g. water) and the enzyme complex. Bound to the enzyme *4* possesses an
non-coordinated basic N in the imidazole ring in contrast to *8*. During the transition from the
solvent to the enzyme the polar contacts of the nitrogens to the solvent molecules have to be
stripped off. The energy required for this release has to be compensated by the formation of
new and comparable interactions inside the enzyme. Whereas in all other cases this balance is
leveled out for *4* remains one non-coordinated nitrogen. This fact explains the pronounced
reduction in binding constant and shows that transportation phenomena are equally important
as a geometrical fit for the biological activity of a drug.

2-Phenyl-imidazole is also bound in the active site but no coordination to iron is detected (*Fig.
14*). The phenyl ring is oriented towards iron and a water molecule remains as coordinative
ligand at the sixth position at iron. This situation is very similar to that of the substrate free
enzyme, thus, the closely related spectroscopic data are explained. The inhibitor is bound
through an H-bond to TYR 96 and an additional water molecule in the active site transmits an
H-bond from the second nitrogen to ASP 251 which reorients in the active site. Analog to
camphor *5* binds to the enzyme while replacing some water molecules in the active site. This
release of water results in a substantial increase of entropy (driving force!). A second
contribution to entropy originates from the transfer of a molecule with lipophilic side chains
from the hydrophilic solvent environment into the more hydrophobic binding pocket of the
enzyme. Entropic effects often give an important if not the determining contribution to the
driving force of drug/receptor binding.

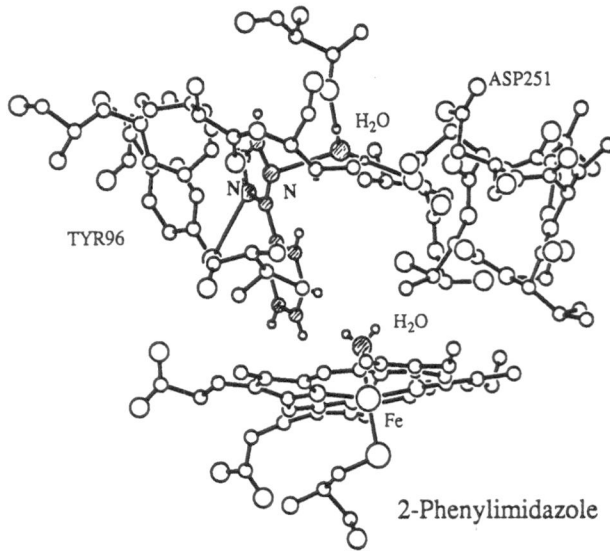

Fig. 14. Active site of Camphor-P450 with the
inhibitor 2-phenyl imidazole **5** co-crystallized

6. Conclusion

The structure determination of enzyme/substrate or inhibitor complexes gives indication on how an enzyme reaction might proceed. Due to the limited resolution of crystal data obtained from these large molecule structures geometrical and electronic details are not resolved. To describe more precisely the geometrical transitions along the reaction coordinate the structure correlation method could be used together with the results obtained from complementary computation methods. The enzyme catalyst perturbs the ground state structure of the substrate towards the geometry of the transition state to facilitate the reaction and to lower the activation barrier. To get some ideas on the electronic nature of these perturbations the results of deformations density studies of related model compounds could be consulted. The inhibition of an enzyme reaction could be achieved through several ways. If the inhibitor binds to the active site where normally the substrate molecule is modified it is structurally related to the natural substrate. In the first case discussed in this paper the inhibitor is distinct by its charge distribution which make its reaction turn-over very unlikely. In the second example the reaction could be blocked by a molecular framework which closely resembles the transition state of the reaction. In the last case the reaction center is inhibited by small molecules which compete with oxygen on the coordination site at iron. Furthermore, this example demonstrates the importance of entropic effects to be considered.

References

[1] "There is no more basic enterprise in chemistry than the determination of
the geometrical structure of a molecule. Such a determination, when it is
well done, ends all speculation as to the structure and provides us with
the starting point for the understanding of every physical, chemical and

biological property of the molecule..." R. Hoffmann, Foreword to "Determination of Geometrical Structure of Free Molecules", (1978), engl. transl. (1983), L.V. Vilkov, V.S. Mastryukov and N.I. Sadova, Mir Publ., Moscow

[2] R.B. Wilson, Y.S. Chen, I.C. Paul and D.Y. Curtin, J. Am. Chem. Soc., (1983) 105 1672

D. Kobelt and E.F. Paulus, Acta Cryst., (1974) B30 232

[3] M. Cohen, Angew. Chem., (1975) 87 439

[4] I. Grenthe and E. Nordin, Inorg. Chem., (1979) 18 1109 and 1869

Y. Ohashi, Acc. Chem. Res., (1988) 21 268

[5] T. Kurihara, Y. Ohashi, Y. Sasada and Y. Ohgo, Acta Cryst., (1983) B39 243

[6] C. Riekel, Prog. Solid State Chem., (1980) 13 89

C. Riekel and E.O. Fischer, J. Solid State Chem., (1979) 29 181

[7] K.P.C. Vollhardt and T.W. Weidman, J. Am. Chem. Soc., (1983) 105 1676

[8] G. Chioccola and J.J. Daly, J. Chem. Soc. A, (1968) 568

[9] S.E. Rasmussen and N.C. Broch, J. Chem. Soc., Chem. Comm., (1965) 298

[10] O.S. Mills and G. Robinson, Proc. Chem. Soc., (1964) 187

K. Miki, Y. Kai, N. Kasai and H. Kurosawa, J. Am. Chem. Soc., (1983) 105 2482

[11] P. Goodford, J. Med. Chem., (1984) 27 557

C.R. Bedell, Chem. Soc. Rev., (1984) 13 279

W. Hol, Angew. Chem., (1986) 98 765

[12] G. Marschall and I. Motoc, (1986) in "Molecular Graphics and Drug Design", eds. A.S.V. Burgen, G.C.K. Roberts and M.S.Tute, p.116, Elsevier Sci. Publ.

G. Marschall, Ann. Rec. Pharmacol. Toxicol., (1987) 27 193

[13] T.L. Poulos and A.J. Howard, Biochem., (1987) 26 8165

J. Badger, I. Minor, M.J. Kremer, M.A. Oliveira, T.J. Smith, J.P. Griffith, D.M.A. Guerin, S. Krishaswamy, M. Luo, M.G. Rossmann, M.A. McKinlay, G.D. Diana, F.J. Dutko, M. Fancher, R.R. Rueckert and B.A. Heinz, Proc Natl. Acad. Sci. USA, (1988) 85 3304

W. Bode, E.F. Meyer Jr. and J.C. Powers, Biochem., (1989) 28 1951

[14] M. Makinen and A.L. Fink, Annu. Rev. Biophys. Bioeng., (1977) 6 301

[15] J. Hajdu, K.A. Acharya, D.I. Stuart, P.J. McLaughlin, D. Barford,
H. Klein and L.N. Johnson, Biochem. Soc. Trans., (1986) 14(3) 538

J. Hajdu, K.A. Acharya, D.I. Stuart, P.J. McLaughlin, D. Barford,
N.G. Oikonomakos, H. Klein and L.N.Johnson, L. N., EMBO J., (1987) 6(2)
539

J. Hajdu, K.A. Acharya, D.I. Stuart, P.J. McLaughlin, D. Barford,
TIBS, (1988) 13 104

[16] R. Wolfenden, Annu. Rev. Biochem. Bioeng., (1976) 5 271

R. Wolfenden and L. Frick, (1987) in "Enzyme Mechanisms", eds. M.I. Page
and A. Williams, p. 97, Royal Society of Chemistry, London

[17] I.M. Kovach, J. Mol. Struct. (Theochem), (1988) 170 159

W. Bode and R. Huber, (1986) in "Molecular and Cellular Basis of Di-
gestion", p. 213 eds. P. Desnuelle, H. Sj str m and O. Nor n, Elsevier
Sci. Publ., Amsterdam New York Oxford

[18] J.L. Finney and H.F.J. Savage, J. Mol. Struct., (1988) 177 23

F.H. Allen, O. Kennard and R. Taylor, Acc. Chem. Res., (1983) 16 146

P. Murray-Rust and J.P. Glusker, J. Am. Chem. Soc., (1984) 106 1018

R. Taylor and O. Kennard, Acc. Chem. Res., (1984) 17 320

E.N. Baker and R. Hubbard, Prog. Biophys. Molec. Biol. (1984) 44 97

[19] A. Vedani and J.D. Dunitz J. Am. Chem. Soc., (1985) 107 7653

N.L. Allinger, R.A. Kok and M.R. Imam, J. Comput. Chem., (1988) 9 591

A. Vedani, J. Comput. Chem., (1988) 9 269

A. Glodblum, J. Comput. Chem., (1987) 8 835

D.N.A. Boobbyer, P.J. Goodford, P.M. McWhinnie and R.C. Wade, J. Med.
Chem., (1989) 32 1083

[20] M. Tintelnot and P. Andrews, J. Comp.-Aided Mol. Des., (1989) 3 67

[21] W.A. Hendrikson and J.H. Konnert, (1981) in "Biomolecular
Structure, Function, Conformation and Evolution", ed. R.
Srinivasan, Vol. 1, p. 43, Oxford: Pergamon

A.T. Brünger, J. Kuriyan and M. Karplus, Science (1987) 235 458

J. Kuriyan, A. T. Brünger, M. Karplus and W. A. Hendrickson,
Acta Cryst., (1989) A45 396

[22] H.B. Bürgi and J.D. Dunitz, Acc. Chem. Res., (1983) 16 153

H.B. Bürgi, J.D. Dunitz and E. Shefter, J. Am. Chem. Soc.,
(1973) 95 5065

[23] K.N. Trueblood and J.D. Dunitz, Acta Cryst., (1983) B39 120

J.D. Dunitz, V. Schomaker and K.N. Trueblood, J. Chem. Phys.,
(1988) 92,856

E.F. Maverick and J.D. Dunitz, Mol. Phys. (1987) 62 451

[24] J.H. Ammeter, H.B. Bürgi, E. Gamp, V. Meyer-Sandrin and W.P. Jensen,
Inorg. Chem., (1979) 18 733

[25] K. Chandrasekhar and H.B. Bürgi, Acta Cryst., (1984) B40 387

[26] M.E. Newcomer, A.T. Jones, J. Åquist, J. Sundelin, U. Eriksson, L. Rask
and P.A. Peterson, EMBO Journ., (1984) 3 1451

[27] T.L. Poulos, B.C. Finzel and A.J. Howard, Biochem., (1986) 25 5314

[28] F.H. Allen, (1982) in "Molecular Structure and Biological Activity",
p. 117, eds. J.F. Griffin, W.I. Duax, Elsevier Biomed., New York

[29] E. Bye, W.B. Schweizer and J.D. Dunitz, J. Am. Chem. Soc., (1982) 104
5893

[30] H.B. Bürgi, Inorg. Chem., (1973) 12 2321

[31] J.D. Dunitz, (1979) "X-Ray Analysis and Structure of Organic Molecules",
Cornell Univ. Press, Ithaca, New York

[32] H.B. Bürgi and J.D. Dunitz, Acta Cryst., (1988) B44 445

C.P. Brock and R.P. Minton, J. Am. Chem. Soc., (1989) 111 4586

[33] H.B. Bürgi and K. Dubler-Steudle, J. Am. Chem. Soc., (1988) 110 4953

[34] H.B. Bürgi and K. Dubler-Steudle, J. Am. Chem. Soc., (1988) 110 7291

[35] G. Binsch and H. Kessler, Angew. Chem., (1980) 92 445

[36] R. Freeman, J. Mol. Struct., (1988) 173 17

K. Wüthrich, (1986) "NMR of Proteins and Nucleic Acids", Wiley, New York

K. Wüthrich, Science, (1989) 243 45

K. Wüthrich, Acc. Chem. Res., (1989) 22 36

[37] F.L. Hirshfeld, Acta Cryst., (1984) B40 484

F.L. Hirshfeld, (1986) in "Struct. Dyn. Mol. Syst.", Vol. 2,p. 57-67., ed. R. Daudel, Reidel: Dordrecht, Netherlands

[38] D.N Nanda and K. Jug, Theor. Chim. Acta, (1980) 57 95

K. Jug, R. Iffert and J. Schulz, Int. J. Quantum Chem., (1987) 32 265

[39] M.J.S. Dewar and W. Thiel J. Am. Chem. Soc., (1977) 99 4899

[40] M.J.S. Dewar, E.G. Zoebisch, E.F. Healy and J.J.P. Stewart, J. Am. Chem. Soc., (1985) 107 3902

[41] J.J.P. Stewart, J. Comp. Chem., (1989) 10 209 and 221

[42] R.C. Bingham, M.J.S. Dewar and D.H. Lo, J. Am. Chem. Soc., (1975) 97 1285

[43] J.A. Pople and G.A. Segal, J. Chem. Phys., (1966) 44 3289

[44] S. Schröder and W. Thiel, J. Am. Chem. Soc., (1985) 107 4422

M.J.S. Dewar, Int. J. Quant. Chem., Quant. Chem. Symp. (1988) 22 557

[45] D.B. Boyd and K.B. Lipkowitz, J. Chem. Educ., (1982) 59 269

[46] T. Clark, (1985) "Handbook of Computational Chemistry", New York: John Wiley & Sons

[47] C.L. Brooks III, M. Karplus and B. M. Pettitt, (1988) in "Proteins: A Theoretical Perspective of Dynamics, Structure, and Thermodynamics", Adv. in Chem. Phys., Vol. 71, eds. I. Prigogine and S. A. Rice, New York: John Wiley & Sons

J.A. McCammon and S.C. Harvey, (1987) in "Dynamics of Proteins and Nucleic Acids", Cambridge Univ. Press

H.S. Northrup and J.A. McCammon, Biopolymers, (1980) 19 1001

[48] H.W. Hoeffken, S.H. Knof, P.A. Bartlett, R. Huber, H. Moellering and G. Schumacher, J. Mol. Biol. (1988) 204 417

[49] M. Coll, S.H. Knof, Y. Ohga, A. Messerschmidt, R. Huber, H. Moellering L. Rüssmann and G. Schumacher, J.Mol. Biol. (1990) in press

[50] F.H. Allen, S. Bellard, M.D. Brice, B.A. Cartwright, A. Doubleday, H. Higgs, T. Hummelink, B.G. Hummelink-Peters, O. Kennard, W.D.S. Motherwell, J.R. Rodgers and D.G. Watson, Acta Cryst., (1979) B35 2331

[51] G. Klebe and U. Schneider, unpublished

[52] K. Hermansson, Acta Chem. Scand., (1987) A41 513

K. Hermansson, Acta Cryst. (1985) B41 161

[53] K. Hermansson and R. Tellgren, Acta Cryst., (1989) B45 252

[54] R.G. Delaplane, R. Tellgren and I. Olovsson, Acta Cryst., (1989) submitted

[55] O.E. Taurian, S. Lunell and R. Tellgren, J. Chem. Phys., (1987) 86 5053

[56] M. Häser and R. Ahlrichs, J. Comp. Chem., (1989) 10 104

[57] G. Klebe, Struct. Chem. (1990), in press

[58] A. Wlodawer, Acta Cryst., (1980) B36 1826

L. Sjölin and A. Wlodawer, Acta Cryst., (1981) A37 594

[59] N. Borkakoti, D.S. Moss and R.A. Palmer, Acta Cryst., (1982) B38, 2210

[60] A. Wlodawer, M. Miller and L. Sjölin, Proc. Natl. Acad. Sci, USA, (1983) 80 3628

B. Borah, Chi-wan Chen, W. Egan, M. Miller, A. Wlodawer and J.S. Cohen, Biochemistry (1985) 24 2058

[61] N. Borkakoti, R.A. Palmer, I. Haneef and D.S. Moss, J. Mol. Biol., (1983) 169 743

N. Borkakoti, Euro. J. Biochem., (1983) 132 89

[62] D. Findlay, D.G. Herries, A.P. Mathias, B.R. Rabin and C.A. Ross, Biochem. J. (1962) 85 152

[63] G. Klebe, J. Organomet. Chem., (1985) 293 147

[64] P. Murray-Rust, H.B. Bürgi and J.D. Dunitz, Acta Cryst., (1979) A35 703

[65] G. Klebe, J.W. Bats and H. Fuess, J. Am. Chem. Soc., (1985) 106 5202

[66] G. Klebe, J. Organomet. Chem., (1987) 332 35

[67] T.L. Poulos, B.C. Finzel, I.C. Gunsalus, G.C. Wagner and J. Kraut, J. Biol. Chem., (1985) 260 16122

[68] T.L. Poulos, B.C. Finzel and A.I. Howard, J. Mol. Biol., (1987) 195

[69] G.B. Jameson, F.S. Molinaro, J.A. Ibers, J.P. Collman, J.I. Brauman, E. Rose and K.S. Suslick, J. Am. Chem. Soc., (1980) 102 3224

[70] G. Klebe, unpublished

[71] T.L. Poulos, (1986) in "Cytochrome P-450", p. 505, ed. P.R. Ortiz de Montellano, Plenum Press, New York & London

[72] S.C. Tang, S. Koch, G.C. Papaefthymiou, S. Foner, R.B. Frankel, J.A. Ibers and R.H. Holm, J. Am. Chem. Soc., (1976) 98 2414

[73] N. Li, J. Landrum and P. Coppens, Inorg. Chem., (1988) 27 482

[74] P.R. Mallinson, T. Koritsanzsky, E. Elkaim, N. Li and P. Coppens, Acta Cryst., (1988) A44 336

[75] C. Lecomte, R.H. Blessing, P. Coppens and A. Tabard, J. Am. Chem. Soc., (1986) 108 6942

[76] C. Lecomte, D.L. Chadwick, P. Coppens and E.D. Stevens, Inorg. Chem., (1983) 22 2982

[77] P. Coppens, ACS Symp. Series, (1989) 394 39

[78] B.W. Griffin and J.A. Peterson, Biochemistry, (1972) 11 4740

[79] J.D. Lispcomb, Biochemistry, (1980) 19 3950

CHARGE DENSITY STUDIES OF DRUG MOLECULES

Edwin D. Stevens

Department of Chemistry
University of New Orleans
New Orleans, Louisiana 70148

Cheryl L. Klein

Department of Chemistry
Xavier University of Louisiana
New Orleans, Louisiana 70125

INTRODUCTION

The role of single-crystal x-ray diffraction in structural studies is well recognized. Much of our understanding of the details of molecular structure, structure-activity relationships, and even chemical bonding has been drawn from the many structures which have been determined by x-ray diffraction. Parameterization of molecular mechanics models is also largely dependent on x-ray structural data.

Inherent in the x-ray experiment, however, is the potential for providing direct information on the electronic structure of drug molecules. Since x-rays are scattered by the electrons in a crystal, the electron density distribution may be obtained experimentally from accurate, high-resolution x-ray intensity measurements. Like the molecular wavefunction, the experimental electron distribution provides detailed 3-dimensional information on chemical bonding. In addition, other one-electron molecular properties such as net atomic charges, dipole moments, electric field gradients, and electrostatic potentials may also be derived from the experimental electron density distribution. For recent reviews, see Angermund, Claus, Goddard, and Kruger, (1985), Klein and Stevens (1988), and Coppens (1989).

Over the past 20 years, experimental techniques and methods of analysis have been developed to the point where experimental charge density studies are possible on most materials that can be crystallized. When compared with large basis set ab inito calculations on small molecules, theoretical and experimental results agree to within the estimated error in the experimental density. Smaller basis set calculations and other less rigorous methods are easily discriminated against using experimental charge densities.

The Application of Charge Density Research to Chemistry and Drug Design
Edited by G. A. Jeffrey and J. F. Piniella, Plenum Press, New York, 1991

319

Since the difficulty of experimental electron density studies increases only moderately with the size of the molecule, compared with the fourth power dependence of theoretical calculations, experimental studies offer a significant advantage for molecules of moderate size (50-200 atoms). Over the past several years we have been measuring experimental electron density distributions of drugs and other molecules of biological interest, and of transition metal complexes. In both cases, obtaining the same information by rigorous theoretical methods would be extremely difficult.

Many drugs are believed to exhibit their pharmacological activity by binding to specific receptors. While the stereochemical requirements of drug-receptor binding has received much attention, it is reasonable to assume that a complimentarity must also exist between the charge distribution of drug and receptor. Thus the x-ray experiment has the potential of providing information on both the stereochemical and charge distribution requirements of drug-receptor bindings.

According to density functional theory, the energy (and from it all other properties) of a system are a unique functional of the electron density distribution. At present, however, very little is known about how to proceed from the experimental electron density distribution. In this paper, we will describe the methods we use for obtaining experimental electron distributions, present some examples of experimental distributions we have recently obtained in New Orleans, and describe those predictions of properties and reactivity we have been able to make.

METHOD

Deformation Densities

A periodic function such as the electron density distribution can be represented as a Fourier series,

$$\rho(r) = \frac{1}{V} \sum_H F_H \, e^{2\pi i H \cdot r}$$

where the coefficients, F_H, are x-ray structure facators. Given accurate x-ray intensity measurements, calculation of the electron density distribution should be straight forward. Two major difficulties must be considered, however. First, the structure factors are, in general, complex numbers, and only the magnitude and not the phase can be determined from the experiment. Secondly, only a finite number of measurements are possible, leading to series termination errors in the result. The phase problem can be overcome by using the phase calculated for a model structure. This requires special consideration for acentric structures which will be discussed later.

To avoid series termination errors, a difference density is calculated by subtracting from the total density the density of a reference molecule,

$$\Delta \rho(r) = \rho_{obs} - \rho_{ref} = \frac{1}{V} \sum_H (F_{obs} - F_{ref}) \, e^{2\pi i H \cdot r}$$

The reference density, often refered to as the "promolecule" or IAM (independent atom model), is usually taken as the sum of the densities of isolated spherical Hatree-Fock atoms placed at the nuclear positions with the observed thermal motions. The resulting density is termed the "deformation" density to distinguished it from the usual difference density of x-ray crystallography and because it displays the deformations that occur in the atomic electron distributions as a result of chemical bonding. Series termination is greatly reduced because the very high order terms in the series which are omitted correspond to the scattering from the core densities where the actual and reference densities closely coincide. In some cases, some series termination effects may still exist for sharp features of the valence density such as lone pairs on oxygen atoms and especially non-bonding d-electron density on transition metal atoms.

To substract the reference density, accurate positions and thermal motion parameters are required. As the conventional least-squares refinement method attempts to minimize the difference, the resulting parameters will be biased by the valence electron distribution and cannot be used. Unbiased parameters can be obtained by a separate neutron diffraction experiment, or by refinement of only high angle x-ray data, where scattering is due largely to the core electrons.

Since the x-ray intensity measurement has some statistical uncertainly in addition to possible systematic errors, the deformation density calculated by Fourier summation will contain some random noise, typically with $\sigma(\Delta\rho) = 0.04 - 0.06$ e/Å^3. And as more observations are included in the calculation, the noise level increases. For molecules of biological interest, the increased size leads to a large number of weak reflections, yielding higher R-factors and higher noise levels in the experimental deformation density. To reduce the noise, experimental measurements are now more commonly fit with a multipole refinement which filters out much of the experimental noise.

Multipole Refinement Model

As an alternative to obtaining the experimental density by Fourier series summation, the least-squares refinement model can be modified to include parameters which describe the distortions of the atomic electron distributions. Rather than describing the atoms as spherical, each atom ("pseudoatom") is described by the following model,

$$\rho_{atom}(r) = \rho_c(r) + P_v \rho_v(\kappa r) + \Sigma P_{\ell m} R_\ell(\kappa' r) y_{\ell m}(\theta, \phi)$$

where $\rho_c(r)$ and $\rho_v(r)$ are the spherically symmetric Hartree-Fock core and valence densities, P_v and P_m are refinable populations, κ and κ' are refinable expansion/contraction coefficients, and the $y_{\ell,m}(\theta, \phi)$ are spherical harmonic angular functions in real form. The radial dependence of the multipole deformation functions is given by

$$R_\ell = N r^n e^{-\zeta r}$$

where N is a normalization constant, and n and ζ can be chosen for each value of ℓ.

The use of a multipole refinement model offers several advantages. All of the x-ray data (and neutron data, if available) can be included in the refinement with proper weights. If the model itself is plotted, the experimental noise is effectively filtered out. Such a density map, given by

$$\Delta \rho(r) = \frac{1}{V} \sum_{H} (F_{c, \text{ multipole}} - F_{c, \text{ promolecule}}) e^{2\pi i H \cdot r}$$

is properly designated a "model" deformation density. The multipole parameters are defined in terms of coordinate systems local to each atom, making it easy to impose noncrystallographic symmetry constraints on the refinement. For example, the density in an aromatic ring is commonly constrained to have a mirror plane of symmetry in the plane of the ring. The validity of any constraints, and the success of the fit in general, can be obtained by plotting a residual density.

$$\rho(r) = \frac{1}{V} \sum_{H} (F_{\text{obs}} - F_{c, \text{ multipole}}) e^{2\pi i H \cdot r}$$

which should be featureless except for random experimental noise.

For acentric structures, the uncertainty in the phases of the observed x-ray structure factors leads to a lowering of peak heights and loss of resolution in the features of the experimental deformational density. With a multipole refinement model, however, the correct amplitude and phase of ΔF can be obtained, since the phases of both structure factors in the expression for the model density are known.

The multipole model may also be plotted directly without thermal motion parameters, yielding a static model density. Convolution of the thermal motion with the molecular electron density greatly limits the information present in the experimental data on sharp features of the valence density. The reliability of static electron distributions is therefore still open to debate. Our preference is to stay closer to the experiment by reporting dynamic densities in most cases.

Electrostatic Potentials

The molecular electrostatic potential is a property of the charge distribution which has been used by theoretical chemists with some success as a predictor of chemical reactivity (Politzer and Truhlar, 1981, and references therein). The molecular electrostatic potential gives the net potential experienced by a point positive charge at position r due to electrons and nuclei of the system,

$$V(r) = \sum_{A} \frac{Z_A}{|R_A - r|} - \int \frac{\rho(r') \, dr'}{|r' - r|}$$

where Z_A is the charge on nucleus A located at R_A. In a crystal, the electrostatic potential is periodic, and as with the electron density, can be calculated using a Fourier series summation (Spackman and Stewart, 1981),

$$V(r) = \frac{1}{4\pi V} \sum_{H} F_H' e^{2\pi i H \cdot r} / (\sin \theta / \lambda)^2$$

where F_H' is the charge (nuclei + electrons) structure facator given by

$$F_H' = \sum_i (Z_i - f_i)\, e^{-2\pi^2 (U_{ij}h_ih_j)}\, e^{-2\pi i\, H \cdot r}$$

and f_i is the pseudoatom scattering factor. Note that this expression for the electrostatic potential is of the same form as the expression for the charge density except for the factors of $4\pi^{-1}$ and $(\sin\theta/\lambda)^{-2}$. Thus exisiting programs can easily be modified to calculate the electronstatic potential.

The series for $V(r)$ converges more quickly than the series for $\rho(r)$ because of the $(\sin\theta/\lambda)^{-2}$ term. Unfortunately, calculation of the potential of the static density still suffers from series termination errors at the resolution limit of most experiments. To avoid series termination effects, we have chosen to calculate the electrostatic potential of the dynamic density, where the term containing the temperature factor U_{ij} further dampens the high order terms, and to extend the calculation to terms beyond the resolution of the experiment.

The electrostatic potential of a molecule in a crystal, as calculated above, may differ significantly from the potential of an isolated molecule. To approximate the molecular electrostatic potential of an isolated molecule, we include the provision for computing the potential in an expanded cell. This has the effect of increasing the separation between molecules.

EXPERIMENTNAL RESULTS

Chemical Carcinogens

We have determined experimental electron density distributions of several chemical carcinogens with the goal of using the experimental information to help elucidate the role of electronic factors in directing the steps of metabolic activation and reaction with cellular targets.

Polyaromatic hydrocarbons are important environmental carcinogens which undergo metabolic activation by mixed function oxidases (Harvey, 1981). In several cases, the ultimate carcinogenic metabolic has been identified as a diol epoxide. As a model for these metabolites, we have determined the experimental deformation density of anti-3,4-dihydroxy-1,2,3,4-tetrahydronaphthalene-1,2-oxide (NDE) a diol epoxide derivative of naphthalene using high-resolution x-ray intensity measurements at 100K. Experimental details are published elsewhere (Klein and Stevens, 1986). Although not carcinogenic itself, NDE has been identified as an in vivo metabolite of naphthalene.

The model deformation density of NDE in the plane of the aromatic ring is plotted in Fig. 1. Since the density in the saturated portion of the molecule is non-planar, the map has been constructed as a composite of planes passing through the bonds.

Of greatest interest is the electron distribution of the epoxide ring (Fig. 2), since this is the functional group believed responsible for the reactivity of the diol epoxides of carcinogenic hydrocarbons. The bond density peaks lie outside the straight lines connecting the atom centers indicating "bent" bonding as expected for a three-membered ring. Large peaks observed directly above and below the oxygen atom indicating largely unhydbridized s and p lone-pairs on the oxygen. The C1-C2 bond peak is polarized toward C2 indicating that

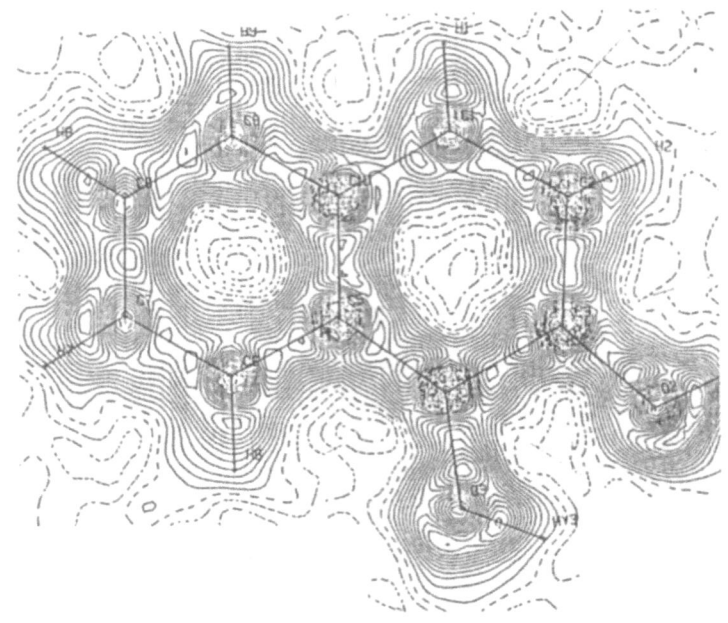

FIG 1. Composite plot of the model deformation density of
anti-3,4-dihydroxy-1,2,3,4-tetrahydronaphthalene-1,2-oxide
in the plane of the aromatic ring. Contours are plotted at
0.05 e/Å^3 intervals with zero and negative contours dashed.

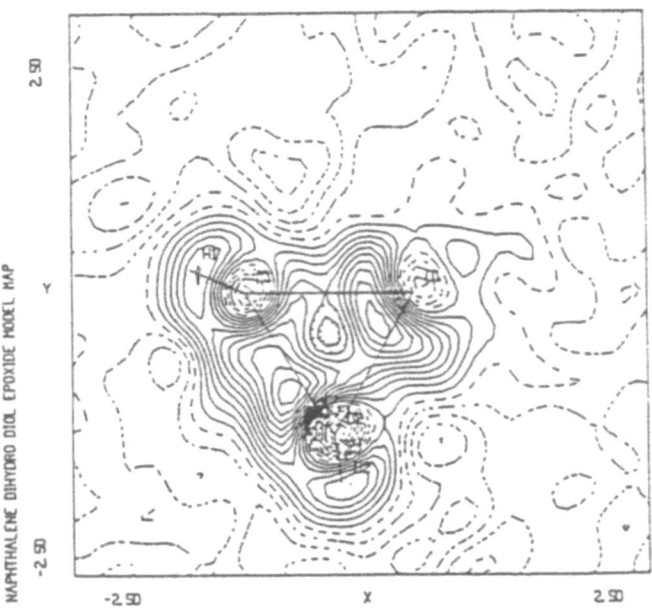

FIG 2. Model deformation density of NDE in the plane of the
epoxide ring. Contours are ploted as in Fig. 1. Tick marks
on the contour lines indicate a depression in the center of
the ring.

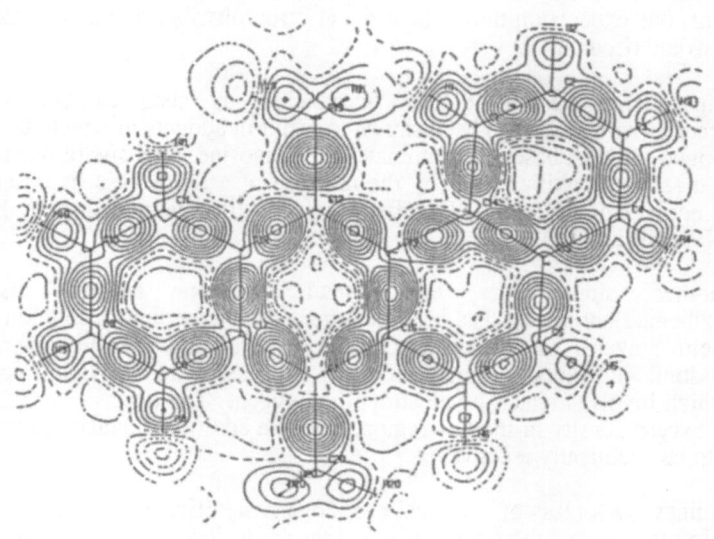

FIG 3. Composite plot of the model deformation density of
7,12-dimethylbenz(a)anthracene in the molecular plane.
Contours are plotted as in Fig. 1.

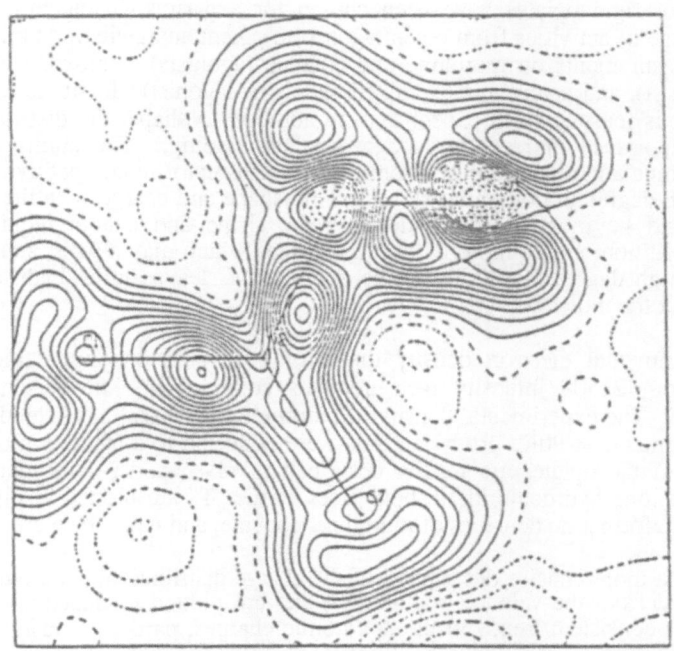

FIG 4. Model deformation density of N-nitrosodiphenylamine
in the plane of the N-nitroso group. Contours are ploted as
in Fig. 1.

C2 is more electronegative than C1. This is confirmed by net atomic charges of C1 = +0.37 and C2 = -1.38 calculated from the multipole valence populations. Aside from steric considerations, one expects on the basis of the electron distribution that chemical attack by a nucleophile would occur at C1.

Reaction of epoxide metabolites of carcinogens such as benzo(a)pyrene and benz(a)anthracene with DNA has been shown to result in adducts in which the carcinogens is covalently bound to DNA bases, predominately at guanosine. The site of reaction, which can be considered as nucleophilic attack by the nitrogen of a DNA base, is observed to be the carbon atom equivalent to C1 in NDE (Jeffrey, Jennete, Blobstein, Weinstein, Bland, Harvey, Kasci, and Nakanishi, 1976).

In another study, the experimental electron density distribution of 7,12-dimethylbenz(a)anthracene has been determined from medium resolution data collected at 180K (Klein, Stevens, Zacharias, and Glusker, 1987). Studies of the parent hydrocarbon should give some indication of preferred sites of reaction in the first step of metabolic activation which involves attack by electrophilic oxygen. The experimental density (Fig. 3) shows some excess density in the bay region, but little additional density in the K-region, an area known to be chemically reactive.

Nitrosamines are a class of chemical carcinogens significant because of their occurence in the diet. In many cases, the require metabolites activation. Because the intermediates are generally unstable and have not been isolated, relatively little is known about the mechanistic details of nitrosamine carcinogenesis. The experimental electron density distribution of N-nitrosodiphenylamine has been determined from x-ray data collected at 100K (Foss, 1986). In the deformation density calculated in the plane of the nitroso group (Fig. 4), large lone-pair peaks are observed on the oxygen and nitroso nitrogen atoms. The N-N and N-O bond peaks fall off the bond axes, apparently a result of repulsion by the large lone-pair densities.

Opiates

A series of rigid opiates have been chosen for experimental electron density studies spanning a range of activities from potent agonist to potent antagonist. At this point we have completed measurements on morphine and codeine (agonists), naloxone hydrochloride (a potent antagonist), and nalorphine (a mixed agonist-antagonist). In studies of drug-receptor interactions, it is reasonable to expect that, in addition to shape, the distribution of charge will be of importance in receptor recognition and binding. The ability of naloxone to displace morphine and other agonists from the opiate receptor in competitive binding studies suggests a high degree of complimentarity in both shape and charge with the receptor. Thus naloxone should be most useful in defining the shape and charge requirements of the receptor, In addition, unlike more flexible molecules which may also be active at the opiate receptor, the high degree of rigidity in the opiate drugs leaves little doubt as to the active conformation of the molecule.

The experimental electron density distribution of naloxone hydrochloride has been determined from 21509 intensity measurements collected at 90K (Klein, Majeste, and Stevens, 1987). The experimental density of morphine (free base) has been determined from 16826 measurements at 90K. Both data sets were collected to a resolution of 1.08Å^{-1}. The conformation of the opiates are readily described as "T-shaped". The model deformation density of naloxone hydrochloride in both faces of the "T" are plotted in Fig. 5 and Fig. 6. The observed deformation densities of morphine, codeine, and nalorphine are very similar.

To obtain a more concise description of the charge distribution, net atomic charges have been calculated from the valence populations, and are plotted for naloxone in Fig. 7. It is difficult to see consistent trends in the net atomic charges, partially a result of the arbitrary nature of attempting to partition a continuous electron distribution into discrete fragments. To get a less localized description of the charge on an atom and the net atomic charges of all

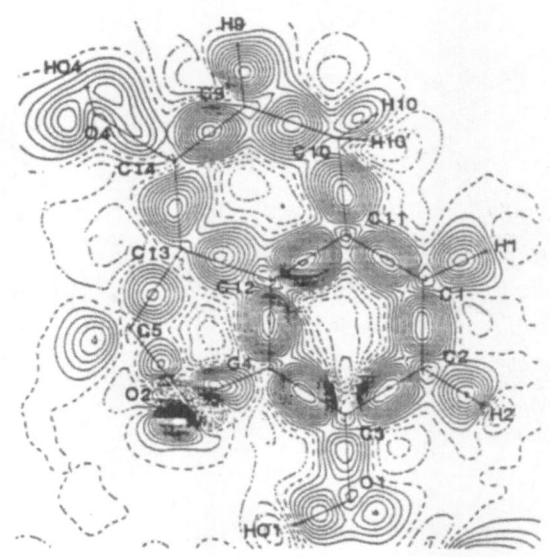

FIG. 5. Model deformation density distribution in
naloxone hydrochloride dihydrate at 90K plotted in the
A, B and E rings. Contours are plotted as in Fig. 1.

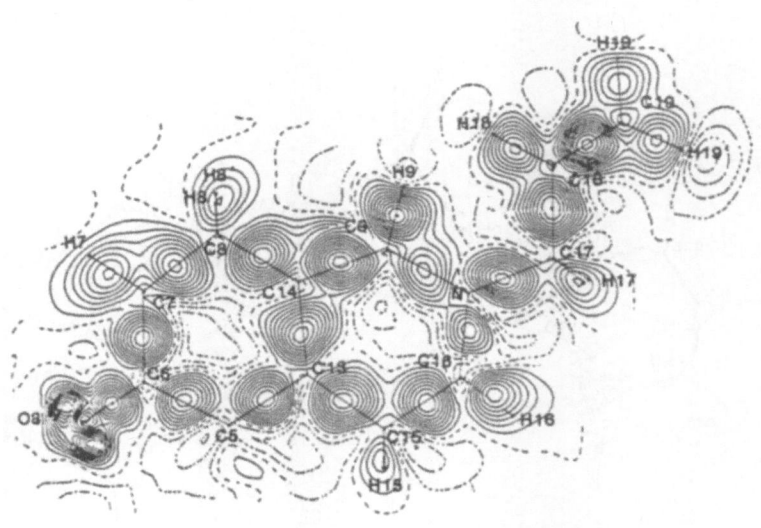

FIG 6. Model deformation density of naloxone hydrochloride
in the C and D rings. The map represents a composite of 9
planes. Contours plotted as in Fig. 1.

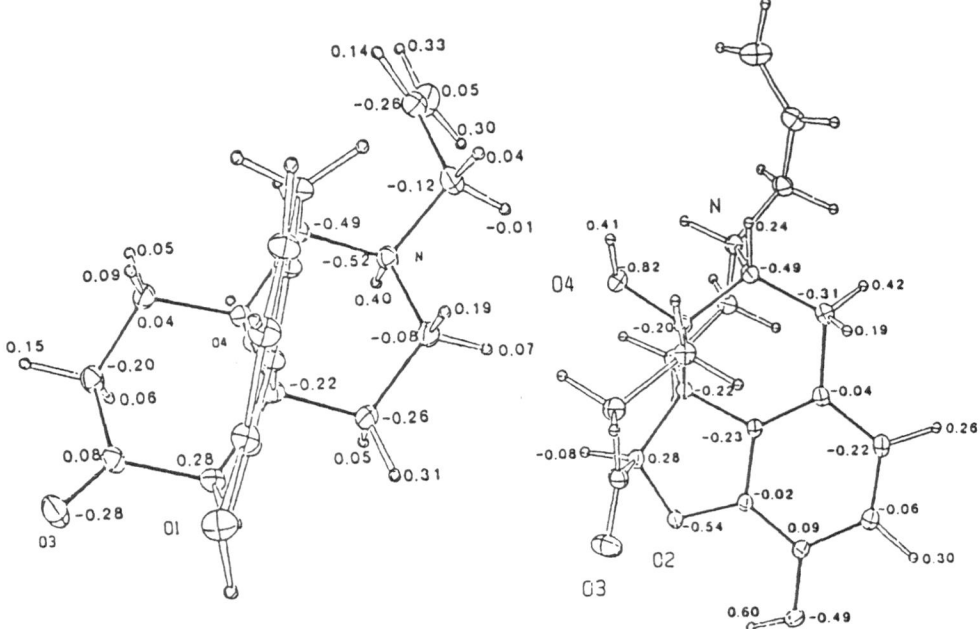

FIG. 7. Distribution of net atomic charges in naloxone hydrochloride dihydrate as determined by multipole refinement of the x-ray data.

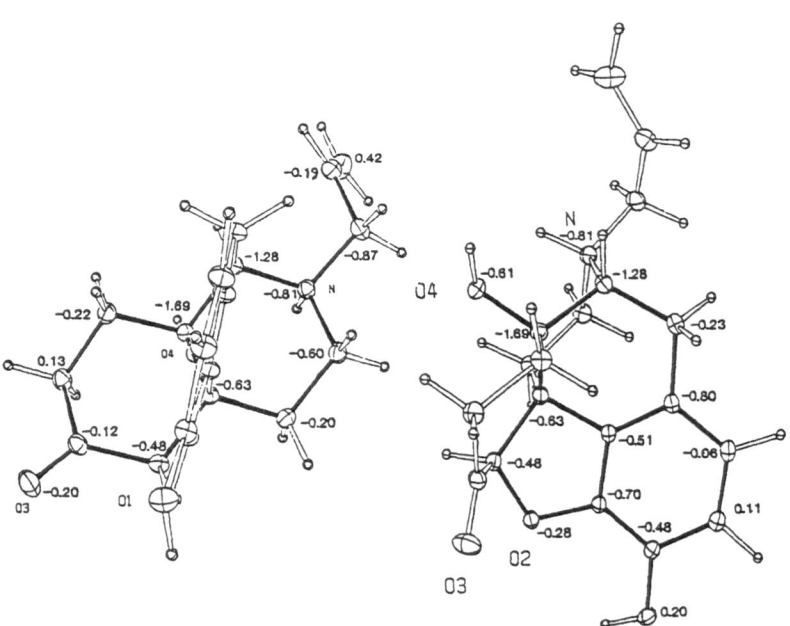

FIG 8. Distribution of "local area charges" in naloxone hydrochloride. Local area charges are the sum of the net atomic charge on an atom and the charges of its neighbors.

other atoms bonded to it as the "local area charge". Local area charges for naloxone are shown in Fig. 8. The local area charges show a gradual variation from positive charges at the edge of the aromatic ring to negative charges at the nitrogen and O_{14}. Most surprising is the negative charge observed on the nitrogen, despite the fact that it is protonated and formally carries a positive charge. The local area charge distribution of morphine is similar, with the unprotonated nitrogen being even more negative.

The similarity of the charge densities among these compounds, which bind to a common receptor but produce widely different pharmocological responses, suggests that the charge distribution is most important in determining receptor recognition and binding. This is reasonable in view of the fact that the long-range interactions must be electrostatic in nature. Steric fit and specific interactions such as hydrogen bonding may be expected to be of equal or greater importance at shorter distances, and may be responsible for the differences in drug activity.

Neurotransmitters

The experimental electron density distribution of an important neurotransmitter, dopamine, has been determined with the goal of using the information to help elucidate the electronic charge density and its role in the biological processes.   Dopamine, a decarboxylation product of DOPA, is a precursor for the hormones norepinephrine and epinephrine and has its own biological function. Excess dopamine in the central nervous system results in depression, schizophrenia, and paranoia, while insufficient levels result in the rigidity and tremors associated with Parkinsonism.

A high-resolution data set was collected for dopamine hydrochloride at 90K (Klein, 1990).   A composite model electron density distribution (EDD) map for a molecule of dopamine HCl is shown in Fig. 9. Both hydroxyl oxygen atoms show peaks of electron density corresponding to nonbonded lone pairs. The lone pairs have merged together into a single elongated peak with maximum in the plane plotted. These hydroxyl groups are the most electron-rich regions of the protonated dopamine molecule, which is consistent with the negative net atomic charges on the oxygen atoms of -0.68e and -0.66e for O_1 and O_2, respectively.

γ-Aminobutyric acid (GABA) is also a neurotransmitter in the mammalian central nervous system. The experimental electron density distribution for the zwitterionic form of GABA was determined by Craven and Weber (1983) at 122K.

Phenothiazines

Phenothiazine and many of its derivatives belong to a class of compounds that are generally used to treat the symptoms of psychosis. Many of these compounds have been used as tranquilizers, antidepressants, and antischizophrenics. It has been suggested that many of these antipsychotic drugs function as competitive antagonists of dopamine and that the phenothiazines block postsynaptic dopamine receptors in the brain and peripheral nervous system.

A study of the experimental electron density distribution using medium resolution data for chlorpromazine hydrochloride (Klein and Stevens, 1990), a potent transquilizer, and levomepromazine sulfoxide (Klein and Wilson, 1990), the sulfoxide metabolite of levomepromazine (which is used in Europe as a neuroleptic with pronounced sedative properties) shows interesting trends of charge throughout the molecules.

As in the opiates, the net atomic charges do not give consistent information concerning areas of the molecules that are electron rich or electron deficient. However, the "local area charges" for both chlorpromazine hydrochloride and levomepromazine sulfoxide show that the substituted aromatic rings are more negative than the unsubstituted rings (Fig. 10). Additionally, the charge in the region around the amino nitrogen atoms appears to become more negative despite the protonation of the nitrogen in chlorpromazine hydrochloride.

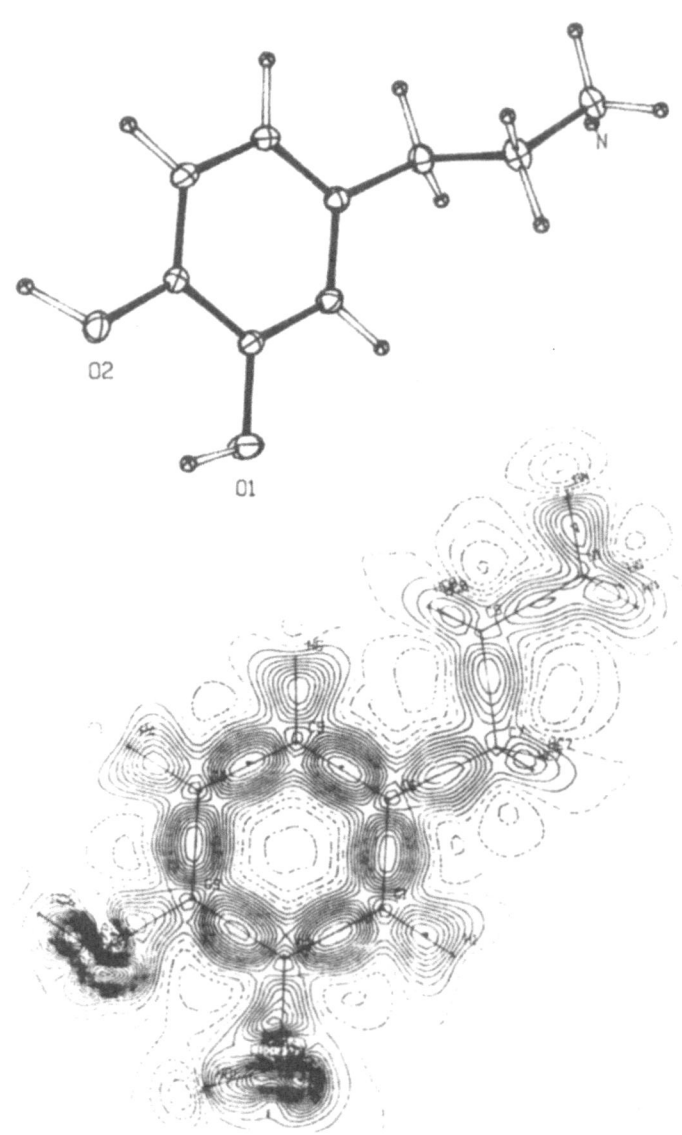

FIG. 9. Molecular structure of the neurotransmitter dopamine
hydrochloride at 90K and a contour map of the model deformation
density of dopamine plotted in planes passing through all
nonhydrogen atom positions. Contours are plotted as in Fig. 1.

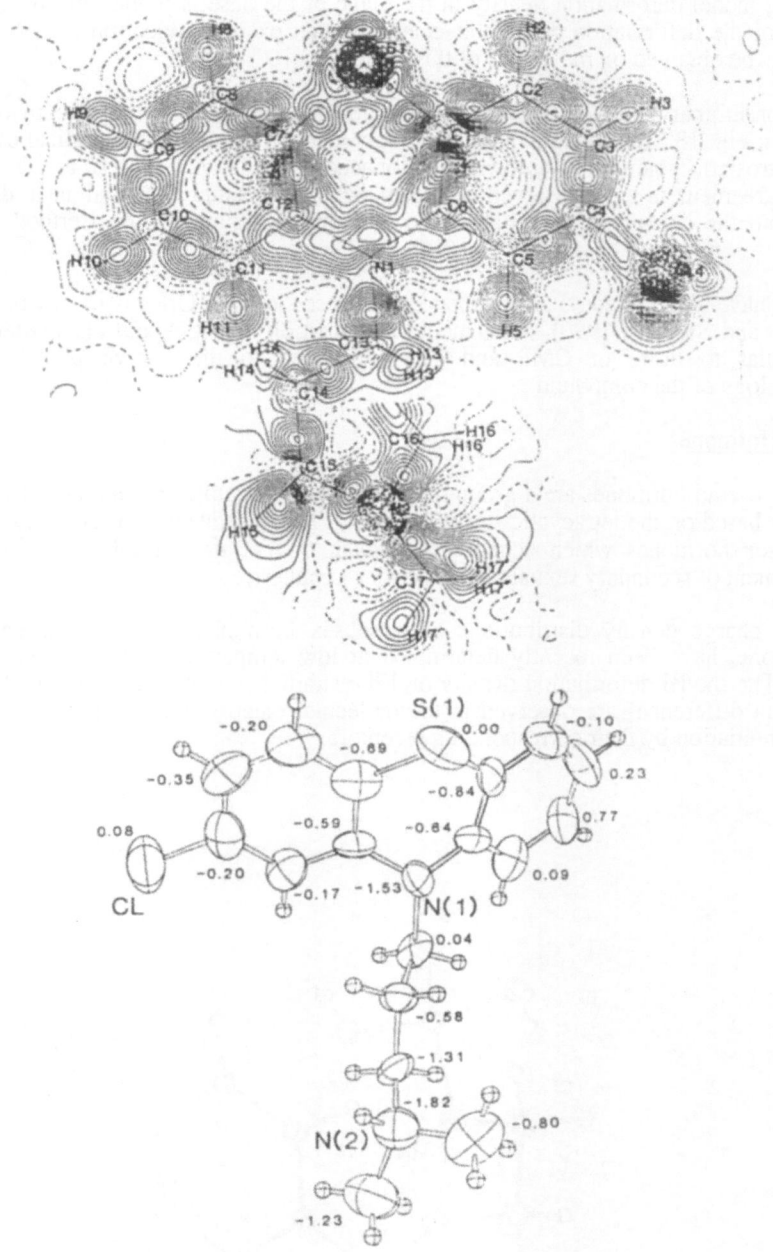

FIG. 10. Contour map of the model deformation density measured
in chlorpromazine hydrochloride at 90K and a plot of the molecular
structure including "local area charges." The structure contains
two independent molecules in the asymmetric unit. Contours are
plotted as in Fig. 1.

Nucleotides

We have recently reanalyzed x-ray data collected at 123K by Pearlman and Kim (1985) on 2'-deoxycytindine-5'-monophosphate. A plot of the structure is given in Fig. 11 and the resulting model deformation density in the plane of the base is plotted in Fig. 12. From the quality of the deformation density, it is evident that excellent electron density distributions may now be obtained on molecules of this size.

A preliminary plot of the electrostatic potential in the plane of the cytosine base in plotted in Fig. 13. Reasonable qualitative agreement is found with theoretical calculations of the electrostatic potential of cytosine by Bonaccorsi, Pullman, Scrocco, and Tomasi (1972). Good agreement is also observed between the experimental potential near the phosphate group and the calculated potential of the dimethylphosphate ion by Berthod and Pullman (1975).

We have also collected an independent data set in New Orleans on thymidine at 100K (Nai-Jue and Stevens, 1990). The model deformation densities and electrostatic potentials are similar to those of CMP and can be related to the known photochemistry and photobiology of the compound.

Steroid Hormones

The steroid hormones are a group of physiologically active compounds closely related in structure based on the tetracyclic steriod nucleus. The androgens and estrogens are male and female sex hormones which control the physiology of the reproductive system and the development of secondary sexual characteristics at puberty.

The charge density distributions of three sex hormones, estradiol, progesterone, and testosterone, have been recently determined at low temperature (Wu, Stevens, and Klein, 1990). The model deformation density of 17β-estradiol is plotted in Fig. 14. In this series, significant differences are observed in the molecular electrostatic potentials, as is necessary for differentiation by their corresponding receptors.

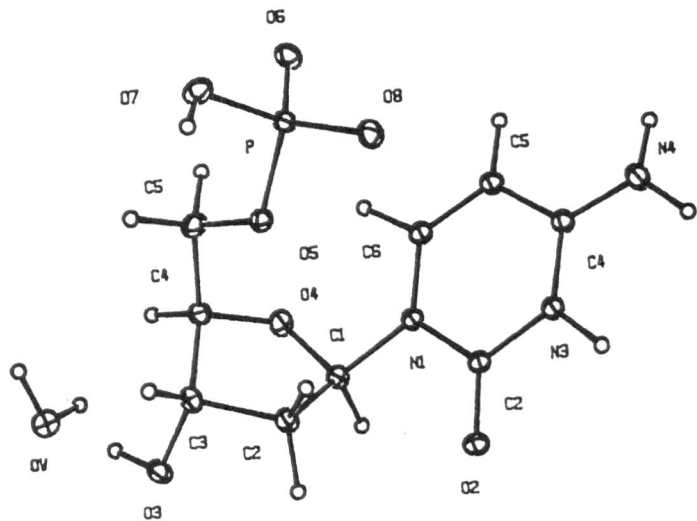

FIG. 11. ORTEP plot of the structure of 2'-deoxy-cytidine-5'-monophosphate at 123K. Ellipsoids are plotted at the 50% probability level.

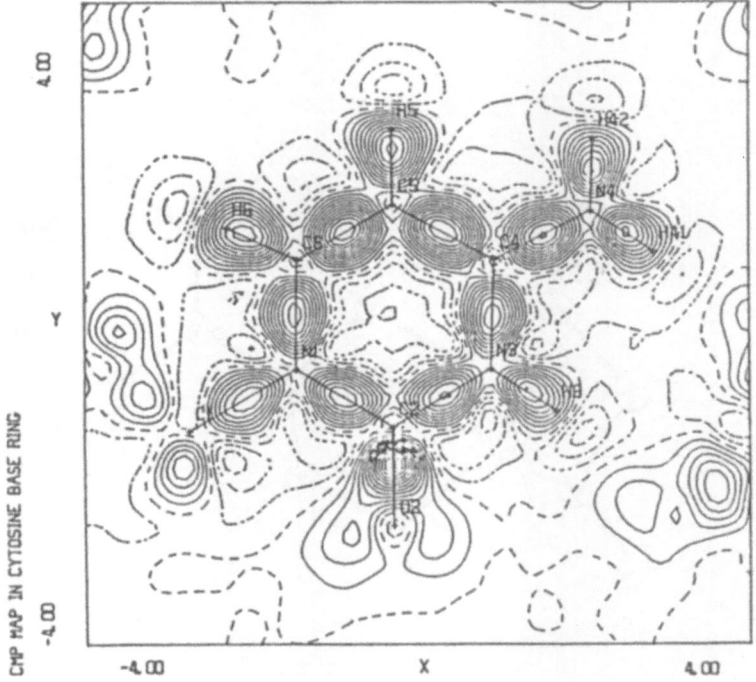

FIG. 12. Model deformation density of 2'-deoxycytidine-5'-monophosphate at 123K in the plane of the cytosine ring. Contours plotted as in Fig. 1.

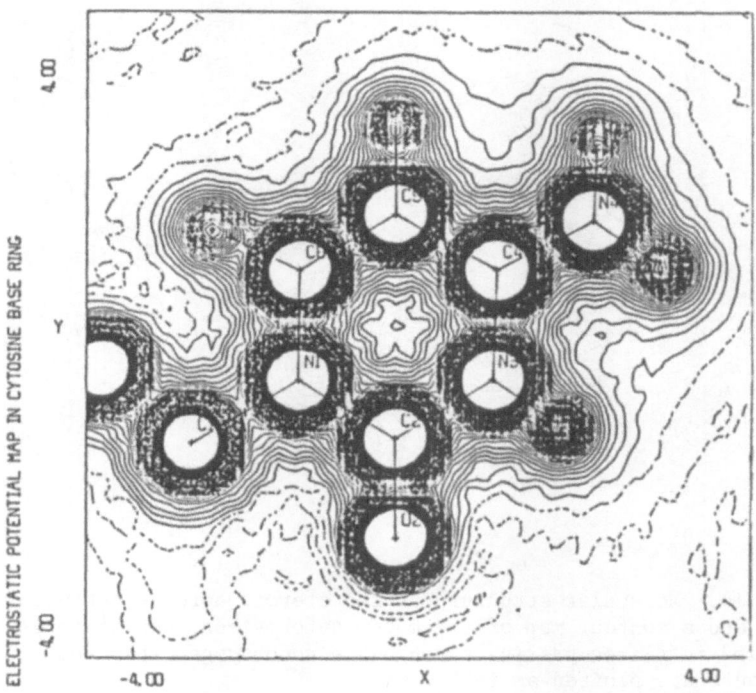

FIG. 13. Experimental electrostatic potential of 2'-deoxy-cytidine-5'-monophosphate at 123K. Contours are plotted at 0.12 e/Å intervals (40 kcal/mole).

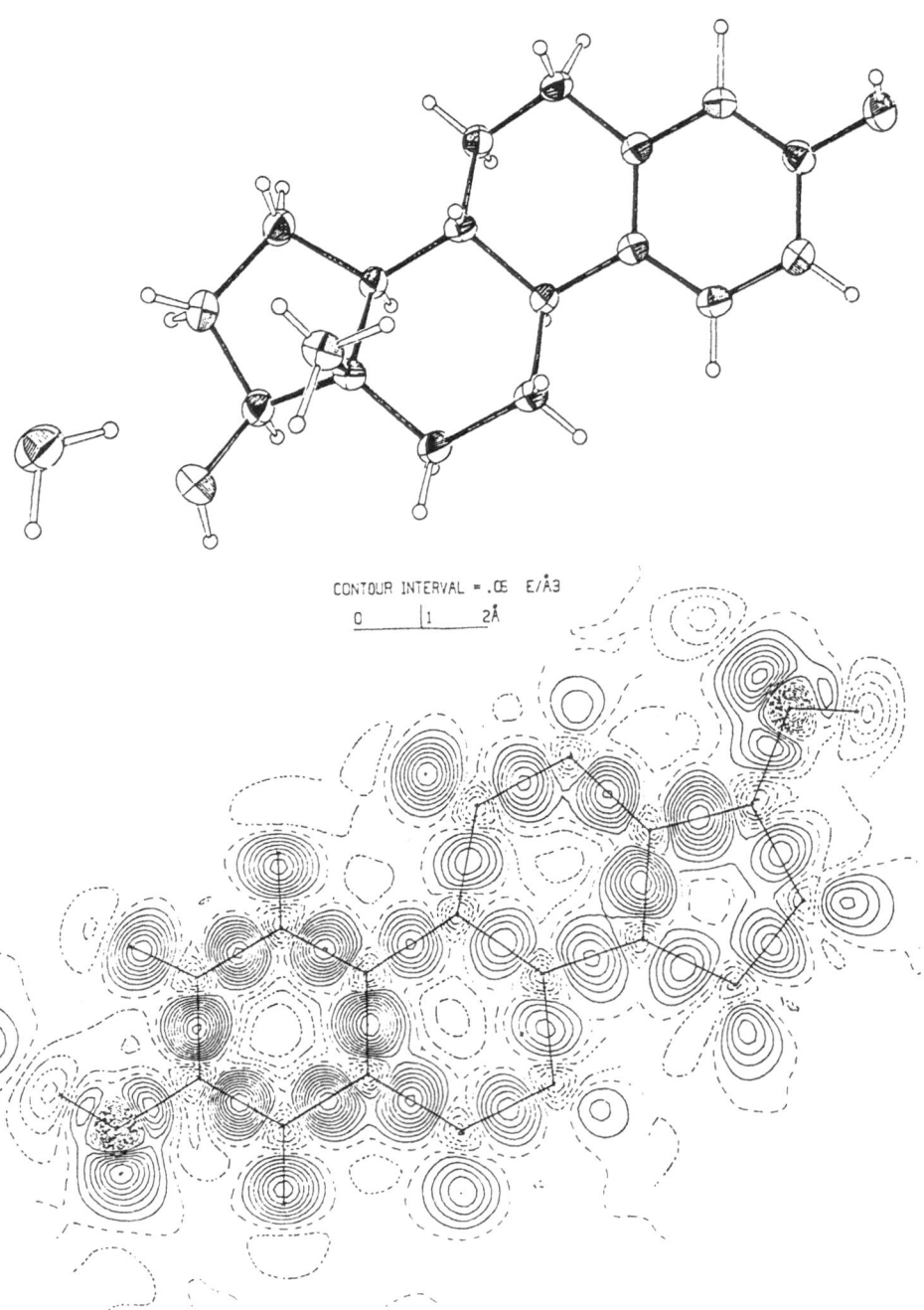

CONTOUR INTERVAL = .05 E/Å3

0 | 1   2Å

FIG. 14. Molecular structure of the steroid hormone estradiol at
105K and a contour map of the model deformation density in estradiol
plotted in planes passing through the nonhydrogen atom positions.
Contours are plotted as in Fig. 1.

SUMMARY

The examples provided and other recent work show that experimental electron density distributions may be obtained without unusual difficulty on moderately large molecules of biological interest. The experimental molecular electrostatic potential also appears to be a property well determined by the x-ray data, although experience with this property is much more limited. Essential to the utilization of such high-resolution x-ray data, however, is the continued development of methods for obtaining predictions of chemical reactivity from the molecular charge distribution.

ACKNOWLEDGEMENTS

Support of this research by the National Institutes of Health (NS-19789 and NS-24828 to E.D. Stevens and RR-08008 to C.L. Klein) is greatly ackowledged. Also a grant from the Petroleum Research Fund sponsored by the American Chemical Society (to C.L. Klein) is greatly acknowledged. We also wish to thank the many people who have collaborated on these projects including Professors Yong-ji Li, Richard J. Majeste, Sung-Huo Kim, Arun K. Pant, and Zhu Nai-jue and Drs. L.I. Hannick (Foss) and J.-C. Wu. Special thanks is due to Mrs. Frances Caldwell for assistance in preparing this manuscript.

REFERENCES

ANGERMUND, K., CLAUS, H.K., GOODARD, R. and KRÜGER, C. (1985). Angewandte Chemie, International Edition in English, **24**, 237-247.

BERTHOLD, H. and PULLMAN, A. (1975). Chemical Physics Letters, **32**, 233-235.

BONACCORSI, R. PULLMAN, A., SROCCO, E., and TOMASI, J. (1972). Theoret. Chim. Acta., **24**, 51-60.

CRAVEN, B.M. and WEBER, H.-P. (1983). Acta Crystallographica, Sect. B. **39**, 743-748.

CLINTON, W.L. and MASSA, L. (1972). Physical Review Letters, **29**, 1363-1366.

COPPENS, P. (1989). Journal of Physical Chemistry, **93**, 7979-7984.

FOSS, L.I. (1986). Ph.D. Thesis, University of New Orleans, New Orleans, Louisiana.

HARVEY, R.G. (1981). Accounts of Chemical Research, **14**, 218-226.

JEFFREY, A.M., JENNETTE, K.W., BLOBTEIN, S.H., WEINSTEIN, I.B., BELAND, P., HARVEY, R.G. KASAI, H., MIURA, I. AND NAKANISHI, K. (1976). Journal of the American Chemical Society, **98**, 5714-5715.

KLEIN, C.L. and STEVENS, E.D. (1990). Unpublished results.

KLEIN, C.L., and STEVENS, E.D. (1988). Experimental Measurements of Electron Density Distributions and Electrostatic Potentials. In Structure and Reactivity, (ed. J.F. Liebman and A. Greenberg). pp. 25-64. VCH Publishers, New York.

KLEIN, C.L. and STEVENS, E.D. (1986). Acta Crystallographic, Sect. B., **44**, 50-55.

KLEIN, C.L., STEVENS, E.D., ZACHARIAS, D.E. and GLUSKER, J.P. (1987). Carcinogensis, **8**, 5-18.

KLEIN, C.L. (1990). Unpublished results.

KLEIN, C.L., and WILSON, (1990). Unpublished results.

KLEIN, C.L., MAJESTE, R.J. and STEVENS, E.D. (1987). Journal of the American Chemical Society, **109**, 6675-6681.

NAI-JUE, Z. and STEVENS, E.D. (1990). Unpublished results.

PEARLMAN, D.A. and KIM, S.-H. (1985). Biopolymers, **24**, 327-357.

POLITZER, P. and TRUHLAR, D.G. (1981). Chemical Applications of Atomic and Molecular Electrostatic Potential. Plenum Press, New York.

SPACKMAN, M.A. and STEWART, R.F. (1981). Electrostatic Potentials in Crystals. In Chemical Applications of Atomic and Molecular Electrostatic Potentials (ed. P. Politzer and D.G. Truhlar), pp. 407-425. Plenum Press, New York.

WU, J.-C., STEVENS, E.D. and KLEIN, C.L. (1990). Unpublished results.

CHARGE DENSITY DISTRIBUTION OF DIMETHYL-1,3-AMINO-4-URACIL

F. Baert, A. Guelzim and V. Warin

Lab. Physique Fondamentale
U. d Sciences et Techniques de Lille
59655 Villeneuve d'Ascq, FRANCE

In the field of research aimed at providing new antitumor drugs we have worked on 5-acyl-6-amino-1,3-dimethyluracils. Among these compounds, 6-amino-5-cinnamoyl-1,3-dimethyluracil (NSC 290115) has revealed interesting antileukaemia activity. The aromatic groups are stabilized in an extended planar structure whose size fits with the base planes of the DNA.

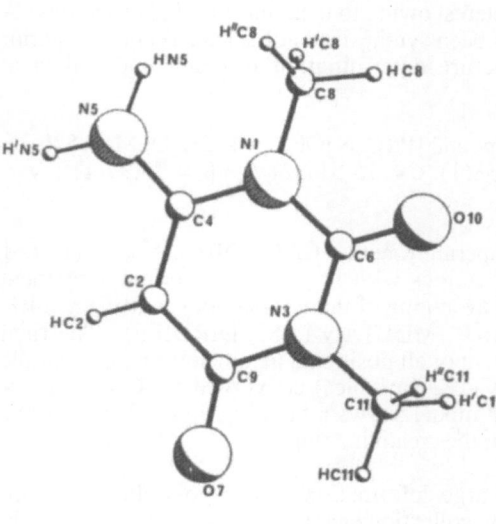

Fig. I

This suggests that the antitumor activity may be a consequence of an intercalative process, the affinity for the DNA being enhanced by the presence of a uracil moiety. The study of the charge density distribution of the title compound which is part of all the drug molecules of this family should provide information about the process of intercalation of the molecule inside the DNA.

Dimethyl-1,3-amino-4-uracil, (Fig. 1) $C_6H_9O_2N_3$, crystallizes in the space group Pnma with a = 14.800, b = 6.519 and c = 7.129 Å. The number of molecules in the cell is four, consequently molecules are lying in the mirror planes. The experimental electron density distribution has been determined from high resolution single crystal X-ray diffraction measurements at 120 K ($2° < \theta < 56°$, with MoK_α = 0.71069 Å).

10,981 intensities have been recorded giving 4626 independent reflections. The deformation density calculated with X-XHO formalism at a resolution of 0.90 Å^{-1} shows well-resolved peaks in the bonds and in the lone pair regions of the oxygen atoms. We observe a peak between O(10) and HC(8) which corresponds to an intramolecular hydrogen bond. Approximate net charges were obtained by least-squares refinement of the occupancy of the valence electron density function. Maps from multipole refinements using the Stewart formalism, modified by Coppens and Hansen, are compared to the experimental ones. The electrostatic potential around the molecule and atomic net charges deduced by direct integration of the experimental structure factors are presented.

STRUCTURE AND ELECTRON DENSITY OF TRANS-DICHLORO-BIS(CREATININE) PLATINUM(II) DI-HYDRATE

Ana Matos Beja and J. A. Carvalho Paixão

Centro FC1, INIC, Departamento de Física
Universidade de Coimbra, PORTUGAL

J. Martin Gil and M. Aragon Salgado

Escuela Técnica Superior de Ingenieros Industriales
Universidad de Valladolid, ESPAÑA

Platinum compounds have biological interest owing to their anti-tumoral properties. A series of creatinine platinum compounds has been synthesized at Escuela Técnica Superior de Ingenieros Industriales, and their structure determination is being undertaken at Departamento de Física, Coimbra.

We report the structure of the title compound [$Pt(C_4N_3OH_6)_2Cl_2 \cdot 2H_2O$, $M_r = 526.25$, monoclinic, $P2_1/n$, a = 6.584(1), b = 7.538(1), c = 15.511(2) Å, β = 93.92(1)°, V = 768.0 Å^3, Z = 2].

X-ray data were collected at room temperature with a CAD4 diffractometer (13,661 reflections), yielding 2014 independent reflections with I > 3σ. The internal agreement factor after ψ-scan absorption correction and averaging of these reflections is $R(F^2)$ = 1.4%. The structure was solved using direct methods (MULTAN 11/82, DIRDIFF). The final R-factor after spherical least-squares refinement of all positional parameters and anisotropic thermal parameters of non-hdyrogen atoms (122 variables) converged to R = 1.7%. A difference Fourier map against the spherical model shows a building up of charge on the covalent bonds and on the lone pair regions of the creatinine ring.

Our results suggest that the study of charge deformation may be possible on organo-metallic compounds with heavy atoms, if data collection enables good statistics and mainly if the heavy atom occupies a special position. That is the case for the present compound, where Pt atoms are at (0,0,0), (1/2,1/2,1/2), so that all reflections with h+k+l = 2n+1 have no contribution from the Pt atoms that could mask small effects in the intensities of these reflections due to asphericities of the charge density.

A LOW-TEMPERATURE (23 K) STUDY OF L-ALANINE: TOPOLOGICAL

PROPERTIES OF EXPERIMENTAL AND THEORETICAL CHARGE DISTRIBUTIONS

R. Bianchi, R. Destro, C. Gatti and F. Merati

Centro di Studio delle Relazioni tra Struttura e Reattivita' Chimica
C.N.R., and
Dipartimento di Chimica Fisica ed Elettrochimica
Universita' degli Studi di Milano
via Golgi 19
I-20133 Milano, ITALY

The experimental charge density of L-alanine has been derived from single-crystal x-ray diffraction data collected at 23 K and interpreted with a multipole (pseudoatom) formalism. The topological properties of the experimental density are compared with those obtained from ab-initio computations of the alanine monomer at the geometry of the molecule in the crystal.

RHF computations with basis sets of increasing quality (STO-3G, 6-31G, 6-31G*, D95*, 6-31G**) have been performed and the effect of the inclusion of Coulomb correlation has been partly accounted for through GVB-PP/6-31G* (18 pairs) and frozen core SD-CI/6-31G computations on the monomer. The electron distribution of three dimers, two trimers and one quadrimer has also been computed, in order to compare the properties of the hydrogen bonds observed in the crystal and to partly mimic the crystal field effect.

Experimental charge densities display the same set of *bond critical points* (bcp) exhibited by theoretical densities for both intramolecular and intermolecular interactions. Although a qualitative agreement between experimental and ab-initio results is always observed, a good quantitative agreement is only recovered for C–C and hydrogen bonds, provided a good quality basis set (from 6-31G* on) is used in the ab-initio computations. The observed discrepancies mainly derive from small differences in the location of bcps. The shift in their position, which is determined by differences in the portrait of the charge density gradient, turns out to be greater, as expected, for ionic than for covalent bonds. Other factors which prevent a good quantitative agreement between experimental and theoretical densities are investigated and pointed out.

Topological properties of the charge density are demonstrated to be a sensitive tool to determine inadequacies in model, basis set of theoretical computations and limitations in accuracy, precision, multipole analysis of experimental data.

ELECTROSTATIC PROPERTIES AND TOPOLOGICAL ANALYSIS OF THE CHARGE DENSITY OF *SYN*-1,6:8,13-BISCARBONYL[14]ANNULENE DERIVED FROM X-RAY DIFFRACTION DATA AT 16 K

R. Bianchi, R. Destro and F. Merati

Dipartimento di Chimica Fisica ed Elettrochimica e
Centro del CNR per lo Studio delle Relazioni tra Struttura e
Reattivita' Chimica
Universita' degli Studi di Milano
via Golgi 19
I-20133 Milano, ITALY

Single crystal X-ray diffraction data of *syn*-1,6:8,13-biscarbonyl[14]annulene, collected at 16 K and interpreted with a multipole formalism[1], have been employed to yield very accurate estimates of the electrostatic properties for this compound.

A value of 10(1) D has been derived for the molecular dipole moment. The values of the electric field gradient at the oxygen nuclei yielded quadrupole coupling constants of 9.1(5) and 9.5(5) MHz, in excellent agreement with NQR measurements reported in the literature for carbonyl compounds.

A topological analysis of the experimental electron charge density distribution showed the expected set of *critical points* for CC, CO and CH bonds. In addition, an unprecedented bonding interaction has been found between the two carbonyl carbons, as indicated by the presence of the corresponding *bond critical point*. Accordingly, two five-membered rings, both of which involve the transannular OC-CO bond, have been found in addition to the two seven-membered rings involving the bridging bonds and the butadienyl fragments. The observed structure of the charge density distribution closely resembles that calculated[2,3] for 1,6-methane[10]annulene and provides a tentative explanation of the aromatic character of the title compound, as compared to the corresponding *anti*-isomer.

Finally, six *bond critical points* have been located in connection with short –CH····O intermolecular distances, which are usually taken as an indication of intermolecular interactions. The values of the Laplacian of charge density at these points are positive, with higher values corresponding to shorter interatomic distances.

[1] R. F. Stewart, *Acta Cryst.* **A32** (1976) 565.
[2] C. Gatti, M. Barzaghi and M. Simonetta, *J. Am. Chem. Soc.* **107**, 3185 (1985).
[3] M. Simonetta, C. Gatti and M. Barzaghi, *J. Mol. Struct. (Theochem)* **138**, 39 (1986).

ELECTRON DENSITY ANALYSIS OF THE ANTIMICROBIAL DRUG AND

RADIOSENSITIZER DIMETRIDAZOLE AT 105 K

Hendrik L. De Bondt [1], N.M. Blaton, O.M. Peeters and C.J. De Ranter [2]

Laboratorium voor Analytische Chemie en Medicinale Fysicochemie
Instituut voor Farmaceutische Wetenschappen, Katholieke Universiteit Leuven
E. Van Evenstraat 4, B-3000 Leuven, Belgium

I. Kjøller Larsen

Department of Chemistry BC
Royal Danish School of Pharmacy
Universitetsparken 2, DK-2100 Copenhagen, Denmark

INTRODUCTION

Nitroimidazoles are used in the treatment of anaerobic infections. Reduction of the nitro group is believed to induce their antibacterial and antiprotozoal activity (Edwards, 1981). Recently, several studies have been undertaken to establish their activity against hypoxic tumours and have opened exciting prospects for their use as radiosensitizing anticancer agents. π-electrons determine the electron affinity (Smeyers et al., 1981) which is highly correlated with radiosensitization efficiency.

Investigation of the total π-electron system in selected nitroimidazoles will hopefully reveal conditional features for antiprotozoal and radiosensitizing activity. Electron distributions are evaluated by calculation of difference densities in sections perpendicular to bond axes and in the molecular plane. Multipole refinements, necessary to evaluate the bond features quantitatively, are being performed and will be published later when comparable data on a second crystal structure will have been obtained.

ABSTRACT OF THE EXPERIMENTAL PART

An extensive X-ray dataset of dimetridazole, 1-2-dimethyl-5-nitroimidazole $C_5H_7N_3O_2$ (Fig. 1), was collected at 105.0(4) on a Nonius CAD4 single-crystal diffractometer at H.C.Ørsted Institute in Copenhagen. Full details are included in a comparitive study of the title compound at room temperature and at 105 K (De Bondt H. et al.). Space group Pmcn, $\lambda(MoK\alpha) = 0.71073$ Å, a = 6.389(1), b = 9.218(2), c = 10.784(2) Å. For all reflections, $\sin\theta/\lambda$ up to 1.23 Å^{-1}, peak profiles were stored. The profile analysis programs REFPK and BGLP of the data reduction package DREAM (Blessing, 1987) were used to distinguish background from peak intensity. The peak limits of weak reflections are centered around the calculated peak

---

[1] Research assistent of the National Fund for Scientific Research (Belgium).

[2] To whom correspondence should be addressed.

The Application of Charge Density Research to Chemistry and Drug Design
Edited by G. A. Jeffrey and J. F. Piniella, Plenum Press, New York, 1991

341

positions. These peak positions are obtained from a new orientation matrix, obtained from least-squares refinement using as orientation angle observations the ω values of the intensity-weighted centroids of the strongest 3421 reflections, along with the diffractometer values of φ and χ. Gaussian absorption corrections were applied using the program ABSORB (DeTitta, 1985). Internal consistency of the multiple observed reflections was 1.8%. All non-hydrogen atoms lie in the mirror plane m.

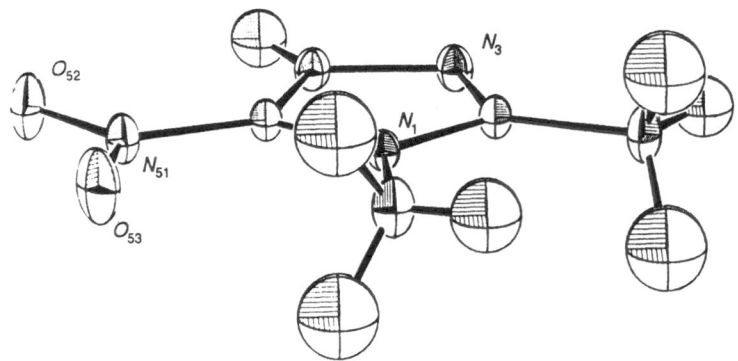

Fig. 1. ORTEP (Johnson 1976) view of the molecule with the atom numbering scheme. Non-hydrogen atoms are represented by thermal ellipsoids at 50% probability level.

REFINEMENTS

Full-matrix least-squares refinements on F were performed using the program package SDP installed on a PDP 11/73 computer and on the MicrovaxII of the H.C.Ørsted Institute in Copenhagen. Each observation was assigned a weight $1/\sigma^2(F_o)$ as determined by the data reduction program DREAM (Blessing, 1987). Unobserved reflections ($F_o < 4\sigma(F_o)$) were excluded from the refinements. No single reflections were omitted. Scattering factors for non-hydrogen atoms (Cromer & Waber, 1974) and corrections for anomalous scattering were taken from the *International Tables for X-ray Crystallography (1974)*. For hydrogen, the spherical scattering factors of Stewart, Davidson & Simpson (1965) were used.

Full Angle (FA) refinements

The first refinement was a FA refinement. Hydrogen atoms were refined isotropically. *NO* (Number of Observations) = 3099, *NV* (Number of Variables) = 79, *NO/NV* = 39.2. Convergence (maximal shift/error < 0.005) was obtained at $R(F) = 0.036$, $R_w(F) = 0.051$. $S = GOF = 1.964$ is considerably higher than 1.0. This may be attributed to the neglect of bonding density assuming the spherical atom model. The scale factor, which is defined as the coefficient by which the relative values of F_o are multiplied to bring them to the scale of F_c, refined to 0.6511(9). Max. $|kF_o-F_c|/\sigma = 19.4$. For 13 reflections $|kF_o-F_c|/\sigma$ exceeded 10.0. Isotropic extinction coefficient $g = 2.04(6).10^{-5}$ was defined as $F_{c, corrected} = F_{c, uncorrected}/(1 + gLp\ F^2_{c, uncorrected})$.

High Order (HO) refinements

As hydrogen atoms do not scatter very well, their parameters were fixed on the FA values. Several refinements with varying lower limits were performed to choose the right cutoff limit. Cutoff was considered at sin θ/λ of 0.7, 0.8, 0.9 and 1.0 Å^{-1}. The corresponding *NO/NV* values were 36.6, 30.4, 23.3 and 15.5; the R values were 0.029, 0.032, 0.036 and 0.043 respectively. Sin $\theta/\lambda = 0.8$ Å^{-1} seemed to be the optimal cutoff because deformation peak heights tended to be constant beyond this limit.

Table 1.
Atomic HO positional parameters with e.s.d.'s.
Hydrogen atomic parameters were obtained from FA refinements.
For non-hydrogen atoms the isotropic equivalent displacement
parameter, U_{eq}, is defined as: $U_{eq} = {}^{1}/_{3} \Sigma_i \Sigma_j U_{ij} a_i^* a_j^* \mathbf{a_i} \cdot \mathbf{a_j}$

Atom	x	y	z	U_{eq} (10^{-2} Å^2)
N1	0.250	-0.04977(4)	0.86900(3)	1.142(4)
C2	0.250	-0.16829(4)	0.94356(4)	1.232(4)
N3	0.250	-0.13281(4)	1.06372(4)	1.378(4)
C4	0.250	0.01475(4)	1.06782(4)	1.264(5)
C5	0.250	0.06792(4)	0.94856(3)	1.160(4)
C11	0.250	-0.05314(6)	0.73351(4)	1.893(7)
C21	0.250	-0.31935(5)	0.89505(6)	1.983(7)
N51	0.250	0.21500(4)	0.91152(4)	1.560(5)
O52	0.250	0.30687(4)	0.99401(6)	2.372(9)
O53	0.250	0.24394(6)	0.80006(5)	2.795(11)
H11A	0.250	-0.147(12)	0.7093(9)	4.4(2)
H11B	0.137(1)	-0.002(11)	0.7011(6)	6.0(2)
H21A	0.250	-0.382(14)	0.9585(8)	3.1(2)
H21B	0.361(1)	-0.335(10)	0.8395(6)	5.8(2)
H4	0.250	0.066(12)	1.1461(9)	2.7(2)

The isotropic extinction coefficient needed fixation due to excessive correlations. Refinement of this parameter caused an increase from $2.04(6).10^{-5}$ to $3.0(4).10^{-4}$ which was responsible for a decrease of the scale factor from 0.6742(7) to 0.6666(7). Therefore, the extinction coefficient obtained from the FA data set was held constant in the HO refinements. As expected, the exclusion of the isotropic extinction corrections did not influence any HO parameter significantly.

Convergence (maximal shift/error < 0.005) was obtained at $R(F) = 0.032$, $R_w(F) = 0.030$ and $S = GOF = 0.828$ for $NO = 1854$, $NV = 61$, $NO/NV = 30.4$. The S value is closer to 1, reflecting that the spherical atom model fits better the HO data with mainly core scattering. Max $| kF_o - F_c | / \sigma = 4.6$. Six reflections had $| kF_o - F_c | / \sigma > 3.0$. The highest correlation coefficients was observed between k and U and did not exceed 0.527. Consequently, the scale factor could be refined easily and did not need to be fixed on its FA value.

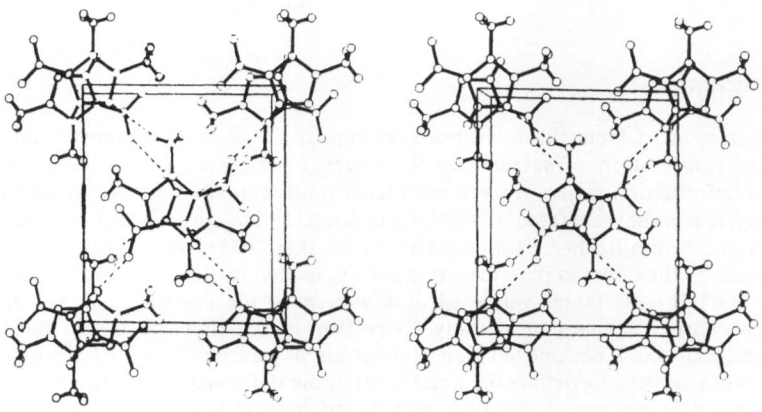

Fig. 2. Molecular packing diagram showing the hydrogen bonds. View along a axis. PLUTO (Motherwell *et. al.*,1978)

DISCUSSION

HO positional parameters (Table 1) are not influenced by the deformation density. The N-O bonds and the N3-C4 bond (Table 2) are significantly shorter in the HO results as FA parameters of the nitro oxygens and of N3 include a slight movement towards their lone pairs. Since there is no aspherical shift, the C2-N3-C4 angle (Table 2) is about 5σ more obtuse. The difference in HO bond length of the two N-O bonds is not significant. Table 3 shows a slight systematic decrease of the HO-atomic vibrational parameters in comparison with the FA ones, the greatest differences being about 8σ. This effect is attributed to enlargement of the FA vibrational parameters due to the bonding density, only observed when spherical atom scattering factors are used and low order reflection are enclosed *i.e.* only for FA refinements (Hirshfeld, 1976). This decrease of the thermal parameters causes a marked increase 0.6511(9) to 0.6742(7) of the scale factor in the HO-refinement. Fig. 2 reveals the packing forces N3(-δ)... N51'(+δ).

Table 2. Bond features: distances (Å), deformation peak heights (e Å^{-3}) and bond angles (°). Numbers in parentheses are estimated standard deviations in the least significant digit(s).
i = 1/2-X, Y, Z

Bond	Distance	Peak height		Angle
N1 -C2	1.3566(6)	0.58	C11 -N1 -C2	125.14(4)
C2 -N3	1.3365(7)	0.53	C11 -N1 -C5	129.57(4)
N3 -C4	1.3609(6)	0.22	C2 -N1 -C5	105.30(3)
C4 -C5	1.3764(6)	0.69	N1 -C11 -H11A	108.07(5)
C5 -N1	1.3831(6)	0.55	N1 -C11 -H11B	111.5(4)
N1 -C11	1.4615(6)	0.43	H11A-C11 -H11B	112.1(6)
C2 -C21	1.4876(7)	0.50	H11B-C11 -H11B$_i$	101.7(5)
C4 -H4	0.968(9)	0.47	N1 -C2 -C21	123.05(5)
C5 -N51	1.4134(6)	0.37	N1 -C2 -N3	112.19(4)
N51-O52	1.2283(9)	0.42	C21 -C2 -N3	124.76(5)
N51-O53	1.2312(9)	0.49	C2 -C21 -H21A	109.6(6)
C11-H11A	0.901(12)		C2 -C21 -H21B	111.3(6)
C11-H11B	0.928(8)		H21B-C21 -H21A	113.0(6)
C21-H21A	0.895(9)		H21B-C21 -H21B$_i$	98.1(7)
C21-H21B	0.938(7)		C2 -N3 -C4	106.03(5)
N3 (Lone Pair)		0.63	N3 -C4 -C5	109.00(5)
O (L.P. Aver.)		0.42	N3 -C4 -H4	121.21(5)
			C5 -C4 -H4	129.80(4)
			N1 -C5 -C4	107.48(4)
			N1 -C5 -N51	125.24(4)
			C4 -C5 -N51	127.28(4)
			C5 -N51 -O52	117.17(5)
			C5 -N51 -O53	118.93(5)
			O52 -N51 -O53	123.89(5)

Difference densities

Because, apart from the hydrogen parameters, HO X-ray refinements do not differ significantly from neutron refinements, X-X difference density maps resemble fairly well those of the X-N deformation maps . $\Delta\rho$ was calculated *via* difference synthesis using the F_c from HO-parameters and the scale factor (0.6641(6)) obtained by least-squares refinement on the full X-ray data set with all other parameters fixed on their HO values. This scale factor was preferred because the integrated deformation density tended to differ too much from zero when the FA or the HO scale factor was used. If k was set to the FA value, 0.6511(9), the total deformation charge became considerably lower than zero. However, when the HO value, 0.6742(7) was assigned, it became too high. Reflections beyond $\sin\theta/\lambda = 1.1$ Å^{-1} and unobserved reflections were omitted to reduce the noise level in the deformation density maps (Fig. 3-6). The difference densities were calculated from the differences between kF_o and F_c of the 2577 remaining reflections. Cutoff at $\sin\theta/\lambda = 1.1$ Å^{-1} in the difference Fourier synthesis is justified as $\Delta F / \sigma(\Delta F)$ reaches very low values.

Table 3. Thermal parameters ($\mathring{A}^2$) of non-hydrogen atoms.
The form of the anisotropic displacement parameter is:
$T = -2\pi^2 \Sigma_i \Sigma_j U_{ij} h_i h_j a_i^* a_j^*$
$U(1,2) = U(1,3) = 0.0$.
First line : FA-refinement. Second line : HO-refinement.
$\Delta/\sigma = (U_{FA}-U_{HO}) /(\sigma^2(U_{FA}) + \sigma^2(U_{HO}))^{1/2}$.

Atom	U(1,1)	Δ/σ	U(2,2)	Δ/σ	U(3,3)	Δ/σ	U(2,3)	Δ/σ
N1	0.0144(1)	5.3	0.0106(1)	2.4	0.0109(1)	5.8	-0.0006(1)	-2.2
N1	0.01372(8)		0.01032(6)		0.01022(6)		-0.00034(6)	
C2	0.0149(2)	0.0	0.0103(1)	7.1	0.0131(2)	2.1	-0.0000(1)	2.0
C2	0.0149(1)		0.00943(7)		0.01264(8)		-0.00025(7)	
N3	0.0187(2)	1.3	0.0119(1)	3.5	0.0122(1)	5.6	0.0011(1)	-2.5
N3	0.0184(1)		0.01147(7)		0.01152(7)		0.00141(7)	
C4	0.0154(2)	0.4	0.0123(1)	4.1	0.0117(1)	7.0	-0.0012(1)	-1.9
C4	0.0153(1)		0.01178(8)		0.01080(8)		-0.00097(7)	
C5	0.0150(2)	4.5	0.0095(1)	1.1	0.0120(1)	4.8	-0.0001(1)	-0.2
C5	0.0140(1)		0.00937(7)		0.01142(7)		-0.00007(7)	
C11	0.0303(3)	1.9	0.0172(2)	3.6	0.0107(2)	-0.2	-0.0006(2)	0.9
C11	0.0296(2)		0.0164(1)		0.01075(9)		-0.0008(1)	
C21	0.0278(3)	2.2	0.0111(2)	0.5	0.0229(2)	4.9	-0.0034(2)	0.0
C21	0.0270(2)		0.01100(9)		0.0215(2)		-0.0034(1)	
N51	0.0203(2)	-0.4	0.0105(1)	4.7	0.0176(2)	5.4	0.0012(1)	-1.6
N51	0.0204(1)		0.00995(6)		0.0164(1)		0.00140(7)	
O52	0.0366(3)	-0.7	0.0113(1)	5.0	0.0250(2)	4.9	-0.0044(1)	-6.8
O52	0.0369(3)		0.01069(7)		0.0236(2)		-0.00348(9)	
O53	0.0482(3)	-2.4	0.0174(2)	4.5	0.0183(2)	0.9	0.0072(1)	1.6
O53	0.0494(4)		0.0164(1)		0.0181(1)		0.00699(9)	

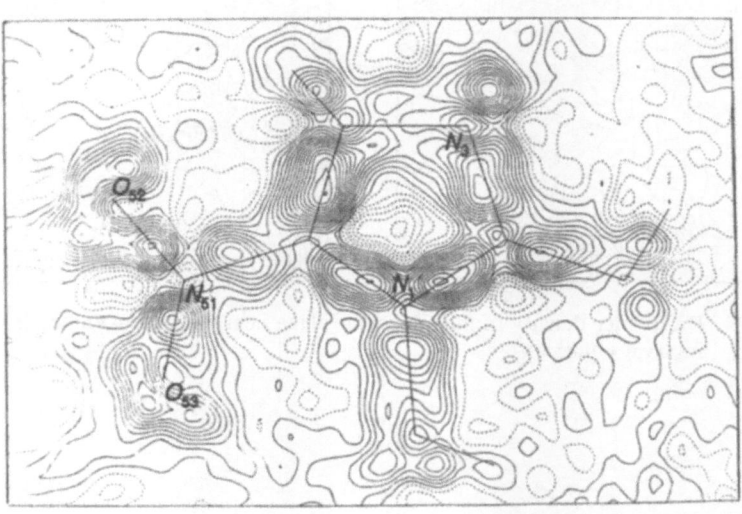

Fig. 3. Difference density map in the molecular plane. Level interval = 0.05 e $\mathring{A}^{-3}$

$$\Delta \rho_{obs}(\bar{r}) - \frac{1}{V}\sum_{\bar{h}} (kF_{obs}(\bar{h}) - F_{c,IAM}(\bar{h}))e^{-2\pi i\bar{h}.\bar{r}}$$

Table 4.   Hydrogen bonds. Distances obtained by elongation of X-H bond to
            theoretical values

```
C11-H11A    H11A....N3(i)    C11...N3(i)       C11-H11A...N3(i)
1.08        2.402(9)         3.425(12)         157.6(7)

C4-H4       H4...O53(ii)     O53(ii)...C4      C4-H4...O53(ii)
1.08        2.306(9)         3.350(13)         162.1(7)

i  = X, -1/2-Y, -1/2+Z
ii = X,  1/2-Y,  1/2+Z
```

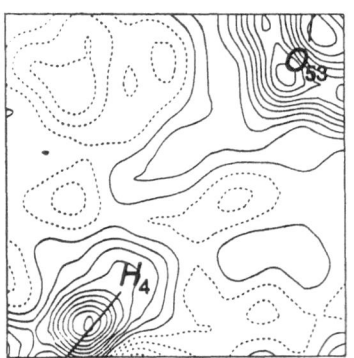 

Fig. 4. Difference density in the molecular plane near the hydrogen bonds. Level
        interval = 0.05 e Å^{-3}.

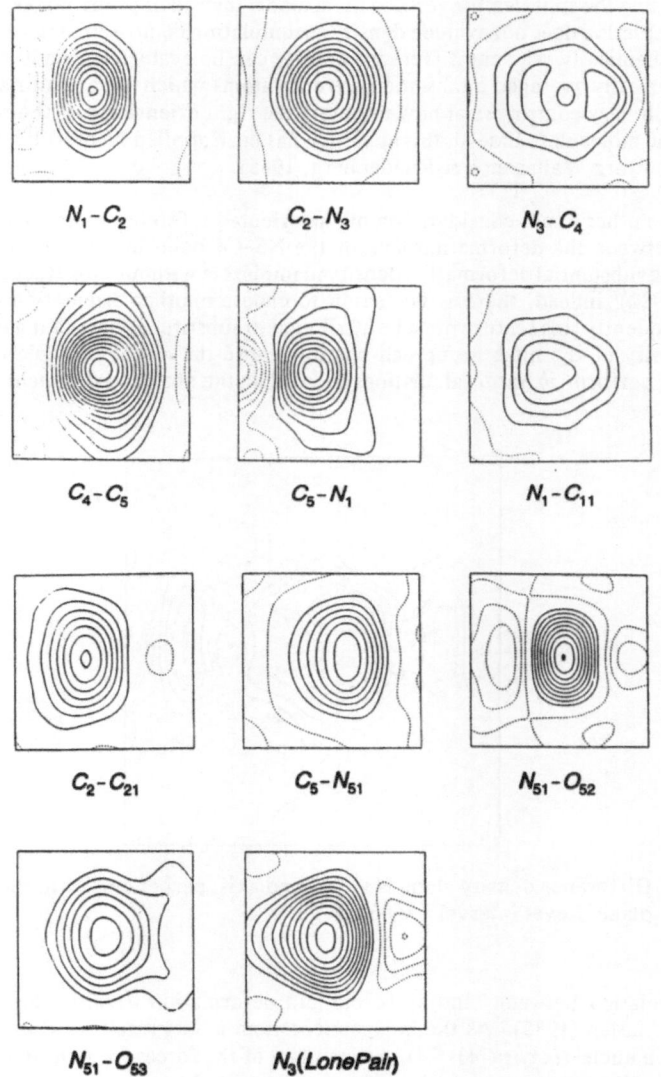

Fig. 5.   Difference density in sections perpendicular to bond axes. Sections at half bond lengths. Left, right, top and bottom are defined by view direction, (from first atom to second), and by orientation of the molecule, (as in Fig. 1). Level interval = 0.05 e Å^{-3}.

H-bonding (Table 4) perturbs the electron density near the lone pair O53. This perturbation can be observed in the Δρ map in Fig. 3 and Fig. 4. The bond peaks (Fig. 5) are extended perpendicular to the molecular plane, indicating partial π-character. This character is also observed in the C2-C21 bond.

As the N3-C4 bond shows a marked depletion of deformation density a section along the bond and perpendicular to the mirror plane was calculated (Fig. 6). The observation of an unexpected polarisation towards the C4 atom may suggest a substraction of too many electrons close to N3, using the spherically averaged atom model instead of prepared fragments (Coppens, 1982). This depletion does not exclude density accumulation compared to the oriented reference atoms. This chemically relevant difference density can be evaluated by adding the difference between spherically averaged atoms and reference atoms which are appropriately oriented. It is however still very controversial how to obtain the right orientation of the reference orbitals for polyatomic molecules . Indeed, this approach has been applied succesfully only to very small molecules (Schwarz, Valtazanos & Ruedenberg, 1985).

On the other hand, consideration of the oriented reference atoms does not explain the difference between the deformation lack of the N3-C4 bond in the title compound and the corresponding substantial deformation density in imidazole without substituents (Epstein, Ruble & Craven, 1982); indeed, there is no reason to choose another orientation of the reference atoms. Consequently this feature must be attributed to substituent or packing effects. The high electron affinity of the nitro group can possibly cause the observed depletion. It would be interesting to perform *ab initio* calculations to validate the observed deficient bonding density.

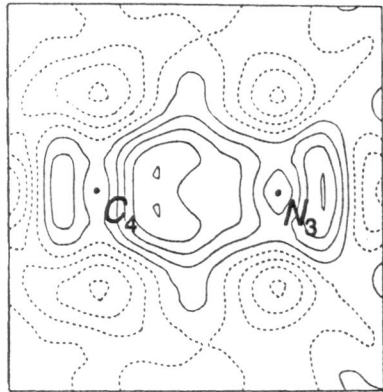

Fig. 6. Difference density along N3-C4 bond axis, perpendicular to the molecular plane. Level interval = 0.05 e Å^{-3}.

The relation between binding forces and deformation densities has been studied by Spackman & Maslen (1985). As the independent atom model results in repulsive electrostatic forces between nuclei (*in casu* N3-C4), the resultant of the forces from the deformation density charge distribution must be attractive. In most bonds, this attractive force is realised by a deformation accumulation around the midpoint of the bond. Sometimes, however, there is a polarisation of this deformation density towards the nuclei. The attracting forces become stronger (F is proportional to $1/r^2$) and less deformation density is required. The considerable experimental errors near the nuclei due to thermal smearing and due to the lack of the resolution of the X-ray diffraction experiment, make the Δρ maps inaccurate to evaluate deformation density peaks close to the nuclei. Consequently, theoretical calculations will be needed to investigate polarisation towards the nuclei. This hypothetical polarisation should be attributed to the nitro-substituent or to the packing as such a lack of electron density was not observed in the unsubstituted imidazole (Epstein *et al.*, 1982).
Deformation and total density studies on multipole refinement models on several compounds of this series will presumably reveal the origin of the deficiency of the observed bonding deformation electron density.

ACKNOWLEDGMENT

Dr. Sine Larsen is thanked for the given opportunity to collect this high resolution data in H.C.Ørsted Institute in Copenhagen. The authors also thank Mr. Flemming Hansen for collecting the X-ray data. The diffractometer and an X-ray generator were acquired by means of grants 11-1837, 11-2360 and 11-3531 from the Danish Natural Science Research Council.

REFERENCES

BLESSING, R.H. (1987). *Cryst. Rev.*, **1**, 3-58

COPPENS, P. (1982). *Elecron density distribution and the chemical bond.*, 61-92. New York Plenum Press. Edited by Coppens, P. and Hall, M.B.

CROMER, D.T. & WABER, J.T. (1974). *International Tables for X-ray Crystallogaphy*, Vol IV Table 2.2B, Birmingham: Kynoch Press. (present distributors: Kluwer Academic Publishers, Dordrecht.)

DE BONDT, H., BLATON, N.M., PEETERS, O.M., DE RANTER, C.J. & LARSEN S. in preparation for *Acta Cryst. B.*

DETITTA, G. (1985). *J. Appl. Cryst.*, **18**, 75-79

EDWARDS, D.I. (1981). *Progress in Medicinal Chemistry*, **18**, 87-116

ENRAF-NONIUS (1982). *Structure Determination Package.* Enraf-Nonius, Delft, the Netherlands.

EPSTEIN, J., RUBLE, J.R. & CRAVEN, B.M. (1982). *Acta Cryst. B*, **38**, 140-149

HIRSHFELD, F.L. (1976). *Acta Cryst A*, **32**, 239-244

JOHNSON, C.K. (1976). ORTEP. *A Fortran Thermal-Ellipsoid Plot Program for Crystal Structure Illustrations. Report ORNL-5138 (Third Revision).* Oak Ridge National Laboratory, Tennessee, USA.

MOTHERWELL, W.D.S. & CLEGG, W. (1978). *PLUTO. Program for plotting molecular and crystal structure.* Univ. of Cambridge, England.

SCHWARZ, W., VALTAZANOS, P. & RUEDENBERG, K. (1985). *Theor. Chem. Acta*, **68**, 471-506.

SMEYERS, Y.G., DE BUEREN, A., ALCALA, R. & ALVAREZ, M.V. (1981). *Int. J. Radiat. Biol.*, **39**, 649-653

SPACKMAN, M.A. & MASLEN, E.N. (1985). *Acta Cryst. A*, **41**, 347-353

STEWART, R.F., DAVIDSON, E.R., SIMPSON, W.T. (1965). *J. Chem. Phys.*, **42**, 3175-3187.

THE INFLUENCE OF PACKING AT A MOLECULAR LEVEL: CONFORMATION, GEOMETRY, SPECTROSCOPY (IR, RAMAN)

Ben Bracke, B. van Dijk, C. Van Alsenoy and A. T. H. Lenstra

Department of Chemistry
Universitaire Instelling Antwerpen
Universiteitsplein 1
B-2610 Wilrijk, BELGIUM

Ab initio calculations are widely applied to optimize the geometry of isolated molecules. To include intermolecular interactions is a necessary prerequisite to optimize a molecular geometry in its crystal environment. We do this using a 'crystal field' approach. For full details we refer to Popelier et al. (*J. Am. Chem. Soc.* 1989, **111**, 5658-5660). In our calculations we apply a 4-21G basis set. This choice is dictated by a purely practical argument; it is the only basis set for which a list of empirical corrections δ exists, which allows us to extrapolate from the calculated r_e geometry to the r_g geometry via $r_g = r_e(4\text{-}21G) + \delta(4\text{-}21G)$. This allows a direct comparison between the theoretical result and the experimental evidence. Analysing a series of carbonylhydrazide derivatives we noted that for 12 products the position of the NH_2 group in the solid state is practically equal to the position typical for the conformation of the isolated minimum energy structure. However, in oxalyldihydrazide (Quaeyhaegens et al., *J. Mol. Struct.*, accepted for publication) and cyanoacetohydrazide (Nanni et al., *J. Mol. Struct.* 1986, **147**, 369-380), the NH_2 group is rotated over an additional 120°. This solid state configuration is stabilised by the formation of intermolecular H-bonds. It is gratifying to note that these discrepancies between solid state and free molecule are reproduced by our 'crystal field' approach, in which we always started our calculations with the minimum energy conformer derived for the isolated form.

Another discrepancy between the solid state structure and its isolated model is the C=O bond length. The calculated shifts suggest that H-bonding (typical for the solid state only) increases the C=O distance. They fit with the equation:

$$d(C=O)_{obs} = 1.225 \text{ Å} + 0.011n \text{ Å} \qquad (eq.\ 1)$$

where n represents the number of H-bonds in which O participates as H-acceptor. This equation is typical for the peptide sequence N–C=P (Popelier et al., *Struct. Chem.*, submitted).

In the isolated molecule 2,3-diketopiperazine the two C=O are identical. In the solid state, however, this equivalence is absent, because only one O acts as acceptor in the H-bonding scheme. As a consequence its bond length is 0.015 Å longer than its unaffected one (C=O is 1.22 Å, which compares favourably with C=O expected from eq. (1) with n = 0). This asymmetry is reproduced by our 'solid state' modelling procedure. So we went one step further, calculating the force constants for the two distinct C=O's. We calculated a frequency separation of 54 cm^{-1}, which is in line with the observed value of 50 cm^{-1}.

THE EXPERIMENTAL ELECTRON DENSITY DISTRIBUTION

OF A C–H→Co BRIDGED COMPLEX

Lee Brammer

Chemistry Department
Brookhaven National Laboratory
Upton, NY 11973, U.S.A.

Edwin D. Stevens

Department of Chemistry
University of New Orleans
New Orleans, LA 70148, U.S.A.

INTRODUCTION

C-H→M bridged species have generated particular interest in recent years[1] with a view to their serving as models for the intermediate stages of oxidative addition of C-H to transition metals, a reaction of great importance in organometallic chemistry. The complex studied here, bis(cyclo-octane-1,5-diyl-dipyrazolylborate)Co(II), has $\bar{1}$ symmetry about the central cobalt atom. The *pseudo*-octahedral coordination sphere involves ligation of 4 pyrazolyl nitrogen atoms in a square-plane and is completed by two agostic[1] C–H→Co interactions (Figure 1). This coordination is particularly unusual as the parent bis(pyrazolyl) complex is tetrahedral with no agostic interactions.

11465 X-ray intensity data were measured at reduced temperature [94(5)K] to a resolution of $\sin\theta/\lambda < 1.0\text{Å}^{-1}$, primarily in one hemisphere of reciprocal space, for space group $P\bar{1}$. Loss of the crystal due to an instrument failure prevented further data collection, though additional X-ray data as well as neutron data will be collected with new crystals. Results obtained using this initial data set are described here.

SUMMARY OF RESULTS

A spherical-atom refinement based on low angle data [$\sin\theta/\lambda < 0.71\text{Å}^{-1}$, $F > 3\sigma(F)$] confirmed the presence of the agostic interaction [C-H 1.03(3)Å and Co-H 2.15(3)Å] and converged with $R = 0.040, S = 1.67$. Current refinements using a multipole model[2] to describe the aspherical features of the electron density have yielded $R = 0.046, S = 1.17$, for 6544 data [$F > 3\sigma(F)$] and 262 variables.

Deformation density in the C–H→Co region, whilst requiring cautious interpretation

The Application of Charge Density Research to Chemistry and Drug Design
Edited by G. A. Jeffrey and J. F. Piniella, Plenum Press, New York, 1991

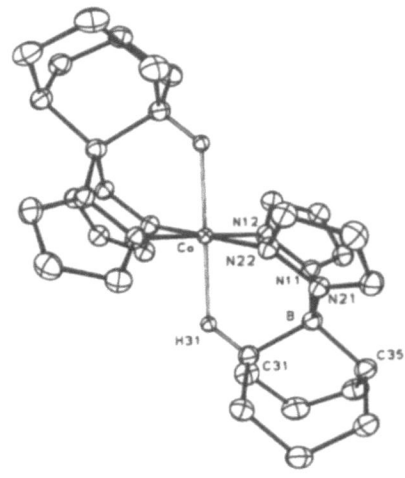

Figure 1. Molecular structure of $[Co\{(N_2C_3H_3)_2B(C_8H_{14})\}_2]$ shown with 75% probability ellipsoids. Hydrogen atoms other than agostic H31 omitted for clarity.

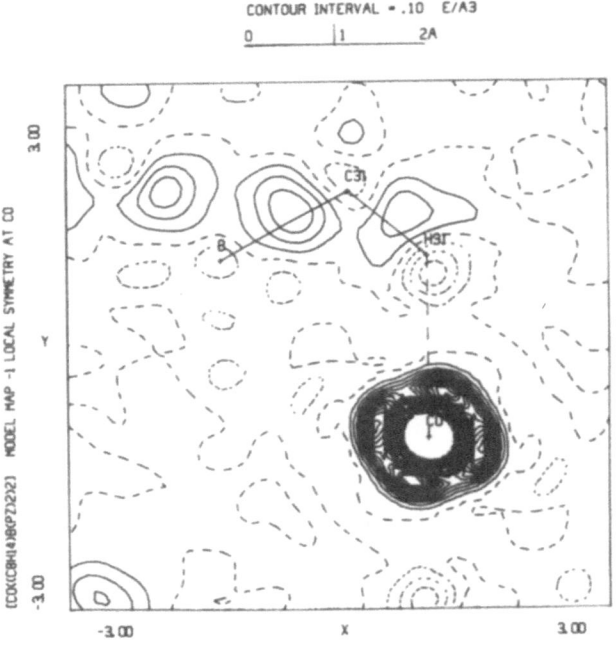

Figure 2. Model deformation density in the C31-H31-Co plane. Positive contours are solid, negative are dotted and zero is broken. Contour interval 0.1e/Å^3.

in the absence of neutron diffraction data, suggests that this interaction is quite weak, *e.g.* there is no significant evidence of incipient Co–H bond formation (Figure 2). Consistent with this is the long Co–H distance of 2.06Å,† *cf.* 1.744(7)Å (neutron diffraction) in the β-C–H$\rightarrow$Co interaction of $[Co(\eta\text{-}C_5Me_5)(C_6H_9)]BF_4$,[3] and 1.56(2)Å (electron diffraction) in $HCo(CO)_4$.[4]

Monopole charges indicate a net transfer to Co of 0.68 electrons from each of the two formally uninegative bis(pyrazolylborate) ligands. Deformation density in the mean plane of the pyrazolyl ring shows interatomic peaks with decreasing magnitude in the sequence C–C > C–N > N–N, analogous to the trend observed by others[5] for formal *single* bonds between these atom types (when using a spherically averaged reference state). Two refinement models using different local symmetry constraints (D_{2h} and C_i) for the Co *multipole populations* have been tested. Resulting deformation density maps in the CoN_4 plane are shown in Figures 3 and 4. Asymmetry of the deformation density in the vicinity of Co is indicative of crystal field effects in both cases. On relaxing the symmetry constraints to C_i (the crystallographic site symmetry of Co) the 4 positive in-plane features of similar magnitude (Fig. 3) become two large and two small accumulations of electron density (Fig. 4). This is consitent with the crystal field effect of the off-axis (N-Co) nitrogen lone-pair features, and results in significant population of the now symmetry-allowed Co d-orbital mixing terms derived[6] from the Co multipole populations [most notably d_{xz}/d_{xy} -1.2(1), d_{z^2}/d_{xy} -1.0(1), $d_{yz}/d_{x^2-y^2}$ -0.9(1)]. However, no significant change in the derived populations[6] of the 5 Co d-orbitals is noted in going from the D_{2h} model to the C_i model. [d_{z^2} 1.41(7). $d_{x^2-y^2}$ 0.48(7), d_{xy} 1.21(7), d_{xz} 1.16(8). d_{yz} 1.79(8). total Co d-orbital population 6.05(9) electrons].

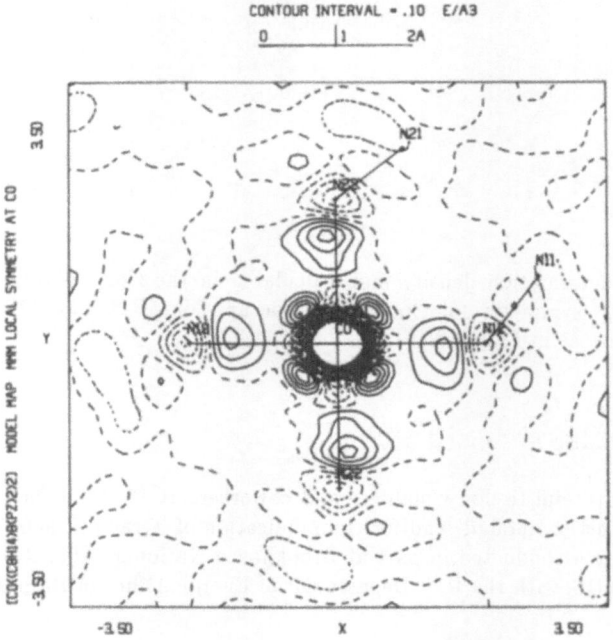

CONTOUR INTERVAL = .10 E/A3

Figure 3. Model deformation density map calculated in the CoN_4 plane, using refinement model with local D_{2h} symmetry constraint on Co multipole occupation. N11 and N21 are out of plane atoms. Contours are as in Figure 2.

---

† Hydrogen position fixed at a typical neutron diffraction determined separation of 1.15Å from C31. along the C–H vector derived from the spherical-atom refinement.

Refinements are still in progress. The approach being taken is to gradually reduce local symmetry constraints on the multipole populations associated with different atomic sites. In particular, the effects of such constraints on the charge density model around the pyrazolyl rings is being investigated. For example, we believe that local C_{2v} symmetry imposed upon the nitrogen multipole occupancies may be contributing to the off-axis nature of the observed nitrogen lone pair features, and hence perhaps to the nature of the electron density distribution observed around the cobalt atom. A full report of these results will be given elsewhere, including refinements based on the further X-ray and neutron diffraction data being collected.

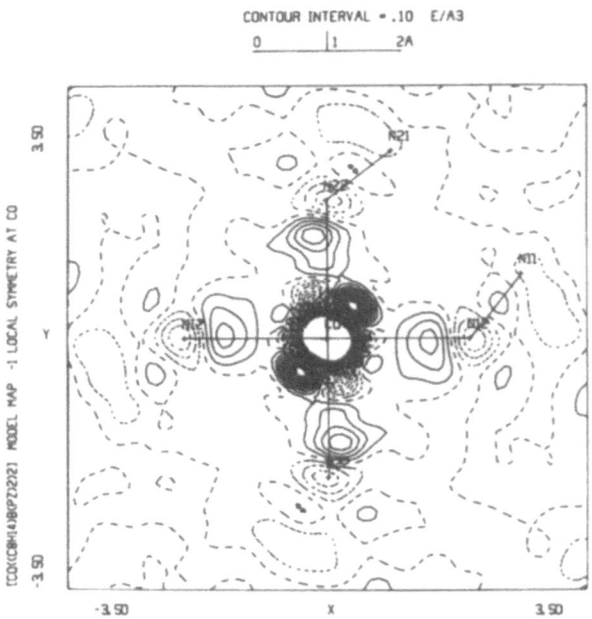

Figure 4. Model deformation density map calculated in the CoN_4 plane, using refinement model with local C_i symmetry for Co multipole occupation. N11 and N21 are out of plane atoms. Contours are as in Figure 2.

ACKNOWLEDGEMENT

We are very grateful to Joe Calabrese and coworkers at Du Pont for the synthesis and crystallization of this compound, and for communication of X-ray diffraction results prior to publication. Research conducted in part at Brookhaven National Laboratory under contract DE-AC02-76CH00016, with the U.S. Department of Energy, Office of Basic Energy Sciences.

REFERENCES

1. M. Brookhart, M. L. H. Green and L.-L. Wong, Prog. Inorg. Chem., 36:1 (1988).
2. N. K. Hansen and P. Coppens, Acta Crystallogr., A34:909 (1978).
3. T. F. Koetzle, L. Brammer, A. G. Orpen, R. B. Cracknell, and J. L. Spencer, American Crystallographic Association Meeting, Abstract PA68 (1989).
4. E. A. McNeill and F. R. Scholer, J. Am. Chem. Soc., 99:6243 (1977).
5. For example, see J. D. Dunitz and P. Seiler, J. Am. Chem. Soc., 105:7056 (1983).
6. A. Holladay, P. Leung and P. Coppens, Acta Crystallogr., A39:377 (1983).

TRANSFERABLE ATOM EQUIVALENTS. MOLECULAR ELECTROSTATIC POTENTIALS FROM THE ELECTRIC MULTIPOLES OF *PROAIMS* ATOMIC BASINS

Curt M. Breneman

Department of Chemistry
Rensselaer Polytechnic Institute
Troy, NY 12180 USA

ABSTRACT

The *ab-initio* electrostatic potential fields surrounding several heteroatom-containing molecules are compared to the corresponding potential fields generated by atom-centered multipole expansions of the electron density distributions within PROAIMS-defined atomic basins. When the electron populations of a set of atomic basins are subtracted from the associated nuclear charges, the resulting "atomic charges" can be used as the monopole terms of a three-term expansion (monopole + dipole + quadrupole). The monopole terms were found to inadequately represent the atomic electron distribution when considered alone, but when the dipole and quadrupole terms are included, the molecular electrostatic potential field can be reconstructed with a high degree of accuracy. These multipolar terms were found to be transferable between molecules within defined limits. This result has allowed the construction of accurate electrostatic potential fields around molecules which are much too large for direct *ab-initio* treatment. These results are compared to some popular "spherical atom" approaches currently in use for approximating intermolecular electrostatic interactions.

INTRODUCTION

Electrostatic forces are generally recognized to be an important component of non-covalent molecular interactions. Consequently, molecular mechanics and dynamics force fields must take these terms into account. Since most molecules of biological interest are outside the realm of accurate *ab-initio* treatments, approximations must be used which facilitate the evaluation of electrostatic terms. To this end, Williams has advocated the use of transferable bond dipoles, both with and without atom-centered terms.[1] This approach has been adopted by Allinger in the MM2 program, while Kollman utilizes atom-centered point charges to compute electrostatic interaction energies in the AMBER force-field.[2] In light of the increased computational resources now available, it has become feasible to use a larger number of terms to define the electrostatic interaction potential. To date, extraction of the required atomic multipole components has been an exercise in approximations, albeit useful ones. With the advent of Bader's PROAIMS technique, however, molecular electron density can now be uniquely partitioned between atoms. The PROAIMS approach has been successfully used to explain a number of molecular properties, including the origin of rotational barriers, and the nature of "resonance" interactions. In this work, we have used the PROAIMS technique to examine not only the electron populations of the atomic "basins", but also the electrostatic potentials arising from the non-spherical distribution of electron density within those regions of space.

RESULTS AND DISCUSSION

In order to analyze the electron density distribution within a defined region of space containing an atom, the density can be expanded about the nuclear origin in a power series in 1/R. Since the ultimate goal of this investigation is to use these terms in a computationally-intensive inner loop, it was decided that unless the higher terms (such as the octupole or hexadecapole terms) provided significant additional information, they would not be included. In this work, we found that the density need only be expanded as far as the second electric moment (or quadrupole) to reproduce the molecular electrostatic potential field to within ±3 kcal/mol at the molecular VDW surface.

The molecules included in this study (methanol, methylamine and nitromethane, among others) were chosen since they have "lone-pairs" of electrons which are expected to be the preferential locations for electrophiles to bind. The shape of the electrostatic potential well in the vicinity of the heteroatom is of value in predicting the directionality of these interactions. In this work, both two- and three-dimensional techniques were used to examine the electrostatic potential fields of the model compounds and to evaluate the approximate potential fields generated by using point-charge approximations and multipolar expansions. In each case, the analytically-derived potential field is used as a standard. The two-dimensional analysis was done by calculating the electrostatic potential of a plane passing through each molecule, while the three-dimensional analysis was accomplished by color-coding the electrostatic potential values at the 0.002 e/au^3 electron density surface. This value of electron density has been used in earlier investigations to define the molecular van der Waals surface.

Examination of the coded molecular surfaces by visual and numerical means allowed a comparison of the values of the analytical potential on this surface with the potential calculated by several approximate methods including atom-centered charges and multipoles. The results obtained by constructing the molecular electron density from reference atoms calculated using the PROAIMS method gave results visibly superior to the CHELPG atom-centered charge model and Ritchie's alternative multipolar scheme, even though the competing methods were based upon fitting of the charges or multipoles to the exact case in question, rather than selecting the data from a generalized library of values, as in the PROAIMS cases. Thus, the PROAIMS reference atoms have the added advantage of being uniquely defined, rather than being a part of a more arbitrary parameter set. An automated topology-driven version of this technique for generating electronic properties of large molecules from a library of atomic fragments is currently undergoing development.

REFERENCES

[1]Williams, D. E. *J. Comp. Chem.* 1988, **9**, 745.
[2]Wiener, P. K. and Kollman, P. A. *J. Comp. Chem.* 1981, **2**, 287.

AN AB INITIO STUDY OF CLAVULANIC ACID AND THE RELATION

TO ITS CHEMICAL REACTIVITY

Berta Fernández and Christian Van Alsenoy

Universitaire Instelling Antwerpen
Universiteitsplein 1
B2610 Wilrijk, BELGIUM

Clavulanic acid (CA) has a wide β-lactamase inhibition spectrum[1], especially when combined with a β-lactam-like amoxycillin or ampicillin. The biological activity of this compound and its derivatives being related to their respective structures, it was deduced from a number of empirical structure-activity relationships that a structure close to that of CA is needed for a potent β-lactamase inhibitor.

We performed 'ab initio' gradient optimizations of six conformations of CA. In these conformations, all possible intramolecular hydrogen bonds are taken into account. In particular, the H-bond between the alcoholic H-atom and the intra-annular O-atom and the H-bond between the ethylene H-atom and the carbonyl O-atom from the acid group, both close a six-membered ring. The geometry optimizations are performed by standard ab initio gradient procedures[2] using the 4-21G basis set[3]. Because of the size of the system, the Multiplicative Integral Approximation[4] is used to perform the SCF calculation. A general assessment of the accuracy of 4-21G geometries for organic molecules can be found in the literature[3,5].

When calculated parameters are compared with available experimental data from an X-ray study of the p-nitrobenzyl ester derivative[6], agreement is excellent. The largest differences found in bond lengths and valence angles are 0.02 Å and 2° respectively.

REFERENCES

1. S. M. Roberts and B. J. Price, *The Role of Organic Chemistry in Drug Research*, in "Medicinal Chemistry," Academic Press, London (1985).
2. P. Pulay, *Mol. Phys.* **17** (1969) 197. P. Pulay, *Theor. Chim. Acta* **50** (1979) 299.
3. P. Pulay, G. Fogarasi, F. Pang and J. E. Boggs, *J. Am. Chem. Soc.* **101** (1979) 2550.
4. C. Van Alsenoy, *J. Comp. Chem.* **9** (1988) 620.
5. L. Schâfer, *J. Mol. Struct.* **100** (1983) 51.
6. A. G. Brown, D. F. Corbett, J. Goodacre, J. B. Harbridge, T. T. Howarth, R. J. Ponsford, I. Stirling and T. J. King, *J. Chem. Soc. Perkin Trans. 1*, 635 (1984).

UHF CALCULATIONS OF THE g TENSOR IN METAL-CARBONYLES RADICALS

André Grand

Laboratoires de Chimie
D.R.F., CENG, 38041 Grenoble, FRANCE

Paul J. Krusic

Central Research and Development Department
E. I. Du Pont de Nemours, Wilmington, DE 19898 USA

Robert Subra

LEDSS - Université J. Fourier, 53X
38041 Grenoble, FRANCE

An experimental and theoretical study of the transient $Fe(CO)_4^{\cdot-}$ radical is presented. This radical, isoelectronic with $Co(CO)_4$, is obtained for the first time in solution, by γ-irradiation of dilute $Fe(CO)_5$ at 77 K. The general features of its ESR spectrum are similar to those of $Co(CO)_4$, i.e., a nearly axial g tensor ($g_\parallel = 2.0039$, $g_\perp = 2.0707$), but it was impossible to get further information on ^{13}C couplings. On the basis of the experimental results for $Co(CO)_4$, we present a theoretical study by both Extended Huckel and $X\alpha$-methods for $Fe(CO)_4^{\cdot-}$ in trigonal pyramidal geometry.

The SOMO is found of A_1 symmetry with a LUMO of E symmetry in both methods. In both calculations, the SOMO is made up mostly of a highly directional part centered on the metal which uses only metal s, pz and dz^2 orbitals and of a second part which is strongly delocalized over the equatorial CO ligands and uses almost exclusively the pz orbitals of the equatorial atoms. The apical CO's contribute negligibly to the SOMO.

The only carbon giving a substantial coupling with the unpaired electron is the apical one, in accordance with ESR results on $Co(CO)_4$ ($a_{13C} = 27$ Gauss). These calculations are consistent with the experimental work and allow a complete understanding of the structure of the radical.

MAGNETOSTRICTION IN NiF$_2$: COMBINED γ-RAY AND NEUTRON DIFFRACTION

W. Jauch and A. Palmer

Hahn-Meitner-Institut
Glienicker Str. 100
D-1000 Berlin 39, WEST GERMANY

A. J. Schultz

Chemistry Department
Argonne National Laboratory
Argonne, IL 60439, USA

NiF$_2$ has a tetragonal rutile type structure with fluorine at $(x,x,0)$. According to a prediction deduced from the Ewald-Born theory of structural birefringence, a very small shift in the fluorine positional parameter x due to magnetic ordering may be the cause for a change in the optical birefringence near the Néel temperature [$T_N = 73.2$ K]. The predicted value of $\Delta x = x(15$ K$) - x(298$ K$) = -4.5 \times 10^{-4} \cong 2.96 \times 10^{-3}$ Å demands measurements of very high accuracy. The experiments were performed using the Berlin γ-ray diffractometer [$\gamma = 0.0392$ Å] and time-of-flight neutrons [$\gamma = 0.5$–3 Å] from the pulsed spallation source IPNS, Argonne. The same single crystal plate of about $3 \times 3 \times 1.5$ mm^3 was investigated by both methods. The results obtained from the neutrons are in good agreement with those from γ-ray diffraction which confirm the theoretical prediction:

	298 K	15 K	Δx
γ-ray diffraction	$x = 0.30371(9)$	$x = 0.30325(7)$	$-4.6(1.1) \times 10^{-4}$
Neutron diffraction	$x = 0.30366(4)$	$x = 0.30333(3)$	$-3.3(0.5) \times 10^{-4}$

THE CONFORMATION OF SOME NEUROMUSCULAR BLOCKING AGENTS

H. Kooijman, J. A. Kanters and J. Kroon

Laboratorium voor Kristal- en Structuurchemie
Vakgroep Algemene Chemie, Rijksuniversiteit
Padualaan 8, 3584 CH Utrecht, THE NETHERLANDS

A wide range of compounds shows neuromuscular blocking activity, *e.g.* tubocurarine, chandonium, atracurium, gallamine, pipecuronium and pancuronium. All these compounds have one or more quaternary nitrogen atoms in common.

We have performed molecular mechanics calculations on three neuromuscular blocking agents of the pancuronium type: vecuronium and protonated vecuronium with one quaternary nitrogen atom, and pancuronium, which possesses two quaternary nitrogen atoms. The steric energy of two conformers of these molecules was calculated using Allinger's MM2 force field[1]. The calculations were performed using standard bond dipole moments as well as atomic point charges. The influence of various charge distributions, calculated with MOPAC[2], on the conformational energy map is compared.

References

[1] N. L. Allinger MM2, QCPE, Indiana University (1985).
[2] J. P. Stewart, MOPAC 5.00, QCPE 455, Indiana University.

A STRUCTURAL COMPARISON OF POTENTIAL HIV INHIBITORS:

2',3'-DIDESOXINUCLEOSIDES AND 2',3'-DIDESOXICARBONUCLEOSIDES

R. A. Mosquera, B. Fernández, S. A. Vázquez and M. A. Ríos y E. Uriarte

Departamentos de Química Física y Química Orgánica
Universidad de Santiago
Santiago de Compostela, E-15708 SPAIN

2',3'-Didesoxinucleosides, such as 3'-azido-3'-desoxitymidine (AZT), are known as powerful and selective inhibitors of the HIV replication. These compounds are phosphorylated inside the infected cell (in the cytoplasm) yielding the 2',3'-didesoxinucleo-side-5'-phosphate that is able to bind to the virus DNA polymerases. Lack of the 3'-OH group prevents the 5'-3'-phosphodiester bonding which converts these compounds in DNA terminators. The structural similarity between 2',3'-didesoxicarbonucleosides and 2',3'-didesoxinucleosides led to the postulation that the former could display a similar anti-replicative HIV activity. In this work we present a comparative structural study between 2',3'-didesoxiuridine (I) and its carbocyclic derivatives (II-IV). The study includes several semi-empirical calculations (AM1 and MNDO), aimed at a concrete description of the structural proximities between these groups of systems through charge distributions and distances between functional groups.

EXPERIMENTAL OBSERVATION OF INTERMOLECULAR π-ELECTRON INTERACTIONS IN PHOTOREACTIVE CRYSTALS

I. Ortmann, St. Werner, C. Krüger

Max-Planck-Institut für Kohlenforschung
Kaiser-Wilhelm-Platz 1
D-4330 Mülheim (WEST GERMANY)

INTRODUCTION

We present the results of electron density determination studies on two prospective photoreactive solids, 4-Acetoxy-6-styryl-α-pyrone (1) and the isomorphous, isoelectronic 4-Acetoxy-2-styryl-1,3-oxazin-6-one (4).

The crystal packing of these compounds is illustrated in Figure 1. Irradiation (λ= 360 nm) of 1 leads to the formation of topochemical [2+2]-photodimers 2 and 3, whereas 4 is photoinactive (Scheme 1) although the distances between the midpoints of the photoreactive C=C double bonds leading to the two possible dimers are similar in both cases - 4.20 (unsym. reaction) and 3.97 Å (sym. reaction) for compound 1 and 4.08 and 4.03 Å for compound 4 (Fig. 1) -.

An explanation for this unusual behavior was sought in the electron density distributions.

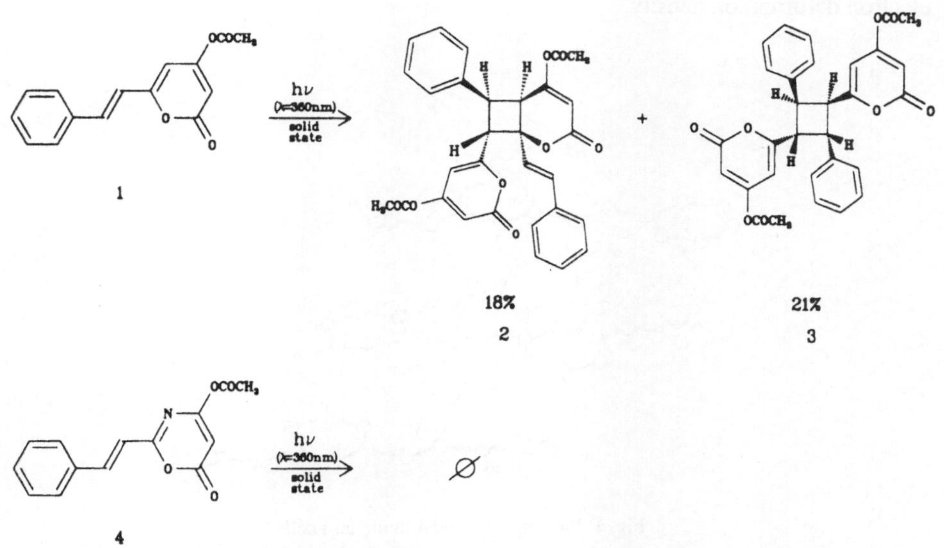

Scheme 1. Different solid state photoreactivity of 1 and 4

RESULTS

Accurate single-crystal X-ray data were refined with an aspherical multipole expansion of the electron density using the program VALRAY (Stewart, Spackman, 1983). Positional and thermal parameters were obtained from two refinements based on the independent atom model:

1)   high-angle refinement of the non-hydrogen atoms
2)   low-angle refinement of the hydrogen atoms with the
     heavy atom parameters fixed at their high-angle
     values.

The multipole refinement included exponential functions up to the octopole level for the non-hydrogen atoms.

The electron deformation density was calculated as the Fourier transform of the difference between structure factors obtained from the multipole model and structure factors obtained from the independent atom model. The results are summarized in Table 1.

Figures 2-5 show Fourier cuts mapping the π-bond region of the photoreactive C=C double bonds.

The deformation density of the photoreactive compound **1** is distorted away from the direction of the interaction so that the major axis of the bonding electron density makes an angle of about 70° with the plane of the molecules. In contrast, the EDD of the photoinactive compound **4** is as expected for the isolated molecule.

The results show that intermolecular repulsion between the electrons in the π-systems of the photoreactive substance **1** in the solid state can be observed in the electron deformation density.

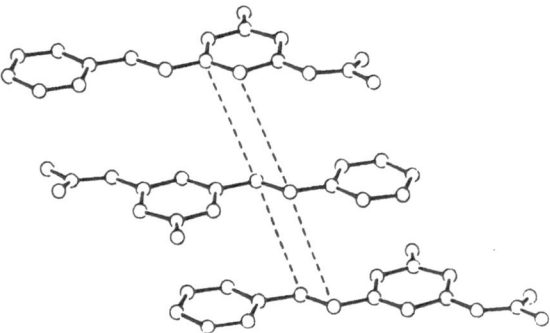

Fig. 1: Packing of **1** and **4** in the unit cell

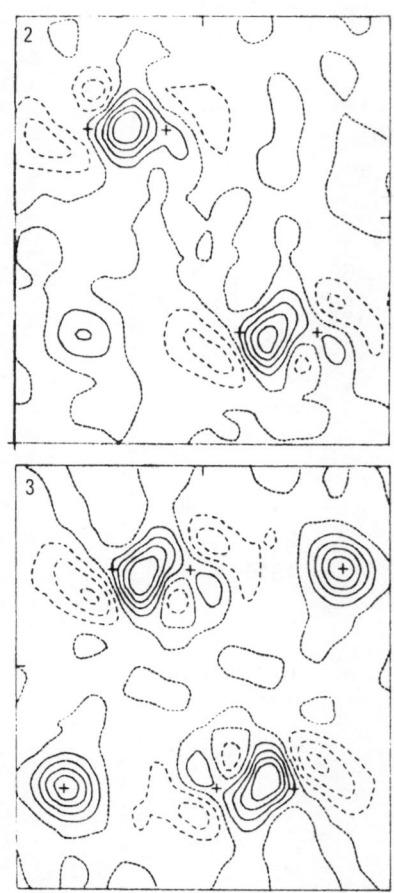

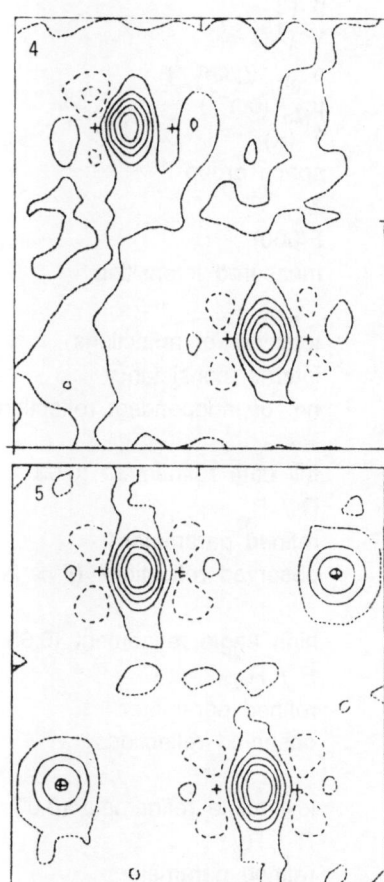

Fig. 2 and 3. Fourier cuts mapping the p-bond region of the photoreactive C-C double bonds of **1**

Fig. 4 and 5. Fourier cuts mapping the p-bond region of the photoinreactive C-C double bonds of **4**

Table 1

	1	**4**
	$C_{15}H_{12}O_4$	$C_{14}H_{11}NO_4$
a (Å)	9.383(1)	9.186(1)
b	13.026(2)	13.233(1)
c	10.028(1)	9.803(1)
β (°)	98.75(1)	96.284(5)
V (Å³)	1211.39	1184.48
$\rho_{calc.}$ (g/cm⁻³)	1.41	1.44
μ_{Mo} (cm⁻¹)	0.96	1.00
T (K)	100	
space group	$P2_1/c$	
Z	4	
F(000)	536	
measured intensities	±h±k+l, 1 < θ < 36°, Ψ = 0,15°	
	±h±k+l, 1 < θ < 20°, Ψ = −45,−30,...,45°	
total no. of reflections	23591	26823
internal consistency	0.034	0.020
no. of independent reflections	5719	5598

full data refinement (0.03 < sin θ/λ < 0.83):

R / R_w	0.065/0.051	0.060/0.040
refined parameter	220	216
observed reflections (F > 2σF)	4225	4919

high angle refinement (0.65 < sin θ/λ < 0.83):

R / R_w	0.054/0.045	0.052/0.033
refined parameter	172	172
observed reflections	1860	2390

low angle refinement (0.03 < sin θ/λ < 0.65):

R / R_w	0.054/0.063	0.052/0.055
refined parameters	49	45
observed reflections	2365	2529

multipole refinement (0.03 < sin θ/λ < 0.65):

R / R_w	0.038/0.031	0.030/0.023
refined parameters	304	304
observed reflections	2365	2529

THE METAL-NITROXYL INTERACTION IN MNO METALLACYCLES

(M = CU, PD): AN AB-INITIO SCF/CI STUDY

Marie-Madeleine Rohmer*, André Grand[†] and Marc Benard*

*Laboratoire de Chimie Quantique
E.R. 139 du CNRS Institut Le Bel
F-67000 Strasbourg, FRANCE

[†]Laboratoires de Chimie (U.A. 1194 du CNRS)
Département de Recherche Fondamentale
Centre d'Etudes Nucléaires de Grenoble
F-38041 Grenoble Cedex, FRANCE

Ab initio SCF and CI calculations have been carried out on the isoelectronic and isostructural complexes $(CH_3)_2NOCuBr_2$ and $(CH_3)_2NOPdClPH_3$, where dimethyl-nitroxide is a model for 2,2,6,6-tetramethylpiperidinyl-1-oxy (TMPO). Calculations show that the $CuBr_2$ and $PdClPH_3$ radicals have a quite different electronic structure, the magnetic orbital being delocalized with major weight on the ligands for $CuBr_2$, and mainly localized on the metal for $PdClPH_3$. The different nature of the metal fragments influences the η^2 coordination of the nitroxyl radical. In the copper complex, the bonding interaction is largely delocalized over the five atoms of the molecular plane. Through-space interactions between the bromine atoms and the nitroxide group contribute to stabilize the molecule. The net charge of the nitroxide is not significantly modified with respect to the free $(CH_3)_2NO$ radical. In the palladium complex, a significant charge transfer occurs from the metal to the NO π^* orbital, resulting in a strong covalent interaction between these orbitals. The other ligands have little contribution in the bonding. Both the Cu and the Pd complexes are computed to be strongly diamagnetic, with respective singlet-triplet energy separations of 5000 cm^{-1} and 17,000 cm^{-1}.

FIRST-PRINCIPLES THEORETICAL METHODS FOR THE CALCULATION

OF ELECTRONIC CHARGE DENSITIES AND ELECTROSTATIC POTENTIALS

Jorge M. Seminario, Jane S. Murray and Peter Politzer

Department of Chemistry
University of New Orleans
New Orleans, LA 70148

INTRODUCTION

Recent years have seen dramatic and exciting advances in computational chemistry, due to important developments in both hardware and software technology. One of these has certainly been the increasing availability of supercomputers for research purposes, as for example at the centers supported by the National Science Foundation and at government laboratories such as the Naval Research Laboratory or the Air Force Supercomputing Center. A complementary factor has been the expanded use of personal computers, workstations and graphics terminals that allow the researcher to interact with supercomputers and mainframes, and to visually monitor and examine the results. At the same time, there has been a mushrooming of software packages at all levels, ranging from *ab initio* to empirical, which permit continually improving computational treatments of systems ranging from individual molecules to molecular aggregates.

The purpose of this contribution is to survey some first-principles approaches to the calculation of the electronic densities and electrostatic potentials of chemical systems. These properties can also be obtained experimentally, by diffraction methods.[1,2] This is accordingly an area in which there is a special opportunity for theory and experiment to complement and support each other, thereby stimulating and facilitating progress in both.

THEORETICAL METHODS

Two first-principles methods will be discussed: (a) *ab initio* procedures,[3] and (b) density functional theory.[4] For both, the initial step is usually to apply the Born-Oppenheimer, or adiabatic, approximation,[5] according to which the movement of the electrons may be treated separately from that of the nuclei; the latter may be viewed as occupying stationary positions.

The Application of Charge Density Research to Chemistry and Drug Design
Edited by G. A. Jeffrey and J. F. Piniella, Plenum Press, New York, 1991

In an *ab initio* treatment, the objective is to solve the Schrödinger equation for the system of interest, $\hat{H}\Psi(\mathbf{r}_1, \mathbf{r}_2, \ldots \mathbf{r}_N) = E_e\Psi(\mathbf{r}_1, \mathbf{r}_2, \ldots \mathbf{r}_N)$, in which $\hat{H}$ is the Hamiltonian operator that corresponds to an N-electron system having total electronic energy E_e and described mathematically by Ψ, which is a function of the positions of the electrons. If Ψ can be determined, then it can be used to obtain any desired electronic property of the system; in particular, its electronic density $\rho(\mathbf{r})$ and electrostatic potential $V(\mathbf{r})$ are given rigorously by eqs. (1) and (2):

$$\rho(\mathbf{r}) = \rho(\mathbf{r}_1) = N\int \Psi^*(\mathbf{r}_1, \mathbf{r}_2, \ldots \mathbf{r}_N)\Psi(\mathbf{r}_1, \mathbf{r}_2, \ldots \mathbf{r}_N)\,d\mathbf{r}_2 \ldots d\mathbf{r}_N \qquad (1)$$

$$V(\mathbf{r}) = \sum_A \frac{Z_A}{|\mathbf{R}_A - \mathbf{r}|} - \int \frac{\rho(\mathbf{r}')\,d\mathbf{r}'}{|\mathbf{r}' - \mathbf{r}|} \qquad (2)$$

Z_A is the charge on nucleus A, located at the fixed position $\mathbf{R}_A$. The total energy of the system is given by,

$$E = E_e + \sum_{A<B} \frac{Z_A Z_B}{|\mathbf{R}_A - \mathbf{R}_B|} \qquad (3)$$

The Hamiltonian operator is the sum of kinetic and potential energy operators, $\hat{H} = \hat{T} + \hat{V}$, where

$$\hat{T} = -\frac{1}{2} \sum_{i=1}^{N} \nabla_i^2 \qquad (4)$$

and

$$\hat{V} = -\sum_i \sum_A \frac{Z_A}{|\mathbf{R}_A - \mathbf{r}_i|} + \sum_{i<j} \frac{1}{|\mathbf{r}_j - \mathbf{r}_i|} \qquad (5)$$

The second term on the right side of eq. (5) represents the instantaneous repulsion between two moving electrons, and it is the presence of this term that has hitherto prevented the exact analytical solution of the Schrödinger equation, in closed form, for $N > 1$.[6]

In order to obtain at least reasonable approximations to Ψ and E_e, two simplifications are frequently introduced. First, Ψ is expressed in determinantal form in terms of one-electron functions called "molecular orbitals", which are usually written as linear combinations of basis functions centered at the various nuclei. During the past twenty years, gaussian-type functions have come to be the most widely used for the latter purpose. The larger the basis set of gaussian functions, the better will be the representation of Ψ; however the time and space requirements of the computations will also increase markedly.

The second simplification is that each electron is treated as moving in the *average* potential of all of the other electrons;

computationally, this is handled by going through an iterative process until self-consistency is achieved. Hence this is known as a "self-consistent-field molecular orbital" (SCF-MO) procedure. By increasing the size of the basis set (or by seeking numerical rather than analytical representations of Ψ), one can approach a limiting case, the Hartree-Fock solution. (The term Hartree-Fock is often applied, loosely, to any *ab initio* SCF calculation.)

The primary deficiency of the Hartree-Fock method is that it overestimates the repulsive interactions between the electrons; by dealing with an average rather than the instantaneous potential, it does not properly reflect the correlation between their movements that tends to diminish the repulsive effect. There are two general "post-Hartree-Fock" approaches that have been developed to take account of electron correlation. In the "configuration-interaction" (CI) procedure, Ψ is expressed as not just a single determinant but rather a linear combination of several, each corresponding to a different distribution (configuration or "excitation") of the electrons among the molecular orbitals.

We normally use the other post-Hartree-Fock method, which is based on many-body, or Møller-Plesset (MP), perturbation theory. Here, the Hartree-Fock wave function is taken to be the zero-order solution, with the perturbation being the difference between the exact Hamiltonian and the Hartree-Fock operator. Solving the perturbation theory equations then produces the first-, second- and higher order corrections to the wave function and correspondingly the second-, third- and higher order corrections to the energy (MP2, MP3, MP4, etc.) (It should be noted that the nth order wave function is sufficient to obtain perturbation energies up to order 2n+1.)

Density Functional Theory[4]

In density functional theory, the focus is on the electronic density, $\rho(\mathbf{r})$, rather than an electronic wave function, $\Psi(\mathbf{r}_1, \mathbf{r}_2, \cdots \mathbf{r}_N)$. Since ρ is a function of only three variables in contrast to Ψ, which depends on 3N, the magnitude of the computation is considerably diminished.

The rigorous basis, for density functional theory is the Hohenberg-Kohn theorem,[7] which states that all of the electronic properties of a chemical system are determined by the electronic density; in particular, the energy is a functional of ρ: $E_e = F[\rho]$. The nature of the functional F is not specified by the theorem, and a great deal of work has gone into its elucidation.[2] In general terms,

$$F[\rho] = T[\rho] + \sum_A \int \frac{Z_A \rho(\mathbf{r})\,d\mathbf{r}}{|\mathbf{R}_A - \mathbf{r}|} + \frac{1}{2} \int \int \frac{\rho(\mathbf{r})(\mathbf{r}')\,d\mathbf{r}d\mathbf{r}'}{|\mathbf{r}' - \mathbf{r}|} + V_{xc}[\rho] \qquad (6)$$

where T is the kinetic energy functional and V_{xc} represents exchange and correlation effects.

Due to the problems associated with determining the natures of the functionals T and especially V_{xc}, an approximate procedure is frequently used, based on the formalism developed by Kohn and Sham.[8] $\rho(\mathbf{r})$ is written as a sum of the densities of non-interacting single particles, $\rho(\mathbf{r}) = \sum_{i=1}^{N} |\phi_i(\mathbf{r})|^2$, and N one-electron equations are obtained:

$$\left[-\frac{1}{2} \nabla^2 - \sum_A \frac{Z_A}{|R_A - \mathbf{r}|} + \int \frac{\rho(\mathbf{r}') d\mathbf{r}'}{|\mathbf{r} - \mathbf{r}'|} + v_{XC}(\mathbf{r}) \right] \phi_i = \varepsilon_i \phi_i \qquad (7)$$

v_{XC} is commonly expressed in terms of the uniform-electron-gas model (local density approximation).

Local density functional (LDF) theory is currently in a period of considerable activity designed to explore its capabilities and limitations.[9] It has the great advantage of requiring less computational time and space than do *ab initio* procedures; for example, the time required for LDF calculations increases with the third power of the number of basis functions, compared to the fourth power for Hartree-Fock and fifth power for MP2. Since the LDF method takes account of electron correlation, the results are normally superior to Hartree-Fock, and frequently comparable to MP2 and MP4.[10]

ELECTRON DENSITIES AND ELECTROSTATIC POTENTIALS

Once the wave function Ψ has been obtained in an *ab initio* calculation, the electron density ρ follows via eq. (1). In the local density functional procedure, solving eqs. (7) gives a set of ϕ_i's and these lead to ρ. The electrostatic potential can then be computed by means of eq. (2).

The time and space requirements for post-Hartree-Fock *ab initio* computations rapidly become prohibitive as the size of the molecule increases. For example, we found that a single run on a Cray Y-MP8 supercomputer for 1,3-dinitro-1,3-diazacyclobutane (I) required 26 CPU hours at the MP4/6-31G* level.

$$O_2N-N \diamond N-NO_2 \qquad \qquad I$$

Accordingly, most *ab initio* treatments of reasonably-sized molecules are SCF or near-Hartee-Fock, and therefore do not reflect electron correlation. While the resulting errors in calculated energies are very small on a percentage basis, they are unfortunately of the same order of magnitude as binding energies, and hence can be very significant. On the other hand, it follows from a theorem due to Møller and Plesset that properties computed from Hartree-Fock wave functions using one-electron operators [as contrasted to the $1/|\mathbf{r}_j - \mathbf{r}_i|$ term in $\hat{V}$, eq. (5)] are correct through first order; i.e. any errors are no more than second order effects.[11] Such properties include $\rho(\mathbf{r})$ and $V(\mathbf{r})$. Thus, if the first-order wave function Ψ_I is written in terms of the Hartee-Fock solution and a correlation Ψ_1, $\Psi_I = \Psi_{HF} + \Psi_1$, so that the expectation value of a one-electron operator $\hat{M}$ is,

$$\left\langle \Psi_I \left| \hat{M} \right| \Psi_I \right\rangle = \left\langle \Psi_{HF} \left| \hat{M} \right| \Psi_{HF} \right\rangle + 2\left\langle \Psi_{HF} \left| \hat{M} \right| \Psi_1 \right\rangle + \left\langle \Psi_1 \left| \hat{M} \right| \Psi_1 \right\rangle, \qquad (8)$$

then the Møller-Plesset theorem states that the quantity $\left\langle \Psi_{HF} | \hat{M} | \Psi_1 \right\rangle$ is zero, so that the error in the Hartree-Fock expectation value is of at least second order. This can be generalized to the true wave function, $\Psi_T = \Psi_{HF} + \Psi_1 + \Psi_2 + \ldots,$ for which,

$$\left\langle \Psi_T \left| \hat{M} \right| \Psi_T \right\rangle = \left\langle \Psi_{HF} \left| \hat{M} \right| \Psi_{HF} \right\rangle + 2\left\langle \Psi_{HF} \left| \hat{M} \right| \Psi_1 \right\rangle$$
$$+ \sum_{i>1} 2\left\langle \Psi_{HF} \left| \hat{M} \right| \Psi_i \right\rangle + \sum_{i>1}\sum_{j>1} \left\langle \Psi_i \left| \hat{M} \right| \Psi_j \right\rangle \qquad (9)$$

The second term on the right is zero, and only second- and higher-order quantities remain.[12] It must of course be recognized that second order corrections may not always be insignificant;[13] nevertheless, it is anticipated on the basis of the Møller-Plesset theorem that one-electron properties will generally be given to relatively good accuracy by Hartree-Fock wave functions.

There is evidence supporting this expectation for a variety of one-electron properties;[3,14-19] however we shall focus on $\rho(\mathbf{r})$ and $V(\mathbf{r})$. A direct comparison of total electron densities shows correlation to have only a small effect; for example, $\rho(\mathbf{r})$ at the C-C bond critical point of ethane changes from 0.252 to 0.242 in going from the Hartree-Fock 6-31G* to a (correlated) generalized valence bond 6-31G* treatment.[15] A more stringent test is to look at density differences (or "deformation densities"), as between a molecule and its constituent unperturbed atoms. This involves comparing small differences between large numbers; thus any discrepancies in the latter assume major proportions. Such comparisons have been made for a number of systems, and confirm the very close resemblance between the Hartree-Fock and the correlated densities.[17-19] A more significant difference is seen in the Laplacian of the electron density, $\nabla^2\rho(\mathbf{r})$, for which the respective values at the aforementioned ethane critical point are -0.660 and -0.592,[15] but this does after all involve taking second derivatives, which are extremely sensitive to the exact form of $\rho(\mathbf{r})$. Two separate studies have concluded that the topological properties of molecular charge distributions are not significantly affected by the inclusion of electron correlation.[15,16]

A more serious consideration is the degree to which an SCF calculation approaches the Hartree-Fock limit. Density differences obtained with minimum-basis-set SCF wave functions may not be even qualitatively correct, sometimes failing to show key regions of charge buildup.[17] It appears that the wave function needs to be of at least double-zeta quality, preferably augmented by polarization functions.

Proceeding to the electrostatic potential and remaining for the moment in the realm of *ab initio* SCF calculations, the general conclusion that emerged from the analysis of a number of studies is that the overall $V(\mathbf{r})$ pattern of a molecule remains qualitatively essentially the same at various computational levels.[20] The major positive and negative regions, and the approximate locations of the most negative potentials (the local minima, V_{min}) are

Table 1. Calculated *ab initio* SCF electrostatic potential
minima (kcal/mole) for formamide, (III), using
3-21G optimized structure.

	Basis Set				
Location	STO-3G	STO-5G	3-21G	6-31G	6-31G*
Oxygen	-65	-67	-79	-78	-66
	-63	-65	-76	-74	-63
Nitrogen	-17	-19	-17	-13	-12
	-16	-18	-16	-12	-11

not greatly affected by varying, for example, the basis set.
When there are several V_{min}, their ordering usually remains unchanged. On the other hand, the magnitudes of $V(\mathbf{r})$, especially
at the minima, may be quite sensitive to the level of the
calculation, and the exact positions of the minima may also be
affected. For example, the V_{min} due to the nitrogen lone pair in
aziridine, II, was found to change from -76.9 to -92.6 kcal/mole
and to move 0.24 Å closer to the nitrogen nucleus in improving
from a small gaussian basis set to a minimal Slater-type.[21] There
is no consistent trend in the direction in which the values of
the potential minima change with the quality of the wave function. This is brought out in Table 1, which lists the minima
found for formamide, III, at several increasing computational
levels. As seen in Figure 1, there are two minima associated
with the oxygen (reflecting its lone pairs) and two with the
nitrogen, because of its near-planarity.[22] The largest basis set
gives essentially the same values for the oxygen minima (but not
the nitrogen) as does the smallest. The magnitudes of the
nitrogen V_{min} are much smaller than is typical for an amine
nitrogen; this is due to the charge delocalization and near-
planarity associated with the nitrogen in formamide.[22,23]

II

III

IV

An early indication of the effect of electron correlation
upon $V(\mathbf{r})$ came from a CI study of formaldehyde, IV.[24] Qualitatively, the potential map remains essentially the same as at the
SCF level; however the magnitudes of the oxygen minima diminish
from -60.4 kcal/mole (SCF) to -50.2 (2474 CI). Nearly all of
this change accompanies the first increase in the number of configurations, from 1 to 261. The subsequent inclusions of additional configurations, while very significantly improving the
energy, affect V_{min} by less than 1 kcal/mole.

A final example, which will be considered in some detail, is
the hydroxide ion, OH^-. This presents the new feature of being a
negative ion, and accordingly having a somewhat spread out electronic charge distribution. It has been pointed out that computational treatments of anions are considerably improved by the inclusions of additional "diffuse" (as contrasted to "polarization")
functions in the basis set.[3,25] Certainly their effect is clearly
seen in our calculated V_{min} values, as will be pointed out.

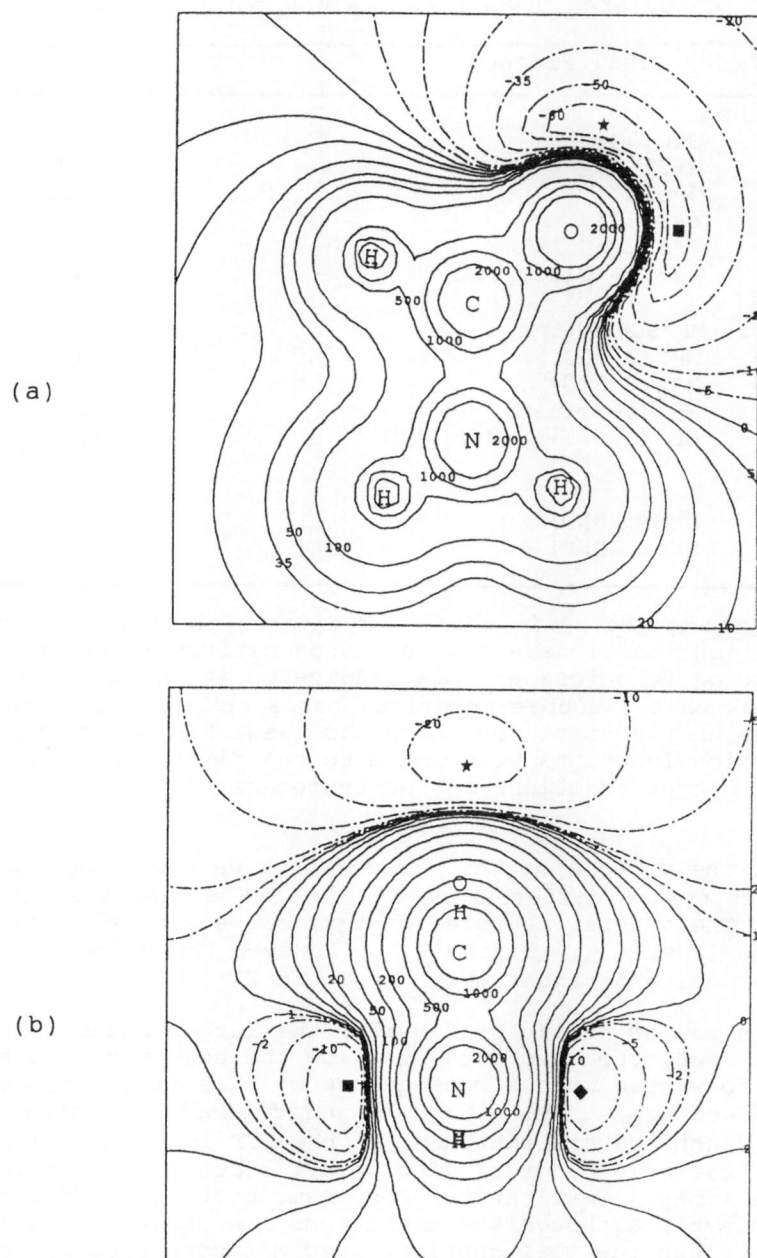

Fig. 1. Calculated electrostatic potential of formamide (III) in
kcal/mole, obtained using an *ab initio* SCF-MO procedure
at the STO-5G level (a) in the plane defined by C, N and
O, and (b) in the plane perpendicular to the C,N,O plane
passing through the C-N bond. The projections of nuclear
positions not in these planes are shown by their atomic
symbols. Dashed contours correspond to negative potentials. The positions of the most negative potentials
are indicated; the values are: (a) ★ -67, ■ -65;
(b) ★ -24, ■ -19, ◆ -18.

Table 2. Calculated Oxygen V_{min} values for hydroxide ion, OH^-.

Computational Procedure	V_{min} (kcal/mole)
ab initio SCF:[a]	
STO-5G	-333
3-21G	-308
6-31G*	-283
6-311G**	-279
6-311++G	-258
6-311++G(3df,3pd)	-248
ab initio MP/SCF:[a]	
MP2/3-21G	-303
MP2/6-31G*	-282
MP2/6-311++G	-241
MP2/6-311++G(3df,3pd)	-233
LDF:[b]	
DMol (DND)	-241
DMol (DNP)	-248

[a]In describing the basis sets, * indicates the presence of polarization functions and + indicates diffuse functions. Polarization functions are also indicated in parentheses.
[b]DND means that a double-numerical basis set with polarization functions on the first row atoms was used; DNP means that polarization functions were added to the first row atoms (as in DND) and p-type functions to the hydrogens.

For the purpose of this discussion, we have computed $V(\mathbf{r})$ for OH^- at twelve different *ab initio* (SCF and MP/SCF) and LDF levels. The general features of $V(\mathbf{r})$ are essentially the same in all cases (see, e.g., Figure 2.). However the values of the V_{min} vary over quite a range, shown in Table 2.

With the *ab initio* procedures, the larger and more flexible the basis set, the more does it allow the computed electron density to spread out; a consequence of this is a decrease in the magnitude of V_{min}. This effect is particularly striking when diffuse functions are included; at the SCF level they bring V_{min} to -258 kcal/mole, whereas with polarization functions it was -279 kcal/mole. Qualitatively the same picture is obtained at the MP2 level, although the magnitudes are somewhat smaller than the SCF. Thus the most sophisticated MP2 computation, for which the basis set contains both diffuse and polarization functions, yields V_{min} = -233 kcal/mole, whereas the corresponding SCF treatment gives -248 kcal/mole. The two LDF results are quite similar, and reasonably close to the highest level MP2. It is interesting to note that inclusion of polarization functions increases the magnitude of V_{min} in the LDF treatment, in contrast to what is observed in the *ab initio* calculations; we have found the same to be true in a computational study of singlet CH_2 .

SUMMARY

For many applications of calculated electron densities or electrostatic potentials, it is the trends in these properties that are important; for example in interpreting or predicting

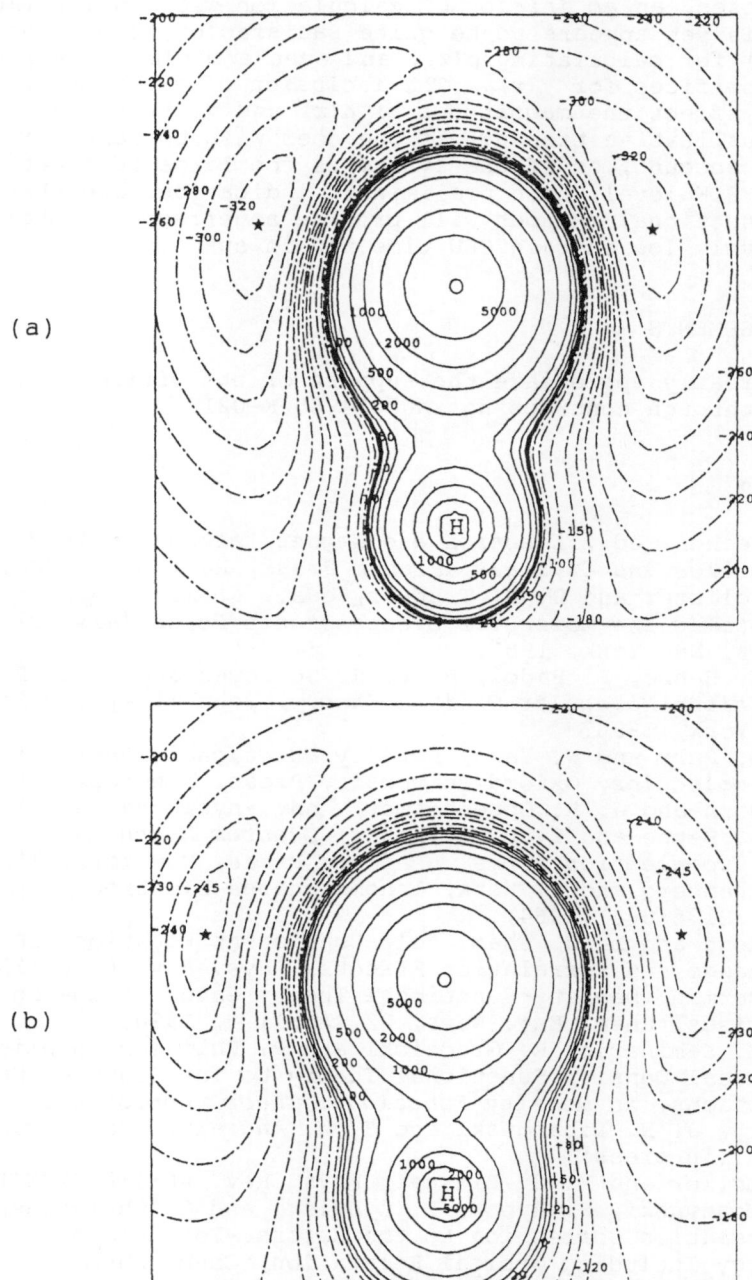

Fig. 2. Calculated electrostatic potential of the hydroxide
ion, OH⁻, in kcal/mole, obtained using (a) an *ab initio*
SCF-MO procedure at the STO-5G level, and (b) a local
density functional procedure (DMol) at the DNP level.
Dashed contours correspond to negative potentials. The
positions of the most negative potentials are indicated;
the values are: (a) ★ -333; (b) ★ -248.

chemical reactivity, one may be comparing different sites in a given molecule, or a series of somewhat related molecules. For such purposes, an *ab initio* SCF calculation with an intermediate level basis set appears to be quite satisfactory (for neutral molecules) for calculating $\rho(\mathbf{r})$, and even a minimal basis set normally suffices for $V(\mathbf{r})$. The inclusion of electron correlation does affect the magnitudes of $\rho(\mathbf{r})$ and $V(\mathbf{r})$, but usually not the qualitative trends. If one does wish to take correlation into account, or if the system is too large to treat at the *ab initio* SCF level, then preliminary indications are that local density functional methods will provide an effective alternative at relatively low cost in CPU time and in space.

ACKNOWLEDGEMENTS

We greatly appreciate the support of the Office of Naval Research through contract no. N00014-85-K-0217.

REFERENCES

1. P. Becker, ed., *Electron and Magnetization Densities in Molecules and Crystals*, Plenum Press, New York, 1980.
2. P. Politzer and D. G. Truhlar, eds., *Chemical Applications of Atomic and Molecular Electrostatic Potentials*, Plenum Press, New York, 1981.
3. W. J. Hehre, L. Radom, P. v. R. Schleyer and J. A. Pople, *Ab Initio Molecular Orbital Theory*, John Wiley and Sons, New York, 1986.
4. R. G. Parr and W. Yang, *Density Functional Theory of Atoms and Molecules*, Oxford University Press, New York, 1989.
5. M. Born and J. R. Oppenheimer, Ann. Physik 84:457 (1927).
6. H. A. Bethe and E. E. Salpeter, *Quantum Mechnics of One- and Two-electron Atoms*, Academic Press, New York, 1957.
7. P.Hohenberg and W. Kohn, Inhomogeneous Electron Gas, Phys. Rev. B136:864 (1964).
8. W. Kohn and L. J. Sham, Self-Consistent Equations Including Exchange and Correlation Effects, Phys. Rev. A140:113 (1965).
9. S. Borman, Density-Functional Theory Gains Following Among Chemists, Chem. Eng. News, 22, April 9, 1990.
10. J. M. Seminario, M. Grodzicki and P. Politzer, Applications of Local Density Functional Theory to The Study of Chemical Reactions, in *Density Functional Theory Approaches to Chemistry*, J. K. Labanowski and J. W. Andzelm, eds, Springer, 1990 (in press).
11. C. Møller and M. S. Plesset, Phys. Rev. 46:418 (1934).
12. R. Moszynski, S. Rybak, S. Cybulski and G. Chalasinski, Correlation Correction to the Hartree-Fock Electrostatic Energy Including Orbital Relaxation, Chem. Phys. Lett. 166:609 (1990).
13. J. A. Pople and R. Seeger, Electronic Density in Møller-Plesset Theory, J. Chem. Phys.62:4566 (1975).
14. B. J. Rosenberg and I. Shavitt, *Ab Initio* SCF and CI Studies on the Ground State of the Water Molecule. I. Comparison of CGTO and STO Basis Sets Near the Hartree-Fock Limit, J. Chem. Phys. *63*:2162 (1975).
15. C. Gatti, P. J. MacDougall and R. F. W. Bader, Effect of Electron Correlation on the Topological Properties of Molecular Charge Distributions, J. Chem. Phys. 88:3792 (1988).

16. R. J. Boyd and L.-C. Wang, The Effect of Electron Correlation on the Topological and Atomic Properties of the Electron Density Distributions of Molecules, J. Comp. Chem. 10:367 (1989).

17. V. H. Smith, Jr., Theoretical Determination and Analysis of Electronic Charge Distributions, Phys. Scr. 15:147 (1977).

18. G. Lauer, H. Meyer, K.-W. Schulte, A. Schweig, and H.-L. Hase, Correlated Electron Density of N_2, Chem. Phys. Lett. 67:503 (1979).

19. L.-C. Wang and R. J. Boyd, The Effect of Electron Correlation on the Electron Density Distributions in Molecules: Comparison of Perturbation and Configuration Interaction Methods, J. Chem. Phys. 90:1083 (1989).

20. P. Politzer and K. C. Daiker, Models for Chemical Reactivity, in *The Force Concept in Chemistry*, B. M. Deb, ed., Van Nostrand Reinhold, New York, 1981.

21. C. Ghio and J. Tomasi, The Protonation of Three-Membered Ring Molecules: The *ab initio* SCF versus the Electrostatic Picture of the Proton Approach, Theoret. Chim. Acta 30:151 (1973).

22. J. S. Murray and P. Politzer, The Effects of Water Upon the Hydrogen Bonding in a Formamide-Ammonia Complex, Chem. Phys. Letters 136:283 (1987).

23. J. S. Murray and P. Politzer, Electrostatic Potentials of Amine Nitrogens as a Measure of the Total Electron-Attracting Tendencies of Substituents, Chem. Phys. Letters 152:364 (1988).

24. R. Daudel, H. LeRonzo, R. Cimiraglia and J. Tomasi, Dependence of the Electrostatic Molecular Potential upon the Basis Set and the Method of Calculation of the Wave Function. Case of the Ground $^3A_1(\pi \to \pi^*)$ and $^1A_1(\pi \to \pi^*)$ States of Formaldehyde, Int. J. Quantum Chem. 13:537 (1978).

25. J. Chandrasekhar, J. G. Andrade and P.v.R. Schleyer, Efficient and Accurate Calculations of Anion Proton Affinities, J Am. Chem. Soc. 103: 5609 (1981).

COMPARISON OF PRECISE ELECTRONIC CHARGE DENSITIES AND ELECTRO-STATIC POTENTIALS OBTAINED FROM DIFFERENT *AB INITIO* APPROACHES

Jorge M. Seminario and Peter Politzer

Department of Chemistry
University of New Orleans
New Orleans, LA 70148 USA

The treatment of molecular systems via standard *ab initio* methods requires the use of a large amount of computational resources. With present *ab initio* techniques, it is possible to obtain electronic charge densities and electrostatic potentials at the HF and MP2 levels of theory, restricted however to about 250 basis functions for the HF and about 100 for the MP2 calculations. The most sophisticated level of theory for computing charge densities and electrostatic potentials within the Gaussian 88 program is the MP2/6-311++G(3df,3pd) procedure. Unfortunately, its use is only feasible for molecules containing up to 2 or 3 atoms. As an alternative solution for calculating precise electronic charge densities and electrostatic potentials, we now have access to methods based on Local Density Functional theory[1]. These schemes produce very accurate electronic charge densities and electrostatic potentials at a relatively low cost in computer resources (N^3 for LDF against N^4 for HF and N^5 for MP2); the results compare in quality with the best standard *ab initio* calculations.

REFERENCE

[1] B. Delley, *J. Chem. Phys.* **92** (1990) 1; Program DMol from BIOSYM Technologies Inc.

THEORETICAL AND EXPERIMENTAL STUDY OF DIMERIZATION

OF SUBSTITUTED PHENYLACETYLENES

F. Sevin and N. Balcioglu

Department of Chemistry, Faculty of Engineering
Hacettepe University
06532 Beytepe, Ankara, TURKEY

Terminal acetylenes dimerize to enzymes under the catalytic effect of Cu(I) salts in glacial acetic acid (Straus Reaction).

$$2RC \equiv C\,Cu \xrightarrow[CH_3COOH]{Cu_2O} RC \equiv C - CH = CHR$$
$$\text{(cis and trans)}$$

In the present work, the effects of substituents on the aromatic ring in the dimerization reaction of p-nitrophenylacetylene (p-NPA), mesitilacetylene (MA) and phenylacetylene (PA) have been investigated and the relative rates of dimerization of these acetylenes have been determined.

Copper(I) acetylides are formed easily in the presence of Cu_2O catalyst by all three acetylenes. While p-NPA and MA remain unchanged at this stage, PA gives the expected enzyme. Cross-coupling Straus experiments result only in the formation of PA-MA products. PA-MA, MA-p-NPA cross-coupling products are not formed. In the results of cross-coupling reactions, existence of the following equilibrium is detected:

In addition, by using the EHM method the theoretical absorption spectrums and dipole moments have been calculated for Cu(I)-phenylacetylide and for Cu(I)-p-nitrophenylacetylide with Cu(I)-mesitilacetylide which do not give the coupling products and have different colors. A good linear relation is not obtained between the experimental (pyridine as solvent) and the theoretical UV-VS spectrums of these copper(I) acetylides. Nevertheless, the dipole moment of Cu(I)-p-nitrophenylacetylide has been found to be greater than the others.

As a consequence, while the non-activity of MA was attributed to the steric effect of the O-methyl groups, the surprising inertness of p-NPA could be due to the increased polarity of the corresponding acetylide anion which results in low solubility and increased stability of this compound. Relative rates of dimerization are increased in the sequence:

$$PA > MA > \text{p-NPA}.$$

ELECTRON DEFORMATION DENSITY OF METAL CARBYNE COMPLEXES

Anne Spasojevic-de Biré[*], Nguyen Quy Dao

Laboratoire de Chimie et Physico-Chimie Moléculaires
U.A. 441 C.N.R.S., E.C.P.
F-92295 Châtenay-Malabry Cedex, France

* present address: Laboratoire Léon Brillouin
  C.E.A. - C.N.R.S., C.E.N. Saclay
  F-91191 Gif-sur-Yvette Cedex, France

Pierre Becker

Laboratoire de Minéralogie Cristallographie
U.A. 9 C.N.R.S., Université P. M. Curie
F-75230 Paris Cedex 05, France

Marc Bénard, Alain Strich, Claudine Thieffry

Laboratoire de Chimie Quantique
E.R. 139 C.N.R.S., Institut Le Bel
F-67000 Strasbourg, France

Niels Hansen, Claude Lecomte

Laboratoire de Minéralogie Cristallographie et
Physique infrarouge
U.R.A. 809 C.N.R.S., BP 239
F-54506 Vandœuvre-lès-Nancy Cedex, France

INTRODUCTION

The carbynic complexes have been discovered in 1973 by E. O. Fischer[1]. Their originality consists in a triple bond between a transition metal and a carbon atom, which makes them very unstable, sensitive to moisture, air and temperature. In relation with this very specific structural character, the chemical bonding in this family of compounds requires special attention in order to understand a new and important area of organometallic chemistry. As a matter of fact, carbyne complexes have been recognized as precursors in the synthesis of organometallic products which take part into catalytic reactions[2]. Our goal was to investigate the electronic deformation density of two members of this family: the *trans*- chlorotetracarbonyl phenylcarbyne chromium (I) and its

The Application of Charge Density Research to Chemistry and Drug Design
Edited by G. A. Jeffrey and J. F. Piniella, Plenum Press, New York, 1991

methylic analog (II). The present paper is designed to show that a joint experimental and theoretical study combining the multipole model refinement and the ab initio CASSCF calculations can lead to informations of interest to the chemist on topics such as backdonation, conjugation or hyperconjugation effects.

EXPERIMENTAL PART

Phenylcarbyne complex (I)

(I) crystallizes in the non centrosymmetric space group $I4_1md$. Using the phases resulting from the refinements of spherical atom models in the calculation of electron deformation density maps are not valid in this case. In order to determine better phases of the observed structure factors, we have used Hansen and Coppens' multipole model[3] (computer program MOLLY)for the least squares refinement of the X-ray diffraction intensities. Crystallographic data and agreement indices obtained in the refinements are summarized in Table 1[4,5]. Positional parameters deduced from the aspherical atom refinement are consistent with those obtained by the neutron diffraction experiment[6].

The experimental electron deformation density is defined as:

$$\rho_{def.}(\mathbf{r}) = (1/V) \sum (\mathbf{F_1}/k - \mathbf{F_2}).exp(-2\pi i\mathbf{H}.\mathbf{r})$$

where $\mathbf{F_1}$ represents an observed structure factor with the phase calculated from a multipole model for the bonded molecule and $\mathbf{F_2}$ is calculated for a promolecule, i.e. using spherical atom form factors; the structural parameters are taken from the converged multipole model.

Methylcarbyne complex (II)

Goddard, Krüger and coworkers.[7,8] have collected the X-ray diffraction data of (II) and studied the electron deformation density by X-X method. In order to compare with the present results, we have also applied[9] the multipole model to their data set (Table 1). Although the space group is centrosymmetric, the use of the multipole model allows a more detailed analysis of the deformation maps than the Fourier methods. The multipole model was constrained so that the four carbonyl groups have identical deformation densities, and the chromium atom a four-fold symmetry. The anisotropy of the chromium d orbital occupations discussed below is only apparent in the static deformation maps.

ORTEP views of the two complexes are shown on Fig. 1.

THEORETICAL CALCULATIONS

Quantum chemical ab initio calculations have been carried out at the CASSCF[10] level on the hypothetical system $Cl(CO)_4Cr \equiv CH$ (III)[11] and on the methyl complex (II)[9]. The observed geometrical parameters have been modified in order to retain a perfect C_{4v} symmetry for the $Cl(CO)_4Cr$ fragment, and C_s symmetry for the complete molecule. Details concerning the Gaussian basis sets selected for these calculations can be found in previous publications[9,11].

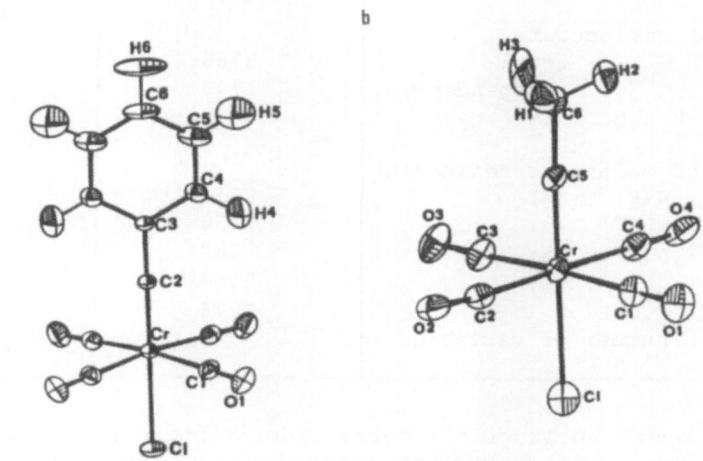

Fig. 1. Molecular structures. a) phenylcarbyne complex (I);

b) methylcarbyne complex (II).

Table 1. Crystallographic data and agreement indices of the refinements.

compound	$Cl(CO)_4Cr \equiv CC_6H_5$	$Cl(CO)_4Cr \equiv CCH_3$
space group	$I4_1md$	$P2_1/c$
cell dimensions		
a (Å)	10.045(3)	6.504(2)
b (Å)	10.045(3)	11.261(5)
c (Å)	12.002(3)	12.756(6)
α (°)	90.0	90,0
β (°)	90.0	105.58(3)
γ (°)	90.0	90.0
V (Å^3)	1211.6	900,0
wavelength, λ (Å)	0.71069	0.71069
$\sin\theta/\lambda_{min}$ (Å^{-1})	0.05	0.05
$\sin\theta/\lambda_{max}$ (Å^{-1})	1.17	0.70
T (K)	110	100
number of reflections		
total	3768	4742
independent	1790	1836
internal R factor (%)	2.5	1.3
results of multipole refinement		
R(F) (%)	2.33	2.08
wR(F) (%)	1.30	2.26
R(F^2) (%)	2.89	2.87
wR(F^2) (%)	2.60	4.95
G.O.F.	2.76	1.36
number of variables	104	91

In order to provide a correct description of the metal-carbon triple bond, the 6 MOs representing the Cr - C bonding and antibonding orbitals (σ, π, π^*, σ^*), altogether populated with 6 electrons, have been included in the active space of the CASSCF calculations, hereafter referred to as CASSCF-6. Only for (III), the highest occupied and lowest unoccupied orbitals of b_2 symmetry have been added to the active space of the CASSCF calculations. Those orbitals are located into the $Cl(CO)_4Cr$ plane and describe the metal to CO backdonation. This latter calculation will be refered to as CASSCF-8, reminding that the CASSCF active space contains 8 orbitals, populated with 8 electrons.

DISCUSSION

The Cr(CO)$_4$ plane

Fig. 2 displays the X-M experimental deformation density and the static density maps of (I) compared to the static density and the theoretical density maps of (II) in the Cr(CO)$_4$ plane. On each map, the characteristic features of the carbonyl bonds are similar. The electronic density pattern around the chromium atom, characterized by positive peaks along the xy directions, agrees with the Chatt and Duncanson model[12] and is consistent with a backdonation from the metal to the carbonyls. This observation is made for molecules (I) and (II), from both the experimental and the theoretical maps.

The density accumulation in the region of the oxygen lone pairs in (I) appears weak and diffuse (Fig. 2a, b) because of the strong thermal smearing in this part of the molecule. The lone pairs do not even show up on the experimental maps of (II) (see for exemple Fig. 2c) due to the lack of high order reflections.

In the theoretical calculations (Fig. 2d), the moderate extension of the atomic basis sets (double-zeta quality for the valence shell of C, O and H atoms, triple-zeta for the 3d shell of the chromium) and the lack of polarization functions somewhat impairs the quantitative accuracy of the maps deduced from the computed wave function. The influence of the basis set truncation on

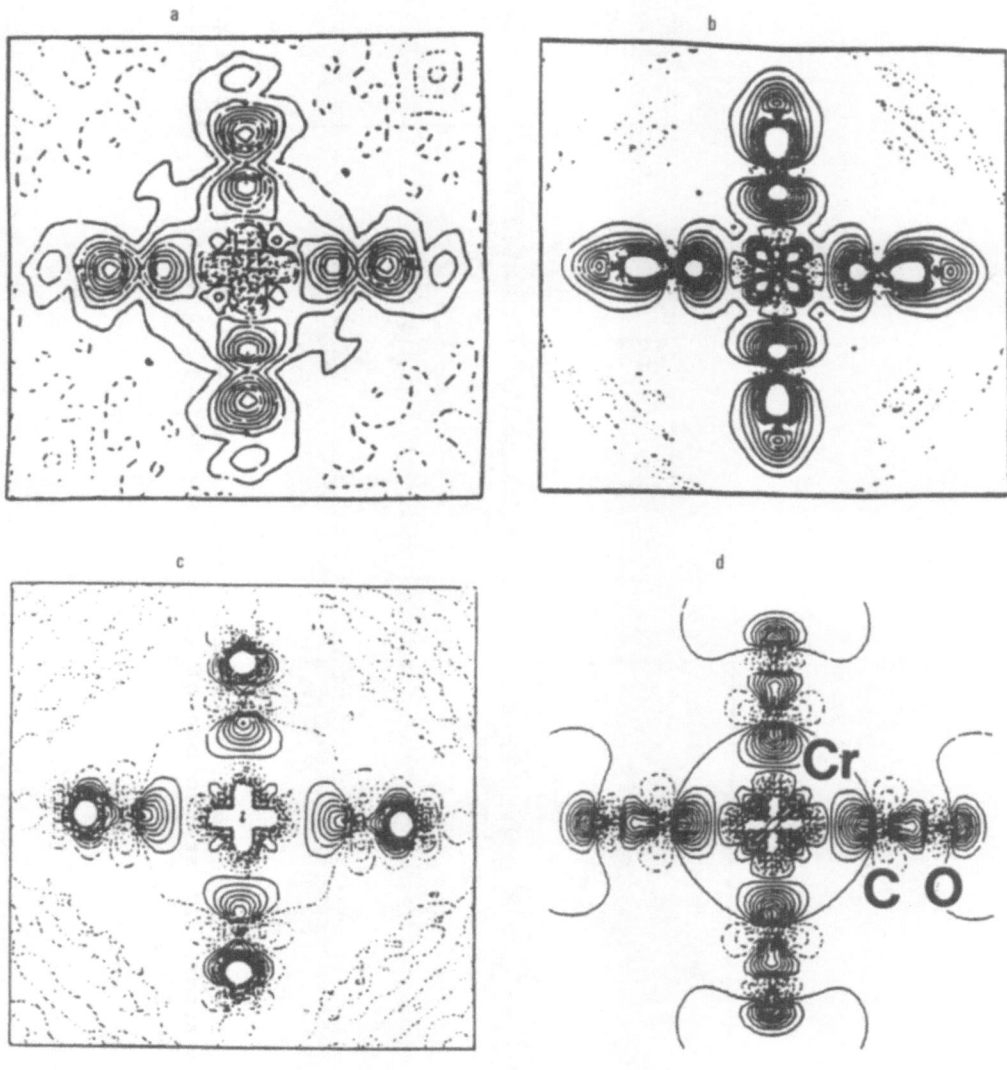

Fig. 2.   The density maps in Cr(CO)$_4$ plane.   a) X-M experimental map of (I);   b) static map of (I);   c) static map of (II);   d) theoretical map of (II). (contour interval 0.1 e.Å^{-3})

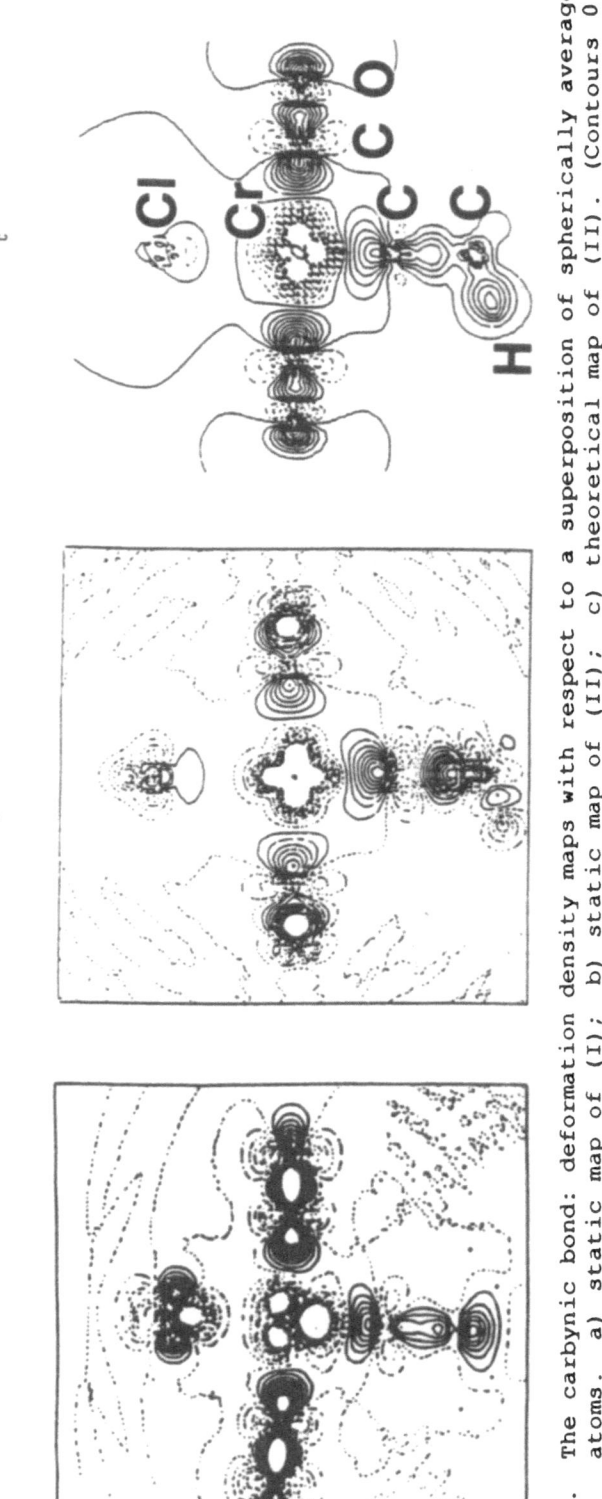

Fig. 3. The carbynic bond: deformation density maps with respect to a superposition of spherically averaged atoms. a) static map of (I); b) static map of (II); c) theoretical map of (II). (Contours 0.1 e.Å^{-3}.)

some specific features of the deformation density are well known: one can for instance expect an underestimation of the interatomic accumulation and an overestimation of the density buildup corresponding to the lone pairs[12]. Nevertheless, the correct agreement obtained in the present work as well as for other organometallic molecules containing carbonyl ligands[14-18] gives confidence into the validity of the obtained pattern.

The carbynic bond

Fig. 3 shows the static density maps of (I) and (II) compared to the theoretical map in a plane containing two carbonyl groups and the carbynic bond. The density accumulation describing the $Cr \equiv C$ bond is shaped as a flat ellipsoïd situated on the bond axis near the carbon atom. Depopulated regions are found along the same axis around the metal on the one hand, and close to the carbynic carbon nucleus on the other hand, with symmetric minima on both sides of the bond axis. It can be noticed that the general features of this bond, obtained either from experiment or from theory, are very similar for the two compounds. In an earlier work, Krüger et al.[7,8] presented experimental deformation density maps characterized by a relatively poor description of the carbyne bond in relation with the use of the spherical atom model. Very recently, Low and Hall[19] have published ab initio calculations at RHF level on (II) and presented theoretical maps very similar to those displayed on Fig. 3c. A quantitative difference can however be noticed: the height of the density peak associated with the carbynic bond reaches 0.8 e.$Å^{-3}$ in the maps obtained at the RHF level, compared to 0.6 e.$Å^{-3}$ in the present work, where the 6 electrons describing the $Cr \equiv C$ bond are correlated through the CASSCF formalism. This 0.2 e.$Å^{-3}$ decrease of the peak height can be reasonably attributed to large correlation effects as documented elsewhere[11].

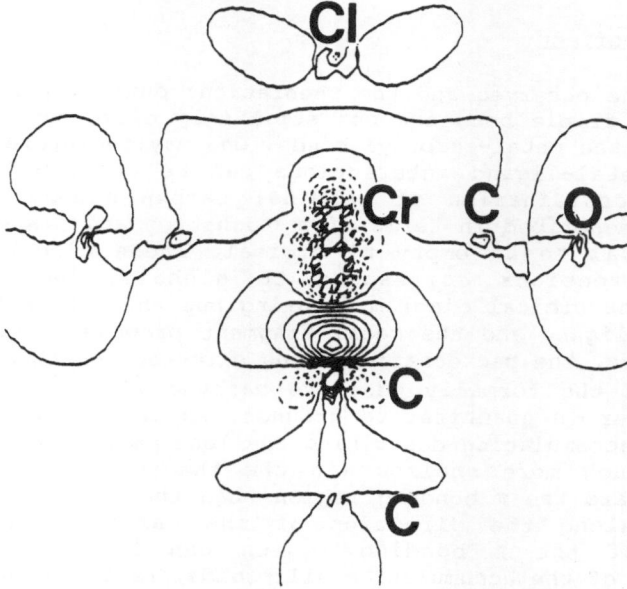

Fig. 4. The carbynic bond: theoretical deformation map of (II) with respect to the superposition of the $ClCr(CO)_4$ and $C-CH_3$ neutral fragments both computed in their quartet state (contours 0.1 e.$Å^{-3}$).

391

Table 2. Eccentricities of the carbyne bond deformation density distribution.

	$Cl(CO)_4Cr \equiv CC_6H_5$ exp.	$Cl(CO)_4Cr \equiv CCH_3$ exp.	CASSCF-6
$Cr \equiv C$	2.6 - 2.0	2.1	2.3
$Cr - C(CO)$	0.5 - 0.8	1.4	1.5

The origin of the accumulation of density, largely extended to the π regions, can be more accurately traced from the theoretical deformation maps obtained with respect to a superposition of two neutral molecular fragments, namely $Cr(CO)_4Cl$ et $C - CH_3$ computed each in their lowest quartet state (Fig. 4). This map is characterized by a depopulation of the metal d_{z^2} orbital to the benefit of a large region corresponding in part to the area of the carbon lone pair, but extending into broad "wings" that evidence the metal-carbon π bonding. The polarization of the density toward the C5 atom indicates that the metal-carbon interaction cannot be strictly considered as a covalent triple bond between the two fragments in their quartet states. This is particularly obvious in the region of the σ coupling, which retains much of the character of a carbon lone pair. The π electrons appear to be more equally distributed between the metal and the carbynic carbon. The distribution of the density around C5 is therefore consistent with a $\sigma^2\pi^2$ formal electronic configuration for the carbynic carbon, thus explaining the origin and the localization of the negative net charge that theoretical treatments consistently assign to that atom[11,20-22].

Backdonation effect

Both the observed and the theoretical density pattern for the metal-carbon triple bond are not strikingly different from the one obtained for the metal-carbonyl bonds. One may therefore expect that these two metal-ligand interactions can be described in similar terms. The coordination of terminal carbon monoxyde ligand is classically described in terms of a σ donation and a π backdonation from the metal to an empty π^* orbital. These electron flows in opposite directions correspond to globally balanced charge transfers. The orbital diagram figuring out the interaction between the carbyne ligand and the metal fragment proposes a qualitatively similar scheme, the backdonation being directed now toward the empty π orbital of the formally positive carbyne ligand[7]. A difference exists however on quantitative grounds, as can be seen in Fig. 3. The density accumulation describing the lone pair of the carbynic C5 carbon is much more shallow (in the theoretical map) and more extended toward the π bonding region than the similar peaks facing the metal along the direction of the carbonyl ligands. This extension of the π bonding region can be measured by the eccentricity of the accumulation ellipsoïds, reported in Table 2. In agreement with the results of the Mulliken population analysis of the CASSCF wave functions[9,11], we interpret these qualitative differences in terms of an increased backdonation toward the carbyne ligand leading to metal-carbyne π bonds characterized by a near balance of the population between the metal and the carbon atomic π orbitals. The equilibrium between donation and backdonation charge

392

transfers is not anymore observed: backdonation becomes dominant thus leading to stronger π bonds and to a negative point charge on the carbynic carbon.

Conjugation effect in the phenylcarbyne complex (I)

An examination of the dynamic X-M and static deformation densities along the Cr ≡ C - C axis in two planes respectively containing and perpendicular to the phenyl ring (Fig. 5) shows that the electronic density is not distributed isotropically around the symmetry axis. First, the accumulation between Cr and the carbynic carbon appears slightly more extended in the plane containing the

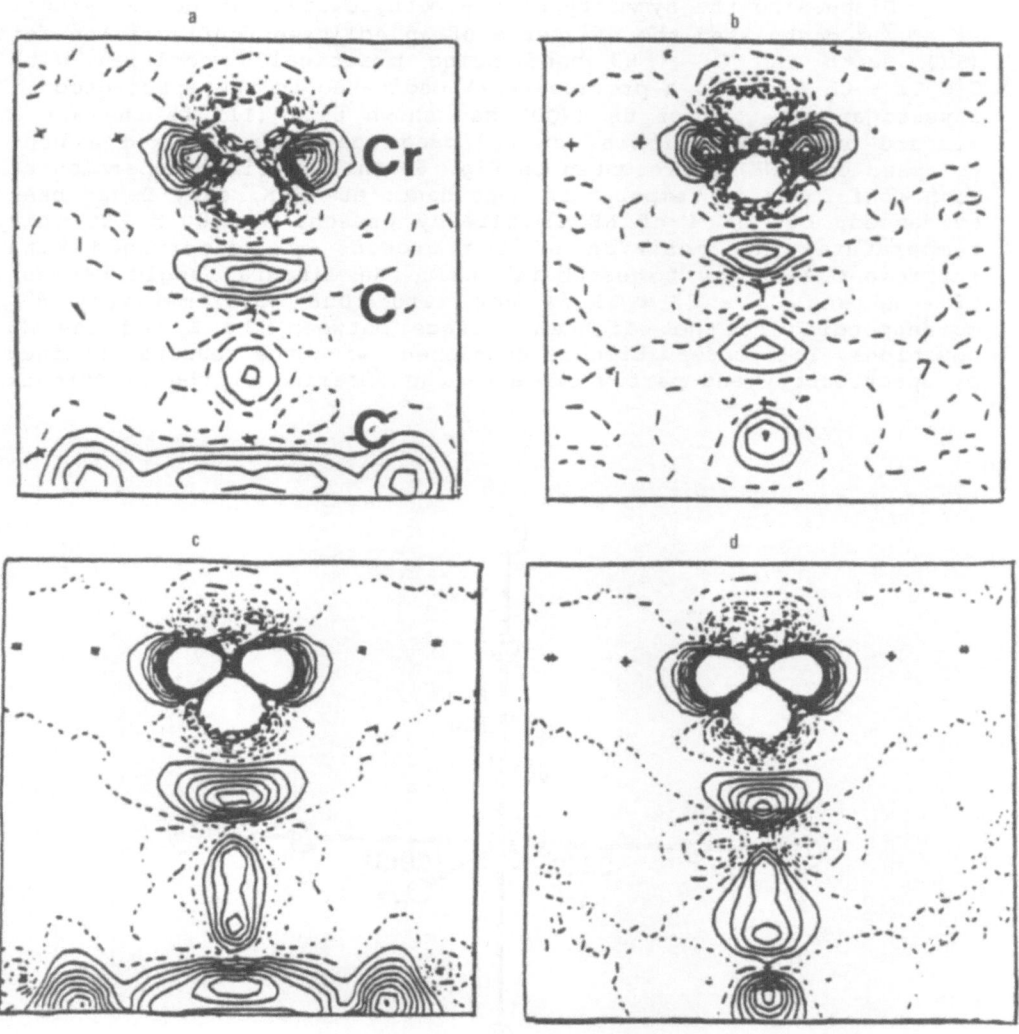

Fig. 5. Conjugation effect in (I)   a) X-M experimental deformation map in the plane containing the phenyl ring;   b) X-M experimental deformation map in the plane perpendicular to the phenyl ring;   c) static deformation map in the plane containing the phenyl ring;   d) static deformation map in the plane perpendicular to the phenyl ring; (contours 0.1 e.Å^{-3})

phenyl. Secondly, an opposite tendency is observed for the accumulation located along the C2 - C3 bond, which appears broader in the plane *perpendicular* to the phenyl. This loss of cylindrical symmetry which is reminiscent of the organic cumulenes[23], can be attributed to the conjugation between the phenyl ring and the triple bond. This interpretation is in agreement with an analysis of the force field obtained from vibrational spectroscopy[24]. The bond orders, as deduced from the force constants, indicated some delocalization of the π bonding from the Cr $\equiv$ C bond (estimated bond order 2.7) toward the next C - C bond, whose bond order was estimated to 1.4.

Existence of two conformers in (II) and hyperconjugation effect

Discussing the symmetry of the methylcarbyne molecule, Krüger et al.[7,8] postulated the existence of an eclipsed configuration for (II), with the C6 - H3 bond being practically coplanar with C $\equiv$ Cr - C3 $\equiv$ O3. In a previous work dedicated to the spectroscopic investigations, one of us (NQD) has shown that (II) is in fact a mixture of two conformers, an eclipsed conformer (E) and a non-eclipsed one (NE) represented on Fig. 6. The relative proportion of both conformers is temperature dependant: at 77 K, only E has been evidenced, but 33 % of NE is already present at 100 K. At this temperature, the position of the protons was determined with sufficient accuracy to establish that the dihedral angle between C6 - H3 and Cr - C3 $\equiv$ O3 is not zero, but aproximatively 8°, corresponding to the weighted average between the E and the NE positions. This observation is consistent with the results obtained by spectroscopy and partly explains the smearing of the electronic

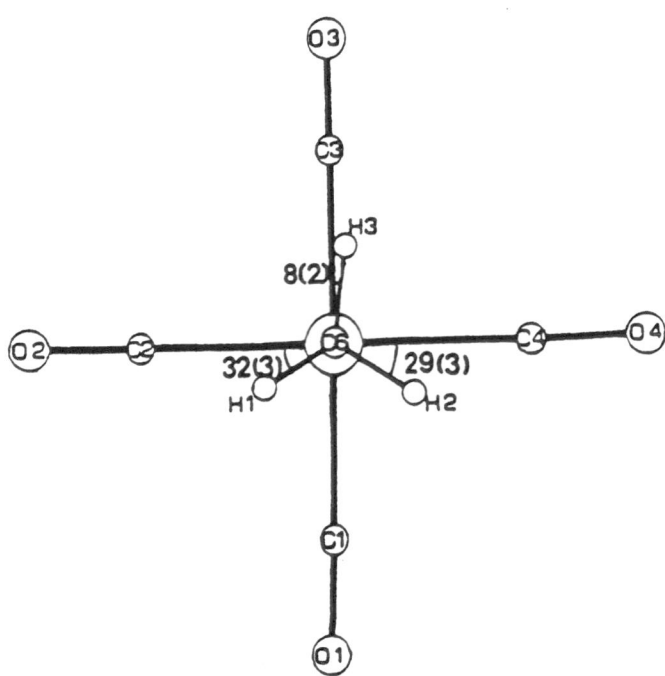

Fig. 6. Dihedral angles of the methyl group.

density in the region of the C - H bonds. At variance from what was observed in the phenyl complex, neither the experimental or the theoretical maps obtained in two perpendicular planes of (II) containing the carbyne bond do display appreciable differences. An information of interest is however obtained from Fig. 4, displaying the theoretical deformation density with respect to a superposition of two molecular fragments. Bonding with the metal fragment leads to an increase of the density along the C5 - C6 bond (0.2 e.$Å^{-3}$) and this accumulation extends toward the π regions. This could be interpreted as an increase of the $C_{carbyne}$ - C_{methyl} bond strength due to hyperconjugation. This interpretation is confirmed by an analysis of the vibrational spectrum of partially deuterated methylcarbyne complex[25-29]. The Cr - C and C - C bond orders, as deduced from the field analysis, appear to deviate from the formal values three and one. The coexistence of two conformers, and the consecutive smearing of the density in the methyl region, could explain why this effect, much weaker than the charge anisotropy observed in (I), was not detected in the experimental maps.

The origin of the Cr - C(CO) bond alternation in the methyl complex

One of the puzzling questions about the structure of (I) concerns the origin of the Cr - C_{CO} bond length alternation that reduces the symmetry of the $Cr(CO)_4$ fragment from C_{4v} to C_{2v} ($d_{Cr-C_{1/3}} = 1.928(3)$ Å, $d_{Cr-C_{1/3}} = 1.928(3)$ Å, $d_{Cr-C_{2/4}} = 1.978(3)$ Å, $d_{C_{1/3}-O_{1/3}} = 1.128(4)$ Å, $d_{C_{2/4}-O_{2/4}} = 1.147(3)$ Å). Krüger et al.[7] tentatively assigned the breaking of the C_{4v} local symmetry to an electronic influence of the methyl group either occuring in the isolated molecule or originating in intermolecular H ⋯ Cl interactions.

The quantum chemical calculations were carried out assuming an eclipsed conformation for the methyl group and an averaged C_{4v} symmetry for the Cr - $(CO)_4$ subunit. However, the point group of the modelized system was no more than C_s due to the presence of the methyl substituent, and the molecular wave function was therefore unconstrained and free to delocalize an anisotropic perturbation into the $Cr(CO)_4$ moiety. No significant influence of the CH_3 dissymetry could be traced however beyond the methyl carbon C6. On the experimental side, the dissymetry observed by Krüger et al.[7,8] in the X-X deformation density maps of the $Cr(CO)_4$ plane vanishes when the multipole refinement is applied to the same set of X-ray diffraction measurement. Both the theoretical and the X-M density distributions therefore seem to rule out the hypothesis of a delocalized electronic effect originating in the methyl substituent and downgrading the symmetry of the $Cr(CO)_4$ fragment. A plausible rationalization for the C_{2v} structure of $Cr(CO)_4$ should therefore rely on intermolecular interactions and induce a minimal dissymetry in the distribution of the electron density.

In a recent theretical study[19], Low and Hall went one step further by fully optimizing the geometry of the molecule at the ab initio RHF level. When the 3-fold axis of the methyl group is assumed to be collinear with the Cr ≡ C axis, the geometry optimization is unable to generate a difference between the two pairs of Cr - C(CO) bond lengths, in agreement with the conclusions of our theoretical study. However, a 12° tilt of the methyl group axis, originating in intermolecular contacts, is sufficient to

Table 3. d orbital populations of (II) and (III).

	$Cl(CO)_4Cr \equiv CH$		$Cl(CO)_4Cr \equiv CCH_3$	
	CASSCF-6	CASSCF-8	CASSCF-6	exp.
d_{z^2}	0.78 (15%)	0.75 (15%)	0.73 (14%)	0.8(4) (15%)
$d_{xz} + d_{yz}$	2.24 (43%)	2.31 (45%)	2.28 (44%)	2.6(4) (47%)
$d_{x^2-y^2}$	0.44 (9%)	0.51 (10%)	0.43 (8%)	0.6(4) (12%)
d_{xy}	1.74 (33%)	1.59 (31%)	1.72 (33%)	1.5(4) (27%)
Total	5.20	5.16	5.16	5.60(8)

generate[19] an alternation of the Cr - C(CO) bond length very similar to what was experimentally reported.

In conclusion, both studies agree to rule out the existence of an electronic intramolecular effect originating in the methyl group and propagating to the $Cr(CO)_4$ plane to induce the observed dissymetry. This dissymetry is rather the consequence of a local effect of the packing forces exclusively affecting the non eclipsed conformer.

Experimental and theoretical populations of the metal d orbitals in the methyl complex

Table 3 displays the metal d orbital populations derived from the multipole refinement of (II) using the formalism due to Holladay et al[16]. These values are compared to the populations obtained from the Mulliken analysis of the CASSCF-6 wave function for the methylcarbyne (II), of the CASSCF-6 and CASSCF-8 wave functions for the hypothetical unsubstitued carbyne (III). Experimental results were obtained assuming a C_{4v} averaging of the electron distribution around the metal, but the use of a C_{2v} symmetry for the chromium atom has shown no significant difference between the d_{xz} and d_{yz} orbital populations.

The agreement between the experimental and theoretical orbital populations appears satisfactory. The only significant discrepancy concerns the d_{xy} orbital (33% for the CASSCF population and 27% only from the multipole model distribution). This can be attributed to the design of the active space in the theoretical treatment, which does not account for the non dynamic correlation effects in the xOy plane. When these effects are accounted for as in the CASSCF-8 treatment of molecule (III), the population of the d_{xy} orbital decreases by 0.15 e, partly benefit of the $d_{x^2-y^2}$ orbital, thus leading in this plane to a charge distribution in better agreement with experiment.

It can be noticed that the deviations from isotropy are smaller in the metal d_σ and d_π orbital facing the carbyne (respectively d_{z^2}, d_{xz}, d_{yz}) than for the corresponding orbitals of chromium facing the carbonyl ligands. This relative trend toward a balance of the d_σ and d_π populations of the metal and of the carbynic carbon orbitals reflects the relative homopolar or even covalent character of the metal-carbyne interactions *as compared to the metal-carbonyl ones*.

CONCLUSION

The joint multipole model refinement and ab initio CASSCF study has given to important chemical information concerning carbyne complexes. The carbyne bond has been characterized by an ellipsoïd of density accumulation originating in the carbon atom and facing a depopulated metal orbital. A qualitative resemblance with the density pattern of the carbon monoxyde interaction was noticed, indicating that both types of interaction can be rationalized in terms of σ donation and π backdonation. The density ellipsoïd associated with the carbynic bond, relatively smooth and more extended toward the π bonding regions, is in keeping with the unbalanced character of the electron flows, more important for the π backdonation than for the σ donation, eventually leading to a large negative net charge for the carbynic carbon. This backdonation interaction gives the carbyne atom an electronic configuration close to $\sigma^2\pi^2$ and then formally corresponding to a negatively charged carbon. The backdonation seems to be even stronger, though anisotropic, in the phenylcarbyne compound, in relation with a conjugation effect between the phenyl ring and the metal-carbon triple bond. A weaker hyperconjugation effect was noticed in the methyl complex. The experimental investigations on the charge density agree with the evidence obtained from vibrational spectroscopy that two conformers of the methylcarbyne, eclipsed and non-eclipsed, do coexist in the crystal at 100 K. The hypothesis of an electronic, intramolecular effect originating in the methyl group of (II) and responsible for the observed breakdown of the C_{4v} symmetry in the $Cr(CO)_4$ moiety was ruled out from an analysis of the CASSCF wave function, and from both the theoretical and the modelized X-M deformation density maps. Finally, the experimental and theoretical orbital d populations are in good agreement for the methyl complex and confirm the homopolar character of the metal-carbyne interactions (espacially for the π bonds) as compared to the similar metal-carbonyl couplings.

ACKNOWLEDGEMENTS

All the singlecrystals were kindly provided by the Anorganisch-Chemisches Institut der Technischen Universität München and we are pleased to express our gratitude to Professsor E. O. Fischer. We are also very much indebted to Professor C. Krüger for providing the set of X-ray diffraction measurement for the methylcarbyne complex. We wish to thank H. Février for collecting the neutron and X-ray data of the phenylcarbyne complex. We also thank F. Baert for constructive discussion.

REFERENCES

1. E. O. Fischer , "Nobel Vortrag", On the way to Carbene and Carbyne Complexes, Angew. Chem. 86:651 (1974)

2. H. Fischer, P. Hofmann, F. R. Kreisl, R. R. Schrock, U. Schubert and K. Weiss, "Carbyne complexes", VCH ed. New York (1988)

3. N. K. Hansen and P. Coppens, Testing Aspherical Atom Refinements on Small-Molecule Data Sets, Acta Cryst. A34:909 (1978)

4. Nguyen Quy Dao, A. Spasojevic - de Biré, E. O. Fischer et P.J. Becker, Cartes de densité électronique de déformation du complexe $Cl(CO)_4Cr \equiv CC_6H_5$ par diffraction des rayons X et des neutrons, C. R. Acad. Sci. Paris 307:II:341 (1988)

5. A. Spasojevic - de Biré, Nguyen Quy Dao, E. O. Fischer and N. K. Hansen, Experimental electronic density deformation of trans-chlorotetracarbonyle phenylcarbyne chromium, J. Am. Chem. Soc. to be submitted to the editor (1990)

6. Nguyen Quy Dao, D. Neugebauer, H. Février, E. O. Fischer, P. J. Becker, and J. Pannetier, Transition metal carbyne complexes, neutron diffraction study on single crystal of trans-chloro (tetracarbonyl) phenylcarbyne chromium, Nouveau J. Chim. 6:359 (1982)

7. R. Goddard and C. Krüger, Deformation Density determinations (X-X, X-N) of Organometallic compounds in: "Electron Distributions and the Chemical Bond", P. Coppens and M. B. Hall eds, Plenum Press, New York (1982).

8. C. Krüger, R. Goddard and K. H. Claus, Die experimentelle Bestimmung der Elektronendichteverteilung in einem Chrom-Carben- sowie einem Chrom-Carbin- Komplex, Z. Naturforsch. 38b:1431 (1983)

9. A. Spasojevic - de Biré, Nguyen Quy Dao, A. Strich, C. Thieffry, and M. Bénard, Tetracarbonylchloroethynechromium: the electron deformation revisited. A joint multipole model refinement and ab initio CASSCF study, Inorg.Chem. in press (1990)

10. P. E. M. Siegbahn, J. Almlöf, A. Heilberg and B. O. Roos, The complete active space SCF (CASSCF) method in a Newton-Raphson formulation with application to the HNO molecule, J. Chem. Phys. 74:2384 (1981)

11. J. M. Poblet, A. Strich, R. Wiest, and M. Bénard, The metal-carbon triple bond in $Cl(CO)_4Cr \equiv CH$: a CASSCF study of neutral carbyne complexes, Chem. Phys. Letters 126:169 (1986)

12. J. Chatt, L. A. Duncanson, and L. M. Venanzi, Transmission of electronic effect through a palladium atom, J. Chem. Soc. 3:3203 (1958)

13. V. H. Smith Concept of Charge Density Analysis: The Theoretical Approach in: "Electron Distributions and the Chemical Bond", P. Coppens and M. B. Hall eds, Plenum Press, New York (1982) See more specifically pp 24-27

14. D. A. Clemente, M. C. Biagini, B. Rees and W. Herrmann, Molecular Structure and Experimental Electron Density of (μ-Methylen)bis[dicarbonyl(η^5cyclopentadienyl)manganese] at 130K, Inorg. Chem. 21:3741 (1982)

15. M. Martin, B. Rees and A. Mitschler, Bonding in a Binuclear Metal Carbonyl: Experimental Charge Density in $Mn_2(CO)_{10}$, Acta Cryst. B38:6 (1982)

16. A. Holladay, P. Leung and P. Coppens, Generalized Relations Between d-Orbital Occupancies of Transition-Metal Atoms and Electron-Density Multipole Population Parameters from X-ray Diffraction Data, Acta Cryst. A39:377 (1983)

17. P. Leung and P. Coppens, Experimental Charge Density Study of Dicobalt Octocarbonyl and Comparison with Theory, Acta Cryst B39:535 (1983)

18. Y. Wang, K. Angermund, R. Goddard and C. Krüger, Redetermination of the Experimental Electron Deformation Density of Benzenetricarbonylchromium, J. Am. Chem. Soc. 109:587 (1987)

19. A. Low, M. B. Hall, Origin of Inequivalent Chromium-Carbonyl Bond Lengths in Chlorotetracarbonylethylidynechromium, Organometallics 9:701 (1990)

20. N. M. Kostic and R. F. Fenske, Molecular Orbital Calculations on Carbyne Complexes $CpMn(CO)_2CR^+$ and $(CO)_5CrCNEt_2^+$. Frontier-Controlled Nucleophilic Addition to Metal-Carbon Triple Bond, J. Am. Chem. Soc. 103:4677 (1981)

21. N. M. Kostic and R. F. Fenske, Molecular Orbital Calculations on Octahedral Carbyne Complexes: Bonding, Structure Deformations, and Frontier-controlled nucleophilic Additions, Organometallics 1:489 (1982).

22. J. Ushio, H. Nakatsuji, and T. Yonezawa Electronic Structures and Reactivities of Metal-Carbon multiple bonds; Schrock-Type Metal Carbene and Metal Carbyne Complexes, J. Am. Chem. Soc. 106:5892 (1984)

23. H. Irngartinger, Electron density distribution in the bonds of cumulenes and small ring compounds, in: "Electron Distribution and the Chemical Bond", P. Coppens and M. B. Hall Eds, Plenum Press, New York, p 36 (1982)

24. H. Février, Contribution à l'étude de la liaison chimique dans les complexes carbyniques neutres des métaux de transition de la famille 6A, Thesis, Université de Paris XIII, (1982)

25. Nguyen Quy Dao, G. P. Foulet – Fonséca, M. Jouan, E. O. Fischer, H. Fischer, J. Schmid, Mise en évidence de deux conformères du complexe $Br(CO)_4CrCCHD_2$ par spectrométrie de diffusion Raman, C. R. Acad. Sci. Paris 307II:245 (1988)

26. Nguyen Quy Dao, H. Février, M. Jouan, N. H. Tran Huy, E. O. Fischer and D. Neugebauer, Sur la non-équivalence des liaisons CH du groupement méthyle dans les complexes cristallisés de trans-bromotetracarbonyle méthyle carbyne de chrome, J. Org. Chem. 241:C53 (1982)

27. Nguyen Quy Dao, H. Février, M. Jouan, E. O. Fischer, Champ de force des complexes trans-bromo (tetra-carbonyle) methyle carbyne de tungstène, Nouv. J. Chim. 7:719 (1983)

28. G. P. Foulet – Fonséca, M. Jouan, Nguyen Quy Dao, N. T. Huy and E. O. Fischer, Etude par spectroscopie vibrationnelle des dérivés partiellement deutériés de trans-bromo (tetracarbonyl) méthylcarbyne de chrome, J. Chim. Phys. 87:13 (1990)

29. G. P. Foulet – Fonséca, M. Jouan, Nguyen Quy Dao,, J. Schmid and E. O. Fischer, Etude vibrationnelle du composé partiellement deutérié pur $Br(CO)_4CrCCHD_2$ et détermination du champ de force des complexes carbyniques de la famille $Br(CO)_4CrCR$ (R=CH_3, CH_2D, CHD_2, CD_3), Spectrochim. Acta 46A:339 (1990)

THE POTENTIAL FUNCTION IN SHORT O–H⋯O BONDS.

EXPERIMENTAL EVIDENCE AND AB INITIO MODELLING

F. Vanhouteghem, C. Van Alsenoy and A. T. H. Lenstra

Department of Structural Chemistry
Antwerp University (U.I.A.)
Universiteitsplein 1, B2610 Wilrijk, BELGIUM

Short hydrogen bonds (D⋯A distance 2.5 Å) are either symmetric (*e.g.* acetamide hydrochloride[1]) or asymmetric (*e.g.* the hydrogen maleate ion in different metal derivatives[2]). Neutron studies at different temperatures of the Mg salt of the hydrogen maleate ion indicated no proton disorder[3], so a double well potential is not self-evident. Moreover, the alternative H-sites (reversing donor and acceptor) are in general situated in a "forbidden" region of the repulsive constraints plots[4].

For an isolated HMal ion, the symmetric double well potential was calculated with a 4-21G basis set, shown in Fig. 1. Inclusion of the coordinating water molecules adjacent to the short intramolecular H-bond changes this potential drastically (Fig. 1). We now find an asymmetric and anharmonic potential. The minimum energy position agrees satisfactorily with the observed H-position. So, the intermolecular H-bonds cause the observed asymmetry in HMal, reducing the role of the counterion to a secondary one. In addition, our analysis is not only in line with the F&S plots, but it also provides an energetic explanation for the excluded region as being simply less attractive. For the symmetric H-bond in acetamide hemi-hydrochloride, we enumerated the potential energy related to proton shifts in the protonated acetamide dimer along O⋯O ($E_{||}$) and perpendicular to it ($E_{\perp}$). We find $E_{\perp}(d) = 82500d^2$ cal/mol, *i.e.* $E_{\perp}$ is an harmonic potential. For $E_{||}$ we find $E_{||}(d) = 933600d^4$ cal/mol, *i.e.* a symmetric, but anhamonic potential. The temperature dependent H-distribution is given by $\exp(-\Delta E/kT)$. This gives at 120 K $\mathrm{rms}_{\perp}(H) = 0.04$ Å and $\mathrm{rms}_{||} = 0.09$ Å. These estimates were experimentally determined as 0.08(1) Å and 0.15(1) Å respectively. So the "intradimer" potential is an important contributor to the observed H-distribution. We recalculated the $\mathrm{rms}_{||}$-value forcing an harmonic approach to match the experimental analysis of Speakman *et al*[1]. This yields a value of 0.14 Å provided the function $E_{||} = Pd^2$ is only fitted to E-values for which the population at 120 K is non-zero (and thus including only models producing a non-zero contribution to the observed intensities). So a re-examiantion of acetamide seems necessary to eliminate the artifacts introduced by the applied harmonic model for $E_{||}$.

It is gratifying to see that an 'ab initio' verification of the thermal parameters is possible, enabling the crystallographer to avoid subsequent errors in *e.g.* deformation analyses.

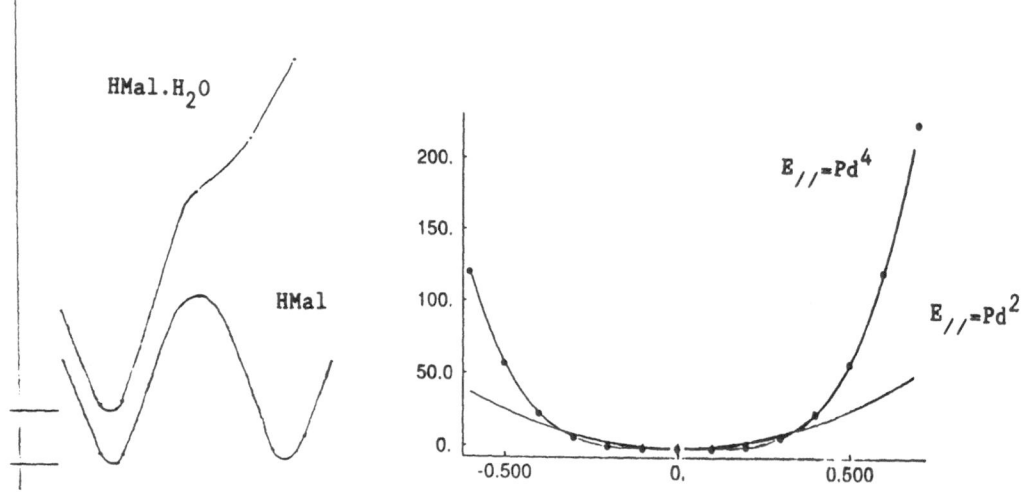

Fig. 1. Potential wells for O–H····O systems, asymmetric and symmetric case.

REFERENCES

1. J. C. Speakman, M. S. Lehmann, J. R. Allibon and D. Semmingsen, *Acta Cryst.* **B37**, 2098-2100 (1981).
2. G. Olovsson, I. Olovsson and M. S. Lehmann, *Acta Cryst.* **C40**, 1521-1526 (1981).
3. F. Vanhouteghem, A. T. H. Lenstra and P. Schweiss, *Acta Cryst.* **B43**, 523-528 (1987).
4. J. L. Finney and H. J. Savage, *J. Mol. Struct.* **177**, 23-41 (1988).

THE EFFECT OF CONFORMATIONAL CHANGES ON THE POLARIZABILITY OF PNA

G.J.M. Velders and D. Feil

Chemical Physics Laboratory, University of Twente
P.O. Box 217, 7500 AE Enschede, The Netherlands

Para-nitroaniline, PNA, is an important material for nonlinear optics: wave mixing, frequency doubling, optical computing etc. It exhibits a large second order hyperpolarizability compared to anorganic materials. PNA is a planar molecule but a crystallographic environment can rotate the NH_2 and NO_2 group. We have investigated with Hartree-Fock calculations the effect of rotation of these groups on the first and second order (hyper)polarizability using an externally applied electric field: the Finite Field method.

Rotation of one or both groups appeared to have only a small effect on the linear polarizability. The second order hyperpolarizability reduces strongly when one or both groups are perpendicular to the benzene ring. Charge analysis for PNA show that the pi-electrons are responsible for the main part of the polarizability but that the contributions from the other electrons can not be neglected. Rotation of a group disturbs the pi-electron system and therefore the polarizability. The second order polarizability appears to be more sensitive to these changes than the first order. Electron density maps of the first and second order polarizability illustrate these changes very clearly.

Perturbation theory can be used to interpret the polarizability in PNA. Electron density maps and polarizability calculations show that the HOMO-LUMO interaction alone does explain most but not all the features in the polarizability. There is a good agreement in the linear polarizability between the perturbation theory and the calculation in the presence of an electric field. For the second order polarizability the agreement is also good qualitatively but quantitatively the perturbation theory gives slightly too small values.

CHARGE DENSITY IN ADENINIUM HYDROCHLORIDE HEMIHYDRATE AND 1H+-ADENINIUMTRICHLOROZINC(II): THE EFFECTS OF METAL BINDING TO A NUCLEOBASE

Louise M. Vilkins and Max R. Taylor

School of Physical Sciences, The Flinders University
of South Australia, Bedford Park SA 5042. Australia

The study of charge density and electrostatic properties in adeninium hydrochloride hemihydrate and 1H+-adeniniumtrichlorozinc(II) is part of a broader investigation into the effects of substituents, protonation and bound metal ions on electron density distributions in some nucleobase derivatives. Any redistribution of electron density might be expected to influence the hydrogen-bonding capability of the base. In this study, we are looking for effects on the electron density and electrostatic potential in protonated adenine caused by the coordination of zinc to N7.

A high resolution X-ray data set has recently been collected on 1H+-adeniniumtrichlorozinc(II) at 123K. A full sphere of data was measured to $\theta=33°$, followed by a half-sphere of selected high-order reflections to $(\sin\theta/\lambda) = 1.32\text{Å}^{-1}$, chosen via calculated E-values (based on structure refined with existing data) such that $I>2\sigma(I)$. A total of 16987 reflections were measured and merging resulted in a unique set of 5348 reflections. Conventional refinement of positional and thermal parameters using high order data $((\sin\theta/\lambda) >0.65\text{Å}^{-1})$ gave R=1.7% and R_w=1.5% (on F). A multipole refinement to octapole level is in progress - R=1.5% and R_w=1.3%. Calculated maps of the experimental static deformation density reveal features in the adeninium moiety such as lone pair electron density at nitrogen and C-N bond polarisation.

X-ray data have also been collected for adeninium hydrochloride hemihydrate (at 123K). The electron deformation density and derived electrostatic properties for the zinc complex will be compared with those for adeninium hydrochloride and other results from the literature.

PARTICIPANTS

ALVAREZ, A.	Dept. de Cristallogr., Univ. Autonoma de Barcelona, 08193 BARCELONA (SPAIN)
ANDRADE, L. C. R.	Dept. Fisica, Univ. de Coimbra, 3000 COIMBRA (PORTUGAL)
ARMAGAN, N.	Selcuk Universitesi, Muhendislik-Mimarlik Fakultesi, 42040 KONYA (TURKEY)
BADER, R. F. W.	Dept. Chemistry, McMaster Univ., Hamilton, ONTARIO L8S4M1 (CANADA)
BAERT, F.	Univ. des Sci. et Tech., EME Etage, 59655 VILLENEUVE d'ASCQ (FRANCE)
BARRY, C. D.	ICI Pharmaceuticals Div., MacClesfield, CHESHIRE SK10 4TG (UNITED KINGDOM)
BLANCHARD, H.	Dept. Chemistry, Queen Mary & Westfield College, E1 4NS LONDON (ENGLAND)
BLESSING. R.	Medical Foundation of Buffalo, BUFFALO, NY 14203 (USA)
BRACKE, B.	Dept. Chemistry, Univ. of Antwerp, B-2610 WILRIJK (BELGIUM)
BRAMMER, L.	Brookhaven National Laboratory, UPTON, NY 11973 (USA)
BRENEMAN, C.	Dept. Chemistry, Cogswell Lab., Rensselaer Polytechnic Inst., Troy, NY 12180 (USA)
CARDUCCI, M.	Dept. Chemistry, Univ. of Arizona, TUCSON, AZ 85721 (USA)
CHRISTIDIS, P.	Lab. Applied Phys., Univ. of Thessaloniki, 54006 THESSALONIKI (GREECE)
COPPENS, P.	Chemistry Dept., State Univ. of New York, BUFFALO, NY 14221 (USA)
DE BONDT, H.	Lab. v. Anal. Chem. en Med. Fysicochem., B-3000 LEUVEN/BELGIE (BELGIUM)
DE MATHEWS, M.	Inst. de Ciencia de Materials, C.S.I.C., 08028 BARCELONA (SPAIN)
DE VRIES, R.	Chem. Phys. Lab., Univ. of Twente, 7500 AE ENSCHEDE (NETHERLANDS)
DOMINGOS, A. M.	LNETI, ICEN, Dept. Chemistry, 2685 SACAVEM (PORTUGAL)
DOYLE, M. J.	BP International Ltd., Sunbury Res. Centre, MIDDLESEX, TW16 7LN (ENGLAND)
DUARTE, M. T.	Centro Quimica. Estrutural, I.S.T., 1096 LISBOA CODEX (PORTUGAL)
FEIL, D.	Chem. Phys. Lab., Twente Univ. of Tech., 7500 AE Enschede (NETHERLANDS)
FERNANDEZ, B.	Quim.-Fisica, Univ. of Santiago, E-15708 SANTIAGO DE COMPOSTELA (SPAIN)
FOCES FOCES, C.	C.S.I.C. Inst. Rocasolano, Uie Cristalografia, 28006 MADRID (SPAIN)
FRAMPTON, C.	Dept. Chemistry, McMaster Univ., HAMILTON L8S 4M1 ONTARIO (CANADA)
FUESS, H.	Inst. Kristallografia, Universitat, D-6000 FRANKFURT/MAIN (GERMANY)
GALVAO, A.	Centro Quimica. Estrutural, I.S.T., 1096 LISBOA CODEX (PORTUGAL)
GATTI, C. E.	CNR Center for Study of Structure Reactivity Relationships, 21033 MILANO (ITALY)
GHERMANI, N.-E.	Lab. de Min. et Crist., Univ. de Nancy, 54506 NANCY CEDEX (FRANCE)
GILLI, G.	Dip. di Chimica, Unvi. di Ferrara, I-44100 FERRARA (ITALY)
GRAND, A.	DRF - Lab. de Chimi, C.E.N.G.-85X, F-38401 GRENOBLE CEDEX (FRANCE)
HABBOU, A.	Lab. de Min. et Crist., Univ. de Nancy, 54506 NANCY CEDEX (FRANCE)
HOWARD, J. A. K.	Dept. Inorg. Chem., Univ. of Bristol, BRISTOL BS8 1TS (UNITED KINGDOM)
HOWARD, S.	Dept. Chemistry, Queen Mary & Westfield College, E1 4NS LONDON (ENGLAND)
JEFFREY, G. A.	Dept. Crystallography, Univ. of Pittsburgh, PITTSBURGH, PA 15260 (USA)
JOSE, J.	Dept. de Cristallogr., Univ. Autonoma de Barcelona, 08193 BARCELONA (SPAIN)
KANSIKAS, J.	Dept. Chemistry, Univ. of Helsinki, 00100 HELSINKI (FINLAND)
KENDI, E.	Dept. Physics Eng., Hacettepe University, 06532 BEYTEPE-ANKARA (TURKEY)
KLEBE, G.	ZHV/W, BASF AG, D-6700 LUDWIGSHAFEN (GERMANY)
KLOOSTER, W.	Chem. Fysica, afdeling Chem. Technologie 7500 AE ENSCHEDE (NETHERLANDS)
KOOIJMAN, H.	Kristal. en Struct.Chem., Rijksuniversiteit, 3584 CH UTRECHT (NETHERLANDS)
LAIDIG, K.	Dept. Chemistry, McMaster Univ., HAMILTON L8S 4M1 ONTARIO (CANADA)
LARSEN, F. K.	Dept. Chemistry, Univ. of Aarhus, DK-8000 AARHUS (DENMARK)
LECOMTE, C.	Lab. Mineralogic-Cristallogrraphie, Univ. de Nancy, 54506 NANCY (FRANCE)

LEHMANN, C.   Inst. f. Kristallographie, Freie Univ. Berlin, 1000 BERLIN 33 (GERMANY)
LEHMANN, M.   Institut Laue-Langevin, 38041 GRENOBLE (FRANCE)
LLAMAS SAIZ, A.L.   C.S.I.C. Inst. Rocasolano, Uei Cristalografia, 28006 MADRID (SPAIN)
LOPEZ CALVO, J. A.   Dept. Quimica Inorg., Univ. de Zaragoza, 50009 ZARAGOZA (SPAIN)
MALLINSON, P. R.   Dept. Chemistry, Univ. of Glasgow, GLASGOW G12 8QQ (SCOTLAND)
MATOS, A.   Dept. Fisica, Univ. de Coimbra, 3000 COIMBRA (PORTUGAL)
MERATI, F.   Dip. di Chimica Fisica ed Elletrochim. Univ. di Milano, 20133 MILANO (ITALY)
MOSQUERA, R.   Quim.-Fisica, Univ. of Santiago, E-15708 SANTIAGO DE COMPOSTELA (SPAIN)
MUTIKAINEN, I.   Dept. Chemistry, Univ. of Helsinki, 00100 HELSINKI (FINLAND)
NARDELLI, M.   Inst. di Chim. Generale, Univ. di Parma, I-43100 PARMA (ITALY)
PAIXAO, J.A. DE C.   Dept. Fisica, Univ. de Coimbra, 3000 COIMBRA (PORTUGAL)
PALMER, A.   Hahn-Meitner-Inst. Berlin GMBH, D-1000 BERLIN 39 (GERMANY)
PARADA, C.   Dept. Quimica Inorg., Univ. Complutense de Madrid, 28040 MADRID (SPAIN)
PEDERSEN, B. F.   Inst. Pharmacy, Univ. of Oslo, BLINDERN, 0315 OSLO 3 (NORWAY)
PINIELLA, J. F.   Dept.Cristallografia, Univ. Auto. Barcelona, 08193 BELLATERRA (SPAIN)
RICHARD, P.   Lab. des Heterocycles, Fac. des Sciences, 21000 DIJON (FRANCE)
ROBERTSON, B. E.   Dept. Physics, Univ.of Reginia, REGINA, SASK., S4S 0A2 (CANADA)
SAWIN, P.   Dept. Chemistry, Univ. of California, LOS ANGELES, CA 90024 (USA)
SCOTT, B.   Dept. Chemistry, Washington State Univ., PULLMAN, WA 99164 (USA)
SEMINARIO, J. M.   Dept. Chemistry, Univ. of New Orleans, NEW ORLEANS, LA 70148 (USA)
SEVIN, F.   Dept. Chemistry, Hacettepe University, 06532 BEYTEPE-ANKARA (TURKEY)
SHERWOOD, P.   Daresbury Laboratory, Daresbury, WARRINGTON WAD 4AD (UNITED KINGDOM)
SOUHASSOU, M.   Medical Foundation of Buffalo, BUFFALO, NY 14203 (USA)
SPASOJEVIC, A.   Lab. Leon Brillouin, CEN-Saclay, F-91191 GIF SUR YVETTE CEDEX (FRANCE)
SPICKETT, R. W.   Laboratoris Almirall, 08024 BARCELONA (SPAIN)
STEVENS, E. D.   Dept. Chemistry, Univ. of New Orleans, NEW ORLEANS, LA 70148 (USA)
STEWART, R. F.   Dept. Chemistry, Carnegie Mellon Univ., PITTSBURGH, PA 15213 (USA)
SWENSON, D.   Medical Foundation of Buffalo, BUFFALO, NY 14203 (USA)
ULKU, D.   Dept. Physics Eng., Hacettepe University, 06532 BEYTEPE-ANKARA (TURKEY)
VAZQUEZ, S.   Quim.-Fisica, Univ. of Santiago, E-15708 SANTIAGO DE COMPOSTELA (SPAIN)
VELDERS, G.   Chem. Phys. Lab., Univ. of Twente, 7500 AE ENSCHEDE (NETHERLANDS)
VILKINS, L. M.   Flinders Univ. of South Australia, BEDFORD PARK, SA 5042 (AUSTRALIA)
WERNER, S.   Max-Planck-Inst. f. Kohlenforschung, D-4330 MULHEIM A.D. RUHR (GERMANY)

INDEX

Ab-initio, *see* Computations
Absorption corrections, 159-161
Active sites
  of camphor P, 307, 330-312,
  of creatine amidinohydrolase, 295
  of ribonuclease A, 302-303
Atomic form factors, 65, 66, 129, 171

Cambridge Crystallographic Data Base, 4
Cell membranes, 243
Charge density, 7, 23
  in B_2F_4, 27, 35
  in B_2H_6, 30
  in $B_6H_6^-$, 32
  comparison of methods for, 139-146
  critical points in, 29
  in diamond, 12
  early work on, 7
  in formamide, 377
  gradient vector fields in, 31
  in the hydrogen bond, 103-121
  in the hydroxyl ion, 379
  in oxalic acid dihydrate, 113-117
  partitioning of, 15
  poor-mans analysis of, 13
  topology of, 26
Charge transfer, 257
Chemical bonds
  bent, 49
  in $C_4H_7^+$, 51
  classification of, 47
  in cyclopropane, 50
  ellipticities of, 50
  paths of, 49
  peak heights of, 145
  in [2.1.1] propellene, 51
Compton scattering, 158
Computations
  ab-initio, 2, 3, 144
    of charge densities, 371-378
    of clavulenic acid, 359
    of metal-nitrosyl interactions, 369
    of short OH$\cdots$O bonds, 401
  molecular mechanics
    blocking agents of, 362
  semi-empirical, 293

Cooling techniques
  apparatus for, 191
  closed-cycle, 196
  cryostream, 193
  displex expansion, 197-199
  gas-stream, 192
  helium flow, 195
Count-rate linearity, 173
Critical points
  3,-1; 3,+1; 3,-3, 35
  in benzene, 98
  central cages of, 32
  of chemical bonds, 38
  classification of, 26
  in imidazole, 87
  in 9-methyladenine, 93
  in molecular charge distributions, 29
  in rings and cages, 43
  signature values of, 29
  in urea, 83
Crystal quality, 169, 171

Data collection methods and analysis, 182
Deformation densities
  of 4-acetoxy-5-styryl α-pyrone, 365-368
  of adeninium HCl·0.5H$_2$O, 404
  of Al_2O_3, 220-223
  of a C–H→Co bridged complex, 353-356
  of chemical carcinogens, 323-325
  of chloropromazine HCl, 331
  of $Cr_2(mhp)_4$, 17
  of $Cr(NH_3)_6Cr(CN)_6$, 219-221
  of 18-crown-6, 234-235
  of Cu_2O, 220-222
  of cyclobutadiene, 11
  of 2'-deoxy-cytidine 5'-monophosphate, 333
  of dimethyl 1,3-amino-4-uracil, 369
  of dimetridazole at 105 K, 341-349
  of DOPA, 329-330
  for drug design, 319
  of estradiol, 334
  of formamide, 15
  function method by, 134-136
  of L-alanine at 23 K, 339
  of metal carbyne complexes, 385-399
  of MgS_2O_3·6H$_2$O, 232

Deformation densities (continued)
  of N-acetyl-α,β-dehydrophenylalanine methyl
      amide, 138, 141
  of N-acetyl-L-tryptophane methylamide, 138
  of $Na_2S_2O_3$, 233
  of $NaHC_2O_4 \cdot H_2O$, 299
  of nascolone HCl, 327-328
  of neurotransmitters, 329
  of nucleotides, 332
  of opiates, 326
  of oxalic acid dihydrate, 113-117
  of peptide link, 146-149
  of phenothiazines, 329
  of pyridinium-1-dicyanomethylide, 139-141
  of pyrrole ring, 124
  standard deviations of, 215-218
  of steroid hormones, 332
  of syn-1,6:8,13-biscarbonyl[14]annulene at 16 K,
      340
  of triazine, 11
  of urea, 107-108
  of water molecule, 105-106
  by X–N method, 10, 225, 231
Diamond, 12
  222 reflection, 9
Difference density, see Deformation densities
DORIS II, 220
Drugs
  affinity of, 242
  as agonists, 242
  as antagonists, 242
  definition of, 241, 251
  design of, 250, 288
  Free-Wilson model for, 269
  Hansch method for, 270-272
  linear free-energy studies of, 272-273
  lyophilicity of, 252
  receptors of, 241, 289
    allosteric models of, 249
    in BDZ, 275-279
    enthalpy of, 262, 266-268
    mapping studies of, 273
    in membranes, 243
    for neurotransmitters, 244
    pharmacology of, 245
    physical chemistry of, 255
    regulatory mechanisms of, 245
    thermodynamics of, 263-265
  structure-activity relationship of, 274
    in benzodiazephines, 279-282
    in β-carbolines, 279
  transport of, 252

Elastic constants, 169
Electron densities
  in Fe(II) and Fe(III) complexes, 308
  models for, 121-155
  in S_N2 transition, 305
  thermal, 156
  in trans dichloro bis(creatinene) platinum II,
      $2H_2O$, 338

Electron densities (continued)
  in water molecule, 105
Electrostatic interactions, 15, 256
  dipole-dipole, 256, 262
  potential maps, 56, 322
    of benzene, 95-96
    from electric multipoles, 357
    of imidazole, 83-87
    of 9-methyladenine, 84-94
    of urea, 74-83
  properties of, 63, 68
Enzyme catalysis, 287
Ewald construction
Extinction corrections, 161-167
  anisotropic, 166
  primary, 162, 164
  secondary, 162, 165

Focal spot scans, 173-174
Fourier sums, properties from, 68
Free-Wilson method, 269
Frozen-core approximation, 121

GaAs, 18
Gamma ray diffraction, 361
Gradient vector field, 31, 35
  phase portrait of, 33
  of B_2F_4, 36
  of B_2H_6, 44
  of $B_6H_6^{-2}$, 45
Gram-Charlier expansion, 64, 131
g-tensor in metal-carbonyl radicals, 360

Hansch method, 270-272
Hessian matrix, 28
Hilbert space, 23
Hydrogen bonds
  charge distribution of, 103-120
  C=O bond lengths, influence on 297
  definition of, 59
  interaction density of, 112-112
  in 9-methyladenine, 91
  short OH···O, 401
  theoretical calculations of, 109, 401
  theory of, 257-260

Instrument performance, 171
  stability, 180
Interaction density, 112
  of oxalic acid dihydrate, 113
Interatomic surface, 37
Ion channels, 243
Isotropic β values
  temperature dependence of, 190
  vs $\sin\theta/\lambda$, 203
IUCr oxalic acid project, 14

K absorption frequencies, 66
κ parameter, 13
  refinement of, 16, 122, 132

Laplacian of ρ, 24
    in benzene, 98
    in imidazole, 88
    in 9-methyladenine, 94
    in N_2, 58
    in NaCl, 58
    in urea, 81- 82
LSEXP method, 134-136

Magnetic densities, 236
    superexchange of, 237
Magnetostriction, in NiF_2, 361
Mathieson scans, 171, 180
Microcrystallography, 211
Molecular graphs, 38, 40
    of bicyclooctane, 40
    of CH_3-CH_2-CH_2^+→CH_2^+-CH_2-CH_3, 48
    of CH_3–CH_2^+→CH_2^+-CH_3, 48
    equivalent class of, 46
    of [1.1.1]-propellene, 40-42
    of [2.2.2]-propellene, 40-42
Molecular mechanics (see also Computations)
    flexibility of, 290
    form factors in, 69
Molecular tensor analysis, 170
MOLLY method, 132
Monochromators, 175
Mosaic block orientation, 165
    anisotropy, 166
    scans, 176
    size, 165
Multiple reflections, 158
Multipole refinement, 122-126, 321-322
    effect of phases in, 136

Neurons, 247
    inhibitory synapsis, 248
    neuronal synapsis, 247
    neurotransmitters, 248
    synaptic contact, 249
Neutron scattering, 155, 228-230
    energy dispersion of, 227
    lengths, 229
    properties, 227
    sources
        Brookhaven National Laboratory, 10
        Institute Laue-Langevin, 225
Nuclcophilic attack, 294
    in amide hydrolysis, 300
    in creatinase, 294

Oxalic acid dihydrate, 10, 110, 113-117
Oxalic acid project, see IUCR

Pendellosung measurements, 15, 18
Polarization density, of water, 111
POP method, 131
Population parameters, 122

Protein Data Bank, 4
Pulse energy scan, 172

Radial density functions, 122
    for benzene, 95
    for imidazole, 83
    for 9-methyladenine, 89
    for urea, 73
Radiation damage, 182
Renninger reflection, 159
Rigid-body test, 170, 205
Roux density, 10
Russell-Saunders, coupling, 24

Scan ranges, 202
Scheringer principle, 60
Silicon, 12, 15
Spectral dispersion scans, 178-179
Spherical harmonic functions, 126-128
Stockholder method, 16
Structural stability, 46-47, 51
Structure-correlation method, 291
Surface crystallography, 211

Take-off angle scans, 176-177
Temperature measurement, 200
Thermal diffuse scattering (TDS), 67, 167
    corrections for, 169
    intensity profiles of, 168
Thermal vibration averages, 67
    Debye-Waller factors, 157, 160, 190
Thomson cross-sections, 65
Transmission factors, 160

Valence/core ratios, 171
Valence shell charge concentration (VSCC), 55
VALREY, 133
Vibrational wave functions, 67
VSEPR model, 55

X–N method, 10, 225, 231
X-ray elastic scattering, 64
    α_1,α_2 doublets
    beam divergence, 176
    generalized scattering factors, 70
        of benzene, 97
        of imidazole, 84-87
        of 9-methyladenine, 91
        of urea, 72
    synchrotron counter deadtime, 211
    synchrotron measurement
        at CHESS, 211
        at HASYLAB, 220
        at national sources, 200
        at Stanford facility, 210

Zero-flux surface, 38
    boundary condition, 53